INTERNATIONAL ASTRONOMICAL UNION

UNION ASTRONOMIQUE INTERNATIONALE

SYMPOSIUM No. 47

HELD AT THE UNIVERSITY OF NEWCASTLE-UPON-TYNE, ENGLAND, 22–26 MARCH, 1971

THE MOON

EDITED BY

STANLEY KEITH RUNCORN

School of Physics, University of Newcastle-upon-Tyne, England

AND

HAROLD C. UREY

Chemistry Department, University of California at San Diego, La Jolla, Calif., U.S.A.

D. REIDEL PUBLISHING COMPANY

DORDRECHT-HOLLAND

1972

Published on behalf of
the International Astronomical Union
by
D. Reidel Publishing Company, Dordrecht, Holland

Softcover reprint of the hardcover 1st edition 1972

Library of Congress Catalog Card Number 73–188005

ISBN 978-94-010-2863-9 ISBN 978-94-010-2861-5 (eBook)
DOI 10.1007/978-94-010-2861-5

Professor H. C. Urey and Academician A. A. Mikhailov outside the School of Physics of the University of Newcastle-upon-Tyne.

THE MOON

PREFACE

The IAU Symposium No. 47 *The Moon* was held in the School of Physics of the University of Newcastle-upon-Tyne, England, from 22 to 26 March 1971. The Meeting was sponsored by Commission 17 and co-sponsored by URSI. The Symposium was supported financially by the IAU. The Scientific Organizing Committee included Prof. S. K. Runcorn, Chairman (University of Newcastle-upon-Tyne), Prof. H. Alfvén (Royal Institute of Technology, Stockholm), Prof. G. Colombo (University of Padova), Prof. A. Dollfus (Observatoire de Paris), Prof. T. Gold (Cornell University), Dr K. Koziel (Jagellonian University, Poland), Prof. G. P. Kuiper (University of Arizona), Dr B. J. Levin (U.S.S.R. Academy of Sciences), Dr A. A. Mikhailov (U.S.S.R. Academy of Sciences), Prof. A. E. Ringwood (Australian National University) and Prof. H. C. Urey (University of California). The Local Organizing Committee included Prof. S. K. Runcorn (Chairman), Dr G. Fielder, Prof. W. R. Hindmarsh, Prof. Z. Kopal and Prof. W. H. McCrea.

This book includes the majority of papers presented at this the second IAU Symposium on *The Moon*: comparison with the previous IAU Symposium on *The Moon* held in Leningrad, a decade earlier, makes clear the great advances made possible in our knowledge of the Moon and solar system by space technology. Academician A. A. Mikhailov's lecture on Luna 16 and 17 and the fascinating Lunakhod and the various lectures by scientists of the National Aeronautics and Space Administration on the Apollo missions, introduced by Dr Homer E. Newell, Associate Administrator, were highlights of the meeting. We believe that the papers in this volume will give to the reader a better appreciation of the unique contribution which each of the sciences has to make to this quest for a new understanding of our satellite and through it insight into the origin of the Earth and solar system.

We would also like to draw attention to the great opportunity which now exists for the postgraduate student in the physical sciences to assist in the understanding of the great amount of new data being obtained on that celestial body which was once in the forefront of astronomical research and is now again.

We would like to express our appreciation of the work of Mr W. F. Mavor and Mrs M. Turner of the School of Physics who very efficiently handled the conference arrangements. We also wish to thank the Vice-Chancellor of the University of Newcastle-upon-Tyne, Dr Henry G. Miller, the Master of University College, Durham, Mr Slater and Prof. G. M. Brown of the Geology Department, University of Durham for hospitality to the participants.

S. KEITH RUNCORN
HAROLD C. UREY

TABLE OF CONTENTS

E. *Lunar Tectonics*

F. *Physical Properties of Lunar Samples*

G. *The Lunar Interior*

H. *The Evolution of the Moon's Orbit*

I. *Origin and Evolution of the Moon*

(O'Keefe's paper is substantially identical with his paper: 'Geochemical Evidence for the Origin of the Moon' which is to appear in *Die Naturwissenschaften* shortly.)

LIST OF PARTICIPANTS

Mr F. Ackfield, Forest Hall Observatory, Newcastle-upon-Tyne, England.

Dr S. O. Agrell, Dept. of Mineralogy and Petrology, University of Cambridge, England.

Miss C. M. Allen, School of Physics, University of Newcastle-upon-Tyne, England.

Dr P. J. Armitage, Lunar Receiving Laboratory, Houston, Texas, U.S.A.

Mr D. W. G. Arthur, University of Arizona, U.S.A.

Dr J. Bastin, Dept. of Physics, Queen Mary College, Mile End Road, London E1, England.

Dr L. Battiston, Laboratorio per lo studie della dinamica delle grandi masse, Consiglio Nazionale delle Ricerche, S. Polo 1364, Venezia, Italy.

Mr A. R. Birks, Nuffield Radio Astronomy Laboratories, Jodrell Bank, Macclesfield, Cheshire, England.

Dr J. Borg, Centre de Spectrométrie de Masse du C.N.R.S., France.

Prof. M. H. P. Bott, Dept. of Geology, University of Durham, England.

Dr Brett, Manned Spacecraft Center, Houston, Tex. 77058, U.S.A.

Prof. G. M. Brown, Dept. of Geology, University of Durham, England.

Dr J. D. Burke, Jet Propulsion Laboratory, Pasadena, California, U.S.A.

Mr P. H. Cadogan, Dept. of Chemistry, University of Bristol, England.

Dr A. Carusi, Istituto di Geologia e Paleontologia, Università degli Studi di Roma, Roma, Italy.

Dr P. E. Champness, Dept. of Geology, University of Manchester, England.

Mr A. J. Cohen, Dept. of Earth and Planetary Physics, University of Pittsburgh, U.S.A.

Dr D. S. Colburn, NASA, Ames Research Center, Moffett Field, Calif. 94035, U.S.A.

Dr D. W. Collinson, School of Physics, University of Newcastle-upon-Tyne, England.

Prof. G. Colombo, Istituto di Meccanica Applicata alle Macchine, Via Marzolo 9, 35100 Padova, Italy.

Dr G. Comstock, General Electric Co., Schnectady, New York, U.S.A.

Mr P. A. Copley, Dept. of Geology, University of Manchester, England.

Dr A. Coradini, Istituto di Geologia e Paleontologia, Università degli Studi di Roma, Roma, Italy.

Mr B. K. Cowan, School of Physics, University of Newcastle-upon-Tyne, England.

Prof. T. G. Cowling, School of Mathematics, University of Leeds, England.

Prof. K. M. Creer, School of Physics, University of Newcastle-upon-Tyne, England.

Mr P. Crofton Sleigh, 5 Ladbroke Gardens, London W11.

Mr B. W. Darracott, School of Physics, University of Newcastle-upon-Tyne, England.

Dr A. de Sa, School of Physics, University of Newcastle-upon-Tyne, England.

Prof. A. Dollfus, Observatoire de Meudon, 92 Meudon, France.

Mr J. Dolman, School of Physics, University of Newcastle-upon-Tyne, England.

Dr J. C. Dran, Centre de Spectrométrie Nucléaire et de Masse, France.

Prof. H. Drever, Dept. of Geology, University of St Andrews, Scotland.

Dr M. Duke, Lunar Receiving Center, Houston, Tex. 77058, U.S.A.

Mr J. P. Durraud, Centre de Spectrométrie Nucléaire et de Masse, France.

Mr Dzhapiashvili, Academy of Sciences, Moscow, U.S.S.R.

Mr S. I. El-Batroukh, School of Physics, University of Newcastle-upon-Tyne, England.

Dr F. El-Baz, Bellcomm. Inc., 955 L'enfant Plaza North, S.W., Washington D.C. 20024, U.S.A.

Prof. W. E. Elston, Dept. of Geology, University of New Mexico, Albuquerque, N.M. 87106, U.S.A.

Prof. M. Ewing, Lamont Geological Observatory, Columbia University, U.S.A.

Mr J. D. Fairhead, School of Physics, University of Newcastle-upon-Tyne, England.

Dr G. Fielder, Dept. of Environmental Sciences, University of Lancaster, England.

Mr B. Fitton, ESTEC, Noordwijk, Holland.

Mr J. G. Fitton, Dept. of Geology, University of Durham, England.

Prof. J. O. Fleckenstein, Institute of Mathematics, University of Basle, Basle, Rheinsprung 21, Switzerland.

Mr R. J. Foster, Dept. of Geology, University of Newcastle-upon-Tyne, England.

Dr P. Fricker, Swiss National Science Foundation, Wildhainweg 20, Berne, Switzerland.

Dr Froeschle, Paris Observatory, France.

Mr R. J. Fryer, Dept. of Environmental Sciences, University of Lancaster, England.

Dr M. Fulchignoni, Laboratorio di Astrofisica Spaziale, PB 67, 00044 Frascati, Italy.

Mr R. M. Gallet, ERL-NOAA, Radio Building, Boulder, Colo. 80302, U.S.A.

Prof. G. F. J. Garlick, Dept. of Physics, University of Hull, England.

Dr T. F. Gaskell, British Petroleum, Britannic House, Moor Lane, London EC2, England.

Dr J. E. Geake, Dept. of Physics, University of Manchester, Inst. of Technology, Manchester, England.

Dr Gibb, Dept. of Geology, University of Manchester, England.

Dr C. Gilbert, School of Mathematics, University of Newcastle-upon-Tyne, England.

Mr H. Giles, Dept. of Geology, University of Manchester, England.

Dr R. W. Girdler, School of Physics, University of Newcastle-upon-Tyne, England.

Prof. T. Gold, Cornell University, N.Y. 14850, U.S.A.

Mr P. M. Goodwin, BBC External Service, Bush House, London WC2, England.

Prof. C. L. Goudas, Dept. of Mechanics, University of Patras, Greece.

Mr A. G. Green, School of Physics, University of Newcastle-upon-Tyne, England.

Dr D. H. Green, Dept. of Geophysics and Geochemistry, Australian National University, Canberra, Australia.

Prof. N. N. Greenwood, Dept. of Geology, University of Newcastle-upon-Tyne, England.
Mr A. J. Grindrod, Dept. of Geology, University of Newcastle-upon-Tyne, England.
Mr M. J. Gross, School of Physics, University of Newcastle-upon-Tyne, England.
Dr J. E. Guest, University of London Observatory, Mill Hill Park, London NW7, England.
Mr A. E. Hailwood, School of Physics, University of Newcastle-upon-Tyne, England.
Mr J. D. H. Hair, c/o School of Physics, University of Newcastle-upon-Tyne, England.
Mr S. A. Hall, School of Physics, University of Newcastle-upon-Tyne, England.
Prof. J. E. Hemingway, Dept. of Geology, University of Newcastle-upon-Tyne, England.
Prof. W. R. Hindmarsh, School of Physics, University of Newcastle-upon-Tyne, England.
Mrs S. Hofmann, School of Physics, University of Newcastle-upon-Tyne, England.
Miss B. Horsfield, BBC Television, Villiers House, Haven Green, Ealing, London W5, England.
Mr M. J. Ince, c/o Dept. of Geology, University of Newcastle-upon-Tyne, England.
Dr J. O. Isard, Dept. of Glass Technology, University of Sheffield, England.
Mr D. A. Jones, c/o School of Physics, University of Newcastle-upon-Tyne, England.
Prof. P. van de Kamp, Sproul Observatory, Swarthmore, Pa. 19081, U.S.A.
Dr G. N. Katterfeld, Dept. of Geomorphology, Leningrad State University, Leningrad C 124, U.S.S.R.
Dr W. M. Kaula, Inst. of Geophysics and Planetary Physics, University of California, Los Angeles, U.S.A.
Mr P. Kazantzis, Dept. of Mechanics, University of Patras, Greece.
Mr J. W. Kent, Dept. of Botany, University of Newcastle-upon-Tyne, England.
Mrs M. Kimhall, Cavendish Laboratory, Cambridge, England.
Dr L. Kopecký, Central Geological Survey, Prague 1, Czechoslovakia.
Prof. K. Koziel, Observatorium Astronomiczne, Krakow, Poland.
Dr N. A. Kozyrev, The Pulkovo Observatory, Prague, Czechoslovakia.
Dr L. Křivský, Astronomical Institute, Ondřejov, Czechoslovakia.
Mr J. K. Landon, Nuffield Radio Astronomy Laboratories, Jodrell Bank, Macclesfield, Cheshire, England.
Mr J. Leck, Forest Hall Observatory, Newcastle-upon-Tyne, England.
Miss F. Lister, Langdale Hall, Rusholme, Manchester, England.
Dr Long, Dept. of Mineralogy and Petrology, University of Cambridge, England.
Dr F. J. Lowes, School of Physics, University of Newcastle-upon-Tyne, England.
Dr R. C. Maddison, Dept. of Physics, University of Keele, England.
Mr A. J. Manson, School of Physics, University of Newcastle-upon-Tyne, England.
Prof. R. G. Mason, Dept. of Applied Geophysics, Imperial College, London, England.
Dr M. Maurette, Centre de Spectrométrie de Masse du C.N.R.S. France.
Dr J. R. Maxwell, Dept. of Chemistry, University of Bristol, England.
Mr B. J. Mays, Dept. of Chemistry, University of Bristol, England.

Dr J. A. M. McDonnell, Dept. of Electronics, University of Kent at Canterbury, England.

Prof. D. H. Menzel, Harvard College Observatory, Smithsonian Astrophysical Observatory, 60 Garden Street, Cambridge, Mass. 02138, U.S.A.

Dr W. H. Michael, Langley Research Centre, Virginia, U.S.A.

Dr A. Michaelis, Science Correspondent, Daily Telegraph, London, England.

Miss B. M. Middlehurst, Encyclopedia Britannica, 2452 N. Avenue, Michigan, U.S.A.

Dr A. Mills, Dept. of Geology, University of Leicester, England.

Dr J. G. Mitchell, School of Physics, University of Newcastle-upon-Tyne, England.

Dr L. Molyneux, School of Physics, University of Newcastle-upon-Tyne, England.

Mr I. Morrison, Nuffield Radio Astronomy Laboratories, Jodrell Bank, Macclesfield, Cheshire, England.

Mr L. V. Morrison, Royal Greenwich Observatory, Herstmonceux Castle, Hailsham, Sussex, England.

Dr M. Moutsoulas, Dept. of Astronomy, University of Manchester, England.

Mr P. M. Muller, Jet Propulsion Laboratory, Pasadena, California, U.S.A.

Mr J. B. Murray, University of London Observatory, Mill Hill Park, London NW7, England.

Dr H. E. Newell, NASA, Washington, DC 20546, U.S.A.

Mr J. B. O'Donovan, School of Physics, University of Newcastle-upon-Tyne, England.

Prof. M. O'Hara, Dept. of Geology, University of Edinburgh, Scotland.

Dr J. A. O'Keefe, Goddard Space Flight Center, Greenbelt, Maryland, U.S.A.

Dr Orszag, Paris, France.

Mr R. S. Osbond, University of Newcastle-upon-Tyne, England.

Mr H. I. Otulana, School of Physics, University of Newcastle-upon-Tyne, England.

Mr S. J. Pandya, Dept. of Physics, Queen Mary College, London SW1, England.

Dr D. Parkinson, ESRIN, C. P. 64, 00044 Frascati, Rome, Italy.

Dr J. H. Parry, School of Physics, University of Newcastle-upon-Tyne, England.

Dr C. T. Pillinger, Dept. of Chemistry, University of Bristol, England.

Mr H. Pinkerton, Dept. of Environmental Sciences, University of Lancaster, England.

Dr J. E. B. Ponsonby, Nuffield Radio Astronomy Laboratories, Jodrell Bank, Macclesfield, Cheshire, England.

Prof. A. T. Price, 23 Butlers Court Road, Beaconsfield, Bucks., England.

Mrs B. Read, 1 Meadow Prospect, Oxford, England.

Mr J. C. W. Richards, School of Physics, University of Newcastle-upon-Tyne, England.

DrM . L. Richards, School of Physics, University of Newcastle-upon-Tyne, England.

Mr H. Rickman, Stockholm Observatorium, 13300 Saltsjöbaden, Sweden.

Mr B. Riddle, Dept. of Geology, University of Durham, England.

Prof. A. E. Ringwood, Dept. of Geophysics and Geochemistry, Australian National University, Canberra, Australia.

Dr L. B. Ronca, Lunar Science Institute, Houston, Tex. 77058, U.S.A.

Mr B. R. Rosen, School of Physics, University of Newcastle-upon-Tyne, England.

Prof. S. K. Runcorn, School of Physics, University of Newcastle-upon-Tyne, England.

Prof. G. S. Rushbrooke, School of Physics, University of Newcastle-upon-Tyne, England.

Prof. C. Sagan, Cornell University, Ithaca, N.Y. 14850, U.S.A.

Dr A. San Migual, Dept. of Petrology, University of Barcelona, Spain.

Capt. L. Scherer, NASA, Washington DC, U.S.A.

Dr L. Schultz, Eidg. Technische Hochschule, Institut für Kristallographie u. Petrographie, Sonneggstrasse 5, Zürich, Switzerland.

Prof. P. Signer, Eidg. Technische Hochschule Institut für Kristallographie u. Petrographie, Sonneggstrasse 5, Zürich, Switzerland.

Dr G. Simmons, Manned Spacecraft Center, Houston, Tex. 77058, U.S.A.

Dr P. R. Simpson, Institute of Geological Sciences, Prince's Gate, London SW7, England.

Mr W. B. Simpson, Organic Geochemistry Unit, University of Newcastle-upon-Tyne, England.

Mr G. Sisson, Sir Howard Grubb Parsons Optical Works, Fossway, Newcastle-upon-Tyne, England.

Dr A. A. Smailes, A.E.R.E., Harwell, England.

Mr L. Srnka, Culham Laboratory, Abingdon, Berks., England.

Dr A. Stephenson, School of Physics, University of Newcastle-upon-Tyne, England.

Mr F. R. Stephenson, School of Physics, University of Newcastle-upon-Tyne, England.

Mr K. Stevens, Oaklands Observatory, Littleover, Derby, England.

Mrs M. P. Strens, School of Physics, University of Newcastle-upon-Tyne, England.

Dr R. G. J. Strens, School of Physics, University of Newcastle-upon-Tyne, England.

Dr R. G. Strom, University of Arizona, Tucson, Arizona, U.S.A.

Mr R. Thompson, School of Physics, University of Newcastle-upon-Tyne, England.

Mr L. Thorning, School of Physics, University of Newcastle-upon-Tyne, England.

Mr D. C. Thornton, Dept. of Natural Philosophy, University of Aberdeen, Scotland.

Dr C. W. Titman, School of Physics, University of Newcastle-upon-Tyne, England.

Mr D. J. Titman, School of Physics, University of Newcastle-upon-Tyne, England.

Dr D. C. Tozer, School of Physics, University of Newcastle-upon-Tyne, England.

Dr R. Trigila, Università di Roma, Istituto di Mineralogia, Rome, Italy.

Prof. H. C. Urey, Dept. of Chemistry, University of California, San Diego, U.S.A.

Mr G. Walker, Dept. of Physics, University of Manchester, Inst. of Technology, Manchester, England.

Dr T. Weimer, Paris Observatory, France.

Prof. T. S. Westoll, Dept. of Geology, University of Newcastle-upon-Tyne, England.

Mr D. Williams, School of Physics, University of Newcastle-upon-Tyne, England.

Dr I. Wilkinson, School of Physics, University of Newcastle-upon-Tyne, England.

Dr L. Wilson, Dept. of Environmental Sciences, University of Lancaster, England.

Mr W. R. Wollenhaupt, NASA/Manned Spacecraft Center, Houston, Tex. 77058, U.S.A.

Mr C. A. Wood, University of Arizona, U.S.A.

Mr B. J. Wood, School of Physics, University of Newcastle-upon-Tyne, England.

Prof. J. Zussman, Dept. of Geology, University of Manchester, England.

PART I

INTRODUCTION

REPORT BY THE CHAIRMAN

G. FIELDER
Univ. of Lancaster, Lancaster, England

1. Formation of Working Group 3 of Commission 17

A relevant observation regarding the First Lunar Science Conference at Houston, Texas, in January, 1970, was made by Dr A. Dollfus as Chairman of a Special Working Session on lunar geology, geophysics and returned lunar samples, held at Brighton, England, in August 1970 on the occasion of the Fourteenth General Assembly of the International Astronomical Union: Dollfus remarked that very few astronomers were present at Houston. Clearly, the highest efficiency in the study of the Moon will be achieved only if a suitable forum is created for the exchange of views of scientists from a variety of disciplines bearing on the lunar problem. Accordingly, Commission 17 (The Moon) of the International Astronomical Union set up a Working Group to assist with the geological and geophysical interpretation of the Moon in close connexion with the relevant astrophysical topics. Two meetings of this Working Group 3 were held during the IAU Symp. 47 'The Moon', at Newcastle-upon-Tyne, in April, 1971. The following recommendations and problems derive from these meetings and from correspondence with IAU members not present at the meetings.

2. Recommendations of Working Group 3

(i) Working Group 3 stresses the utility of interdisciplinary, international collaboration in future lunar exploration. Joint studies should be encouraged between Commissions 17, 7 (Celestial Mechanics), 16 (Planets and Satellites), 19 (Rotation of the Earth), 22 (Meteorites) and 40 (Radio Astronomy).

(ii) Working Group 3 should assist the ICSU's Inter-Union Commission for Study of the Moon (IUCM) to provide a forum for discussing the application of lunar sample analyses to the advancement of lunar science in general. Such a forum has the potential of defining the lunar problem on an international level by scientists from different, but interconnected, fields. Through such a forum, nations can extract a balanced view on the magnitude and nature of the effort to involve in future lunar exploration, with the aim of achieving the highest scientific rewards.

(iii) The IAU should foster the development of cosmogenic theories which draw on modern space data as well as the older, astronomical data pertaining to the Moon. Commission 17 should bring this problem to the attention of Commission 7. Should there be a revived interest in this field, it would provide an opportunity to bring scientists in some of the smaller astronomical institutes into the main current of modern thought.

(iv) The IAU should foster the international exchange of lunar reports. Investi-

Urey and Runcorn (eds.), The Moon, 3–5.

gators should send copies of their reports to the World Data Centres. Astronomers, in particular, should have access to the principal geological results deriving from the analysis of returned lunar specimens. Catalogues and charts of lunar photographs, and supporting data such as bibliographies stemming from the lunar exploration programme, should be distributed to those members of the international scientific community who would wish to participate in cooperative studies of the Moon. The Working Group hopes that the U.S.A. and U.S.S.R. World Data Centres will be in a position to provide interested members of the IAU with lunar reports, photographs and scientific supporting data.

(v) The IAU should foster the international exchange of returned lunar specimens and the collaborative study of them. Working Group 3 considers it of high scientific importance that lunar samples for analysis be drawn, in the future, from at least one, non-flooded highland site typical in terms of albedo, radar reflectivity, thermal behaviour, polarisation and crater number-density.

(vi) Study of the lunar samples has been more thorough than the study of many terrestrial rocks. Together with the experiments which have now been placed on the Moon, the results are adequate to shed light on some basic problems of lunar history and processes; but these base-line studies are not sufficient to enable geologists to describe with any certainty the chemical and petrological evolution of the Moon. More intensive sampling is required, the use of roving vehicles is to be commended; and every effort should be made to identify the precise location (and orientation) of lunar material sampled in the future. Nevertheless, the space programmes have already provided a backlog of data and have raised questions which can lead to productive theoretical and other Earth-based studies, and Working Group 3 considers that it would not be undesirable to consolidate interpretative studies and thought on lunar problems over a period of several years.

3. Questions that Need to be Answered

A. THE MOON'S FIGURE AND HISTORY

Many well placed laser retroflectors are required to provide (a) a net of points of accurately known relative altitude; (b) modern data for rediscussion of the Moon's physical libration (following the treatment being developed by A. Cook); and (c) the value of the Moon's secular acceleration in recent time (over a sufficiently long period).

Treatments of past evolution of the Moon's orbit are particularly prone to weakness. At what stage of the Moon's evolution did the terrestrial oceans come into being and to what extent and for how long did each contribute to the dissipation of the Earth's angular momentum? Is dissipation in the Earth's core negligible? How have the shallow seas, the distribution and area of continents, and the internal structure of the Earth (and Moon) changed over the ages? Is it clear that hamonics higher than the second should be neglected in discussing tidal friction? R. Hipkin has raised the question "Was the Moon over trapped in a resonant orbit?" Have there been major fluctuations in the secular acceleration of the Moon; and is it even possible, at the present time, to produce a reliable model of the evolution of the Moon's orbit?

B. MASCONS AND THERMAL HISTORY OF MOON

Topographic profiles are required across mascon areas. Used in conjunction with the gravity measurements they will enable closer definitions of mascons to be made.

Many heat flow measurements using instruments above and beneath the Moon's surface are desired; an accurate determination of the amount of heat flowing from the Moon's interior is required, at least by remote sensing methods. Improved magnetic observations are required to provide better conductivity and temperature profiles in the Moon. There is a need for more sophisticated theoretical work on the melting and thermal history of the Moon, taking full regard of the lunar sample analyses as well as of the heat flow and the magnetic measurements.

C. LUNAR VOLCANISM

Samples of lunar highland material are required to test for defluidisation. The distal zones of lunar flows should be examined with a view to studying the behaviour of materials moving under vacuum conditions. The Straight Wall should be examined for evidence of layering; and a means should be found for the mapping of subsurface layering. A close-up study should be made of lunar domes; and of craters related to, and in, rilles. The further study of tektites should be encouraged.

PART II

SCIENTIFIC PAPERS

A. LUNAR MECHANICS

DYNAMICS OF THE MOON

SIR HAROLD JEFFREYS

St. John's College, Cambridge, England

Abstract. Koziel's results on the librations have been rediscussed. Some serious departures from independence of the errors were detected and have been allowed for. The results are

$$\beta = 0.0006271 \pm 0.0000010$$
$$\gamma = 0.0002362 \pm 0.0000082$$

These do not differ much from Koziel's values but the uncertainties are larger. Both are consistent with results previously derived by comparison of results of different authors, and γ is consistent with a reinterpretation of a result of Yakovkin.

The effect of elasticity is considered. It is shown that the elastic strain does not contribute to the librations but would affect the perturbations of a satellite. Allowance for this difference reduces Michael's estimate of 0.4015 for C/Ma^2 to 0.4001 ± 0.0030, which would be consistent with either uniform density or with the value I found in 1936 after allowance for compressibility and a possible thin surface layer, namely 0.3971 ± 0.0007.

Since last year's meeting on the Moon I have carried out some revision of the results described then. A serious mistake had been found by Habibullin and Schrutka-Rechtenstamm in Yakovkin's estimate of the term in the libration in longitude with a period close to 3 years. I was not altogether satisfied with Koziel's analysis of four series of observations, since possibility of correlation of the errors had not been checked. It was possible however to solve his separate sets of normal equations, and the results differed by more than random errors would explain. This was allowed for in a revised treatment. Available values of β and γ are now as follows.

$$\begin{aligned}
\beta &= 0.0006279 \pm 0.0000015 && \text{(Jeffreys, 1961, by comparison of 10 determinations)}\\
&= 0.0006294 \pm 0.0000006 && \text{(Koziel)}\\
&= 0.0006271 \pm 0.0000010 && \text{(Koziel revised)}\\
\gamma &= 0.0002398 \pm 0.0000092 && \text{(Yakovkin, corrected)}\\
&= 0.0002274 \pm 0.0000088 && \text{(Jeffreys, 1961, by comparison of 20 determinations)}\\
&= 0.0002310 \pm 0.0000032 && \text{(Koziel)}\\
&= 0.0002362 \pm 0.0000082 && \text{(Koziel revised).}
\end{aligned}$$

The revised values are within the standard errors of the others.

Elasticity has a well known effect on the free nutation of the Earth, and it seemed possible that it might have one on the Moon's librations. The treatment turned out to be easy. The elastic deformation is always in the direction of the disturbing body, and the couples are consequently unaffected by it. Hence the values of β and γ estimated from the librations do not include the parts contributed by elastic deformation. But these parts do affect the motion of a satellite travelling near the Moon. Now comparison of its perturbations, on the supposition that the Moon is rigid, with the librations leads to a determination of C/Ma^2 for the Moon. The best determination so

Urey and Runcorn (eds.), The Moon, 11–12.

far is by W. H. Michael, announced at the 1970 conference of the IAU, namely

$$C/Ma^2 = 0.4015 \pm 0.0030\,.$$

Correction for elasticity reduces this to 0.4001 ± 0.0030. This would agree with either uniform density or with 0.3971 ± 0.0007, which I derived in 1936 by allowing for compression and the possibility of a granitic layer. At any rate Eckert's result from the motion of the Moon's node, leading to a structure like a tennis ball, must have some other explanation.

Reference

Jeffreys, H.: 1971, *Monthly Notices Roy. Astron. Soc.* **153**, 73.

ON THE INCLINATION OF THE LUNAR AXIS

M. MOUTSOULAS
Department of Astronomy, University of Manchester, Manchester, England

Since the days of Cassini (1693), study of the position of the lunar axis relative to an inertial system of reference has been based on the assumption that "the inclination of the Moon's equator to the plane of the ecliptic is constant". And, although the actual value of that inclination has been subject to continuous changes and modifications, reduced from the originally suggested $4\frac{1}{2}°$ to $1°32'4''$ (Koziel, 1967), no sufficient attention has been paid to the fact that, as the lunar globe moves within the field of varying external forces, the inclination of its axis cannot remain constant.

Moreover, certain confusion seems to have been involved in works dealing with the subject, and several 'axes' of the Moon appear to be interchanged with each other and used quite inconsistently. It must be understood that the shortest inertial axis of the lunar dynamical configuration, the instantaneous axis of rotation of the Moon, and the rotation axis which the Moon would possess if it could obey precisely Cassini's laws of motion, do not coincide with each other (Habibullin, 1968). Therefore, whenever measurements of reference points of the lunar surface are reduced for the determination of the position of the lunar 'axis', either the selenographic coordinates of those points or the inclination of the axis they are referred to, should be expressed as time-dependent functions. Which one would be considered as constant depends on the definition; but they cannot both possess constant values.

The Eulerian equations of motion provide the required relations between the system of principal axes of the Moon, $Oxyz$, and a system fixed in space. If we adopt as our fixed system the ecliptic system of coordinates and we represent the longitude of the descending node of the Moon's equator, the angular distance of the direction Ox from the descending node of the Moon's equator, and the inclination of the lunar equator to the ecliptic, with ψ, φ and θ, respectively, the kinematic equations take the form:

$$\omega_x = -\frac{d\psi}{dt}\sin\theta\sin\varphi - \frac{d\theta}{dt}\cos\varphi \tag{1}$$

$$\omega_y = -\frac{d\psi}{dt}\sin\theta\cos\varphi + \frac{d\theta}{dt}\sin\varphi \tag{2}$$

$$\omega_z = \frac{d\psi}{dt}\cos\theta + \frac{d\varphi}{dt}. \tag{3}$$

where ω_x, ω_y and ω_z stand for the components of the angular velocity along the x, y, z axes.

Urey and Runcorn (eds.), The Moon, 13–21. *All Rights Reserved.*

The system of Equations (1)–(3) solved for $d\psi/dt$, $d\varphi/dt$ and $d\theta/dt$ gives:

$$\frac{d\psi}{dt} = -\frac{\omega_x \sin\varphi + \omega_y \cos\varphi}{\sin\theta} \tag{4}$$

$$\frac{d\varphi}{dt} = \frac{(\omega_x \sin\varphi + \omega_y \cos\varphi)\cos\theta}{\sin\theta} + \omega_z \tag{5}$$

$$\frac{d\theta}{dt} = \omega_y \sin\varphi - \omega_x \cos\varphi\,. \tag{6}$$

We can introduce into the right-hand side of the Equations (4)–(6) the components σ, τ and ϱ of the physical libration, which, as is known, represent deviations in the position of the $Oxyz$ system from that prescribed by Cassini's laws, and therefore are connected to the angles ψ, φ and θ by means of the relations:

$$\psi = \Omega + \sigma \tag{7}$$

$$\varphi = 180^\circ + l - \psi + \tau \tag{8}$$

$$\theta = I + \varrho\,. \tag{9}$$

where Ω is the longitude of the ascending node of the lunar orbit, l the mean longitude of the Moon and I the mean value of the inclination of the lunar equator to the ecliptic.

Retaining the first-order terms for the small quantities σ, τ and ϱ, and taking into account the relation:

$$l = g + \omega + \Omega \tag{10}$$

which expresses the mean longitude of the Moon in terms of the mean anomaly of the Moon, g, the angular distance of the Moon's perigee from the ascending node of the orbit, ω, and the longitude of the ascending node of the lunar orbit, Ω, we find the system of equations:

$$\frac{d\psi}{dt} = \frac{1}{\sin I + \varrho\cos I} \times \times \{[\omega_x \cos(g+\omega) - \omega_y \sin(g+\omega)](\tau - \sigma) + \omega_x \sin(g+\omega) + \omega_y \cos(g+\omega)\} \tag{11}$$

$$\frac{d\varphi}{dt} = -\frac{\cos I}{\sin I + \varrho\cos I} \times \times \{[\omega_x \cos(g+\omega) - \omega_y \sin(g+\omega)](\tau - \sigma) + \omega_x \sin(g+\omega) + \omega_y \cos(g+\omega)\} + \frac{\sin I}{\sin I + \varrho\cos I}[\omega_x \sin(g+\omega) + \omega_y \cos(g+\omega)]\,\varrho + \omega_z \tag{12}$$

$$\frac{d\theta}{dt} = [\omega_x \sin(g+\omega) + \omega_y \cos(g+\omega)](\sigma - \tau) + \omega_x \cos(g+\omega) - \omega_y \sin(g+\omega) \tag{13}$$

or the equivalent to that system:

$$\frac{d\sigma}{dt} = \frac{1}{\sin I + \varrho \cos I} \times \times \{[\omega_x \cos(g + \omega) - \omega_y \sin(g + \omega)] \times \times (\tau - \sigma) + \omega_x \sin(g + \omega) + \omega_y \cos(g + \omega)\} - \frac{d\Omega}{dt} \tag{14}$$

$$\frac{d\tau}{dt} = \frac{1 - \cos I}{\sin I + \varrho \cos I} \{[\omega_x \cos(g + \omega) - \omega_y \sin(g + \omega)] \times \times (\tau - \sigma) + \omega_x \sin(g + \omega) + \omega_y \cos(g + \omega)\} + \frac{\sin I}{\sin I + \varrho \cos I} \times \times [\omega_x \sin(g + \omega) + \omega_y \cos(g + \omega)]\, \varrho + \omega_z - \frac{dg}{dt} - \frac{d\omega}{dt} - \frac{d\Omega}{dt} \tag{15}$$

$$\frac{d\varrho}{dt} = [\omega_x \sin(g + \omega) + \omega_y \cos(g + \omega)] (\sigma - \tau) + \omega_x \cos(g + \omega) - \omega_y \sin(g + \omega). \tag{16}$$

Solution of the system of Equations (11)–(13), or its equivalent (14)–(16), will give the value of the true inclination, θ, of the Moon's axis to that of the ecliptic, as a function of time.

The velocity components ω_x, ω_y and $\tilde{\omega}_z$ are derived from Euler's dynamical equations:

$$A \frac{d\omega_x}{dt} = (B - C)\left(\omega_y \omega_z - \frac{3Gm_\oplus}{R^5} y_E z_E\right) \tag{17}$$

$$B \frac{d\omega_y}{dt} = (C - A)\left(\omega_z \omega_x - \frac{3Gm_\oplus}{R^5} z_E x_E\right) \tag{18}$$

$$C \frac{d\omega_z}{dt} = (A - B)\left(\omega_x \omega_y - \frac{3Gm_\oplus}{R^5} x_E y_E\right) \tag{19}$$

where A, B, C are the principal moments of inertia of the Moon, G is the gravitational constant, $m_\oplus$ is the mass of the Earth, R the distance between the centre of mass of the Moon and that of the Earth, and x_E, y_E, z_E the rectangular coordinates of the centre of the Earth in the selenocentric system $Oxyz$, which are related to the true selenocentric longitude of the Earth, υ, the true geocentric latitude of the Moon, $B_☾$, and the Eulerian angles, by means of the expressions:

$$x_E = R \cos B_☾ [\cos(\upsilon - \varphi) - \sin\upsilon \sin\varphi (1 - \cos\theta) + \tan B_☾ \sin\varphi \sin\theta], \tag{20}$$

$$y_E = R \cos B_☾ [\sin(\upsilon - \varphi) - \sin\upsilon \cos\varphi (1 - \cos\theta) + \tan B_☾ \cos\varphi \sin\theta], \tag{21}$$

$$z_E = R \cos B_☾ [\sin\upsilon \sin\theta - \tan B_☾ \cos\theta]. \tag{22}$$

As is known, the true geocentric latitude of the Moon, $B_☾$, can be expressed in terms of the inclination of the lunar orbit to the ecliptic, i, and the true selenocentric longitude of the Earth, υ, by means of the formula:

$$\tan B_☾ = -\tan i \sin \upsilon . \tag{23}$$

We can, moreover, express the coordinates x_E, y_E, z_E in terms of the true geocentric longitude and latitude of the Moon, L and $B_☾$, its mean anomaly, g, the angular distance of the lunar perigee from the ascending node of the orbit, ω, and the physical libration components σ, τ, ϱ, taking into account the relations (7)–(9) and the fact that

$$\upsilon = 180° + L + \psi . \tag{24}$$

Then the system of Equations (17)–(19) takes the form:

$$\begin{aligned}\frac{d\omega_x}{dt} + \alpha\omega_y\omega_z = {} & \frac{3Gm_{\oplus}\alpha \cos^2 B_☾}{R^3} \{\sigma [\sin (L-l)\cdot\cos (L-l+g+\omega) - \\ & - (1-\cos I + \tan i \sin I) \{\tfrac{1}{2} \sin 2 (L-l+g+\omega) \cos (g+\omega) + \\ & + \sin (L-l+g+\omega) \cos (L-l+2g+2\omega)\}] \sin (I+i) + \\ & + \tau [\cos (L-l) \sin (L-l+g+\omega) - (1-\cos I + \tan i \sin I) \times \\ & \times \sin^2 (L-l+g+\omega) \sin (g+\omega)] \sin (I+i) + \varrho \{[(1-\cos I + \\ & + \tan i \sin I) \sin^2 (L-l+g+\omega) \cos (g+\omega) - \sin (L-l) \times \\ & \times \sin (L-l+g+\omega)] \cos (I+i) + (\sin I + \tan i \cos I) \sin (I+i) \times \\ & \times \sin^2 (L-l+g+\omega) \cos (g+\omega)\} + [(1-\cos I + \tan i \sin I) \times \\ & \times \sin^2 (L-l+g+\omega) \cos (g+\omega) - \sin (L-l) \times \\ & \times \sin (L-l+g+\omega)] \sin (I+i)\}\end{aligned} \tag{25}$$

$$\begin{aligned}\frac{d\omega_y}{dt} - \beta\omega_z\omega_x = {} & -\frac{3Gm_{\oplus}\beta \cos^2 B_☾}{R^3} \{\sigma [\cos (L-l) \cos (L-l+g+\omega) - \\ & - (1-\cos I + \tan i \sin I) \{\tfrac{1}{2} \sin^2 (L-l+g+\omega) \sin (g+\omega) + \\ & + \sin (L-l+2g+2\omega) \sin (L-l+g+\omega)\}] \sin (I+i) - \\ & - \tau [\sin (L-l) \sin (L-l+g+\omega) - (1-\cos I + \tan i \sin I) \times \\ & \times \sin^2 (L-l+g+\omega) \cos (g+\omega)] \sin (I+i) + \\ & + \varrho [\{(1-\cos I + \tan i \sin I) \sin^2 (L-l+g+\omega) \sin (g+\omega) - \\ & - \cos (L-l) \sin (L-l+g+\omega)\} \cos (I+i) + (\sin I + \tan i \cos I) \times \\ & \times \sin (I+i) \sin^2 (L-l+g+\omega) \sin (g+\omega)] + \\ & + [(1-\cos I + \tan i \sin I) \sin^2 (L-l+g+\omega) \sin (g+\omega) - \\ & - \cos (L-l) \sin (L-l+g+\omega)] \sin (I+i)\}\end{aligned} \tag{26}$$

$$\begin{aligned}\frac{d\omega_z}{dt} + \gamma\omega_x\omega_y = {} & \frac{3Gm_{\oplus}\gamma \cos^2 B_☾}{2R^3} \{2\sigma (1-\cos I + \tan i \sin I) \cos 2 (g+\omega) - \\ & - 2\tau [(\cos I - \tan i \sin I) \cos 2 (L-l) - (1-\cos I + \tan i \sin I) \times \\ & \times \cos 2 (g+\omega)] - \varrho (\sin I + \tan i \cos I) [\sin 2 (L-l) + \sin 2 (g+\omega)] + \\ & + (\cos I - \tan i \sin I) \sin 2 (L-l) - \\ & - (1-\cos I + \tan i \sin I) \sin 2 (g+\omega)\}\end{aligned} \tag{27}$$

where α, β and γ represent the ratios of mechanical ellipticities of the Moon:

$$\alpha = (C - B)/A, \quad \beta = (C - A)/B, \quad \gamma = (B - A)/C. \tag{28}$$

We are therefore led to the simultaneous solution of the Equations (14), (15), (16), (25), (26) and (27), which will produce the values of the corrections ϱ for the inclination of the Moon's shortest axis of inertia to the ecliptic, as well as the libration components σ and τ.

The variation in the inclination of the lunar axis during the year 1971, obtained by numerical integration of this system of equations, is presented in Table I. It has been assumed that at the beginning of the year the librations were zero and the inclination possessed the mean value of 1° 32′ 4″. In order to establish an ephemeris for the inclination, it now only remains to obtain by observational detection a set of measurements which will provide the correct initial conditions for the solution, while the periodic nature and other general characteristics of the nutational motion of the lunar axis are apparent in the values presented here.

TABLE I

Variation of the inclination of the lunar axis to the ecliptic over the period of a year

Day	Inclination	Day	Inclination
1	1° 32′ 4″	28	1° 32′ 5″
2	1° 32′ 1″	29	1° 31′ 59″
3	1° 31′ 57″	30	1° 31′ 55″
4	1° 31′ 51″		
5	1° 31′ 42″	31	1° 31′ 46″
6	1° 31′ 31″	32	1° 31′ 38″
7	1° 31′ 18″	33	1° 31′ 25″
8	1° 31′ 2″	34	1° 31′ 13″
9	1° 30′ 43″	35	1° 30′ 55″
10	1° 30′ 31″	36	1° 30′ 41″
		37	1° 30′ 23″
11	1° 30′ 18″	38	1° 30′ 12″
12	1° 30′ 10″	39	1° 30′ 4″
13	1° 30′ 6″	40	1° 30′ 2″
14	1° 30′ 7″		
15	1° 30′ 10″	41	1° 30′ 5″
16	1° 30′ 21″	42	1° 30′ 14″
17	1° 30′ 34″	43	1° 30′ 26″
18	1° 30′ 52″	44	1° 30′ 42″
19	1° 31′ 10″	45	1° 31′ 0″
20	1° 31′ 30″	46	1° 31′ 19″
		47	1° 31′ 38″
21	1° 31′ 44″	48	1° 31′ 55″
22	1° 31′ 58″	49	1° 32′ 7″
23	1° 32′ 4″	50	1° 32′ 16″
24	1° 32′ 11″		
25	1° 32′ 11″	51	1° 32′ 20″
26	1° 32′ 12″	52	1° 32′ 20″
27	1° 32′ 8″	53	1° 32′ 17″

Table I (continued)

Day	Inclination
54	1° 32′ 13″
55	1° 32′ 6″
56	1° 32′ 0″
57	1° 31′ 52″
58	1° 31′ 43″
59	1° 31′ 33″
60	1° 31′ 21″
61	1° 31′ 7″
62	1° 30′ 53″
63	1° 30′ 38″
64	1° 30′ 22″
65	1° 30′ 9″
66	1° 30′ 0″
67	1° 29′ 57″
68	1° 30′ 2″
69	1° 30′ 12″
70	1° 30′ 28″
71	1° 30′ 46″
72	1° 31′ 8″
73	1° 31′ 27″
74	1° 31′ 49″
75	1° 32′ 5″
76	1° 32′ 21″
77	1° 32′ 29″
78	1° 32′ 33″
79	1° 32′ 31″
80	1° 32′ 28″
81	1° 32′ 20″
82	1° 32′ 12″
83	1° 32′ 2″
84	1° 31′ 53″
85	1° 31′ 41″
86	1° 31′ 30″
87	1° 31′ 17″
88	1° 31′ 4″
89	1° 30′ 48″
90	1° 30′ 34″
91	1° 30′ 20″
92	1° 30′ 10″
93	1° 30′ 0″
94	1° 29′ 58″
95	1° 29′ 58″
96	1° 30′ 8″
97	1° 30′ 21″
98	1° 30′ 45″
99	1° 31′ 5″
100	1° 31′ 31″
101	1° 31′ 50″
102	1° 32′ 12″
103	1° 32′ 25″
104	1° 32′ 39″
105	1° 32′ 41″
106	1° 32′ 44″
107	1° 32′ 35″
108	1° 32′ 31″
109	1° 32′ 16″
110	1° 32′ 8″
111	1° 31′ 52″
112	1° 31′ 43″
113	1° 31′ 26″
114	1° 31′ 15″
115	1° 30′ 56″
116	1° 30′ 43″
117	1° 30′ 25″
118	1° 30′ 14″
119	1° 30′ 1″
120	1° 29′ 57″
121	1° 29′ 53″
122	1° 29′ 57″
123	1° 30′ 4″
124	1° 30′ 20″
125	1° 30′ 38″
126	1° 31′ 3″
127	1° 31′ 27″
128	1° 31′ 52″
129	1° 32′ 11″
130	1° 32′ 28″
131	1° 32′ 38″
132	1° 32′ 47″
133	1° 32′ 46″
134	1° 32′ 44″
135	1° 32′ 34″
136	1° 32′ 24″
137	1° 32′ 9″
138	1° 31′ 57″
139	1° 31′ 41″
140	1° 31′ 28″
141	1° 31′ 10″
142	1° 30′ 55″
143	1° 30′ 36″
144	1° 30′ 20″
145	1° 30′ 2″
146	1° 29′ 53″
147	1° 29′ 44″
148	1° 29′ 45″
149	1° 29′ 48″
150	1° 30′ 1″

Table I (continued)

Day	Inclination	Day	Inclination
151	1° 30′ 12″	200	1° 29′ 50″
152	1° 30′ 36″		
153	1° 30′ 56″	201	1° 29′ 38″
154	1° 31′ 26″	202	1° 29′ 33″
155	1° 31′ 47″	203	1° 29′ 33″
156	1° 32′ 13″	204	1° 29′ 43″
157	1° 32′ 27″	205	1° 30′ 2″
158	1° 32′ 43″	206	1° 30′ 25″
159	1° 32′ 45″	207	1° 30′ 53″
160	1° 32′ 50″	208	1° 31′ 21″
		209	1° 31′ 49″
161	1° 32′ 44″	210	1° 32′ 15″
162	1° 32′ 40″		
163	1° 32′ 26″	211	1° 32′ 37″
164	1° 32′ 16″	212	1° 32′ 54″
165	1° 31′ 58″	213	1° 33′ 5″
166	1° 31′ 46″	214	1° 33′ 6″
167	1° 31′ 27″	215	1° 33′ 3″
168	1° 31′ 14″	216	1° 32′ 52″
169	1° 30′ 53″	217	1° 32′ 40″
170	1° 30′ 38″	218	1° 32′ 23″
		219	1° 32′ 8″
171	1° 30′ 17″	220	1° 31′ 49″
172	1° 30′ 3″		
173	1° 29′ 45″	221	1° 31′ 32″
174	1° 29′ 38″	222	1° 31′ 11″
175	1° 29′ 35″	223	1° 30′ 53″
176	1° 29′ 42″	224	1° 30′ 33″
177	1° 29′ 52″	225	1° 30′ 17″
178	1° 30′ 11″	226	1° 30′ 0″
179	1° 30′ 32″	227	1° 29′ 48″
180	1° 30′ 57″	228	1° 29′ 38″
		229	1° 29′ 32″
181	1° 31′ 23″	230	1° 29′ 33″
182	1° 31′ 50″		
183	1° 32′ 14″	231	1° 29′ 41″
184	1° 32′ 35″	232	1° 29′ 56″
185	1° 32′ 47″	233	1° 30′ 18″
186	1° 32′ 56″	234	1° 30′ 48″
187	1° 32′ 56″	235	1° 31′ 17″
188	1° 32′ 52″	236	1° 31′ 48″
189	1° 32′ 43″	237	1° 32′ 13″
190	1° 32′ 33″	238	1° 32′ 38″
		239	1° 32′ 56″
191	1° 32′ 18″	240	1° 33′ 9″
192	1° 32′ 4″		
193	1° 31′ 46″	241	1° 33′ 14″
194	1° 31′ 30″	242	1° 33′ 13″
195	1° 31′ 12″	243	1° 33′ 2″
196	1° 30′ 56″	244	1° 32′ 49″
197	1° 30′ 37″	245	1° 32′ 30″
198	1° 30′ 21″	246	1° 32′ 13″
199	1° 30′ 4″	247	1° 31′ 51″

Table I (continued)

Day	Inclination	Day	Inclination
248	1° 31′ 33″	296	1° 33′ 19″
249	1° 31′ 10″	297	1° 33′ 12″
250	1° 30′ 51″	298	1° 33′ 3″
		299	1° 32′ 45″
251	1° 30′ 28″	300	1° 32′ 27″
252	1° 30′ 9″		
253	1° 29′ 51″	301	1° 32′ 3″
254	1° 29′ 38″	302	1° 31′ 42″
255	1° 29′ 28″	303	1° 31′ 17″
256	1° 29′ 26″	304	1° 30′ 56″
257	1° 29′ 29″	305	1° 30′ 32″
258	1° 29′ 36″	306	1° 30′ 11″
259	1° 29′ 52″	307	1° 29′ 46″
260	1° 30′ 13″	308	1° 29′ 29″
		309	1° 29′ 10″
261	1° 30′ 41″	310	1° 29′ 4″
262	1° 31′ 11″		
263	1° 31′ 44″	311	1° 29′ 4″
264	1° 32′ 11″	312	1° 29′ 16″
265	1° 32′ 37″	313	1° 29′ 32″
266	1° 32′ 54″	314	1° 29′ 59″
267	1° 33′ 8″	315	1° 30′ 27″
268	1° 33′ 15″	316	1° 31′ 0″
269	1° 33′ 16″	317	1° 31′ 34″
270	1° 33′ 9″	318	1° 32′ 7″
		319	1° 32′ 38″
271	1° 32′ 57″	320	1° 33′ 2″
272	1° 32′ 40″		
273	1° 32′ 19″	321	1° 33′ 18″
274	1° 31′ 58″	322	1° 33′ 27″
275	1° 31′ 36″	323	1° 33′ 28″
276	1° 31′ 14″	324	1° 33′ 22″
277	1° 30′ 51″	325	1° 33′ 9″
278	1° 30′ 29″	326	1° 32′ 56″
279	1° 30′ 6″	327	1° 32′ 35″
280	1° 29′ 45″	328	1° 32′ 14″
		329	1° 31′ 49″
281	1° 29′ 27″	330	1° 31′ 26″
282	1° 29′ 19″		
283	1° 29′ 13″	331	1° 31′ 0″
284	1° 29′ 18″	332	1° 30′ 38″
285	1° 29′ 28″	333	1° 30′ 15″
286	1° 29′ 48″	334	1° 29′ 55″
287	1° 30′ 6″	335	1° 29′ 35″
288	1° 30′ 37″	336	1° 29′ 19″
289	1° 31′ 5″	337	1° 29′ 5″
290	1° 31′ 41″	338	1° 29′ 1″
		339	1° 29′ 3″
291	1° 32′ 9″	340	1° 29′ 17″
292	1° 32′ 39″		
293	1° 32′ 57″	341	1° 29′ 43″
294	1° 33′ 13″	342	1° 30′ 12″
295	1° 33′ 17″	343	1° 30′ 48″

Table I (continued)

Day	Inclination	Day	Inclination
344	1° 31′ 22″	355	1° 32′ 25″
345	1° 31′ 58″	356	1° 31′ 57″
346	1° 32′ 30″	357	1° 31′ 35″
347	1° 32′ 59″	358	1° 31′ 6″
348	1° 33′ 20″	359	1° 30′ 43″
349	1° 33′ 35″	360	1° 30′ 17″
350	1° 33′ 35″		
		361	1° 29′ 57″
351	1° 33′ 33″	362	1° 29′ 35″
352	1° 33′ 20″	363	1° 29′ 22″
353	1° 33′ 6″	364	1° 29′ 10″
354	1° 32′ 44″	365	1° 29′ 3″

References

Cassini, G. D.: 1693, *Traité de l'origine et du progress de l'Astronomie*, Paris.
Habibullin, Sh. T.: 1968, *Soviet Astron.* **12**, 526.
Koziel, K.: 1967, *Icarus* **7**, 27.
Moutsoulas, M.: 1971, 'Librations of the Lunar Globe', in *Physics and Astronomy of the Moon* (2nd edition), Academic Press, New York.

THE SHAPE OF THE MOON

S. K. RUNCORN and S. HOFMANN
Department of Geophysics and Planetary Physics, School of Physics, University of Newcastle upon Tyne, England

Abstract. The determination of the heights of points on the lunar surface by Earth based astronomy using the geometrical librations, although individually of low accuracy, still provides our best method of obtaining the global shape of the Moon. The intrinsic scatter of the results arises from the effects of 'seeing' and simple statistical analysis is required to derive valid conclusions about the shape. Baldwin's method of fitting ellipsoidal surfaces to the points on the maria and uplands, separately by the method of least squares proves to be a valuable tool.

Analyses of the ACIC points and of the Pic du Midi studies of G. A. Mills show that good first descriptions of the global shape of the Moon for both the maria and uplands are triaxial ellipsoids with their long axes within 10° of the Earth direction, the major axis of the maria being about 1.3 km smaller than that of the uplands. Of particular significance is that the ellipticity of these surfaces is about $2\frac{1}{2}$ times greater than the dynamical ellipticity; thus the non-hydrostatic figure of the Moon is not simply the result of distortion from a uniform Moon during its early history. The angular variation in density within the Moon cannot be simply a phenomena within the crust but must extend to a great depth. Convection could provide an explanation.

The departures of the lunar surface from the idealised ellipsoids are also of interest. The circular maria are systematically depressed relative to the maria ellipsoid: can the mascons have adjusted isostatically since their formation? Systematic differences in height between the western and eastern southern uplands are also noted.

1. The Ellipticity of the Lunar Surface

The global shape of the Moon, that is, its shape neglecting local topography, is perhaps its most fundamental property. Yet the external shape of the Moon has been a controversial issue for a century and the major questions at issue are as follows. Can the Moon's overall shape be described by 'an earthward bulge', i.e. is there a strong second harmonic in its external surface? Is the bulge really directed towards the mean position of the Earth? If so, what is the height of the bulge, i.e. the excess of the earthward radius over that in the plane of the sky.

Astronomers, using the method of geometrical librations to determine the heights of points on the lunar surface from the centre of the Moon's figure, found values between 0 and 20 km for the height of this bulge. A major step towards understanding the method was taken by Baldwin (1949) who analysed points on the maria and uplands separately; he showed both that the maria were systematically lower than the uplands and that the spheroid fitted to each set of points had nearly the same ellipticity. However Kopal (1967) and Goudas (1963) analysed the points as a whole, fitting them to a spherical harmonic series and showed that higher harmonics than the second, especially the fourth, were needed to describe the external surface adequately. Therefore they were inclined to regard 'the lunar bulge' as meaningless or at least an oversimplification of the description of the shape of the Moon. Runcorn (1967) however argued that because the maria surfaces were distinctly younger than the uplands and evidently of a different physiochemical origin, the surfaces associated with the Moon which are

Urey and Runcorn (eds.), The Moon, 22–31.

of true physical significance are the surfaces of the uplands and maria taken separately rather than the external surface of the Moon.

Particularly the natural step of comparing the external gravitational field of the Moon with its shape would be more revealing if the two regions of the surface, evidently of different chemical composition (and therefore probably of different densities) were separately considered. So Runcorn and Gray (1967) and Runcorn and Shrubsall (1968), following Baldwin's further analysis (1963), analysed the more numerous data of the ACIC program by methods similar to but rather more general than his. A new set of data has now been obtained by Mills (1968) and this is now in this paper subjected to the same analysis, details of which are given in Runcorn and Gray (1967).

The difficulty of what in principle is a simple problem arises because the scatter of the data exceeds the quantities to be determined. Each height determination is subject to errors arising from the astronomical phenomenon of seeing and from the varying lighting of the peaks and other lunar features studied which, unlike stars, are not point sources. Even the differential displacements of the points arising from the geometrical librations are about the same as the angular resolving power of the telescopes used, multiplied by the Earth–Moon distance. Thus the Rayleigh diffraction limit of earth based astronomy sets a limit to the accuracy of a height determination as emphasized by Kopal (1967). The shape of the Moon by the method of geometrical librations is thus a problem of seeing through noise but Runcorn (1967) showed that, providing the errors are random, a determination of the height of the bulge or the low harmonics of its shape should be possible using, as Baldwin did, about 100 points. The much more accurate methods now available through space technology to determine height do not have anywhere near the overall coverage of the earth-based observations, so that the analysis of the geometrical libration data is still most important. The more numerous data now available enable many aspects of the Moon's shape to be discussed which were not hitherto possible. The only way to ensure that the effects are not the result of systematic errors is to use and compare different sets of measurements. Table I compares the different sets of results obtained by previous workers and the present paper giving the length of the semi-axes a_x, a_y and a_z of the ellipsoids which best fit points on the maria and the uplands separately (z is the polar axis and x the Earth aligned axis). Mills (1968) has listed lunar coordinates which we now analyse and compare with previous data. His data is divided, in Figure 1, into heights of points on the maria and for the uplands and the absolute heights of these lunar features are plotted against the sine of their angular distance (K) from the centre of the lunar disc. This shows clearly that the mean level of the maria points is below that of the uplands and also shows the reality of the earthward bulge at the centre of the disc. The full lines show the heights of the best fitting ellipsoids for each set using Mills' data and the dotted lines those fitting ACIC data, using the analysis of Runcorn and Shrubsall (1968), the a_y and a_z axes being taken as equal. The two pairs of lines show the good agreement between the analyses of the two sets of data.

In Figure 2 the variation is shown in the standard deviation of the mean error with the angle of orientation in the final fit of the ellipsoids to the maria and upland points,

TABLE I

Axes of best fitting ellipsoids to uplands and maria comparing with previous papers

	Uplands ellipsoid				
	Runcorn and Gray rot^n about z axis	Runcorn and Shrubsall rot^n about z axis	rot^n about x axis	Present paper rot^n about z axis	Present paper rot^n about x axis
No. of points	90	581	581	542	542
		a'_x	a''_x	a'_x	a''_x
a_x (in km)	1739.63	1739.91	1739.92	1740.19	1740.17
a_y (in km)	1736.65	1736.59	1736.30	1737.31	1736.97
a_z (in km)	1736.15	1736.90	1737.27	1736.48	1736.20
$a_x - a_y$ (in km)	2.98	3.32	3.62	2.88	3.20
$a_x - (a_z + a_y)/2$	3.23	3.165	3.14	3.30	3.58
Angle between a'_x of best fit and axis towards Earth	0°	7°W	–	5°E	
Angle between a''_z of best fit and polar axis	–	–	40° W	–	32.5° W
$a''_z - a''_y$	–	–	0.97	–	−0.77
	Maria ellipsoid				
No. of points	106	391	391	385	385
a_x (in km)	1738.99	1738.62	1738.56	1739.05	1739.02
a_y (in km)	1735.40	1735.28	1735.67	1735.97	1736.13
a_z (in km)	1734.56	1735.44	1735.42	1734.66	1734.68
$a_x - a_y$ (in km)	3.59	3.34	2.89	3.08	2.89
$a_x - (a_z + a_y)/2$	4.01	3.26	3.02	3.73	3.61
Angle between a'_x of best fit and axis towards Earth	5°E	5°E	–	5°E	–
Angle between a''_z of best fit and polar axis	–	–	30° W	–	5°W
$a''_z - a''_y$	–	–	−0.25	–	−1.45
	Uplands axis minus maria axes				
a_x (in km)	0.64	1.29	1.36	1.14	1.15
a_y (in km)	1.25	1.31	0.63	1.34	0.84
a_z (in km)	1.59	1.46	1.85	1.82	1.52
$a_x - a_y$ (in km)					
$a_x - (a_z + a_y)/2$					
	Fit by limb Goudas (1963)			Fit by limb Watts (1963)	
Angle between a''_z of best fit and polar axis	35° W			35°W	
$a''_z - a''_y$	1.9			1.12	

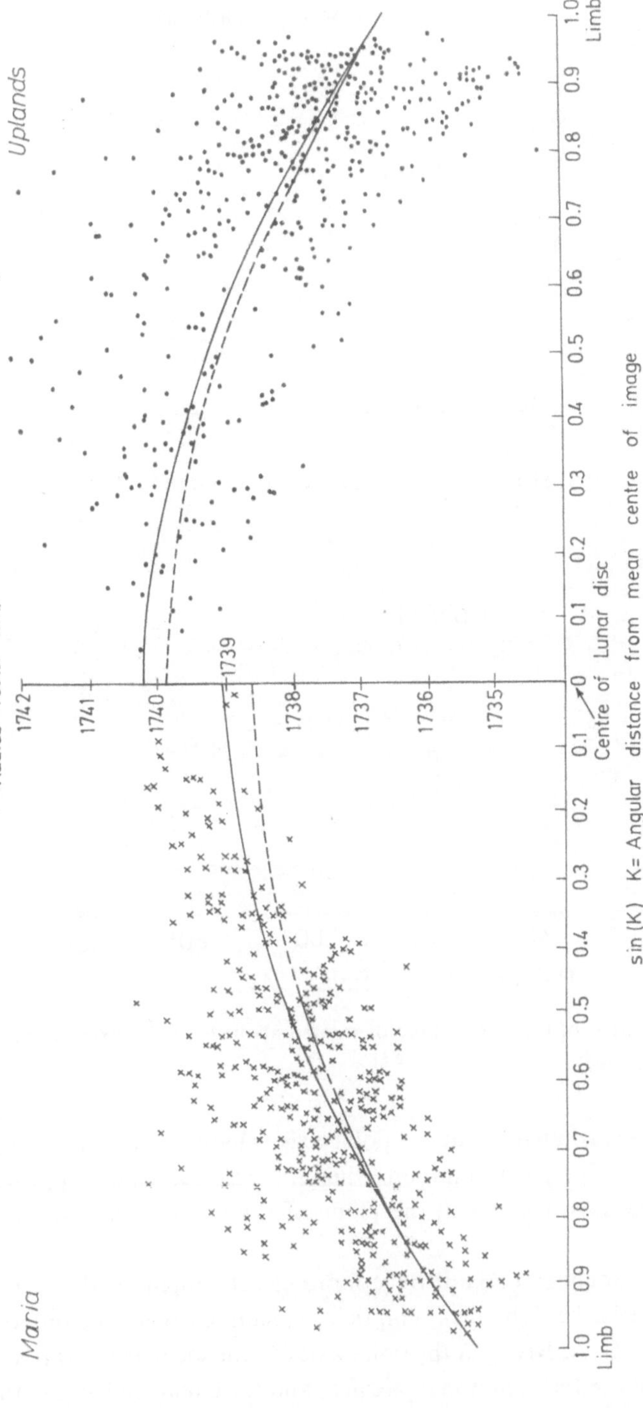

Fig. 1. Radius vector plotted against sine of angular distance from centre of image.

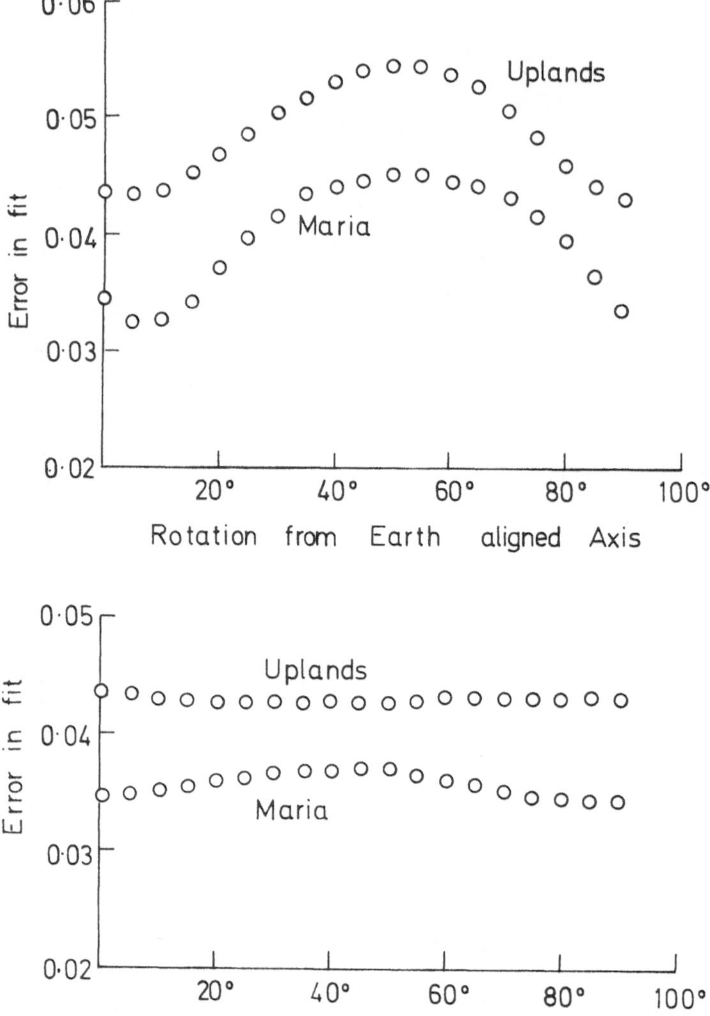

Fig. 2. Variation in the error of the fitted ellipsoids rotated about the polar axis and in the plane of the sky.

first when the ellipsoid is rotated about the polar axis (z) and then in the plane of the sky. For the former there is a clear minimum when the x axis coincides approximately with the Earth aligned axis whereas in the plane of the sky the variation is almost negligible.

The results of these analyses make it once more clearly apparent that the surface ellipticity greatly exceeds the dynamical ellipticity calculated from the differences in the moments of inertia (C, A, B) about the polar axis (z), the mean earthward direction (x) (i.e. the direction of the Earth at nodal passage) and the equatorial axis in the plane of the sky (y) respectively.

As pointed out by Runcorn (1962, 1967) this is the key fact in understanding the cause of the non-hydrostatic condition of the Moon: it cannot simply result from the distortion of a uniform sphere in remote times. The interior of the Moon is not of uniform density: the density varies with angle so that the density below the centre of the disc is about 10^{-3} less than the density below its limb.

2. Heights of the maria

In their analysis, Runcorn and Shrubsall gave two maps on which is plotted the distribution of deviations of the points on the maria from the ellipsoid which best fits them and a similar one for the uplands. From the former it is clear that the surfaces of the circular maria which have the 'mascon' gravity anomalies above them, are systematically lower than those of the other maria by about 1 km. This interesting result has been followed up on the Mills data by separating the maria into 'mascon

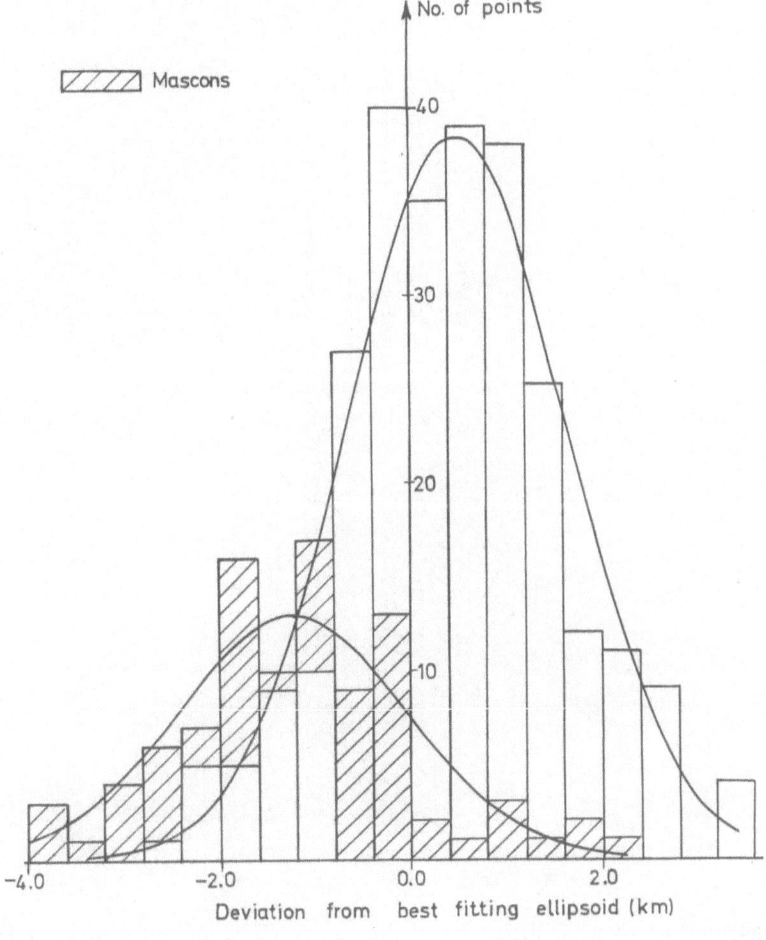

Fig. 3. Histograms showing the heights in the mascon maria and other maria.

TABLE II

Axes of ellipsoids fitted to uplands, all maria, mascon maria and other maria

	Semi-axis toward Earth a_x	Equatorial semi-axis in plane of sky a_y	Polar semi-axis a_z	$a_x - (a_y + a_z)/2$	$a_y - a_z$	Number of points
Uplands	1740.2	1737.3	1736.5	3.3	0.8	532
All Maria	1739.1	1736.0	1734.7	3.8	1.3	385
Mascons	1738.7	1734.5	1734.0	4.5	0.5	95
Other Maria	1739.3	1736.1	1735.2	3.7	0.9	290

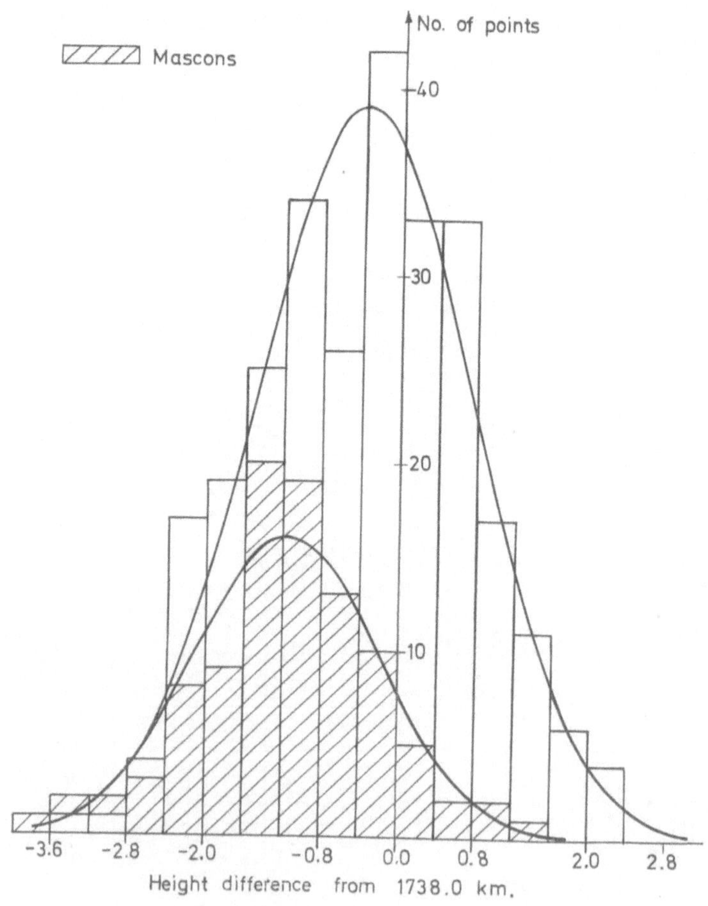

Fig. 4. Histograms showing the departures of the mascon maria and other maria from the best fitting ellipsoid to all maria.

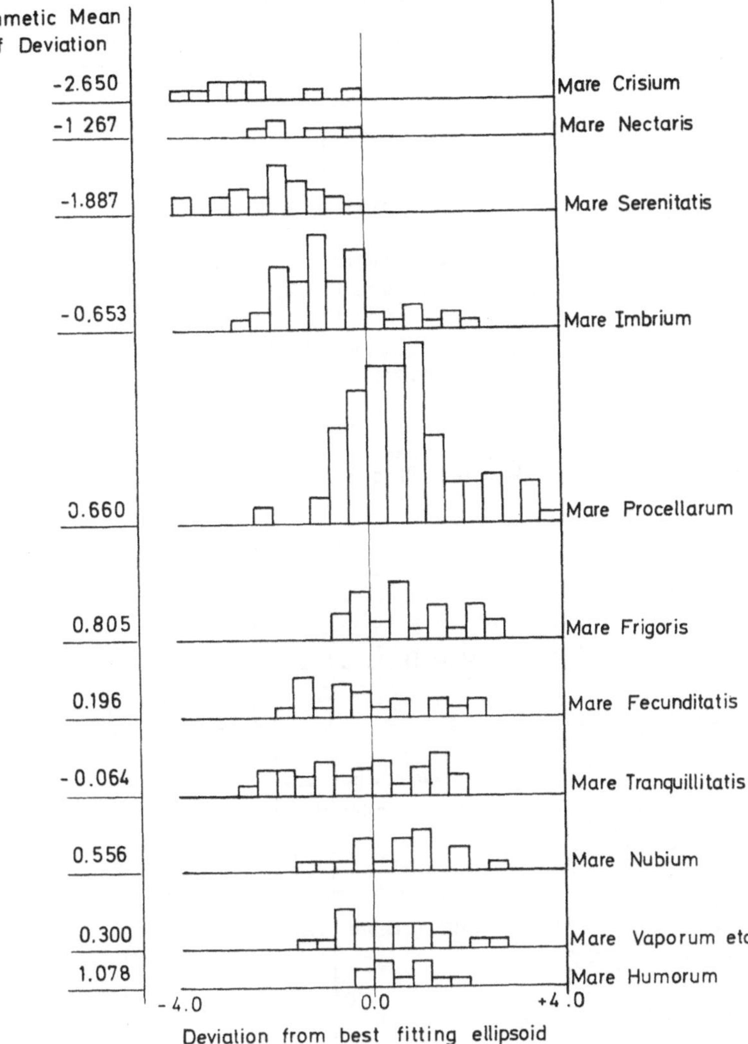

Fig. 5. Histograms showing the deviation of individual maria from the best fitting ellipsoid to all maria.

maria' and 'other maria' and fitting ellipsoidal surfaces to them. Table II shows the axis of the ellipsoids obtained. The 'shape' of the moon ellipsoid is maintained by these further surfaces.

To show further the systematic nature of this effect, Figure 3 is a histogram of the absolute heights of the two groups of maria and Figure 4 a histogram of the deviation of each group from the best fitting ellipsoid to all maria. The relatively small number of 'mascon maria' points available means that they do not contribute much when all maria are taken together.

Finally the deviation of each maria area from the best fitting ellipsoid to all maria

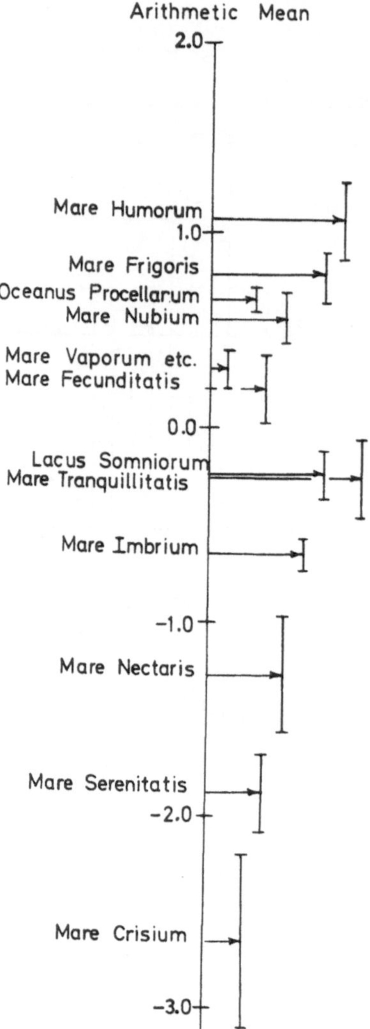

Fig. 6. Arithmetic mean of deviation of individual maria from best fitting ellipsoid showing standard deviation.

was considered separately. In Figure 5 the histogram of these deviations is plotted and in Figure 6 the arithmetic mean of the deviation and the standard mean deviation about this mean is shown. Mare Crisium emerges from this analysis as the lowest surface on the Moon.

References

Baldwin, R. B.: 1949, *The Face of the Moon*, Chicago University Press.
Baldwin, R. B.: 1963, *The Measure of the Moon*, Chicago University Press.
Goudas, C. L.: 1963, *Icarus* **2**, 423.
Kopal, Z.: 1967, *Proc. Roy. Soc. London* **296**, 254.

Mills, G. A.: 1968, *Icarus* **8**, 90.
Runcorn, S. K.: 1962, *Nature*, **195**, 1150.
Runcorn, S. K.: 1967, *Proc. Roy. Soc. London* **A296**, 270.
Runcorn, S. K. and Gray, B. M.: 1967, in *Mantles of the Earth and Terrestrial Planets*, John Wiley, New York.
Runcorn, S. K. and Shrubsall, M. H.: 1968, *Phys. Earth Planet. Interiors* **1**, 317.

SOME DIFFERENCES BETWEEN GEOMETRICAL AND DYNAMICAL FIGURES OF THE MOON

I. V. GAVRILOV

The Main Astronomical Observatory of the Ukrainian Academy of Sciences, Kiev, U.S.S.R.

Abstract. The geometrical shape of the Moon is determined from measurements of absolute heights of the lunar surface, while its dynamical shape is described by means of the Moon's gravity field parameters. All these data are derived from observations of the lunar artificial satellites ('Luna-10', 'Orbiters 1-4') and astronomical measurements.

In the paper differences of the lunar geometrical and dynamical figures are analysed. It is shown, that the homogeneous model of the Moon is not capable of explaining these differences. It is found, that the lunar centre of gravity situated about 0.9 km to the north, and 1.1 km nearer to the Earth, than the centre of its geometrical figure.

The geometrical shape of the Moon is determined from measurements of absolute heights of the lunar surface, while its dynamical shape is described by means of the Moon's gravity field parameters. However there are no reasons to consider, that the smoothed surface of the Moon and its equipotential surface coincide with one another [1]. Therefore it is useful to determine differences between the lunar geometrical and dynamical figures and to compare them with some theoretical lunar models.

Let $\mathbf{R}=\{X, Y, Z\}$ be the position vector of the lunar surface point, and let $\mathbf{r}=\{x, y, z\}$ be the position vector of the equipotential surface point. The absolute heights of the lunar and equipotential surfaces are respectively

$$H = 1738.0\,(\sqrt{X^2 + Y^2 + Z^2} - 1),$$
$$h = 1738.0\,(\sqrt{x^2 + y^2 + z^2} - 1). \tag{1}$$

For the first case the reference surface is a sphere, the centre of which coincides with the origin of the X, Y, Z system, and for the second case–a baricentric sphere.

The dynamical figure of the Moon can be represented by means of spherical harmonics

$$h = \sum_{n=0}^{\infty} \sum_{m=0}^{n} (C_{nm} \cos m\lambda + S_{nm} \sin m\lambda)\, P_{nm}(\sin \beta), \tag{2}$$

where C_{nm}, S_{nm} are harmonic coefficients, that are determined from an analysis of the motion of artificial satellites of the Moon [2, 3]. Analogically we can write for the geometrical figure of the Moon

$$H = \sum_{n=0}^{\infty} \sum_{m=0}^{n} (J_{nm} \cos m\lambda + J'_{nm} \sin m\lambda)\, P_{nm}(\sin \beta). \tag{3}$$

The harmonic coefficients J_{nm} and J'_{nm} can be determined by use of absolute heights of the selenodetic reference points. As at present the reference points on the far side of the Moon are lacking the coefficients J_{nm} and J'_{nm} are determined from various lunar

Urey and Runcorn (eds.), The Moon, 32–34.

models using symmetry and homogeneity hypotheses [4]. However these procedures are open to severe criticism.

We must determine more strictly the approximate absolute heights of the lunar far side from existing relations between heights of the lunar and equipotential surfaces. For example these relations may be expressed by the formula

$$H = \alpha + \beta h , \tag{4}$$

where α and β are parameters determined from measurements of the lunar and equipotential surfaces of the visible side.

The following relations of harmonic coefficients are known from theory

$$\begin{aligned} C_{nm} &= kJ_{nm} , \\ S_{nm} &= kJ'_{nm} . \end{aligned} \tag{5}$$

The parameter k depends on the density distribution along the lunar radius.

All data obtained up to the present show that the homogeneous model of the Moon is not capable of explaining existing differences between the lunar geometrical and dynamical figures. It is very probably, that the real Moon is slightly nonhomogeneous in all three directions (radius, longitude, latitude). Moreover, the geometrical centre of the Moon does not coincide with its centre of gravity.

The nonhomogeneity of the Moon demands an additional analysis.

Comparison of the parameters of the lunar geometrical and dynamical figures permits to obtain the position of the lunar centre of gravity in its body.

For this purpose, the most probable sphere fitting the lunar visible surface can be found from a solution of the equations

$$aX + bY + cZ + d = H , \tag{6}$$

where a, b, c are coordinates of the centre, d is the radius correction.

The most probable sphere is found on the assumption, that a certain correlation exists between the absolute heights of the lunar and equipotential surfaces. Equation (6) should then be written

$$AX + BY + CZ + \beta h + \alpha = H . \tag{7}$$

Here A, B, C are new coordinates of the centre, α is the new radius correction, β is the proportionality coefficient of the heights H and h.

The differences $A\text{-}a = \Delta X$, $B\text{-}b = \Delta Y$, $C\text{-}c = \Delta Z$ show that the centres of the reference spheres of absolute heights h and H do not coincide one with another. So far as the reference sphere of absolute heights h is a barycentric sphere, the vector $M = \{\Delta X, \Delta Y, \Delta Z\}$ can be interpreted as a position vector of the lunar centre of gravity in the X, Y, Z system.

This method was tested using data obtained from the 'Luna-10' and 'Orbiters 1–4' observations. Absolute heights of the visible lunar surface were determined from the lunar hypsometric chart [1], prepared at the Kiev Observatory. The results are represented in Table I.

TABLE I

Coordinates	Position of the lunar centre of gravity in km	
	From 'Luna-10' data	From 'Orbiters 1-4' data
ΔX	-0.33 ± 0.13	-0.20 ± 0.14
ΔY	$+0.48 \pm 0.14$	$+0.36 \pm 0.14$
ΔZ	$+1.45 \pm 0.56$	$+1.05 \pm 0.35$

The weighted mean results (weights 1 and 4, depending on the number of used artificial satellites of the Moon) is

$$\Delta X = -\ 0.26\ km$$
$$\Delta Y = +\ 0.38\ km$$
$$\Delta Z = +\ 1.13\ km\,.$$

Previously during the construction of the Kiev hypsometric lunar chart the following data were used $\Delta X' = +0.26$ km, $\Delta Y' = +0.52$ km, $\Delta Z' = 0.0$ km [1]. Therefore the final results are

$$\Delta X = \quad 0.00\ km$$
$$\Delta Y = +\ 0.90\ km$$
$$\Delta Z = +\ 1.13\ km\,.$$

The obtained data show that the lunar centre of gravity is situated approximately to the north, and nearer to the Earth, than the centre of the lunar geometrical figure. These results do not contradict the data obtained from other investigations [5].

References

[1] Gavrilov, I. V.: 1969, *Figura i rasmery luny po astronomicheskim nabliudeniam*, 'Naukova dumka' press, Kiev.
[2] Akim, E. L.: 1966, *Dokl. Akad. Nauk. S.S.S.R.* **170**, 4.
[3] Lorell, J.: 1970, *The Moon* **1**, 2.
[4] Goudas, C. L.: 1964, *Icarus* **3**, 2; 1965, **4**, 2.
[5] Arthur, D. W. G.: 1966, *Comm. LPL* **5**, 75.

LARGE DISKS AS REPRESENTATIONS FOR THE LUNAR MASCONS WITH IMPLICATIONS REGARDING THEORIES OF FORMATION

P. M. MULLER and W. L. SJOGREN

Jet Propulsion Laboratory, Pasadena, California, U.S.A.

Abstract. It is shown that large (hundred-km diam) near-surface disks are capable of accurately representing the lunar mascons with very few parameters. While this does not 'prove' that the mascons are excess mass in disk form, it is highly suggestive. It is pointed out that virtually every proposed mascon theory is consistent with a disk shape for the excess mass. The tentative hypothesis is advanced that the mascons are, in fact, disk shaped mass excesses. The conditions for mascon formation and preservation are reviewed in the light of this hypothesis.

1. Introduction

The resolution of Doppler tracking data from the Lunar Orbiter Mission to obtain lunar gravity information (Muller and Sjogren, 1968) was not sufficient to differentiate a point mass from a surface disk, as a model for a mascon. This was initially pointed out by Conel and Holstrom (1968). Kane (1969) showed that a low altitude orbit over various mascon models would indeed produce quite different gravity signatures. Apollos 12 and 14 have provided both high- and low- altitude orbits over the Nectaris mascon. Comparison of these gravity profiles with those of theoretical models is presented and some assertions are made.

2. Approach

The raw Doppler data from Apollos 12 and 14 were fit in short arcs (1 revolution or less) with the JPL orbit determination program. This program calculated the theoretical observations accounting for gravitation effects of the Sun, planets and a spherical Moon, the Earth's rotation, signal transit times, and atmospherics. It then did a least-squares fit to the actual observations. The resulting residuals between the observation and theoretical calculations had characteristic signatures which were, in turn, differentiated to provide line-of-sight accelerations. These gravity profiles were plotted and represented the 'real world'.

The next step was to simulate tracking data for two different models. The first model was a point mass buried 100 km which represented possibly the remnants of an impactor. The second model was a surface disk which represented dense surface lava pools. The simulated data was then fit with the theoretical model used for the real Doppler observations (i.e., spherical Moon). The residuals were processed in the same manner as mentioned above and their gravity profiles were over-plotted with the initial 'real world' gravity profiles. Several attempts of modifying the model parameters to best match the real data residuals was done using Apollo 12 data. For

Urey and Runcorn (eds.), The Moon, 35–40. *All Rights Reserved.*

example, the mass and depth of the point model were chosen to match the amplitude of the high- and low-altitude residuals. Likewise, the mass and radius of the surface disk were chosen so as to match these same amplitudes. Apollo 14 then was the independent test for these models. All the simulated data cases placed the perturbing mass at the center of Mare Nectaris (−16 lat. and 34 long.). The point was $9.1 \times 10^{-6}\ GM_{☾}$ at a depth of 100 km, while the 150 and 200 km radius surface disks had masses of $5.1 \times 10^{-6}\ GM_{☾}$ and $9.1 \times 10^{-6}\ GM_{☾}$ respectively.

3. Results

The high-altitude gravity profiles are shown in Figure 1. The amplitudes of the point mass and 150 km disk match the real data fairly well, although the complete shape of the real data curve is somewhat broader. The 200 km disk seems much too large. The low-altitude profile in Figure 2 shows again the fairly good match of the real data amplitude with the point mass and the 150 km disk. However, the shape of both of them do not compare well at all. The point mass is much to narrow with its peak shifted 2° east of the real peak. The 150 km disk is better, but still too narrow, with its peak $1\frac{1}{2}$° east. The 200 km disk does a relatively good job in matching both amplitude and shape. A 200 km disk is physically pleasing, for it is just the radius of the prominent mare basin and the edge of Montes Pyrenaeus.

The Apollo 14 low altitude data provides another comparison of models as shown in Figure 3. Again the point mass is the poorest fit; the 150 km disk is somewhat better, but is still too narrow; and the 200 km disk has about the right shape but is slightly over in amplitude.

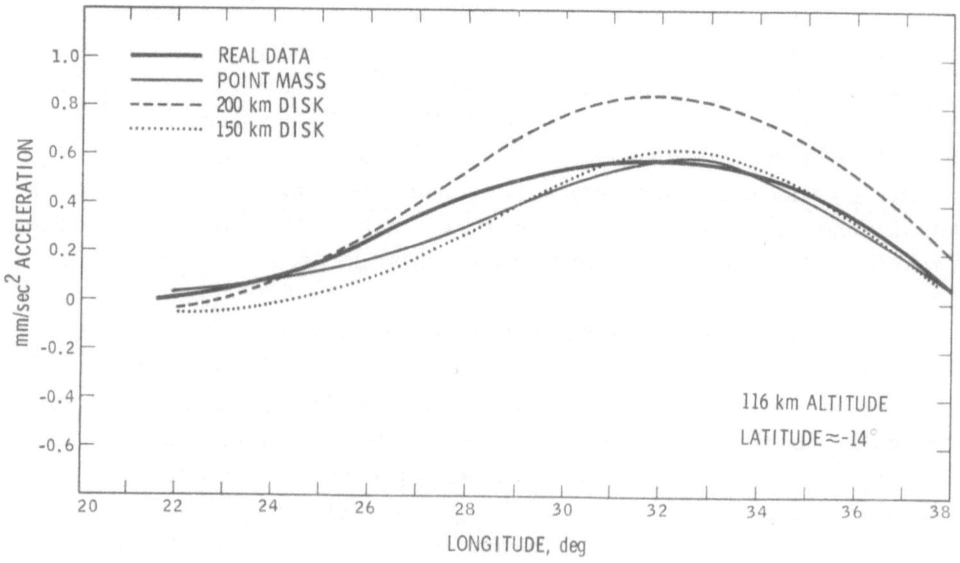

Fig. 1. Nectaris gravity profile by Apollo 12 at 116 km altitude.

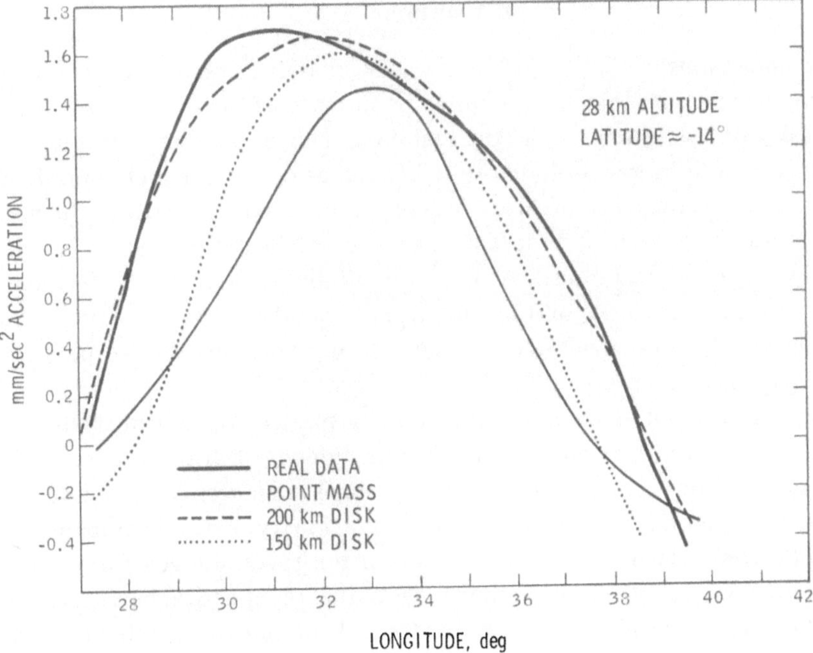

Fig. 2. Nectaris gravity profile by Apollo 12 at 28 km altitude.

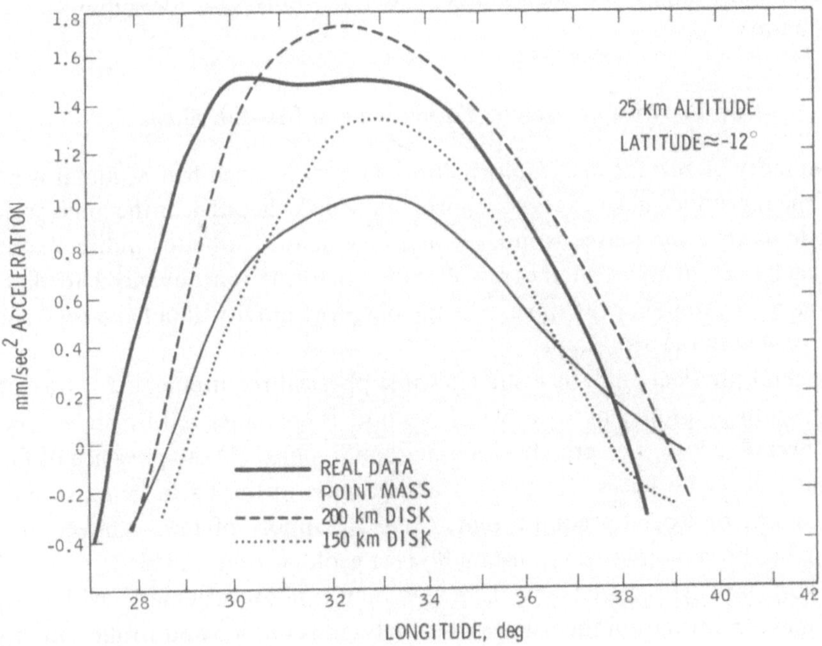

Fig. 3. Nectaris gravity profile by Apollo 14 at 25 km altitude.

4. Conclusion

From the comparisons in Figures 1–3, it is evident that disks do a much better job of representing the real data than does a buried point mass. However, it is not clear that a simple disk can completely model the real data. This is the case with the 200 km disk which does well in low altitude orbits but rather poorly for high altitude data. There may be some compensating negative gravity anomalies adjacent to the mascon area that averages in high altitude data and thus reduces the positive gravity as presented in the theoretical model. A more quantitative answer will be forthcoming with the reduction of all the Apollo 14 data, from which the masses for a dense grid of surface disks will be estimated over a broad area (i.e., not just the Nectaris central basin) in a dynamical least squares fit.

In theory, no amount of above-surface gravity measurements can define a subsurface mass distribution, as there are an infinite number of distributions which can fulfill the observations. In practice, as any geologist can testify, many successful deductions are in fact made from above-surface gravimetry. Few oil companies could make a profit from wildcat drilling were it not for this successful generalization from insufficient observations. It has been shown above that the spherical mass point model does not fit the observations nearly so well as the disk, despite the fact that the disk has 'sharp' edges which the spacecraft overflies. In the hand-working of even some lunar orbiter data, we have noted many cases of 'shoulders' near the edges of mascon seas, and have speculated on disk models. It appears safe to at least advance the hypothesis that the mascons are, in fact, disk shaped mass excesses, and to take note of the indirect but highly suggestive consistency of observations with mass-models of this shape and nature.

5. Theories of Mascon Formation and Mascon Shape

A brief summary of current mascon formation theories is given below, and it is noted that all of them produce mass excess in approximate disk shapes near the lunar surface. The hypothesis that the mascons are, in fact, disk shaped, obtains indirect support from the fact that all physical theories of mascon formation advanced to date do require or produce this shape. Conversely, the observed gravity data have been shown to be *consistent* with the disk shape.

Fundamental physical analysis results in three physical requirements for formation and preservation of a lunar mascon, which we note is characterized by higher gravity over a topographic low. The presence of a mascon implies: (1) the presence of higher density material in the basin; (2) there is added mass in the basin being supported above the point of isostatic equilibrium; (3) this support of mass above isostatic equilibrium has been maintained (presumably) for geologic time. (Table I.)

The authors have synthesized a five-step mascon formation schematic, within which the major mascon formation theories can be analyzed (Gottlieb and Muller, in press). Table II presents this schematic theory of mascon formation. Steps 1 and 2 are

TABLE I

Immediate mascon conclusions

Observation:	Higher gravity over topographic lows (basins).
Requirement:	Presence of higher density material.
Requirement:	Presence of mass supported above isostasy.
Requirement:	That the mass was supported for geologic time (stress 50–400 bars).
Conclusion:	Some mechanism produced very large volumes of significantly higher density material in the ringed sea basins.
Conclusion:	Some process raised very large volumes of lunar mass above the level of isostatic equilibrium (approximately 3 km above this level).
Conclusion:	The Moon retained sufficient strength to a depth of at least 100 km to support a stress of 50 to 400 bars for geologic time.
Conclusion:	The temperature of the lunar material to this depth could hardly have risen to 800 °C and if this high, could only have remained hot for a geologically short time (10^3 to 10^6 yrs).

generally agreed to have taken place – particularly, step 2, which was rather firmly established by Harold Urey after W. C. Gilbert. If step 3 did not take place, then very high density material had to be transported in large quantities. H. C. Urey's suggestion (Table III) of direct formation by impact combines steps 3 and 2, and avoids any problem with steps 4 and 5. It is the simplest explanation, but suffers from apparent theoretical difficulties in accounting for the fact that the asteroidal body is retained after the tremendous release of impact energy. If his hypothesis is simply accepted, however, the resulting mascon will be disk shaped, because a buried sphere is physically impossible, and the shattered asteroidal remains would surely be scattered in the basin bottom. Both the lava-flow and transport models are basically similar, except the latter avoids temperature problems that raise strong doubts in some minds as to

TABLE II

Schematic theory of mascon formation.

Step 1	The Moon forms a crust of lower density than the interior. The thickness must be at least 25 km, 50 km most likely, 100 km likely maximum. This constraint is placed by the next step.
Step 2	An asteroidal-sized body impacts at the future location of a mascon sea, blasting crustal material away to a depth sufficient for the next step.
Step 3	The initial deep crater is back-filled from below with the higher-density material until approximate isostasy is achieved. A shallow basin of 5–10 km depth remains due to the density contrast of the crust versus the interior.
Step 4	The material below the basin floor now achieves and retains sufficient strength to withstand a shear force of 50–400 bars for geologic time.
Step 5	The basin is further filled with material sufficient to give the observed positive gravity anomaly (above isostasy). Any slumping of the floor removes the anomaly, either during or following the deposition. The effective density contrast for this step is approximately 3.0, since vacuum is being replaced with rock.

TABLE III

Basic classes of mascon theories

Lavas: Wise, Yates, and Booker, Kovach, Lu and Rell: Fall into two classes:
Lavas perfuse upward through a crust able to withstand the shear.
Lavas are deposited in the basin by flows from outside the basin.

Physical dynamics of a shrinking Moon: Kaula:
The lunar crust shrinks, does not crack, and forces lavas above isostasy into the basins.

Transport of material into compensated basin: Gilvarry, Gold, others:
Transport of sediments by water in a lunar hydrosphere.
Transport of lunar fines or dust by various mechanisms.
Transport by any other scheme.

Direct formation by impact: Urey
The asteroidal impacting body which formed the seas is of higher density than the Moon and leaves behind more mass than was ejected.

Other theories so far offered fail to explain the super-isostatic condition.

whether the sub-basin materials could simultaneously transmit large quantities of hot lavas and still support the mascon stress of 50–400 bars for even very short geologic time spans.

Several theories of mascon formation have been proposed which fail to account for the super-isostatic condition of the excess mass. It can easily be shown (Gottlieb and Muller, in press) that isostatic mascon models are impossible or impractical for reasonable density contrasts. The three requirements given in Table I are rather firmly advanced as requirements on any theory of mascon formation. When we examine each theory noted in Table III, it is evident that each produces a roughly disk-shaped excess of mass near the surface.

It is, therefore, the strong suggestion of this paper that the mascons are, in fact, near-surface, disk-shaped mass distributions. Since there are no reasonable experiments that can be conducted to firmly establish this hypothesis, and since this model is consistent with the observations and all theories advanced to date, we would appear safe in adopting the hypothesis. Further, we can add a fourth 'requirement' on any future mascon theory; namely, that it either be consistent with the near-surface disk distribution profile, or be carefully tested against the far higher resolution, low-altitude data being currently obtained from Apollo tracking, which does seem capable of discriminating at least between widely different mass distribution models such as point mass at depth, surface disks, and perhaps others.

References

Conel, J. E. and Holstrom, G. B.: 1968, *Science* **162**, 1403.
Gottlieb, P. and Muller, P. M.: in *Proc. of the Lunar Science Institute Symposium on the Geophysical Interpretation of the Moon*, Houston, Texas, June 1970 (to be published).
Kane, M.: 1969, *J. Geophys. Res.* **74**, 6579.
Muller, P. M. and Sjogren, W.L.: 1968, *Science* **161**, 680.

B. THE LUNAR SURFACE AND ITS MORPHOLOGY

THE GEOMORPHIC EVOLUTION OF THE LUNAR SURFACE

L. B. RONCA

Lunar Science Institute, Houston, Texas, U.S.A. *

Abstract. Fundamental parameters of the geomorphology of a lunar surface are (i) the number and size distribution of craters, (ii) the degree of erosion of the craters and (iii) thickness and other characteristics of the regolith. These parameters are not independent of one another; as one changes through time, the other two will also change in a statistically predictable way.

In the *continuous degradation sequence*, the relationship between the number of craters per unit area and the degree of erosion of the craters is continuous and monotonic. This sequence occurs in areas subjected to intermediate to small impacts, and eroded mainly by the impacts themselves plus other small-scale erosional processes. In areas subjected to large impacts and mare flooding the *discontinuous degradation* sequence is predominant..

The relationship of the first two parameters, the number of craters and the degree of erosion, with the third parameter, the regolith, is not simple and is not yet understood. It appears, however, that the geomorphological stage is more important than the mare-versus-highland dichotomy of the lunar surface.

The solution of the function relating craters of the continuous degradation sequence with degree of erosion was defined as the *geomorphic index* of the area. Studies of the geomorphic index of stratigraphic surfaces show that areas covered by considerable ballistic sediments have a geomorphic index which is not a monotonic function of time. On the other hand, areas covered almost exclusively by mare flooding show an index which is a monotonic function of the age of the flooding. As each mare surface shows a considerable range in indices, it is concluded that maria are covered by surfaces formed through a considerable length of time. By using Apollo 11 and 12 radiometric ages it is suggested that the time of mare flooding lasted on the order of one billion (10^9) years.

The geomorphic index of highland surfaces shows a remarkable degree of order; i.e., the farther an area is inland from the mare shores, the higher will be the index. No explanation is given to this phenomenon, but is suggested that lunar erosion is not just a localized phenomenon centered on the locus of an impact, but has lateral trends of regional dimensions. Electrostatic transportation as suggested by Gold is a possible mechanism.

1. Introduction

Fundamental parameters of the geomorphology of a lunar surface are (i) the number and size distribution of craters, (ii) the degree of erosion of the craters, and (iii) thickness and other characteristics of the regolith. At least in first approximation, it is likely that these parameters are not independent of one another; i.e. as the number of craters increases, the degree of erosion of preexisting craters also increases and progressive changes occur in the regolith.

Considering the statistical nature of the phenomenon, it would be worth-while to develop a unit of measure which includes all three of the above parameters. This unit could describe the geomorphology of any area, and, being the result of three independent measurements, could reduce the unavoidable statistical fluctuations. Previous work (Ronca and Green, 1970) has led to the development of a function which relates the number density of craters (excluding ghost craters) to the number of these craters

* Present address: Dept. of Geology, Wayne State University, Detroit, Michigan 48202.

Urey and Runcorn (eds.), The Moon, 43–54.

which are essentially uneroded. The value of this function at different areas was defined as the *geomorphic index* of that area. The relationship between the geomorphic index (that is, the first two parameters of lunar geomorphology) and the regolith (the third parameter) was presented in Ronca (1971a). This relationship appears to be not monotonic, and it is not yet understood. It appears, however, that the geomorphic index is more important, as far as the regolith is concerned, than the mare-versus-highland dichotomy of the lunar surface.

What is the geological meaning of the geomorphic index? It is a description of the geomorphic age of an area. To use an example, a river system on the Earth is variously referred to as being in its youthful, mature, or old stage. This means that there is a sequence of characteristics which a river system displays consecutively from the time of its formation. The same concept can be applied to a lunar surface, from the time of its formation, when presumably no or only endogenous craters are present, to the time when the surface is completely covered with impact craters, many of which are highly eroded. The geomorphic index is the position of the lunar surface in this progression.

It is important to realize the basic difference between geomorphic age and geologic age. Just as on Earth one river may reach its old stage in a much shorter time than another river in a different area, so on the Moon one surface may reach a high geomorphic index sooner than another. Geologic age, on the other hand, is simply the age in years (or its position in the stratigraphic column) of a feature. Only under certain conditions of equal exposure to the modifying agents can the geomorphic age be assumed to be a monotonic function of the geologic age.

Implicit in the concept of geomorphic age is the concept of *rejuvenation*. On the Earth extreme tectonic activity gives plenty of examples of "pushing back the geomorphic clock'. On the Moon, if all the areas were formed at the same time and progressed through essentially the same history, they should all have the same geomorphic index. In reality the index ranges from approximately 5 to 13 for the mare surfaces and 13 to 20 for the terra surfaces. The simplest way to explain the range in indices is to call for rejuvenative processes, which occasionally wipe out most or all craters in an area.

The most evident process of rejuvenation on the Moon is *mare flooding*. This process actually creates a new surface, young both geologically and geomorphologically. Impact craters are soon formed and the geomorphic index begins to increase in value. In general, for mare surfaces, the geomorphic index is a monotonic function of the geologic age. Care must be taken in this assumption, however, as a large impact on or near a mare surface will 'geomorphically age' the mare surface prematurely. For example, the area surrounding Copernicus shows a higher geomorphic index than the mare proper, probably as a result of the Copernicus impact. Detailed analysis show, however, that such areas are covered by considerable amounts of ballistic sediments.

Mare flooding is not the only process of rejuvenation. When a large impact occurs, the surrounding area is subjected to the highly erosive action of the ballistic ejecta and seismic activity (the usage of the adjectives large, intermediate and small, when applied to impact craters follows closely the definitions presented by Hörz and Ronca, 1971). It was calculated that even if only 10^{-4} of the impact energy is converted into seismic energy, a mare-size impact would create moonquakes of approximately magnitude 10 on the Gutenberg-Richter scale, which is considerably larger than any earthquake recorded on Earth. Under the dual attack of the seismic waves and ballistic sediments, crater rims are completely or partially obliterated. From a geomorphological point of view, if no craters or only ghost craters are left, the area will have been rejuvenated. The amount of rejuvenation should decrease progressively as the distance from the impact increases. A clear evidence of this type of rejuvenation is shown by the terrain around Mare Orientale and was discussed in a previous publication (Ronca an Green, 1969).

The absence of craters on slopes surrounding maria (see Gold, p. 55 of these proceedings) gives evidence that rejuvenation may also occur as the result of mass movements produced by the electrostatic action of the solar wind.

The purpose of this paper is to review the assumptions and the methods used to obtain the geomorphic index, to study the relationships between the index and relative time in terms of lunar stratigraphy and to present some conclusions about the length of the interval of time that was necessary to fill up the mare basins with the mare material. It will be shown that this time interval occupies a considerable portion of the lunar geological history. It will also be shown that erosive processes on the highlands are laterally related over large distances, implying a more complicated picture than that offered by erosion by impacts only. Mass movement produced by electrostatic action as described by Gold may be the explanation.

2. The Geomorphic Index

The following is a review of material published in Ronca and Green (1968, 1969 and 1970).

The University of Arizona catalog (Arthur *et al.*, 1963, 1964, 1965, 1966) classifies lunar craters on a scale of 1 to 5 on the basis of their condition. Very sharp fresh-looking craters are classified as 1, craters with blurred rims as 2, craters with extensively broken rims as 3. Craters usually described as ruins are classified as 4, and ghost and fragmentary craters as 5.

No age relationship is intended in the definition of these classes. Intuitively, however, it appears possible that the classes may represent an age sequence. It is possible to perform a test to check the hypothesis that the classes do indeed represent an age sequence. If only those craters larger than a few kilometers are considered (for this size crater saturation is not reached), then, the older a lunar surface is, the more highly cratered it will be. If the classes are a time sequence, then class-5 craters should be common in highly cratered areas, while class-1 craters should be common in areas of low crater densities. This is not actually the case. If we plot the percentage of class-5 craters versus the number of craters per unit area for the craters of the lunar near-side (excluding the limbs) larger than 3.5 km in diam, we can see that the percentage of craters which are of class-5 increases to a maximum very quickly for areas of low-intermediate crater densities and finally decreases for areas of high crater densities. Contrary to the hypothesis, areas of high crater density are relatively low in craters of class-5.

We can make a similar test excluding craters of class 4 and 5. If only craters of class 1, 2, and 3 are considered, then the results fit the hypothesis. The percentage of craters which are of class-3 is low in areas of low crater density and increases monotonically with the crater density. These relationships can be interpreted to indicate that classes 1, 2, and 3 are a time sequence, while classes 4 and 5 are not.

The next step is to check this interpretation by observing a large number of individual craters. For brevity's sake, only the conclusions will be presented here. They are as follows: All the erosional processes operating on lunar craters can be grouped in two categories. The first category produces a degradation by erosion through time from class 1 to class 3, and in some cases, class 4. This can be called the *continuous degradation sequence*. The second category of erosional agents is responsible for the conditions of craters of class 5 and of some of class 4. This is not a continuous process, as it can happen to craters belonging to any class. This category will be called the *discontinuous degradation*. It can also cause rejuvenation, that is, the complete disappearance of craters.

The erosional agents which cause the continuous degradation sequence operate more or less continuously through time (not necessarily at the same rate). Micrometeoritic impact, Gold's electrostatic erosion, and space weathering are likely to be the dominant agents, accompanied by other processes, such as terrace collapse, isostatic recovery and perhaps large-scale tectonics. Specific details of the continuous modification of a crater after its formation have been described by Pike (1967) and Ross (1968).

The erosional agents which cause the discontinuous process are primarily two. Flooding by mare material leaving only a rim or part of a rim above the surface is one. The other is ballistic sedimentation and destruction by seismic waves created by large impacts. If flooding by mare material is so deep as to completely cover the craters of an area or the ballistic sedimentation and seismic waves are so intense as to completely obliterate the craters of an area, the area has been rejuvenated, as previously discussed.

It is evident that if we are interested in a time-related parameter, we must concentrate on the continuous degradation sequence. Craters of class 4 and 5 are, in the great majority, relics of a previous chapter of the geomorphic history of that particular area.

We are now ready to define the geomorphic index. It can be shown that if we plot the logarithm of the percentage of craters of class 1 versus the logarithm of the number of craters per unit area for the craters of the near side larger than 3.5 km in diam, the data distribute themselves along a line of slope -1. The following model fits this observation. Let us start with a newly formed surface. For a very short time, it will be without any large crater. Soon impacts will begin to create more and more craters. At first, all craters will be of class 1, but soon the earliest craters will become class 2. If no large crater and no mare flooding of significance occur to produce any discontinuous degradation, each crater will proceed from class 1 to class 2 and finally to class 3 (a few craters will reach class 4). A newly formed crater will remain in class 1 during the length of time, t, necessary for the crater density of the area to increase by a number, K, of craters per unit area [note that if the impact flux varies through the lunar geological time (Hartmann 1965, 1966), this length of time, t, will not be the same through geologic time]. It can be easily proven that if this model is correct, then the data must distribute themselves on a logarithmic plot on a line of slope -1, for

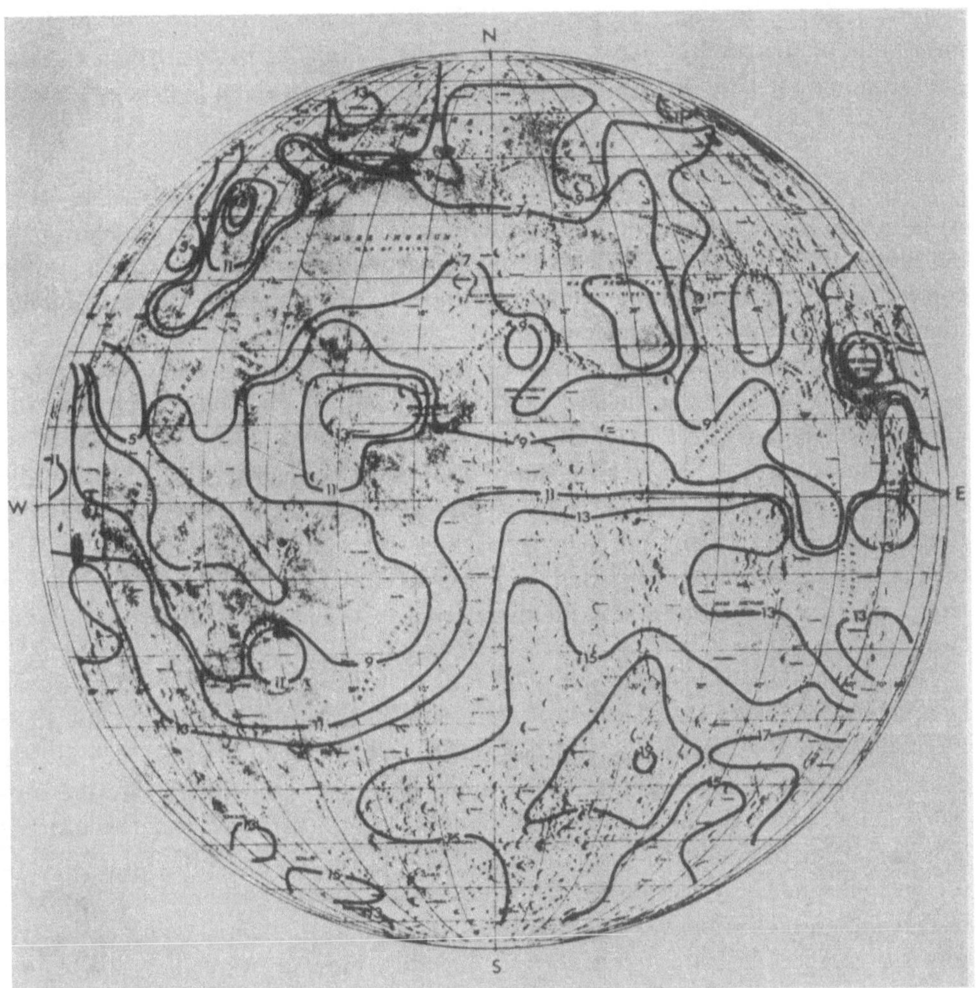

Fig. 1. Countour map of the geomorphic index, calculated for unit areas of 58×10^3 km^2. Apollo 11 landed on 23.49° E, 0.67° N, on an area of geomorphic index equal to 10.3. Apollo 12 landed on 23.34° W, 2.45° S, on an area of index equal to 8.4. Luna 16 landed on 56.30° E, 0.68° S, on an area of geomorphic index equal to 14.3. Although the indices of the landing sites were obtained for areas of 6.4×10^3 km^2, the values fit satisfactorily in the contour map. Also note the trend in the southern highlands.

any value of K. This is actually the case, as discussed above. The geomorphic index is defined as the position, in arbitrary unit, on the line of slope -1.

The geomorphic index of a lunar area is more reliable than the crater density or the average crater class because it combines two independently measured parameters – crater density and crater class. Although the combination of these two parameters could be obtained more simply by calculating their ratios, this procedure would not take into account the scattering of data. The calculation of the geomorphic index is able to eliminate the scattering not in an arbitrary statistical fashion, but as a direct result of a proposed geological model. Figure 1 shows the contour map of the geomorphic index on the lunar near-side, calculated for a unit area of 58×10^3 km^2. At this resolution, most of the mare surfaces have indices ranging from less than 5 to less than 13, and most of the highland surfaces range in index from about 13 to more than 19.

3. The Geomorphic Index of Stratigraphic Units

As discussed in the Introduction, the geomorphology of an area is not necessarily an indication of the age of that area, as certain areas may age faster than others. In the context of our parameters, the geomorphic index is not necessarily the same monotonic function of time in all lunar areas.

A *relative* time scale of lunar areas is presently available in the form of the United States Geological Survey geological maps. For a complete discussion of lunar stratigraphy, the reader is referred to McCauley (1967) and Mutch (1970). Here suffice it to say that lunar surfaces can be grouped, in accordance to their relative ages, into the following stratigraphic systems (starting with the oldest): Pre-Imbrian, Imbrian, Eratosthenian, and Copernican. The Imbrian can be subdivided into an upper part, the Procellarum Group, constituting the mare material, and a lower part, of different names in different localities, here simply referred to as the Pre-Procellarum Imbrian (see Figure 2).

The calculation of the geomorphic index is done from areas arbitrarily drawn by the computer. This means that it would be very difficult to calculate the index for areas displaying only one stratigraphic system on its surface. In most cases, each area is constituted by terrains of several stratigraphic units. In order to compare the geomorphic index with the stratigraphic position, it was necessary to calculate the stratigraphic 'center of gravity' of each area. This was done by measuring, for each area for which the index was calculated, the percentage area occupied by each stratigraphic system and normalize accordingly. Figure 2 shows the stratigraphic 'center of gravity' versus the respective geomorphic index for the areas presently covered by the United States Geological Survey lunar geologic maps. In addition, each datum point is identified by a different symbol indicating the predominant stratigraphic systems of the area.

Although the plot may not be treated rigorously, it is highly indicative of the relationship between geomorphology and time. Areas rich in Copernican and Eratosthenian terrains are mainly constituted by ballistic sediments produced by recent and

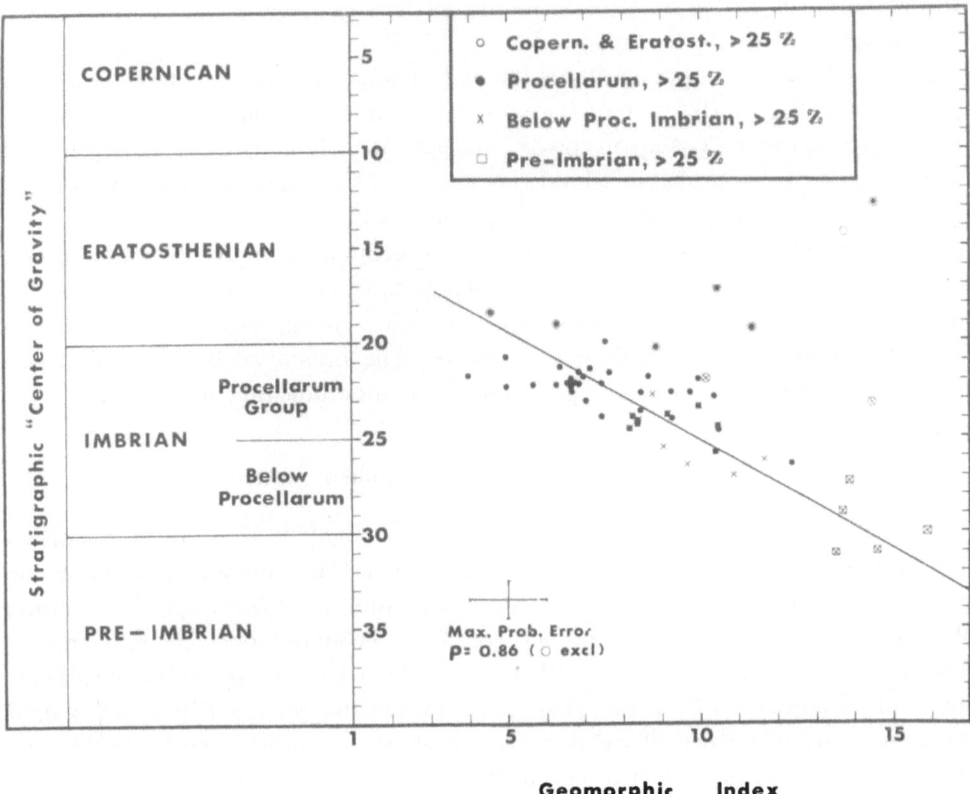

Fig. 2. The stratigraphic 'center of gravity' of areas of 58×10^3 km^2 versus their respective geomorphic index. In addition, the symbols in the upper rectangle indicate the stratigraphic systems which cover more than 25% of each area. When each of two stratigraphic systems occupy more than 25% of the area, a combination of symbols is used. Areas with more than 25% surface covered by Copernican and Eratosthenian deposits (open circles and combinations including open circles) have a larger geomorphic index than other areas. Excluding these areas, the linear correlation coefficient is 0.86 at the 0.1% probability level, indicating a relationship between stratigraphic time and geomorphic index.

almost recent impacts. These areas have been recently subjected to the effects of intermediate to large impacts and are likely to have geomorphologically aged faster than areas where all the recent or almost recent impacts are small. This is clearly shown in Figure 2, where surfaces with more than 25% area composed of Copernican and/or Eratosthenian deposits are shown as open circles or a combination of open circle with another symbol. All the open circles (and combinations with open circles) are higher than the average trend, which means that they have a geomorphic index higher than that to be expected by their relatively younger age. This is also indicated in Figure 1 where the contours nest around Copernicus with a higher value than on the maria.

If we exclude the open circles (and combinations with open circles) in Figure 2, the remaining data points correlate well. The linear correlation coefficient between the stratigraphic 'center of gravity' (excluding terrains of Copernican and/or Eratosthenian age) and the geomorphic index is 0.86. If we assume that the data points represent a random sample of all possible data points obtainable from Imbrian and Pre-Imbrian lunar surfaces, then the null hypothesis that there is no correlation between stratigraphic 'center of gravity' and geomorphic index must be rejected at better than the 0.1% probability level (the chances of making the wrong decision in rejecting the no-correlation hypothesis is less than 0.1%).

We can reach the following conclusion: The geomorphic index of areas rich in ballistic sediment is not a monotonic function of time. The intensity and distance of the impact probably plays the most important role. On the other hand, areas rejuvenated by mare flooding (and with an insignificant amount of ballistic sediments) have a geomorphic index which is approximately a monotonic function of time.

4. The Geomorphic Index of Highland Surfaces

The distribution of geomorphic indices in the lunar highlands has not been worked out yet and only preliminary observations can be given here. Figure 1 shows how the southern highlands appear to offer an amazing amount of regularity in the contours of the geomorphic index. The farther one goes from the mare shores inland, the higher the geomorphic index becomes. In other words, there appears to be a systematic increase in geomorphic age inland. There is no conceivable process that could explain this trend with an equivalent trend in geological age; processes which could be the cause of this trend, equivalent to ocean floor spreading and continental drift on the Earth, must be completely excluded. The only alternative is that the trend is due to different rates of geomorphological ageing. Areas far away from the shores geomorphologically age faster than areas nearer to the shores.

Presently there is no definitive explanation to this phenomenon. We may add one more observation. Runcorn (p. 377 of these proceedings) shows that the areas of higher geomorphic index in Figure 1 (closed contours 17 and 19 in the southern highlands) correspond closely to areas higher in elevation than the surrounding. Is there a relationship between elevation and rate of geomorphic ageing? One possibility is that the electrostatic erosion mechanism proposed by Gold (see p. 55 of these proceedings) is an important agent of the geomorphological ageing process. Lower areas may be regions of transportation or deposition, while higher areas may be areas of degradation.

It is evident that more research must be done on the geomorphology of highlands. The only comment that can be forwarded is that lunar erosion is not just a localized phenomenon centered on the locus of a large impact, but has lateral trends of regional dimensions.

The Geomorphic Index of Mare Surfaces

As discussed above, the geomorphic index of mare surfaces uncovered by substantial amounts of ballistic sediments is a monotonic function of time; i.e., the higher the index of a surface, the older the surface is. For the unit area used in preparing the contours of Figure 1, 58×10^3 km^2, the index for mare surfaces ranges from less than 5 to less than 13. In order to study the maria in more detail, the unit area was decreased to 6.4×10^3 km^2 and Orbiter photographs and United States Geological Survey geological maps were used to exclude any area that showed a cover of ballistic sediments. A more detailed description was presented in Ronca (1971b).

The range of indices for some of the maria is shown in Figure 3 as a vertical histo-

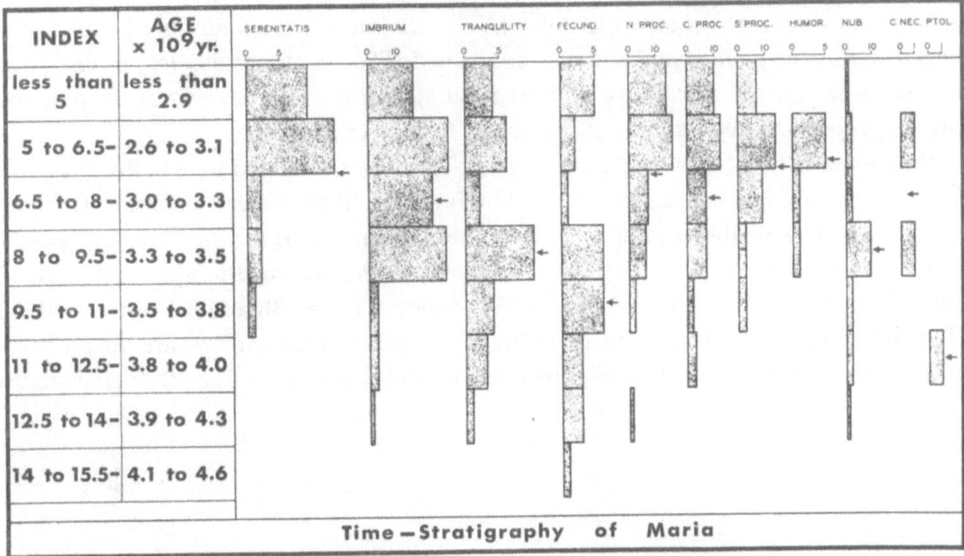

Fig. 3. The distribution of the geomorphic indices for the indicated mare surfaces calculated for unit areas of 6.4×10^3 km^2. The first column to the left shows the values of the geomorphic indices. The second column shows the preliminary range of ages obtained using Figure 4. The scale under the name of the maria is the number of unit areas displaying the corresponding geomorphic index. Arrows point to the average value for each mare surface. Although the age column is preliminary, it seems likely that the span of time between the oldest and youngest flooding is of the order of 10^9 yr.

gram. Each mare displays a wide range of geomorphic indices, from less than 5 to more than 14. Humorum and Nectaris do not have any geomorphologically young areas (geomorphic indices less than 5).

The following interpretations are offered:

(1) The surfaces of the maria are formed by several floodings having a considerable span of geomorphic indices. This suggests that the filling of the mare basins occupied a considerable length of time.

(2) The geomorphologically oldest surface is found in Fecunditatis. It is however, impossible to state that the flooding activity started sooner in Fecunditatis than in the other maria. The absence or paucity of old surfaces in some maria may simply be due to burial by younger floodings.

(3) No geomorphologically young surfaces are found in Humorum, Nectaris and

the floor of Ptolemaeus, and Nubium has only a few. A preliminary interpretation is that flooding stopped in some maria sooner than in others.

(4) A tendency toward bimodality in the flooding activity is apparent. The most common surfaces have indices between 5 and 6.5 and between 8 and 9.5. It is impossible to say whether this is significant or is due to statistical fluctuations.

It would be very important to find the functionality of the index of maria versus time. Several researchers have tried to find the relationship between meteoritic flux and time, for a comprehensive summary of all methods the reader is referred to Mutch (1970). But even if the meteoritic flux was exactly known, it would only be of partial help as far as the geomorphic index is concerned. The geomorphic index, as discussed before, is not only a function of cratering but also of the effectiveness of the flux and other agents in the degradation of previously formed craters.

If we assume that the radiometric ages of the samples collected by Apollo 11 and 12 are also the ages of formation of the landing sites, then we can compare the geomorphic indices of the landing sites with the radiometric ages, and have two points in the relationship between these two parameters. The radiometric ages of Apollo 11 and 12 are respectively $3.65 \pm 0.05 \times 10^9$ yr (Albee *et al.*, 1970) and $3.36 \pm 0.1 \times 10^9$ yr (Papanastassiou and Wasserburg, 1970). The geomorphic indices are respectively 10.3 and 8.4 (Ronca, 1971b). Figure 4 shows these points on the age-versus-index

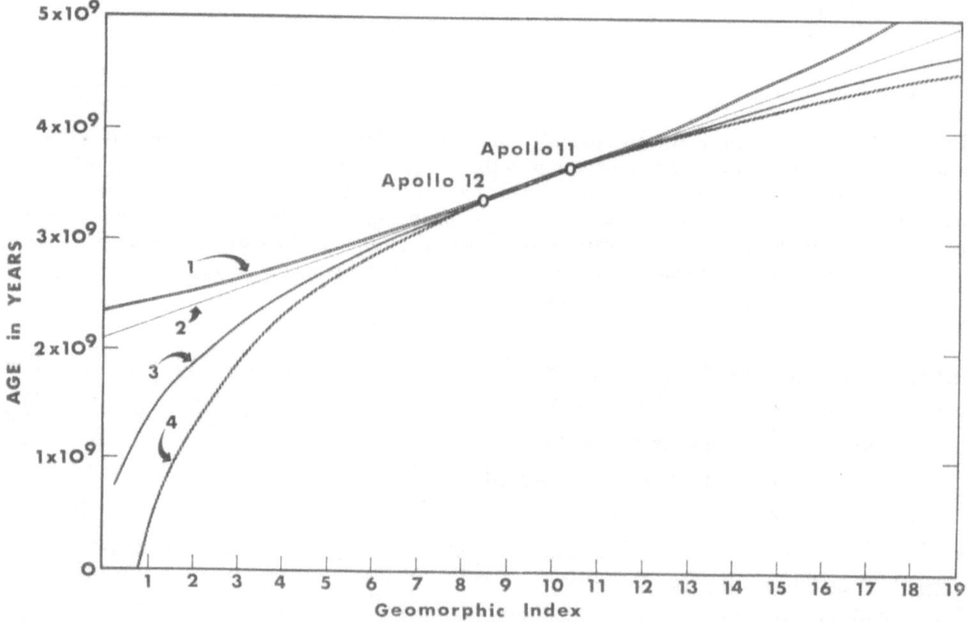

Fig. 4. Geomorphic index versus age, with the values of Apollo 11 and 12 indicated. A linear relationship between the logarithm of age and the index is assumed in curve 1, between age and the index in curve 2, between the logarithm of age and the logarithm of the index in curve 3, and between age and the logarithm of the index in curve 4. If the radiometric age of the basaltic chips collected by Luna 16 is applicable, this age (more than 4×10^9 yr) fits the geomorphic index of the landing site (14.3).

diagram. An infinity of curves can be drawn through the two points, but it is likely that the relationship between index and age is within the limits determined by a linear and logarithmic expression. Four possible curves are drawn: the linear relationship between index and age (curve 2 on Figure 4), between logarithm of the index and age (curve 4), between the index and the logarithm of the age (curve 1) and between the logarithm of the index and the logarithm of the age (curve 3). These four curves were also drawn for the limits of the radiometric ages, but are omitted from the figure for clarity. The second column of the vertical histogram of Figure 3 indicates the age limits corresponding to each index interval.

The preliminary conclusions that can be drawn are:

(1) the youngest mare floodings are between 3×10^9 and 0 yr old,

(2) the oldest mare floodings are between 4×10^9 and 4.5×10^9 yr old,

(3) the flooding activity of the lunar near side lasted for a period of time of the order of one billion years.

Since the drawing of Figure 4 some data have become available from the samples brought back to Earth by the automatic probe Luna 16. The landing area, in Mare Fecunditatis, appears to have a high geomorphic index, approximately 14.3. The radiometric age of the basaltic chips was given to be more than 4×10^9 yr. It is not clear whether this age can be assigned to the soil or the basalt but if it actually refers to the basalt, then there is good agreement between these data and Figure 4.

Acknowledgements

This paper was prepared under the joint support of The Universities Space Research Association and the National Aeronautics and Space Administration Manned Spacecraft Center, Contract No. NSR 09-051-001, and constitutes The Lunar Science Institute paper No. 51.

References

Albee, A. L., Burnett, D. S., Chodos, A. A., Eugster, O. J., Huncke, J. C., Papanastassiou, D. A., Pososek, F. A., Price Russ II, G., Sanz, H. G., Tera, F., and Wasserburg, G. J.: 1970, *Science* **167**, 463.

Arthur, D. W. G., Agnieray, A P., Horvath, R. A., Wood, C. A., and Chapman, C. R.: 1963, 'The System of Lunar Craters, Quadrant I', Lunar Planet. Lab., Univ. Arizona, Commun. 2, p. 1.

Arthur, D. W. G., Agnieray, A. P., Pellicori, R. H., Wood, C. A., and Weller, T.: 1965, 'The Systems of Lunar Craters, Quadrant III', Lunar Planet. Lab., Univ. Arizona, Commun. 3, p. 1.

Arthur, D. W. G., Pellicori, R. H., and Wood, C. A.: 1966, 'The System of Lunar Craters, Quadrant IV', Lunar Planet. Lab., Univ. Arizona, Commun. 5, p. 1.

Hartmann, W. K.: 1965, *Icarus* **4**, 207.

Hartmann, W. K.: 1966, *Icarus* **5**, 406.

Hörz, F. and Ronca, L. B.: 1971; *Modern Geol.* **2**, 65.

McCauley, J. F.: 1967, in S. K. Runcorn (Ed.), *Mantles of the Earth and Terrestrial Planets*, John Wiley & Sons, Inc., New York, p. 431.

Mutch, J. A.: 1970, *Geology of the Moon*, Princeton University Press, Princeton, U.S.A., 324 pp.

Papanistassiou, D. A. and Wasserburg, G. J.: 1970, *Earth Planetary Sci. Letters* **8**, 269.

Pike, R. J. Jr.: 1967 *J. Geophys. Res.* **72**, 2099.

Ronca, L. B.: 1971a, *The Moon* **2**, 202.

Ronca, L. B.: 1971b, 'Ages of Lunar Mare Surfaces. The Geological Society of America Bulletin'.

Ronca, L. B. and Green, R. R.: 1968, *Nature* **218**, 1147.
Ronca, L. B. and Green, R. R.: 1969, *Astrophys. Space Sci.* **2**, 22.
Ronca, L. B. and Green, R. R.: 1970, *Geol. Soc. Am. Bull.* **81**, 337.
Ross, H. P.: 1968, *J. Geophys. Res.* **73**, 1343.

EROSION, TRANSPORTATION AND THE NATURE OF THE MARIA

T. GOLD

Cornell University, Ithaca, U.S.A.

Abstract. Rock dust appears to have been redistributed over the Moon by effects other than impact explosions. A core sample on Apollo 12 showed sharp and distinctive layers and was clearly unmixed. Surface transportation processes that deposit the dust very gently must have been at work. Orbiter pictures confirm that such surface creep has taken place on a very large scale.

The seismic evidence makes clear that there is no continuous sheet of bedrock at a shallow depth in the vicinity of the Apollo 12 site. A deep deposit of powder would match the seismic properties observed. Mascons require for their explanation a surface transportation process that tends to fill in the large impact basins after their formation.

Surface transportation of lunar dust has been demonstrated in the laboratory to occur most readily as a result of electrostatic forces produced by electron bombardment in the energy range of a few hundred volts. Such bombardment happens on the Moon predominantly when it is in the magnetic tail of the Earth, and this may be the reason why mare ground is so remarkably dominant on the hemispere facing the Earth.

There is much evidence that erosion and transportation of material over the surface has taken place on the Moon on a substantial scale. There has been very little discussion of the obvious dilemma "Where is all the eroded material now?". Many of the craters in the uplands have been degraded until they have only a small fraction of the height they must once have had were they to have looked like the younger craters do now. Explosive events could not have generated their present shapes. The amount of material apparently removed from all the overlapping highland craters appears to correspond to more than 1 km of depth averaged over the Moon. Some authors have suggested that all this material has left the Moon altogether, being blasted into space by further impacts. This explanation leads to serious difficulties. Either the infalling stuff removes more from the Moon than the mass falling in, which seems mechanically very unlikely and leads to difficulties for forming the Moon in the first place, or, if less is removed than brought in, the features ought to adopt a generally 'snowed over' appearance, which is certainly not the case.

Where steep mountains meet the plain, the junction line is usually abrupt and there is no indication that all the missing material can be hidden in that vicinity. Some talus slopes exist perhaps on the moon, but certainly they are not a common feature. The only places that can accommodate the large amount of missing material are the surfaces of the maria and the flat interiors of old craters. It was this argument that provided the first reason for suspecting large amounts of finely divided material to cover the mare surfaces (Gold, 1955).

The arguments that surface transportation has taken place on a large scale on th Moon have been greatly strengthened by all the recent evidence. The Moon is indeed covered almost everywhere with a fine powder which could have suffered surface transportation, and not with solid rock which would for the most part be immobile. The depth of this deposit of fine powder is unknown, and while no doubt there will be

Urey and Runcorn (eds.), The Moon, 55–67.

Fig. 1. Rock on the lunar surface, near the Apollo 14 landing site, photographed by the astronauts.

changes of compaction with depth, no one has seen clear evidence of large areas of bedrock anywhere.

On a small scale a surface creep phenomenon must clearly have taken place so that all the many stones scattered over the surface can be neatly imbedded without being snowed over (Figure 1). The random scattering of soil by secondaries from meteorite impacts could not achieve this, but would indeed generate the snowed over appearance. Instead, the rocks are for the most part very clean down to a sharp junction line with the soil, and their surrounds do not show any of the impact scars generated when they were tossed to their present positions. A surface creep phenomenon must be active that is fast enough to achieve this and indeed to counteract the scattering effect of the meteorites.

Plowing over by meteorite impact has been thought of by many authors as being the major surface activity. It has been estimated, for example, that the ground has been plowed over a hundred times to a depth of 40 cm in the lifetime of the mare surface (Shoemaker *et al.*, 1970).

The evidence of the core tube on Apollo 12 makes clear that the ground at that site has not even been plowed over once to a depth of 40 cm. There is clear evidence that the core has striations in height noticeable in optical properties (TLSPET, 1970), in differences of the size distribution of the grains (Gold *et al.*, 1971) and in chemical differences. Anders *et al.* (1971) has found for example, an increase in some trace elements by a factor as large as 10^5 in one layer of the core. This can only be understood by supposing that the surface has been added to at a rate that exceeds the plowing by meteorites. If one supposed a plowing rate to 40 cm depth of once in 40 million yr, one thus requires a deposition rate faster than 1 cm per million yr. The rate required to fill the mare basins in 4 billion yr would be ~1 cm in 10000 yr; however, it seems likely that both infall and transportation were occurring at a much faster rate in earlier epochs.

Mare surface in general is clearly not saturation bombarded since the crater density is regionally quite variable and in general less than in the highlands. This can be understood again as a sign that a deposition process lays down material faster in the mare regions than the plowing by meteorites. Ronca (1971) has studied such regional differences and has mapped them.

The compositional differences between soil and rocks in the same region make clear that the soil is not local bedrock ground up with the rocks being pieces of that same bedrock. This would have been the expected situation if bedrock existed at a shallow depth and if the soil were merely the consequence of its local pulverization.

If instead the soil has suffered some surface creep over big distances, while the rocks are pieces thrown out from major craters and originally represent material at a greater depth, then such compositional differences can indeed be expected.

Lunar Orbiter and Apollo orbital photography show classes of features (Figure 2) on the Moon in fine detail whose origin must be sought in surface transportation mechanisms. Outstanding among those are the 'shoulders' at the junctions of old mountains with flat mare surface. The shoulders are characterized by the following features: at the junction line the ground rises with the steepest gradient that occurs in the entire region. Over distances of some tens or hundreds of meters the gradient flattens, with the level then being some meters or tens of meters above the mare surface. This flattened gradient then meets up in another contact line with the steeper slope of the old mountains (Figure 3).

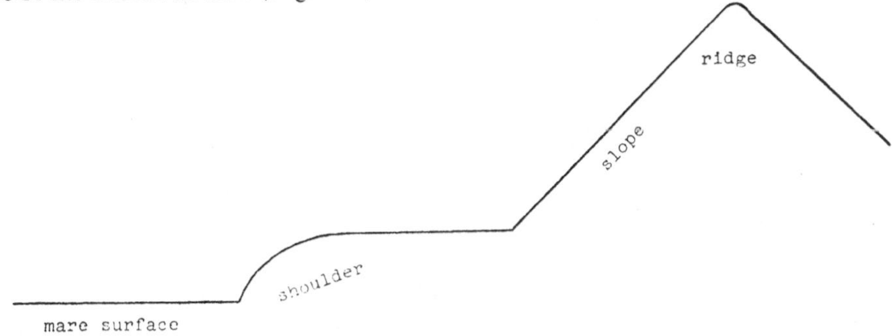

Fig. 2. Illustration of the classes of features studied on Lunar Orbiter photographs.

Fig. 3. Lunar Orbiter photograph showing the abrupt junction line where a mountain meets the plain.

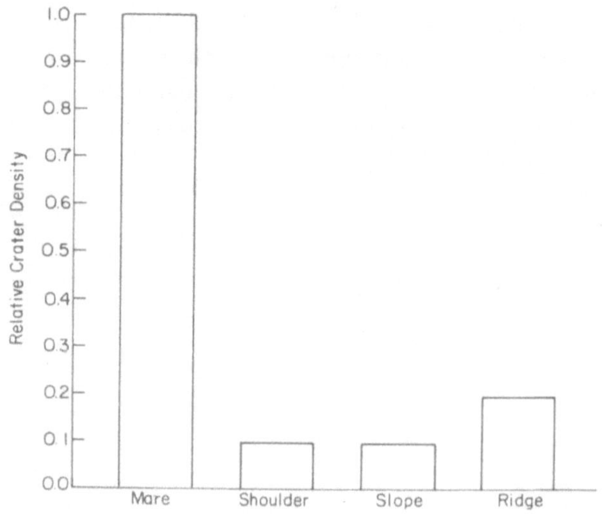

Fig. 4. Average relative crater densities on mountain slopes, shoulders and nearby mare ground, based on Lunar Orbiter photographs.

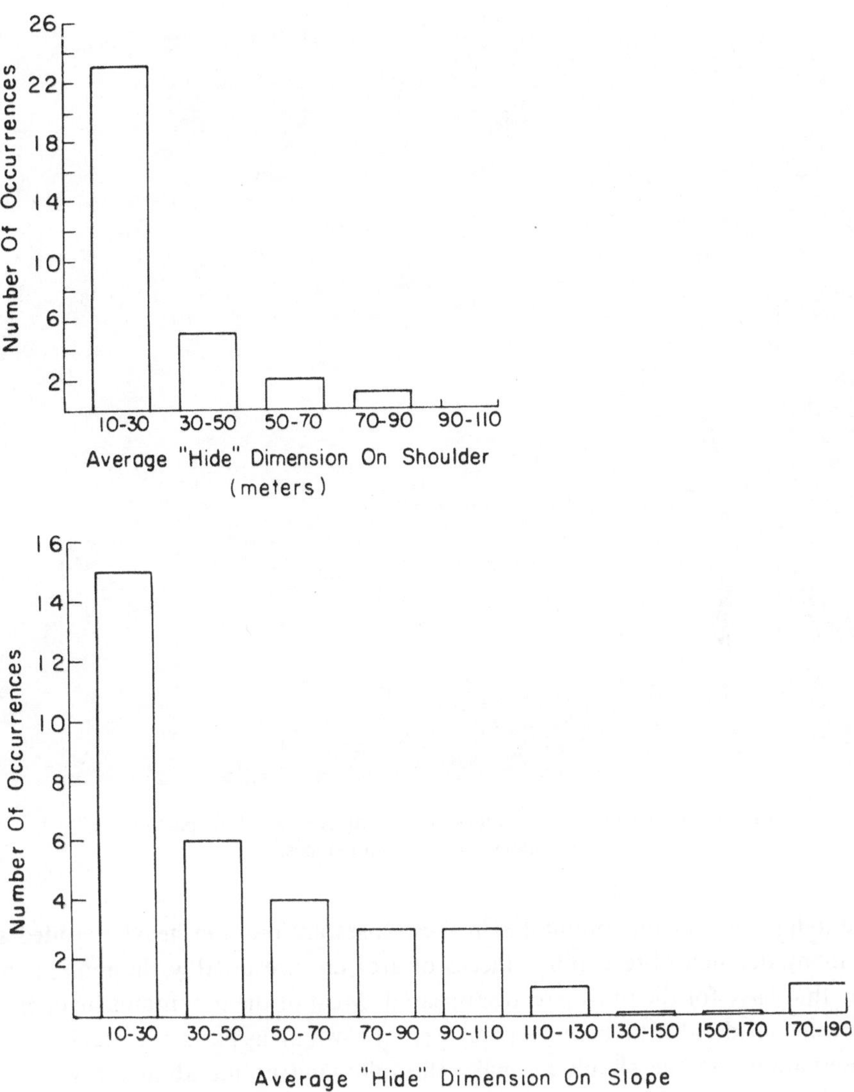

Fig. 5. Histograms of average 'hide' dimensions on mountain slopes and shoulders, based on Lunar Orbiter photographs.

We have investigated more than twenty regions where such shapes are seen and have found that they are quite generally associated with other observable quantities. It is generally true that the crater density in the shoulder is lower than that in the neighboring ground. Also, quite generally, the crisscross pattern seen on most lunar steep slopes ('elephant hide pattern') is usually very pronounced on the old mountain slope and is clearly visible but on a much smaller scale on the shoulders (Figures 4 and 5). This pattern is of course absent on the mare surface. Figures 6 and 7 show features of this kind.

Fig. 6. Lunar Orbiter photograph showing elephant hide patterns on mountain shoulders and slopes.

The judgment that the mountains in these cases are old and heavily eroded stems from many detailed observations. Pieces of arcs are presumably the remains of old craters that have for the most part disappeared. Most of the mountains involved must have been considerably higher when they were thrown up as parts of craters.

Observations on the shoulders make the following points abundantly clear: the material forming the shoulders has come there as a consequence of a surface transportation mechanism. One cannot conceive of any physical process of a general slumping of the shape, nor can one think of lava having poured out into this shape so uniformly along hundreds of miles of old ridges. The relation with the erosion surface patterns strengthens this conclusion. It is further clear that the shoulders have reached their present configuration later than the final shaping of the mare surface. There are many occasions where partly overlapping craters allow a relative chronology to be established. Figure 8 shows such a crater in the flat mare ground partially filled with shoulder material. The nature of the transportation mechanism which has to have been active on quite a large scale is not yet clear. I will discuss a possible mechanism later, and I consider this the outstanding problem for the understanding of the nature of the lunar surface.

Fig. 7. Lunar Orbiter photograph showing elephant hide patterns on mountain shoulders and slopes.

When we have such clear evidence that the lunar soil has been transported on this scale, we may wonder whether the next larger scale phenomenon, namely the filling of the mare basins, did not occur in a similar manner. Two lines of evidence favor this.

The transmission of seismic signals over distances of tens or hundreds of kilometers on the Moon occurs in a manner totally different from the terrestrial case. The seismic evidence cannot be understood except with an absence of widespread bedrock at a shallow depth and instead the presence of a medium of slower sonic velocity over an interval of several kilometers of depth. Lunar soil gradually compacted with depth will account very well for the whole range of seismic phenomena seen (Ronca, 1971).

Mascons (Gold and Soter, 1970) as distributed positive gravity anomalies covering the floors of most large circular mare basins would be the direct consequence of a cold accreted moon suffering large scale surface transportation. An impact large enough to make a mare basin would leave in the first place a concave basin of fully compacted material, while previously the accumulated material subjected to small scale impacts would have been porous down to a depth of as much as 50 km. Such a basin, now made of denser material, must adjust itself hydrostatically immediately

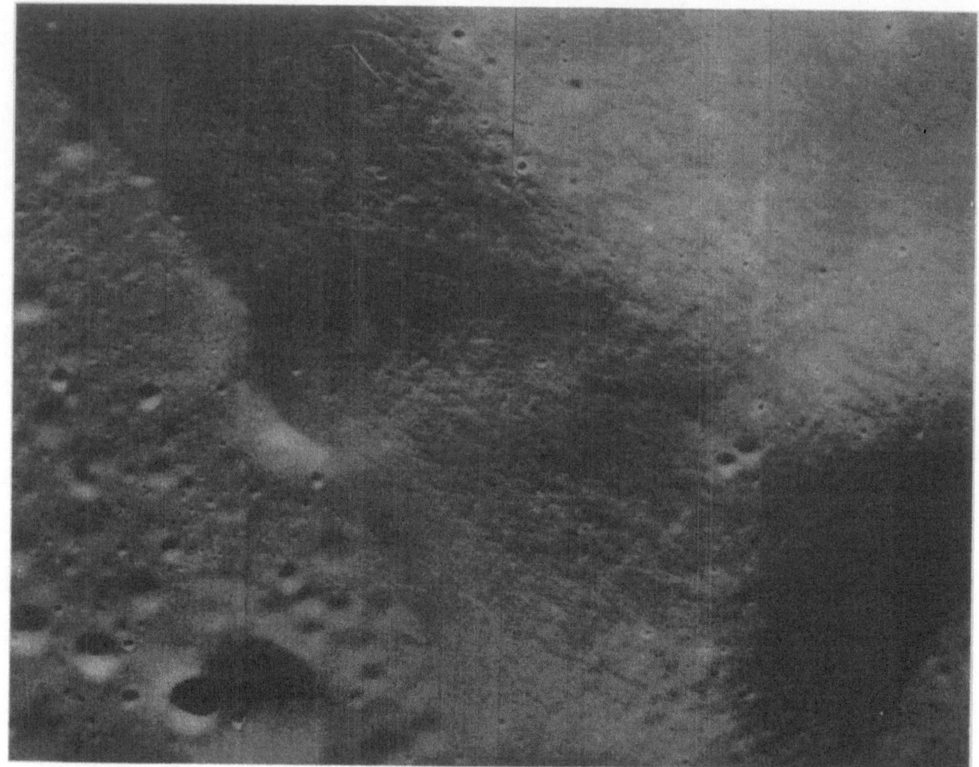

Fig. 8. Lunar Orbiter photograph of a crater in the flat mare ground partially filled with shoulder material.

after the impact, since the strength of no rock can be enough to maintain so large a departure from equilibrium. The floor will thus immediately be forced up to the vicinity of equilibrium, but since the material is denser there than in the surroundings, the equilibrium level will be below the surrounding plain. If porosity was crushed out over a depth of 50 km an equilibrium level 4 km lower would be entirely reasonable.

At this stage it can be shown that to a first order there would be no free air gravity anomaly. Since the basin is lower than the surrounds, surface transportation could cause it to be filled, and when as much as 1 or 2 km has accumulated, a positive gravity anomaly of the observed strength will result. In this case, since the underlying rock can be largely cold (except for a comparatively thin skin of heating and melting caused by the impact) there is no problem of later upholding the extra burden. In theories where this extra burden is due to lava having come up from underneath it is not at all easy to understand how the rock from which it has come can be strong enough to support the imbalance.

I have been concerned for many years now with the problem of discovering the most likely surface transportation mechanism. Although some shapes are suggestive of wind erosion and perhaps a very temporary atmosphere, possibly as a consequence

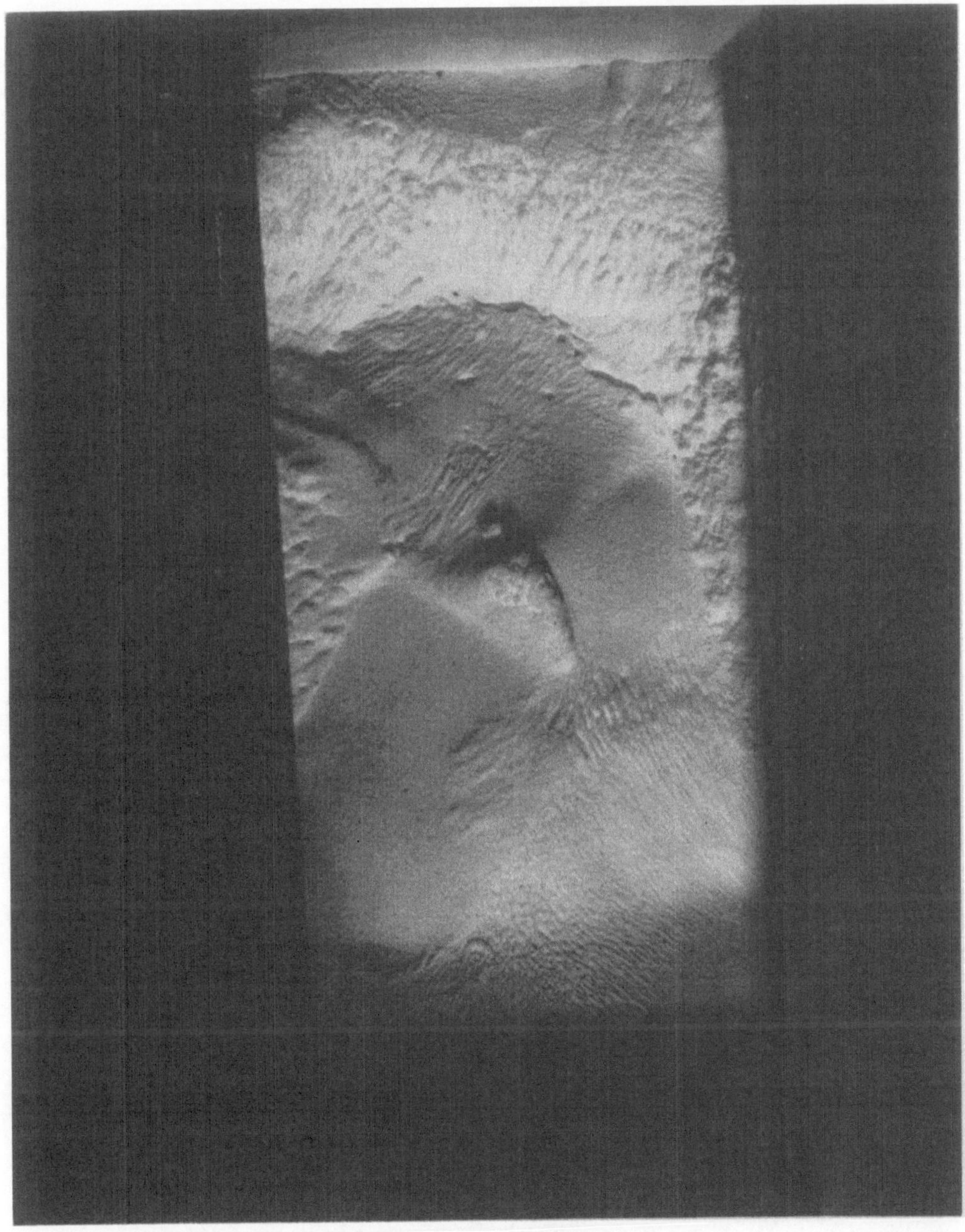

Fig. 9. Surface configuration of an insulating powder (Al_2O_3) after being exposed to an electron beam of ~ 500 V energy. The actual surface area seen on the picture is 6 × 12 cm.

of a large impact, many erosion features such as the shoulders can clearly not be accounted for in that way. Equally seismicity cannot be held responsible. Direct solar radiation pressure is also insufficient and also the corresponding latitute dependence is not seen. Electrostatic effects have thus always seemed the most likely.

Among the many electrostatic effects that I have considered and experimentally investigated, there is one that is quite outstanding in its efficacy. It is an effect related to the secondary electron emission characteristics of the material in the presence of electron bombardment. If the primary bombarding electrons come in with an energy in the vicinity of that for which an equal number of secondaries is produced, then the charging of the surface becomes locally quite unstable. For one spot the secondary emission ratio may be slightly more than unity, and it therefore races to become positive until no other surface can receive the secondary electrons and they fall back upon it. Another spot being slightly below the unity value becomes negative until the energy of the incoming electrons is insufficient for them to reach it. Areas are thus driven apart in potential by as much as the energy of the incoming electron beam, and that, for unity in the secondary emission ratio, is of the general order of a few hundred volts for most materials. If neighboring grains a few microns apart develop differences

Fig. 10. Single frame print of an Apollo 14 close-up stereo photograph of the lunar surface, showing parallel lines of grooves. The area shown here is 8.3×7.6 cm.

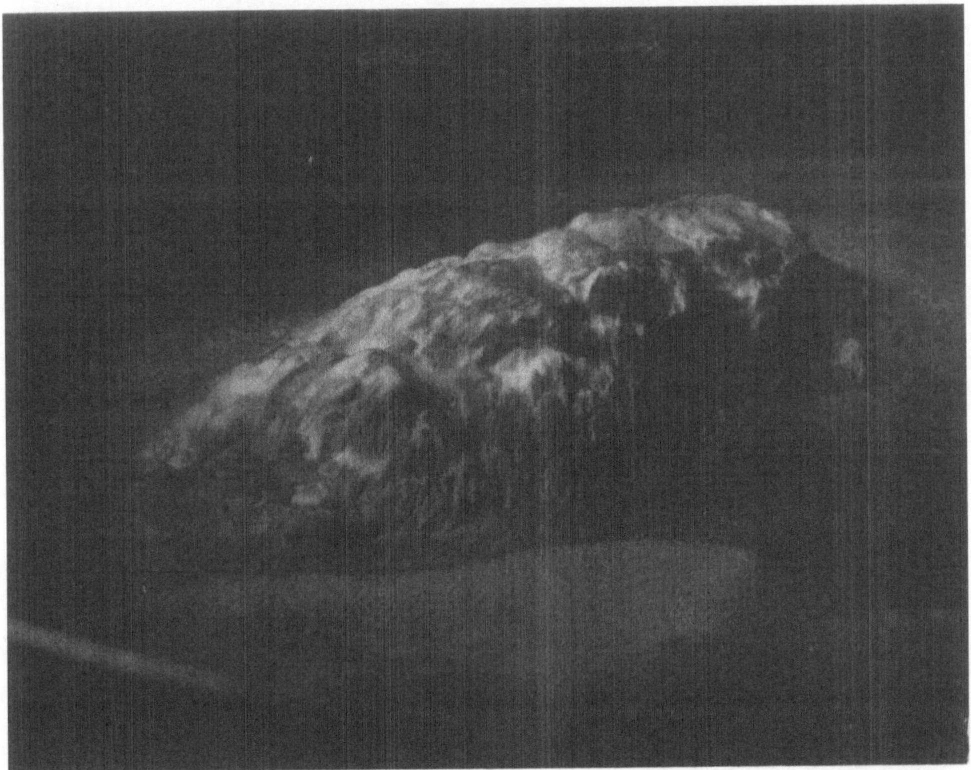

Fig. 11. Configuration of a small pile of insulating rock powder on the top of the flat surface of a chemically different rock powder after electron bombardment in a laboratory vacuum system. (The length of the pile is approximately 6 cm.).

of potential of several hundred volts the electric fields are in the range of millions of volts per centimeter and the forces are then amply sufficient to dislodge and move the grains. This is the effect we have recently investigated extensively in the laboratory and we have observed not only very rapid transportation of grains over the surface but also the development of very particular shapes and features in the surfaces so treated. Parallel lines of grooves and ridges on a millimeter scale are a particularly common type of shape that is so produced. We have seen a remarkable pattern of a similar nature in one of the close-up stereo photographs brought back from the Moon by Apollo 14 (Figures 9 and 10).

Another phenomenon observed in the laboratory is that it is common for the junction line between different materials to remain sharp, and while the junction line itself can move extensively, no mixture of materials is produced. In those cases one material flows over the other retaining a sharp front. This is a phenomenon on a smaller scale but quite reminiscent of the shoulder phenomenon we have described (Figures 11 and 12).

If electrostatic transportation has played a major part in forming the Moon's

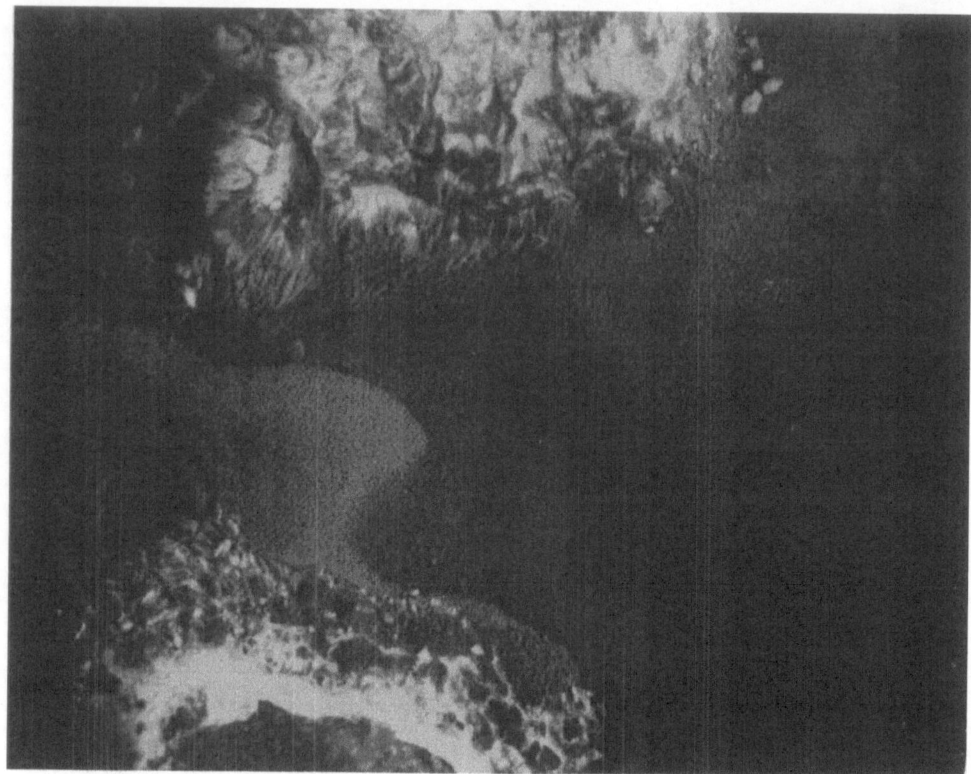

Fig. 12. Junction line formed between different materials subjected to electron bombardment in a laboratory vacuum system. The picture shows an area of approximately 4 × 6 cm.

surface then the difference in the appearance between the front and the back of the Moon can be understood. Electrons in the range of energy of a few hundred volts do not occur very often in the free solar wind. There the electrons are merely transported along with the protons at the solar wind speeds, and this corresponds to very low electron energies. In the magnetic tail of the Earth a different phenomenon takes place. Through the process of magnetoturbulence, suffered by the solar wind as it strikes the magnetosphere, some energy sharing between electrons and protons takes place, and electron energies are thus increased into the range of several hundreds of volts. We believe that this results in electrons in this energy range being able to hit the near side of the Moon but not the far side. Since the Moon is within the magnetic tail of the Earth for approximately four days each month, this may be a large effect. In the past when the Moon's orbit was undoubtedly much smaller, the effect may have been much larger still.

Thus, if electrostatic transportation is a major phenomenon, big basins on the front side of the Moon may have gradually filled in, while similar basins on the back side would merely be subject to further cratering but to much less filling. So long as we have no detailed knowledge of levels on the Moon, we cannot discover such basins when they are not filled. An examination of gravity anomalies should of course then show the mascons to be weaker or absent on the back.

Acknowledgements

The analysis of Orbiter pictures and study of the 'shoulders' and other erosion features was carried out jointly with Mr John Delano. The laboratory work on electrostatic transportation was carried out jointly with Mr Gregory J. Williams. Work on lunar studies at Cornell University is supported by NASA Grant NGL-33-010-005.

References

Gold, T.: 1955, *Monthly Notices Roy. Astron. Soc.* **115**, 585.
Gold, T. and Soter, S.: 1970, *Science* **169**, 1071.
Gold, T., O'Leary, B. T., and Campbell, M.: 1971, Proc. of the Second Lunar Science Conference, *Geochim. Cosmochim. Acta Suppl. 2* **3**, 2173.
Laul, J. C., Morgan, J. W., Ganapathy, R., and Anders, E.: 1971, Proc. of the Second Lunar Science Conference, *Geochim. Cosmochim. Acta Suppl. 2* **2**, 1139.
Muller, P. M. and Sjogren, W. L.: This volume, p. 35.
Ronca, L. B.: 1971, This volume p. 43 and *Geolo. Soc. Am. Bull.*, June 1971.
Shoemaker, E. M., Hait, M. H., Swann, G. A., Schleicher, D. L., Schaber, G. G., Goddard, E. N., and Waters, A. C.: 1970, *Geochim. Cosmochim. Acta Suppl.* 1, **3**, 2399.
The Lunar Sample Preliminary Examination Team: 1970, *Science* **167**, 1325.

RADAR MAPPING OF THE MOON AT 162 MHz

J. E. B. PONSONBY, I. MORISON, A. R. BIRKS, and J. K. LANDON
Nuffield Radio Astronomy Laboratories, Jodrell Bank, Macclesfield, Cheshire, England

Abstract. The result of Lunar radar mapping observations are presented, which were made at Jodrell Bank in January and February 1970 at 162.4 MHz in the depolarized return. The Maps show that most of the depolarized return comes from the highland regions. No significant return has been detected from the Maria. Certain isolated features are particularly prominent, notably the craters Tycho and Theophilus which appear almost equally intense, also the areas around Werner and Fourier. Various other features can be identified with less reliability. These results are discussed in qualitative terms.

The Moon has been observed at Jodrell Bank with an essentially continuous wave (cw) radar operating at 162.4 MHz ($\lambda = 1.85$ m). A circularly polarized wave was used

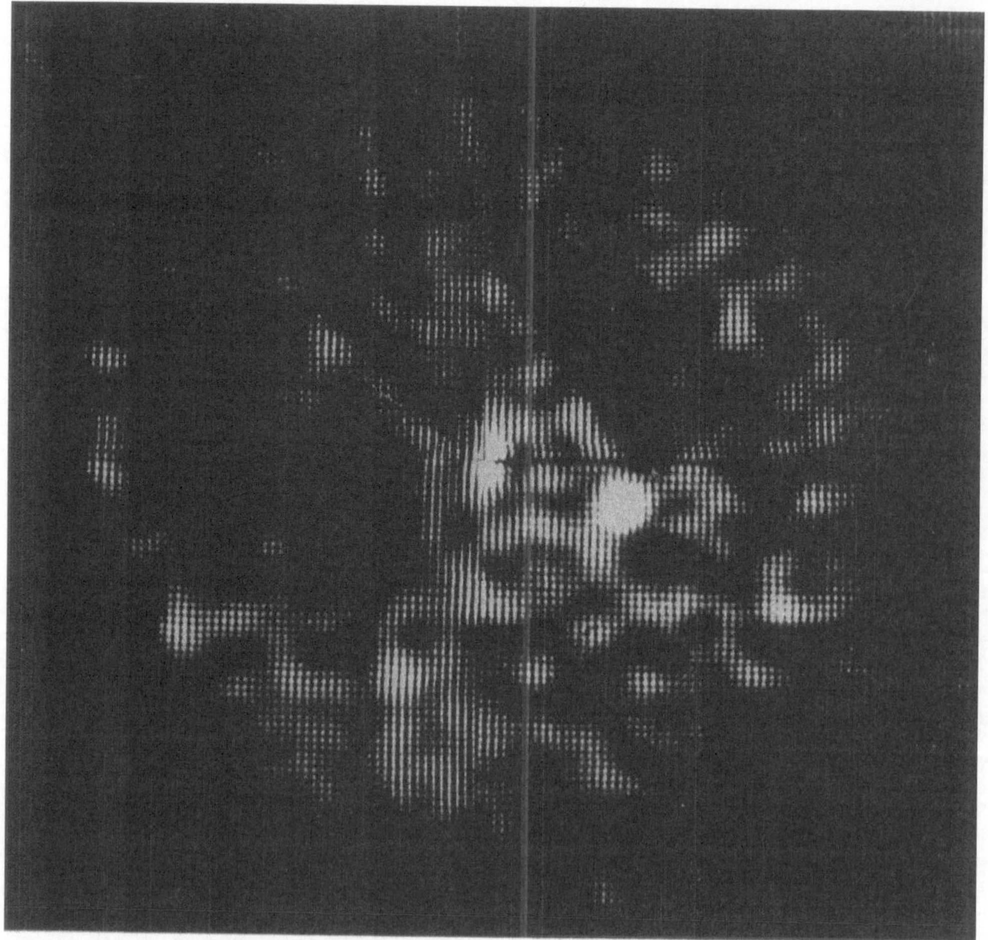

Fig. 1. Moon seen by 162.4 MHz radar with 1.25 arc min. resolution in January 1970.

Urey and Runcorn (eds.), The Moon, 68–71.

to illuminate the Moon, and the component of the reflected signal having the non-specular sense of circular polarization was received and used to make images of the lunar disk.

In this paper we present the results of two independent observations made in January and February 1970. A preliminary map at 2 arc min resolution obtained from some of the January data was presented at the XIV General Assembly of IAU at Brighton (Ponsonby, 1970). Here we present a final map at 1.25 arc min resolution from the January data and one at 2 arc min resolution from the February observations. These are shown in Figures 1 and 2 respectively. North is at the top in both cases so that the disc is seen as it is in the sky in the northern hemisphere.

Two-dimensional mapping has been achieved using the Aperture Synthesis method first described by Thomson (1965) and later by Thomson and Ponsonby (1968) and Hagfors *et al.* (1968). This method is distinct from the better known range-doppler

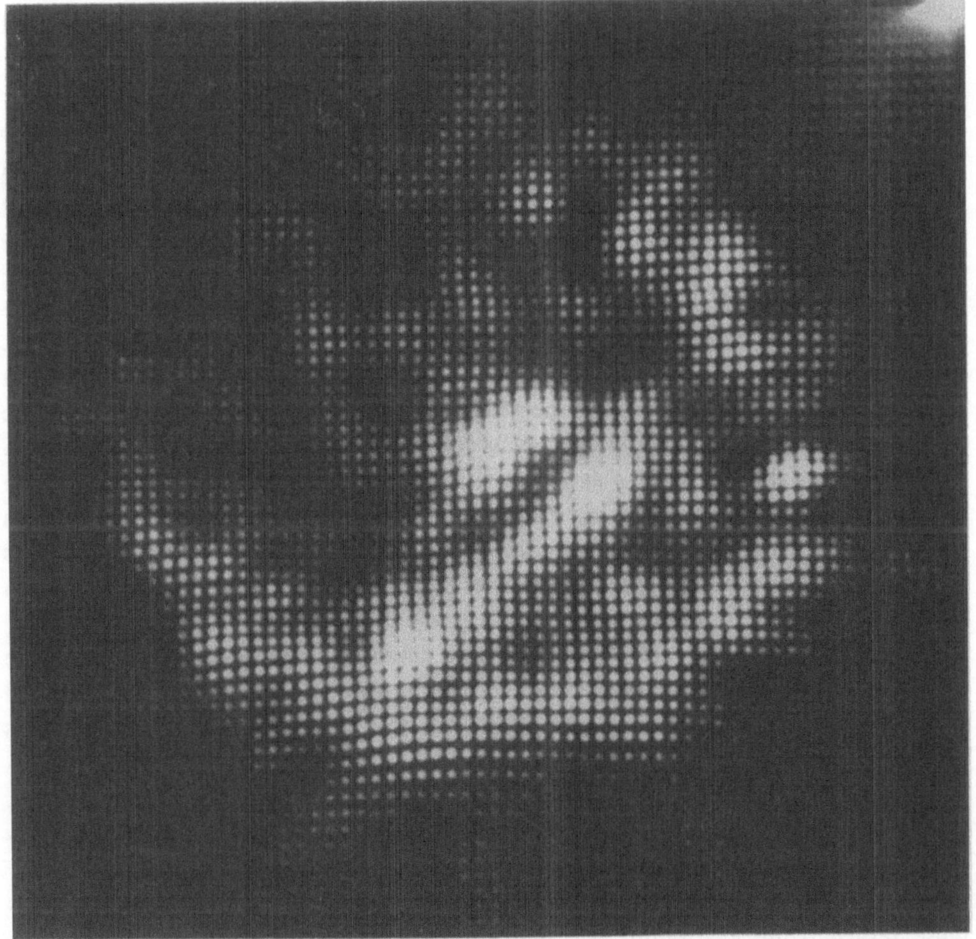

Fig. 2. Moon seen by 162.4 MHz radar with 2 arc min resolution in February, 1970.

mapping technique, and has the advantage of enabling the whole face of the Moon to be mapped simultaneously without ambiguity using a relatively modest radar system.

The present observations have been made using a 1 kW transmitter with a single 50 ft diam. dish having a 7° beamwidth.

The method relies upon the differential doppler shifts caused by the Moon's librations; the peculiar circumstances of the Moon's librations allow the method to be used on two or three favourable days each month. About 12 hr of observations are necessary for a complete map; the total change of Lunar aspect during this period, typically 2° in selenographic coordinates, limits the attainable resolution to about 1 minute of arc. This resolution is equal to what could be achieved directly had one an aerial about 6 km in diam.

At the times of the observations, the mean selenographic coordinates of the disc centre were;

in January 1970, $l = +4{\overset{\circ}{.}}5$, $b = -6°$;

in February 1970, $l = +6{\overset{\circ}{.}}1$, $b = -6°$.

The bright region at the centre of the maps is due to an instrumental malfunction which more seriously effected the February results.

The broad features appear the same on both maps. Thus the highland regions appear bright whilst no significant return has been detected from the extended maria. The dark areas on the left hand sides correspond to maria Imbrium, Procellarum, Nubium and Humorum. On the right hand side Serenitatis and Tranquillitatis form the main dark area, whilst there are isolated dark areas at the position of Crisium, Fecunditatis and Nectaris.

Individual features are best seen on the 1.25 arc. min January map. The most prominent are the craters Tycho and Theophilus which appear almost equally bright. Comparable though lesser returns are seen from the regions of the craters Werner and Fourier. These appear to come from extended areas and are not necessarily to be associated with the craters themselves. Copernicus can be definitely identified in Procellarum on the January map though it appears considerably less bright than Tycho. A comparable return is seen from the region of Proclus. The crater Langrenus coincides with a still weaker bright spot on the January map but surprisingly appears as one of the prominent features on the February map. This may be due to a random fluctuation. We have not detected Aristarchus or Kepler.

We know of no theory of scattering which allows quantitative deductions to be made from the observed intensities of the depolarized return. The depolarized scattering mechanism is however essentially non-specular and one can say that its occurence must be associated with structure on a scale comparable with the wavelength. This structure may be on the surface or be buried at a depth of some meters. Qualitative interpretation of our results can be based on this notion.

That Tycho would appear as a prominent feature was expected from earlier reports that it has appeared unusually bright to radar at short wavelengths. The intense return from Theophilus was not so anticipated. On the map obtained by Hagfors *et al.* (1968) for instance, at $\lambda = 23$ cm using effectively the same mapping technique, Tycho is extremely conspicuous, whilst Theophilus is not prominent at all. Together with Copernicus however, which appears much less intensely at our wavelength, both are rayed craters believed to be of relatively recent origin. We interpret the change in behaviour of Theophilus as indicating a lack of structure on the scale of the shorter wavelength. This view is consistent with the smoother appearance of Copernicus compared to Tycho on the scale of the high resolution photographs taken by the Orbiter space craft. Unfortunately there are no high resolution pictures available of Theophilus. If there were, we suppose they would reveal a degree of roughness intermediate between that seen in Tycho and Copernicus.

References

Hagfors, T., Nanni, B., and Stone, K.: 1968, *Radio Sci.* **3** (*New Series*), 491.

Ponsonby, J. E. B.: 1970 'Mapping of the Moon by Continuous Wave Radar at 162 MHz', presented at XIV General Assembly IAU, Brighton 1970.

Thomson, J. H.: 1965, 'Talk Given at the Symposium on Planetary Atmospheres and Surfaces', Dorado, Puerto Rico.

Thomson, J. H. and Ponsonby, J. E. B.: 1968, *Proc. Roy. Soc. A.* **303**, 477.

C. APOLLO MISSIONS PROGRESS

INTRODUCTORY REMARKS: THE APOLLO 14 MISSION

HOMER E. NEWELL

U.S. National Aeronautics and Space Administration, Headquarters, Washington, D.C.

Abstract. To set the context of our current exploration of the Moon, it is well to recall: Apollo 11 landed near the equator and was capable of affording the astronauts only a few hours on the lunar surface; Apollo 12 checked out pin-point landing which then permitted going to more difficult, non-equatorial sites; in Apollo 14 enlarged propellant tanks permitted some 42 kg of samples to be brought back. Three more Apollo missions follow Apollo 14. The Apollo missions, complemented by unmanned missions to the various planets, form a program to explore the solar system.

Captain Scherer will elaborate on the significant increase in Apollo capability in the last three missions. Of note are the lunar heat flow experiments, the lunar surface gravimeter, and, for Apollo 15, the laser ranging retroreflector which will be three times the area of and four times as efficient as retroreflectors in the previous Apollo missions. In the lunar orbital equipment will be the sub-satellite with instruments for measuring various physical properties of the Moon and its environment. Also noteworthy is the lunar sounder, for Apollo 17, which will use radar for probing a large fraction of the lunar surface to considerable depth.

Hardware changes will allow the astronauts to remain on the lunar surface for up to 66 hrs – an increase of 100%; thus the landed scientific payload may be doubled to approximately 550 kg. Increased range and efficiency of surface operations will result from improved suit mobility, an improved life support system, and a Lunar Roving Vehicle capable of up to 90 km traverses. Changes in the Command and Service Modules will allow for up to 16 days total flight duration.

We will again welcome proposals for analysis of the new samples, but now that the initial exploratory work has been done we intend to be far less duplicative in choosing proposals to support. We are particularly pleased with the growing attention being given by the scientific societies to the results of lunar research.

In the last talk of this series, Mr. Burke will describe some of the future possibilities that have been studied for exploration of the Moon and planets. We will particularly welcome suggestions as to how to improve the value of future missions.

My remarks are intended to be introductory to the more detailed presentations to follow. After my remarks, Captain Lee Scherer, Director, Apollo Lunar Exploration Office, NASA, will discuss the Apollo 14 mission and some of the results. Peter Armitage, Manager, Lunar Receiving Laboratory, NASA Manned Spacecraft Center, will then give some of the results on the examination of the Apollo 14 rocks by the Preliminary Evaluation Team. Then Lee Scherer will return to discuss the remaining Apollo missions. After that, J. D. Burke, Manager, Advanced Technical Studies, Jet Propulsion Laboratory, will peer into the future beyond any currently approved projects or missions.

To set the context of our current exploration of the Moon, it is well to recall a few facts about the earlier missions. Apollo 11 landed on Mare Tranquillitatis near the equator, and was capable of only a limited stay affording the astronauts only a few hours on the lunar surface. Apollo 12, in landing on Oceanus Procellarum, checked out pin-point landing, the success of which then permitted going to more difficult, non-mare, non-equatorial sites on later flights. Also, during the Apollo 12 mission it was observed that the consumables carried by the portable life support system would permit longer extravehicular activities (EVA's) than original appropriately conservative estimates had planned.

Urey and Runcorn (eds.), The Moon, 75–80.

TABLE I

Assignment of approved lunar surface experiments

	Experiment or instrument	Principal investigator	Apollo 11	12	13	14	15	16	17
	Lunar geology investigation	Shoemaker/USGS	A	A					
		Swan/USGS			X	A	A		
		Muehlberger/U. of Texas						A	A
	Laser ranging retro-reflector	Alley/U. of Md.	A						
		Faller/Wesleyan, Mass.				A	A		
	Cosmic ray detector	Fleischer/GE.	A					A	
		Walker/Wash. U. at St. Louis							
		Price/UCLA							
	Lunar surface close-up stereo photo	Gold/Cornell	A	A	X				
	Portable magnetometer	Dyal/ARC				A		A	
	Lunar gravity traverse	Talwani/Columbia							A
	Soil mechanics	Mitchell/U. of Cal., Berkeley				A	A	A	A
	Far UV camera spectroscope	Carruthers/NRL						A	
	Surface electrical properties	Simmons/MSC							A
	Solar wind composition	Geiss/Berne Univ.	A	A	X	A			
ALSEP	Lunar passive seismology	Latham/Columbia	A	A	X	A	A	A	[a]
ALSEP	Lunar active seismology	Kovach/Stanford				A		A	
ALSEP	Lunar tri-axis magnetometer	Dyal/ARC		A			A	A	
ALSEP	Medium energy solar wind	Snyder/JPL		A			A		
ALSEP	Suprathermal ion detector	Freeman/Rice		A		A	A		
ALSEP	Lunar heat flow (with drill)	Langseth/Columbia			X		A	A	A
ALSEP	Cold cathode ionization gauge	Johnson/U. of Texas (Dallas)		A	X	A	A		
ALSEP	Lunar ejecta and meteorites	Berg/GSFC							A
ALSEP	Lunar seismic profiling	Kovach/Stanford							A
ALSEP	Lunar atmos. composition	Hoffman/U. of Texas (Dallas)							A
ALSEP	Lunar surface gravimeter	Weber/U. of Maryland							A[a]
ALSEP	Charged particle-lunar environment	O'Brien/U. of Sidney, Australia			X	A			

A – Assigned
X – Aborted
[a] Lunar passive seismology may be substituted for lunar surface gravimeter

TABLE II

Assignment of approved lunar orbital, in-flight, and pre- and post-flight experiments

Experiment or instrument	Principal investigator	Apollo 13	14	15	16	17
Gamma-ray spectrometer	Arnold/U. of Sal. SD			A	A	
X-Ray fluorescence	Adler/GSFC			A	A	
Alpha particle spectrometer	Gorenstein/Am. Sci. and Eng.			A	A	
S-band transponder (CSM/LM)	Sjogren/JPL	X	A	A	A	A
Mass spectrometer	Hoffman/U. of Texas (Dallas)			A	A	
Far UV spectrometer	Fastie/Johns Hopkins					A
Bistatic radar	Howard/Stanford	X	A	A		
IR scanning radiometer	Low/Rice					A
Apollo window meteoroid	Cour-Palis/MSC			A		
UV Photography – Earth and Moon	Owen/IITRI			A	A	
Gegenschein from lunar orbit	Dunkelman/GSFC			A		
Lunar sounder	Brown/JPL					A
	Ward/U. of Utah					
Bone mineral measurement	Vogel/USPHS San Fran.			A	A	A
Total body gamma spectrometry	Benson/MSC			A*	A*	A*
Subsatellite:						
S-band transponder	Sjogren/JPL			A		A
Particle shadows/boundary layer	Anderson/U. of Cal., Berkeley			A		A
Subsatellite magnetometer	Coleman/UCLA			A		A

X – Aborted
A – Assigned
A*– Assigned pending elimination of quarantine requirements

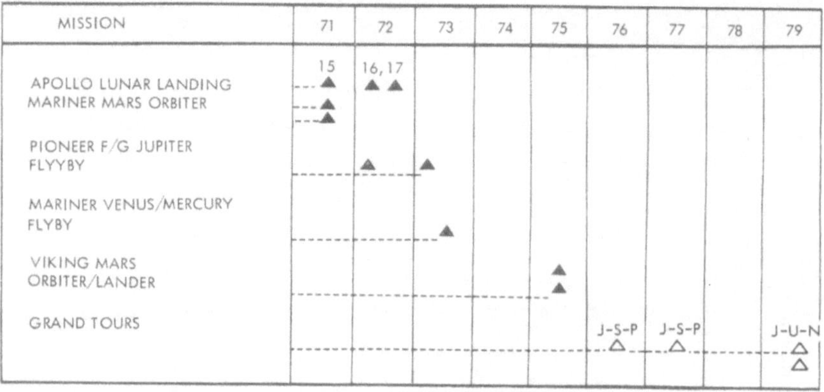

MISSION
71
72
73
74
75
76
77
78
79
APOLLO LUNAR LANDING
MARINER MARS ORBITER
PIONEER F/G JUPITER FLYYBY
MARINER VENUS/MERCURY FLYBY
VIKING MARS ORBITER/LANDER
GRAND TOURS
15
16, 17
J-S-P
J-S-P
J-U-N
▲ CURRENTLY APPROVED
△ PLANNED, FY 72

Fig. 1. Lunar and planetary missions. The currently approved missions to the Moon and the planets are indicated by the solid triangles. New missions included in the FY 1972 budget request are indicated by open triangles.

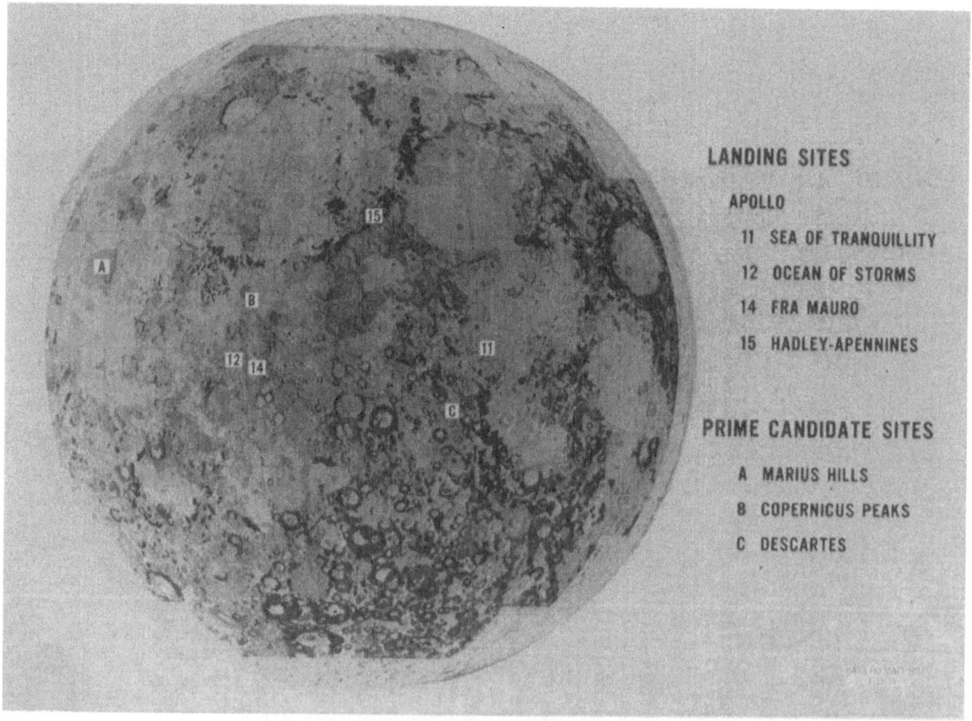

Fig. 2. Geologic map of the Moon with landing sites. Shown are the landing areas for Apollo missions that have been accomplished and candidate sites for the remainder of the Apollo flights. Some of the candidate sites for future missions are: Marius Hills, Copernicus, Hadley/Apennines, and Descartes (NASA Hq MA 71-5037).

Thus, in Apollo 14, which has just been completed, returning safely to Earth on 9 February 1971, more EVA time was achieved. In addition, enlarged propellant tanks permitted a larger payload, and some 42 kg of psamles were brought back.

The completion of the Apollo 14 mission leaves three more Apollo missions to the Moon as shown in Figure 1. The window for Apollo 15 runs from 7–26 July 1971;

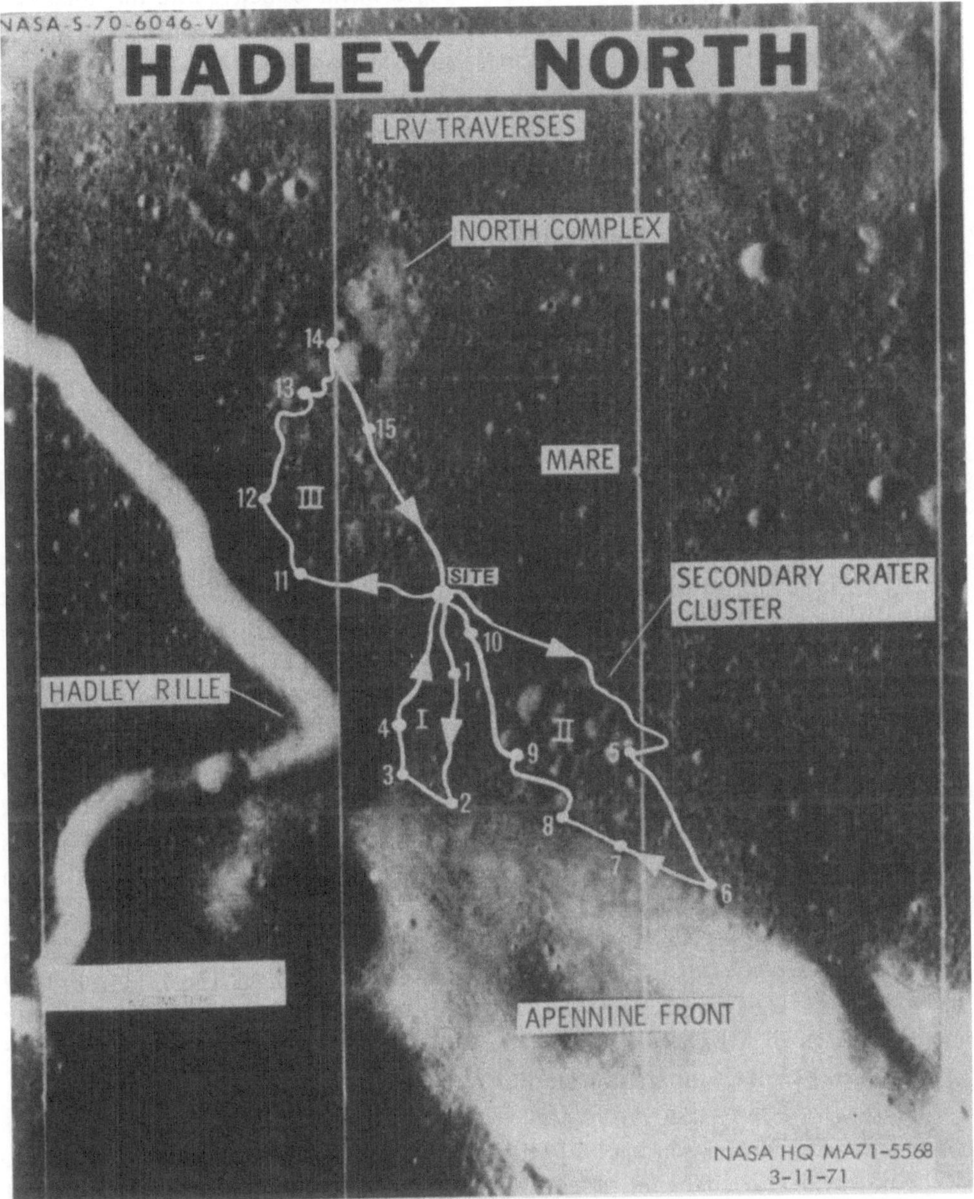

Fig. 3. Hadley/Apennines. The chosen lunar landing site for Apollo 15 (NASA Hq MA 71-5568).

Apollo 16 and 17 are scheduled for 1972. As may be seen from Figure 1, the Apollo missions form part of a program to explore the solar system, being complemented by a number of unmanned missions to the various planets.

As Captain Scherer will elaborate in his second talk, there will be a significant increase in Apollo capability in the last three missions. The instruments to be carried or experiments planned are shown in Tables I and II. Of particular note are the lunar heat flow experiments, in which it is planned to drill three meters into the lunar surface and insert thermocouples for lunar heat flow measurements, and the lunar surface gravimeter which will be carried if it can be ready in time for Apollo 17. Also of note is the laser ranging retroreflector for Apollo 15, which will be three times the area of and four times as efficient as retroreflectors in the previous Apollo missions. Of particular interest in the lunar orbital equipment will be the subsatellite which will carry instruments for measuring various physical properties of the Moon and its environment. Also the lunar sounder, for Apollo 17, which will use radar for probing a large fraction of the lunar surface to considerable depth, is worthy of note.

Changes in hardware will allow the astronauts to remain on the lunar surface for up to 66 hr – an increase of 100%; furthermore, these changes will allow the landed scientific payload to be doubled to approximately 550 kg. Increased range and efficiency of surface operations will result from improved suit mobility, an improved life support system, and a Lunar Roving Vehicle capable of up to 90 km traverses. Changes in the Command and Service Modules will allow for up to 16 days total flight duration.

Candidate landing sites for future missions are shown in Figure 2. We are planning to target Apollo 15 for an area in the vicinity of the Hadley Rille and the Apennine Mountains, some of the features of which can be seen on the Lunar Orbiter photo of Figure 3. Sites for Apollos 16 and 17, however, are yet to be determined.

All three of these missions will, of course, bring back additional lunar samples. We are open to suggestions as to what the astronauts should especially look for, particularly those that investigations of previous samples indicate will be especially illuminating. We will again welcome proposals for analysis of the new samples, but now that the initial exploratory work has been done with the samples from Apollos 11, 12, and 14, we intend to be much more selective, and far less duplicative, in choosing proposals to support.

We are particularly pleased with the growing attention being given by the scientific societies to the results of lunar research, as in this series of meetings sponsored by the I.A.U. In NASA we have raised the question whether NASA should hold any more meetings like the first two lunar symposia in Houston, or should now look to the various societies to sponsor and organize future sessions.

Finally, as we carry out our current program, we are continually studying future possibilities for exploration of the Moon and planets. In the last talk of this series, Mr Burke will describe some of the possibilities that have been studied. We will particularly welcome suggestions as to how to improve the value of future missions.

THE APOLLO 14 MISSION AND PRELIMINARY RESULTS

L. R. SCHERER

Director, Apollo Lunar Exploration, Apollo Program Office, National Aeronautics and Space Administration, Washington, D.C., U.S.A.

Abstract. In February 1971, the Apollo 14 mission successfully landed in the Fra Mauro region of the Moon. Astronauts deployed a sophisticated long-lived geophysical station. While on the surface they performed active seismic and magnetic experiments, deployed a solar wind collector and laser reflector, made an extensive geologic traverse, and performed television, movie, and still camera photography and other minor tasks. They selected a variety of samples of lunar material and returned to Earth over 40 kg for study. In orbit about the Moon, a variety of scientific and operational photography was accomplished.

The initial Apollo missions were directed to smooth mare regions of the Moon. Apollo 14 was the first mission to land in a non-mare area. The Fra Mauro formation is a peculiar geologic unit found in a number of areas around the Imbrium basin. It is thought that this formation represents throw-out resulting from impact of a smaller Moon which formed Imbrium. If this premise is true, material from this region may

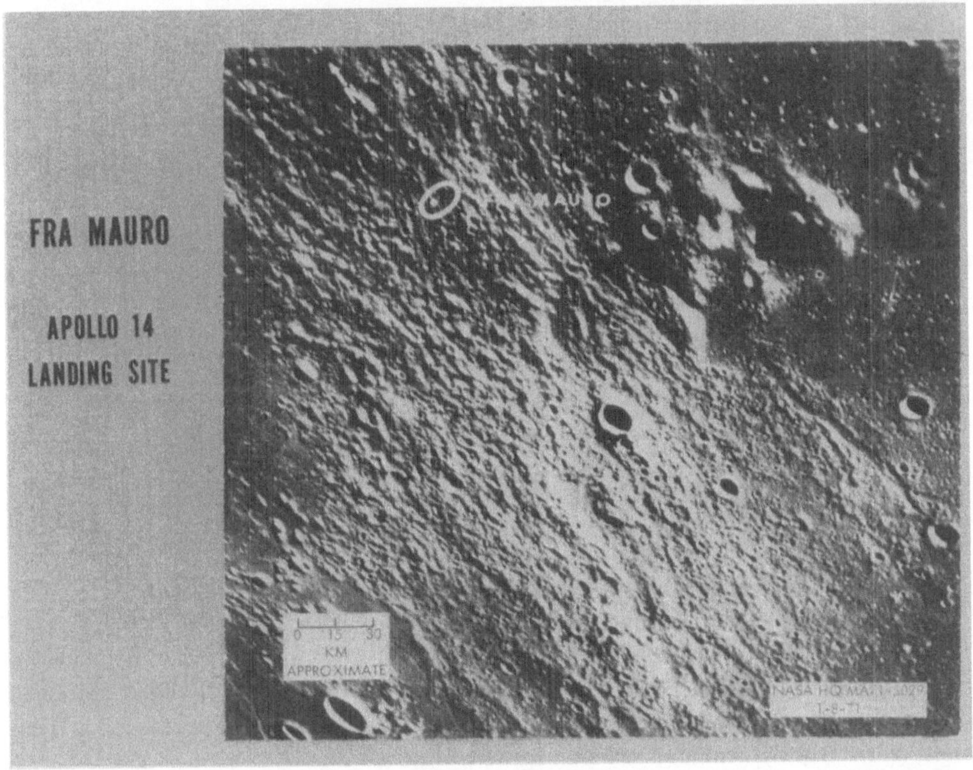

Fig. 1. Medium resolution of Fra Mauro.

Urey and Runcorn (eds.), The Moon, 81–93.

provide a chronology milestone for this major event in the early history of the Moon. The material may represent original crustal material as well as that from deep within the interior. One can see from Figure 1 the apparent radial patterns in the broad area, that seem to come from the Imbrium basin. We selected a specific landing site near a fresh moderate size crater which we concluded would have penetrated the upper surface rubble. It was felt the probability of returning original Fra Mauro material was increased by sampling from the vicinity of the crater rim. The mission proceeded as planned and on February 5, Apollo 14 touched down softly at the prescribed site. For this mission we developed a mobility aid for the crew as shown in Figure 2. As you

Fig. 2. Mobile Equipment Transporter (MET).

can see from the payload mounted on the little cart, the astronauts were able to carry with them a wide variety of cameras, tools, rock bags, and a small portable magnetometer. It proved to be very effective. Figure 3 is a photograph of the geophysical station that was deployed by the astronauts on the first trip to the surface. The geometry of the deployment is shown on Figure 4.

As of this date, March 23, 1971, scientific results from these various experiments must be considered as very preliminary. The passive seismometer in the first 49 days of

Fig. 3. Geophysical station.

operation has recorded some 60 events. Twenty of these were noted also on the seismometer deployed by Apollo 12 about 181 km away. We have been recording seismic signals of distinctive characteristics around the time that the Moon and Earth are in closest proximity. These were thought to be the result of tidal strains. The initial thought of the Principal Investigator is that they might be produced in the Fra Mauro crater region but this is a very preliminary guess. During the first perigee after Apollo 14 two of these events were seen on both seismometers. Further analysis may lead to determination of the more precise origin.

For the active seismic experiment, a geophone line 300 ft long was deployed. A total of thirteen energy impacts were placed into the surface by the astronauts with a thumper. The data returned appear to be excellent. The initial estimate is that the

APOLLO 14 ALSEP LAYOUT
(NOT TO SCALE)

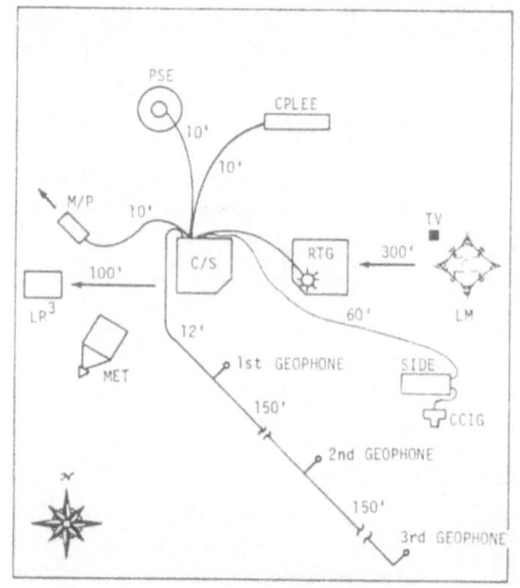

CCIG - COLD CATHODE IONIZATION GAUGE
CPLEE - CHARGED PARTICLE LUNAR ENVIRONMENT EXPERIMENT
C/S - CENTRAL STATION
LR3 - LASER RANGING RETRO-REFLECTOR
MET - MOBILE EQUIPMENT TRANSPORTER
M/P - MORTAR PACKAGE
PSE - PASSIVE SEISMIC EXPERIMENT
RTG - RADIOISOTOPE THERMOELECTRIC GENERATOR
SIDE - SUPRATHERMAL ION DETECTION EXPERIMENT

Fig. 4. Geometry of deployment of the geophysical station.

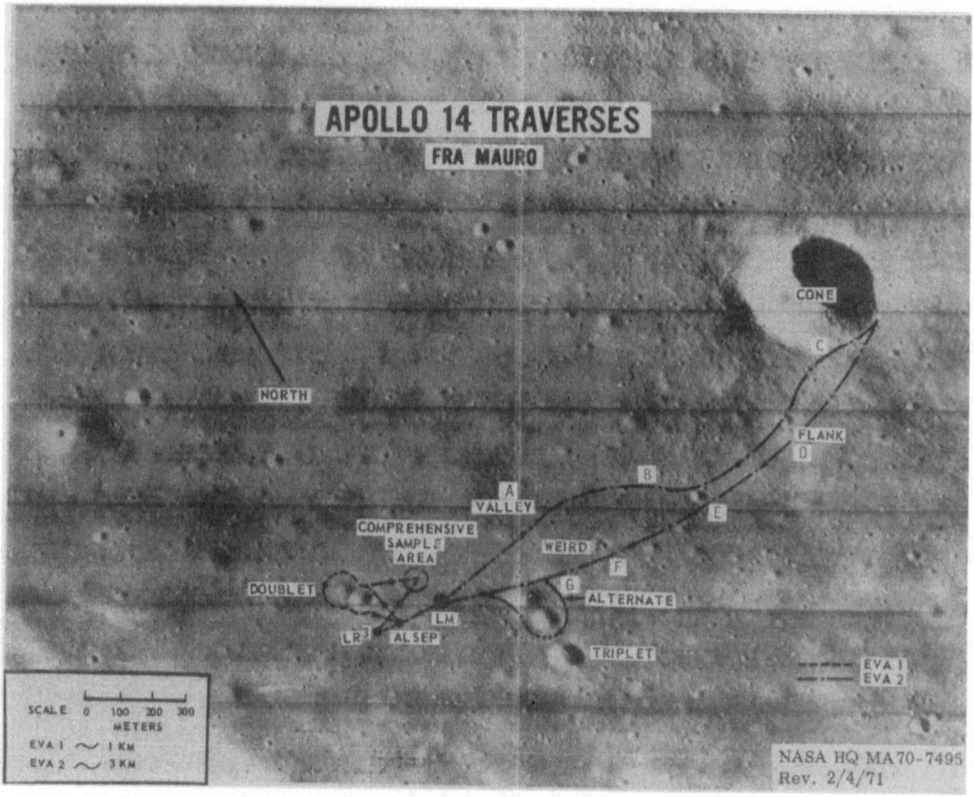

APOLLO 14 TRAVERSES
FRA MAURO

NORTH
CONE
C
FLANK
D
B
A
VALLEY
E
COMPREHENSIVE
SAMPLE
AREA
WEIRD
F
G
DOUBLET
ALTERNATE
LM
LR³
ALSEP
TRIPLET
EVA 1
EVA 2
SCALE 0 100 200 300
METERS
EVA 1 ~ 1 KM
EVA 2 ~ 3 KM
NASA HQ MA70-7495
Rev. 2/4/71

Fig. 5. Apollo 14 traverses.

surface rubble appears to be about 8 m deep. There appears to be another layer at 50–70 m from the data from both seismic instruments.

The Suprathermal Ion Detector has been recording numerous charged particles but no particular analysis has as yet been issued. The Cold Cathode Gauge which measures the intensity of particles has recorded two incidents of sudden increases in pressure which may be indicative of gaseous emissions. On one of these, the pressure increased by a factor of ten for a 1 hr period; on another by a factor of 2–4 for a 9-hr period. It is too early to interpret the significance of these changes. The Charged Particle Instrument has recorded interesting data as it has passed through the Earth's magnetic tail. It has seen extremely high energy particles greater than 50 keV. It has also shown that the electrons seem to be strongly correlated along the tail. The

Fig. 6. White rock.

final instrument deployed by the astronauts was a laser reflector. This was a somewhat improved version over that deployed by the Apollo 11 crew. The McDonald Observatory in Texas was able to bounce signals from the reflector while the astronauts were still on the surface.

After a rest period the astronauts emerged for the second surface period. This was devoted to a variety of tasks on a traverse up to the rim of Cone Crater (Figure 5). Not far from the Lunar Module they took their first magnetometer reading which showed about 100 γ signal. Near the rim of the crater the instrument indicated 40 γ. These readings were considered surprising both in terms of strength and gradient over this short distance. In the vicinity of Cone Crater the astronauts could not determine their precise location due to large boulders and various ridges. When the traverse was

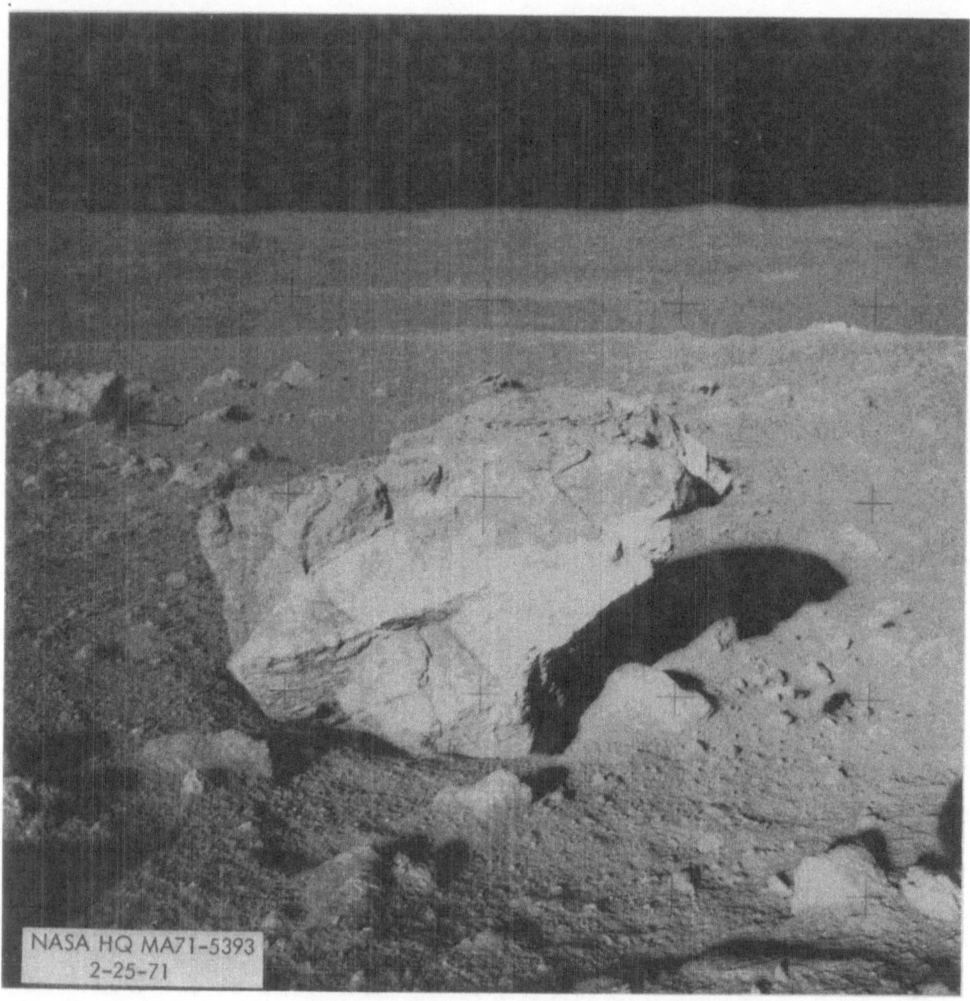

Fig. 7. Multi-colored rock – LM in background.

reconstructed from returned photography it was shown they were within just a few feet of the rim of the crater. Although it would have been interesting if they could have looked down into the crater to see if any stratification could have been observed, this is not thought to have been a serious scientific omission since the prime objective was to return rocks from the edge of the crater, and this was accomplished. Figure 6 is an interesting white rock on the crater rim from which the astronauts removed and returned a large chip. Nearby, Figure 7, is a peculair multi-colored rock. The landing module can barely be seen in the far distance. Within just a few feet, they found the rock shown in Figure 8 which has a completely different appearance and proved to be much harder than the two previously shown. Figure 9 is a trench that the astronauts dug. As can be seen, the sides collapsed easily. This had not been experienced in previous missions and indicates a much lower cohesiveness of the soil at this site than in the previous mare. The astronauts returned a total of 94 lbs of rocks and soil. Most

Fig. 8. Hard rock – Igneous.

Fig. 9. Trench.

of the rocks were extensively documented by photography on the surface before being disturbed. Their highly diverse nature promises to hold an extremely interesting scientific story but one which will be difficult to unravel.

Figure 10 shows the location of the impacts of the spent stage of the launch vehicle and the Lunar Module when its primary job had been completed. These two impacts of known energy and location are important for attempting to understand the lunar interior from seismic signals.

While Captain Shepard and Commander Mitchell were on the surface, Major Roosa, in orbit, had a wide variety of tasks. Photography of potential future landing sites of the Descartes region was most important. In spite of the failure of a large camera, excellent photographs were taken with a back-up camera of the area of prime interest. Numerous other photographs of scientific interest were taken. Figure 11 shows some linear rilles in the Crater Fra Mauro. One can see a line of volcanic domes and apparently a flow of some type which may have obscured the rille in the vicinity. This area might possibly represent the source of seismic signals or gaseous emissions. It is far too early to consider this more than mere conjecture. Another interesting feature on an orbital photograph is shown in Figure 12. It is similar to a sinuous rille but is a positive feature with a peculiar cross section. A large number of additional scientific photographs were taken which will be the subject of further study.

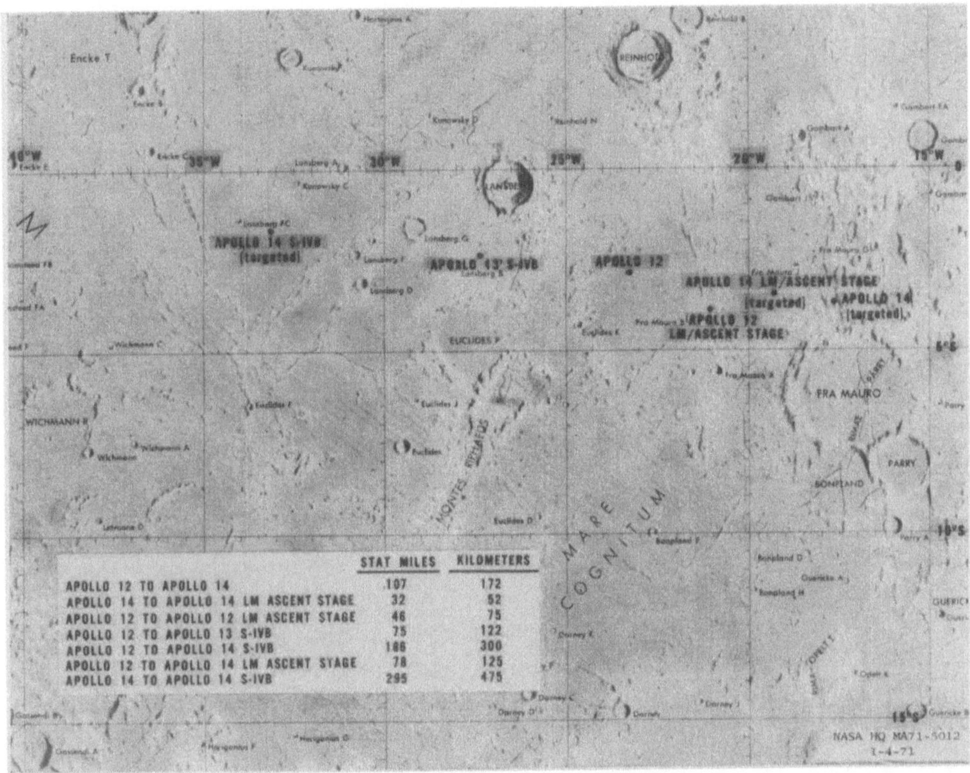

	STAT MILES	KILOMETERS
APOLLO 12 TO APOLLO 14	107	172
APOLLO 14 TO APOLLO 14 LM ASCENT STAGE	32	52
APOLLO 12 TO APOLLO 12 LM ASCENT STAGE	46	75
APOLLO 12 TO APOLLO 13 S-IVB	75	122
APOLLO 12 TO APOLLO 14 S-IVB	186	300
APOLLO 12 TO APOLLO 14 LM ASCENT STAGE	78	125
APOLLO 14 TO APOLLO 14 S-IVB	295	475

Fig. 10. Lunar impact locations.

On this mission, orbital tasks included low light level astronomical photography. The geometry of some of these tasks is indicated in Figures 13, 14, and 15. No results from this photography are as yet available. We know that photographs were made in the proper manner with the proper exposure but careful analysis with micro-densitometry equipment will be required to determine the results.

Another orbital experiment conducted was the recording of signals emanating from spacecraft and reflected from the lunar surface in the S-band and in the VHF frequencies. By comparing direct with reflected signals this bistatic radar data will give us information on the surface and near-surface roughness. With this experiment results are not as yet available but the data appear to be very good.

As a final point regarding the mission of Apollo 14, Figure 16 was constructed to show the progressive growth of scientific capabilities with each of the Apollo missions thus far. What is most significant is that these increases are not the result of major changes in hardware but rather results from increases in confidence and using what we have learned to improve procedures and techniques.

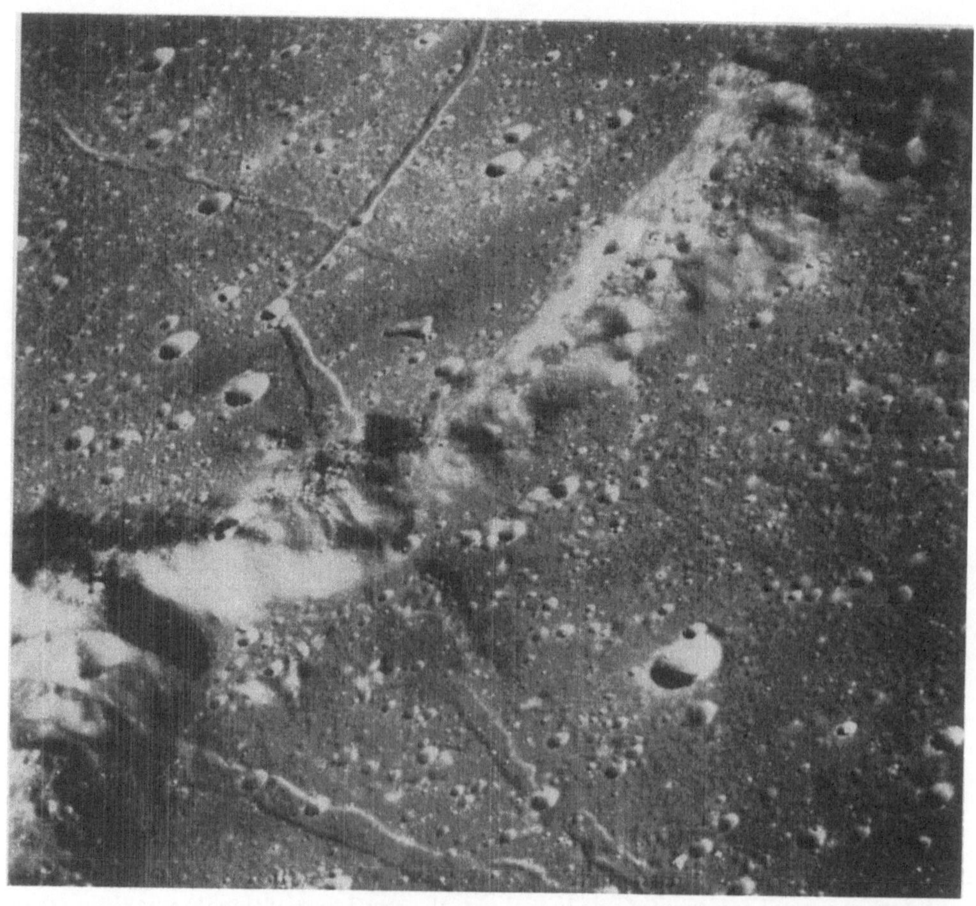

Fig. 11. Linear Rilles in crater Fra Mauro.

Fig. 12. Positive sinuous feature.

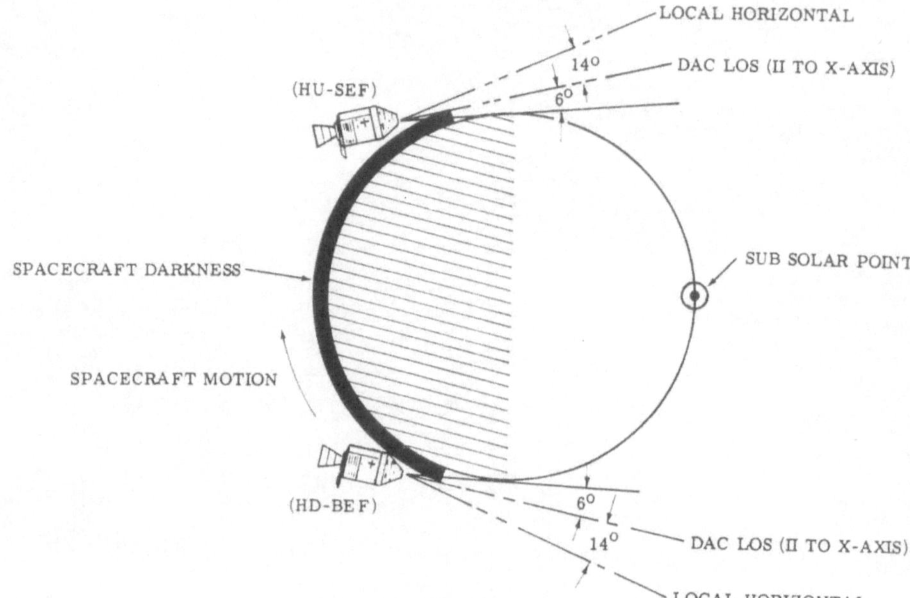

Fig. 13. Zodical light photography.

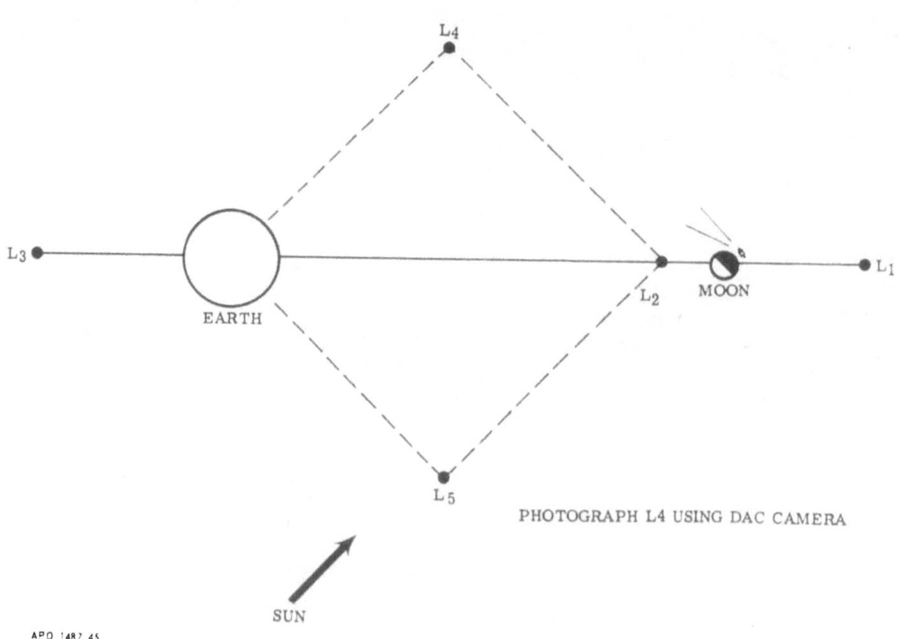

Fig. 14. Apollo 14 libration point photography.

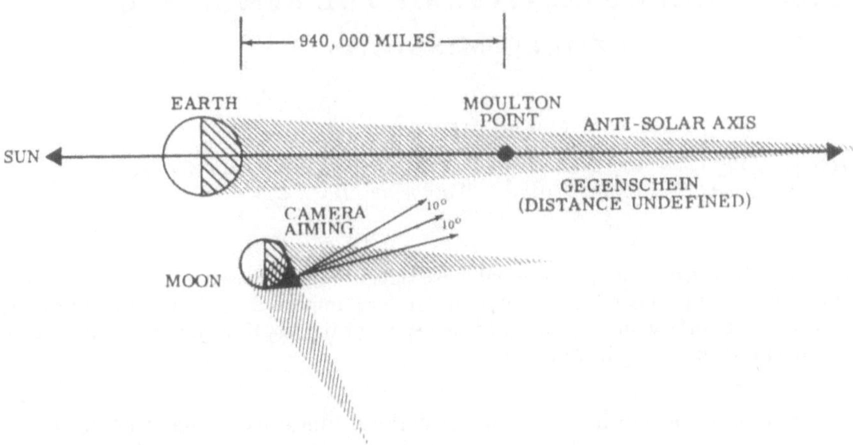

Fig. 15. Gegenschein experiment.

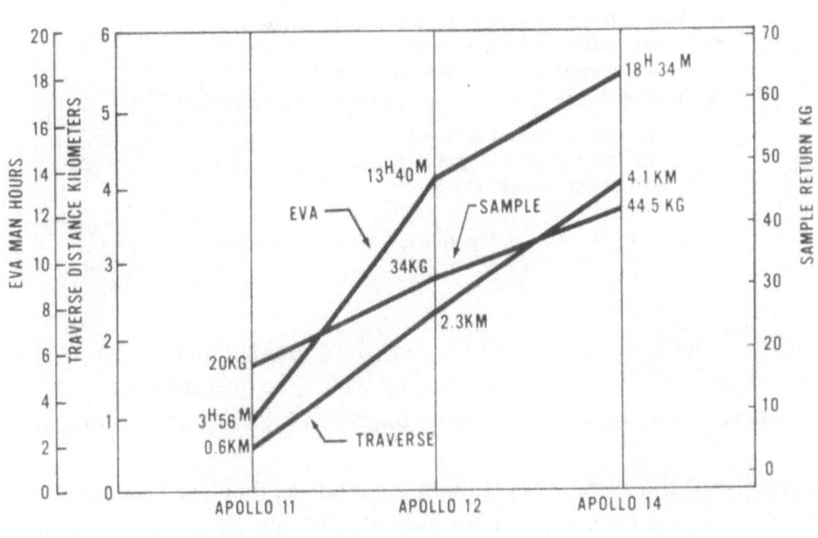

Fig. 16. Mission science growth.

PLANS AND OBJECTIVES OF THE REMAINING APOLLO MISSIONS

L. R. SCHERER

Director, Apollo Lunar Exploration, Apollo Program Office, National Aeronautics and Space Administration, Washington, D.C., U.S.A.

Abstract. The three remaining Apollo missions will have significantly increased scientific capabilities. These result from increased payload, more time on the surface, improved range, and more sophisticated experiments on the surface and in orbit. Landing sites for the last three missions will be carefully selected to maximize the total scientific return.

There are three remaining Apollo missions. For these missions we have been able to incorporate major increases in system capabilities and experimental payloads over that of the earlier missions. This is summarized in Figure 1. The landed payload capability will be greater than 1200 lbs. The astronauts will be able to make three

- Landed scientific payload doubled.
- Addition of service module experiment package.
- Lunar surface stay time doubled.
- Eva manhours increased from 18 to 40.
- Increased range and efficiency of surface operations.
  - Improved suit mobility.
  - Improved life support system.
  - Lunar roving vehicle.

Fig. 1. Increased capabilities Apollo 15 mission.

separate exursions on the surface, each of greater duration than in the past. Major new experiments are planned, particularly for use in orbit. The operational capabilities of the crew have been substantially increased by improved life support systems, suits and mobility.

Figure 2 is a list of the surface experiment payloads as they are now planned. The list is too long to discuss each experiment individually, but let me invite attention to just a couple. Note the traverse type experiments planned for use with the Rover for Apollo 17. These are the surface electrical properties, the traverse gravimeter and the lunar seismic profiling. It is hoped that these complementary experiments will contribute much to our understanding of the near sub-surface of the Moon. On Apollo 17 we do not plan to carry a passive seismometer. In its place we are developing a tidal gravimeter whose data may prove to be of particular significance. The primary function of this instrument is to measure the tidal acceleration of the lunar surface. However, it is so sensitive that it may be able to detect gravitational waves originally predicted by Einstein, if such things actually exist. Experiments searching for such proof conducted on Earth are generally considered controversial. Using the Moon

Urey and Runcorn (eds.), The Moon, 94–103.

	Experiment	Investigator	15	16	17
S-031	Passive seismic	Latham	×	×	(×)
S-033	Active seismic	Kovach		×	
S-034	Lunar surface magnetometer	Sonett/Dyal	×	×	
S-035	Solar wind spectrometer	Snyder	×		
S-036	Suprathermal ion detector	Freeman	×		
S-037	Heat flow	Langseth	×	×	×
S-038	Charged particle lunar environment	O'brien/Reasoner			
S-058	Cold cathode ionization	Johnson	×		
M-515	Lunar dust detector	Freden	×		
S-207	Lunar surface gravimeter	Weber			(×)
S-202	Lunar ejects and meteorites	Berg			×
S-203	Lunar seismic profiling	Kovach			×
S-205	Lunar atmosphereic composition	Hoffman			×
S-201	Far UV camera/spectroscope	Carruthers		×	
S-059	Lunar geology investigation	Swann/Muelberger	×	×	×
S-078	Laser ranging retro-reflector	Faller	×		
S-080	Solar wind composition	Geiss	×		
	Lunar surface close-up camera	Facility			
S-152	Cosmic ray detector	Fleischer/Walker/Price		×	
S-198	Lunar portable magnetometer	Dyal		×	
S-199	Lunar gravity traverse	Talwani			×
S-200	Soil mechanics	Mitchell	×	×	×
S-204	Surface electrical properties	Simmons			×

Fig. 2. Apollo lunar surface science plan.

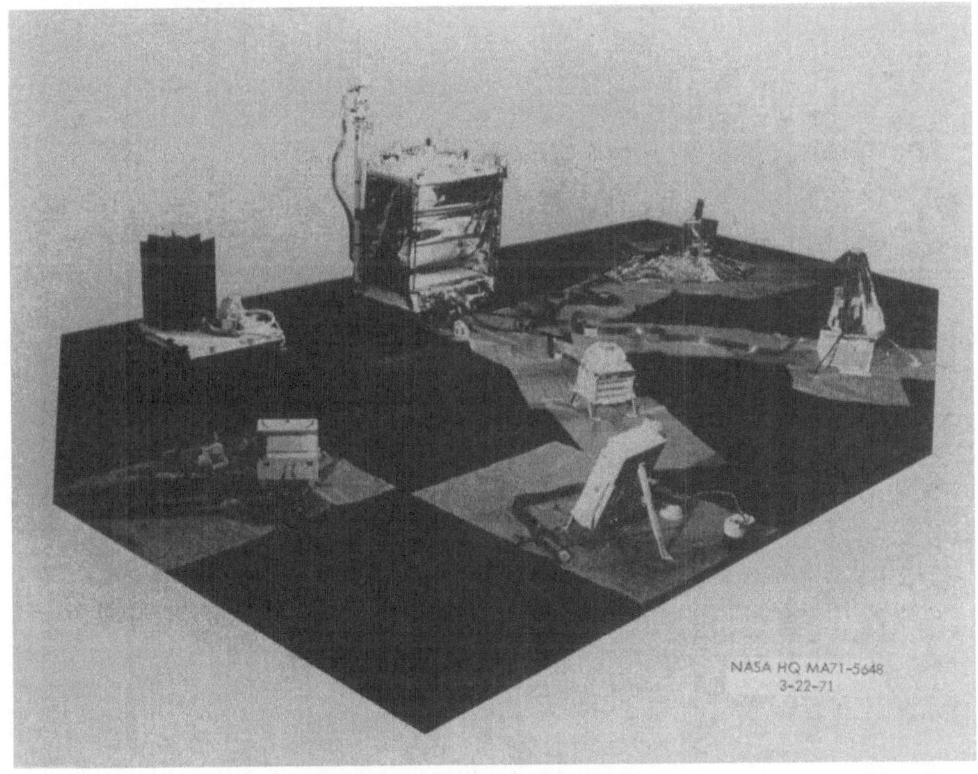

Fig. 3. Apollo 15 geophysical station.

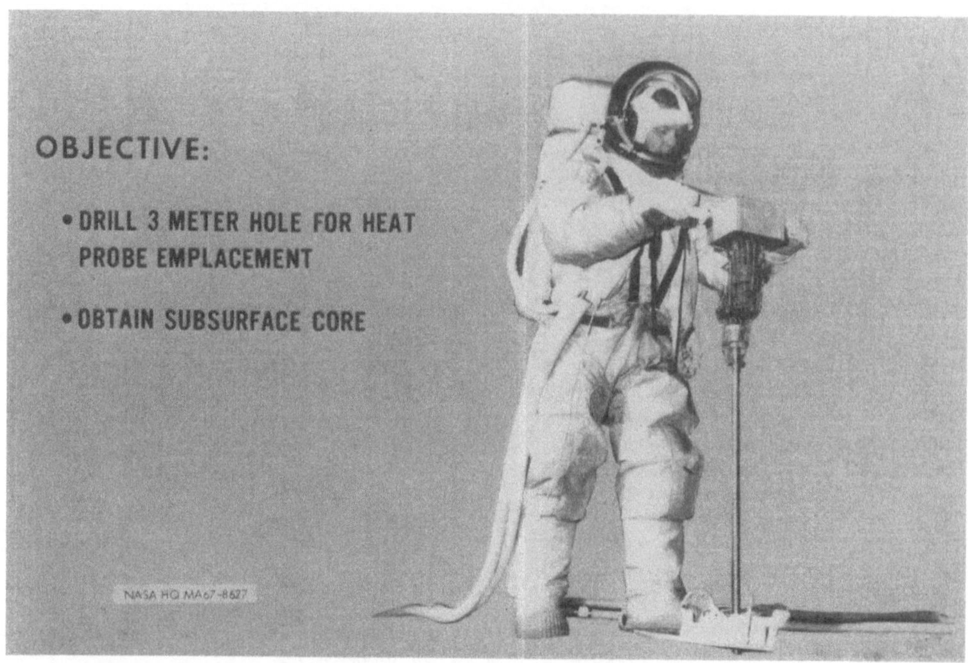

OBJECTIVE:
• DRILL 3 METER HOLE FOR HEAT PROBE EMPLACEMENT
• OBTAIN SUBSURFACE CORE

Fig. 4. Lunar surface drill.

Fig 5. Rover.

	Experiment	Investigator	15	16	17
	Service module:				
S-170	Bistatic radar	Howard	×		
S-160	Gamma-ray spectrometer	Arnold	×	×	
S-161	X-ray fluorescence	Adler	×	×	
S-162	Alpha-particle spectrometer	Gorenstein	×	×	
S-165	Mass spectrometer	Hoffman	×	×	
	SM Photographic tasks: (Photo Team)	(None)			
	24″ panoramic camera		×	×	×
	3″ mapping camera		×	×	×
	Laser altimeter		×	×	×
	Subsatellite:				
S-173	Particle shadows/boundary layer	Anderson	×	×	
S-174	Subsatellite magnetometer	Coleman	×	×	
S-164	S-band transponder	Sjogren	×	×	
S-169	Far UV spectrometer	Fastie			×
S-171	IR Scanning radiometer	Low			×
S-167/S-168	Lunar sounder	Ward/Brown			×
S-164	S-band transponder (CSM/LM)	Sjogren	×	×	×
	Command module:				
	CM Photographic tasks: (Photo Team)	(None)	×	×	×
S-176	Apollo window meteoroid	Flaherty	×		
S-177	UV Photo-Earth and Moon	Owen	×	×	
S-178	Gegenschein from lunar orbit	Dunkelman	×		

Fig. 6. Apollo lunar orbital experiments and photographic tasks.

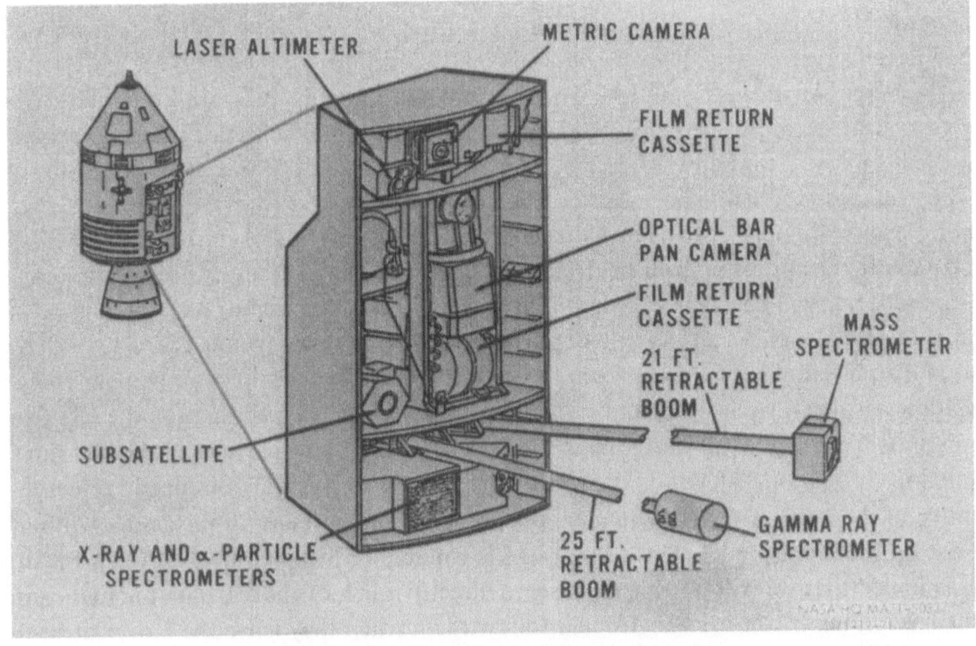

Fig. 7 Apollo orbital science instrument module.

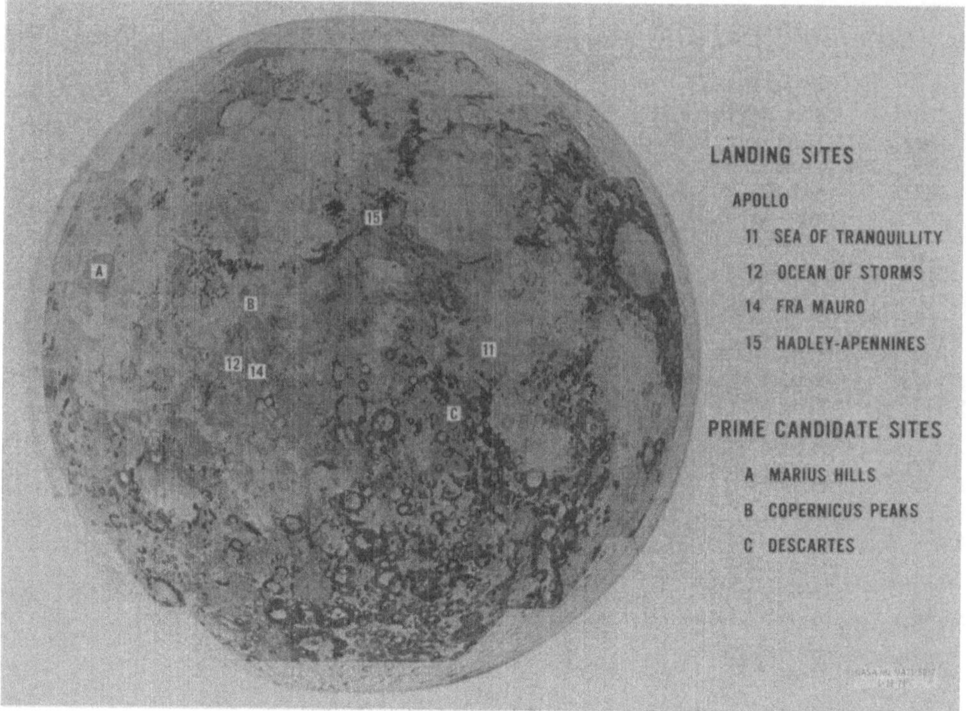

Fig. 8. Geologic map of the Moon with landing sites.

increases the sensitivity many orders of magnitude. The unambiguous discovery of gravitational waves would be an exciting scientific finding.

Our experiment plans for the Apollo 15 mission in July 1971 are quite firm. The geophysical station to be emplaced is shown in Figure 3. The major new experiment is the heat flow. This had been planned for Apollo 13 but, of course, was not deployed when that mission had to be aborted. The drill to be used is shown in Figure 4. Two holes approximately 3 m deep will be drilled and heat probes inserted. It is also planned that a third hole will be drilled and the core returned. On Apollo 15 a larger laser reflector will be carried. This will have 300 cubes compared to the 100 of the two previous reflectors. As a result we hope that smaller telescopes around the world can be used for reflecting laser signals from the Moon. This reflector will complete our three-station network of reflectors and no further instruments are planned for the last two missions. Figure 5 is the very interesting roving vehicle to be carried on the last three missions. The model shown is the trainer used at 1 g so it is not completely representative of the $\frac{1}{6}$ g model. In particular, the structure shown beneath the center will not exist on the flight model. The TV camera is capable of being controlled from Earth. It is hoped that we will be able to observe directly much of the astronaut activities in their very long traverses. The TV will follow them when they leave the Rover at each station. When the surface mission is completed the Rover will be placed in such a position that the actual lift-off from the surface may be observed by the television camera.

As a potential added bonus, by coincidence a lunar eclipse will occur several days after the astronauts leave the Moon. It is hoped that sufficient battery power will remain so the color television can be used to observe the changes in appearance of the lunar surface during the progress of this eclipse.

Figure 6 represents the extensive orbital experiments planned for the final three missions. Excellent panoramic and mapping cameras will be carried and operated in conjunction with the laser altimeter. A set of geochemical sensors is planned for Apollo 15 and 16. These may prove to be very significant for extrapolating the data obtained from specific landing sites over wide regions of the lunar surface. Also on Apollo 15 and 16 we plan to carry small subsatellites. They will remain operative for many months after the manned portion of the mission is completed, providing magnetic and particle fields data. They also should provide greatly improved information on mass concentrations first noted from Lunar Orbiter data. A sophisticated experiment we have under development for the final mission is an electromagnetic sounder which will probe far beneath the lunar surface. Figure 7 is a sketch of the orbital

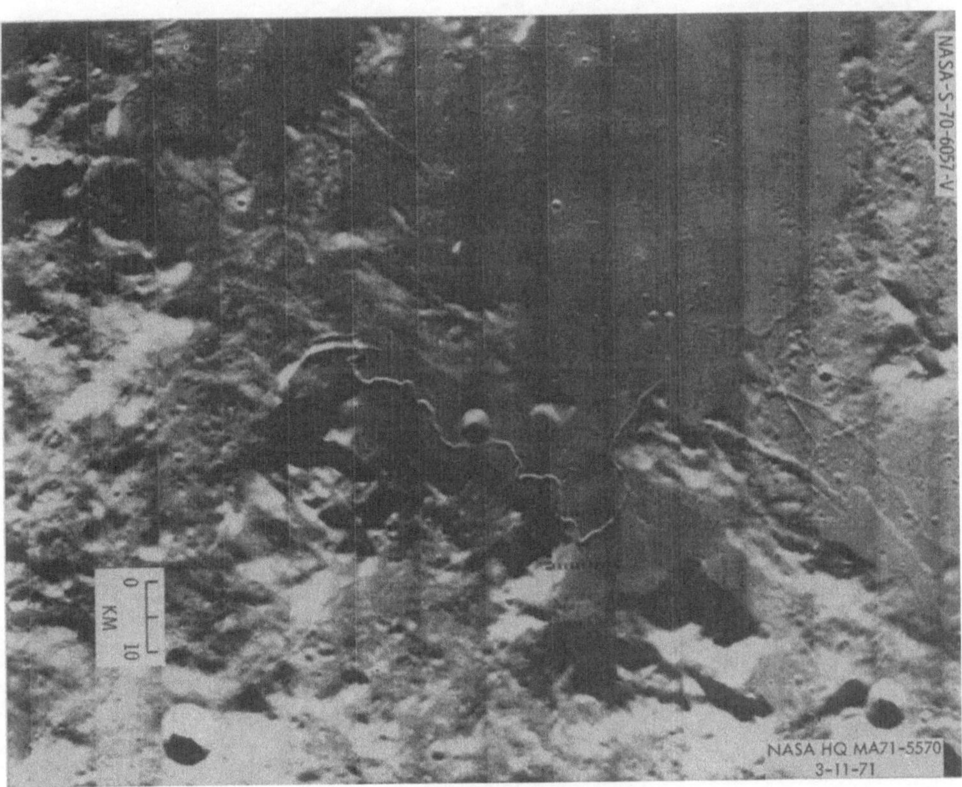

Fig. 9. Hadley Rille.

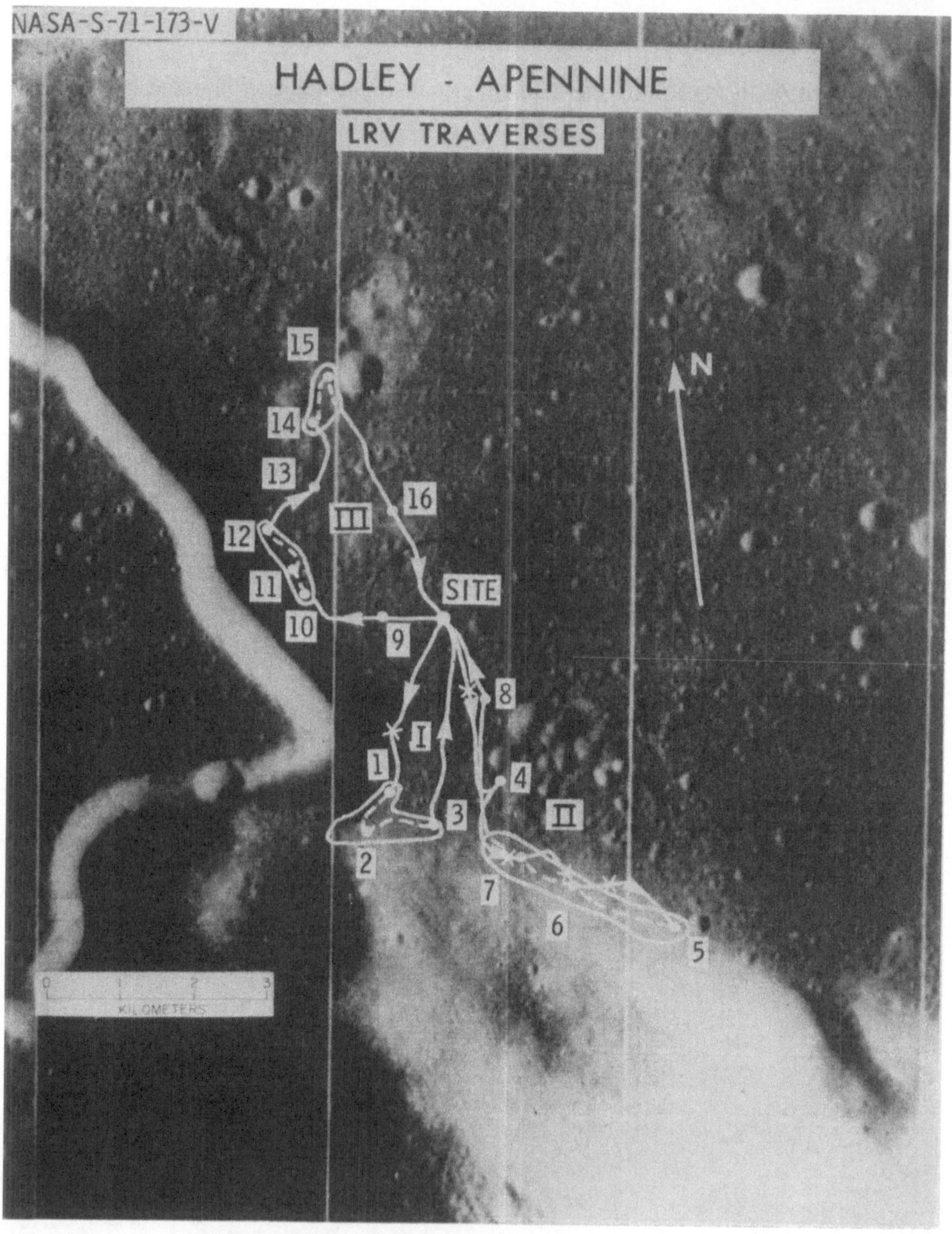

Fig. 10. Close-up view of Hadley-Apennine traverses.

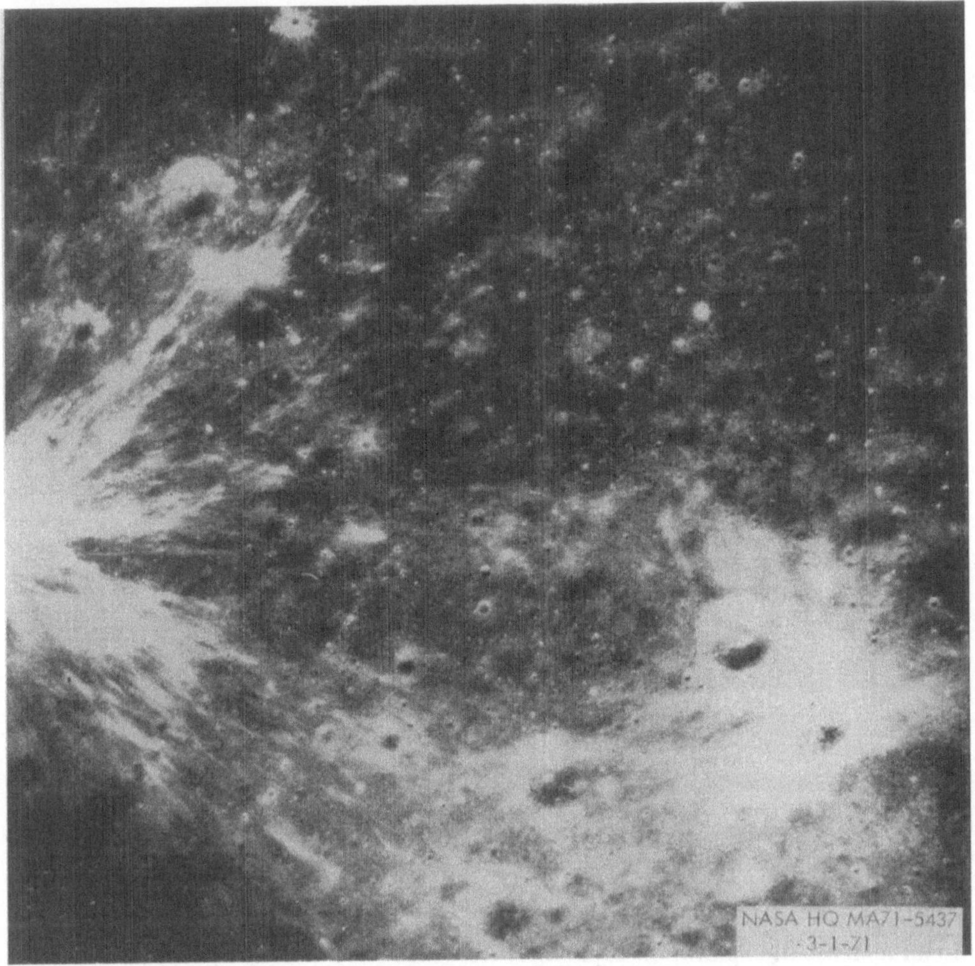

Fig. 11. Descartes.

experiment payload as it is planned for incorporation in the Service Module of Apollo 15. With only three Apollo lunar missions remaining, all possible care is being taken in the selection of the specific landing sites to insure the greatest scientific return. Figure 8 indicates the locations of some prime sites under consideration.

The Apollo 15 site is, of course, firm. The Lunar Module will land between the Hadley Rille and Apennine mountain front as shown on Figure 9. The broad objectives at this site are to study and sample the highland front, the adjacent mare and the rille itself. Figure 10 is a closer view of the site and the tentative traverses as they are now planned. The Apollo 14 astronauts walked a little more than 4 km on the surface. The traverses shown here total 37 km. We feel the various improvements we have been able to make, in particular the roving vehicle, make feasible this excep-

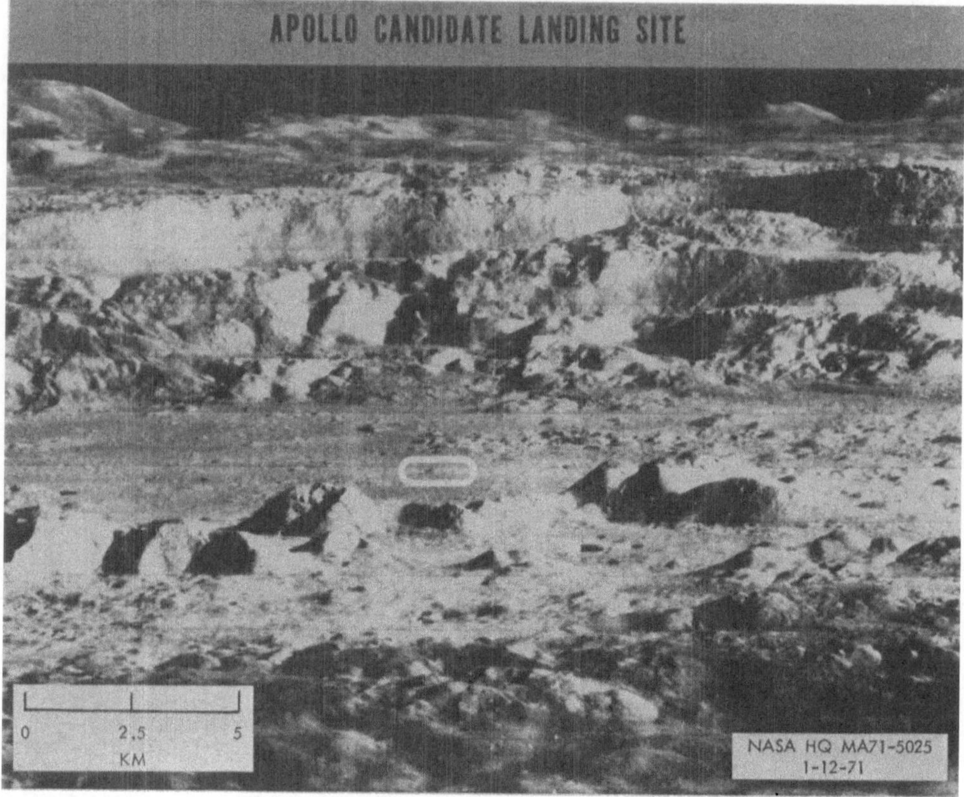

Fig. 12. Copernicus.

tional increase in exploration capability. As with each mission, a geophysical station will be left operating at this site. On the succeeding mission we plan to impact the spent booster out in Mare Imbrium. It is hoped that the resulting seismic data will be significant in unraveling the major mascon that exists in the Imbrium basin.

The last two landing sites have not as yet been selected. A great deal of photography was taken of highland regions around the Descartes site and to the eastward on Apollo 14 (Figure 11). These are now undergoing careful study and it is quite likely that Apollo 16 will be directed somewhere in this highland area. The Apollo 17 site will not be selected until about the end of 1971.

Two of the sites under consideration are shown in Figures 12 and 13. At the spectacular crater Copernicus the landing site would be on the floor near the central peaks. Sampling these peaks would probably provide material from deep in the interior since they represent rebound from the initial impact. Marius Hills is an extensive volcanic region which lies along a major fault line of the moon. Even the Crater Tycho is not excluded from our considerations at this time although it obviously is a very difficult site operationally.

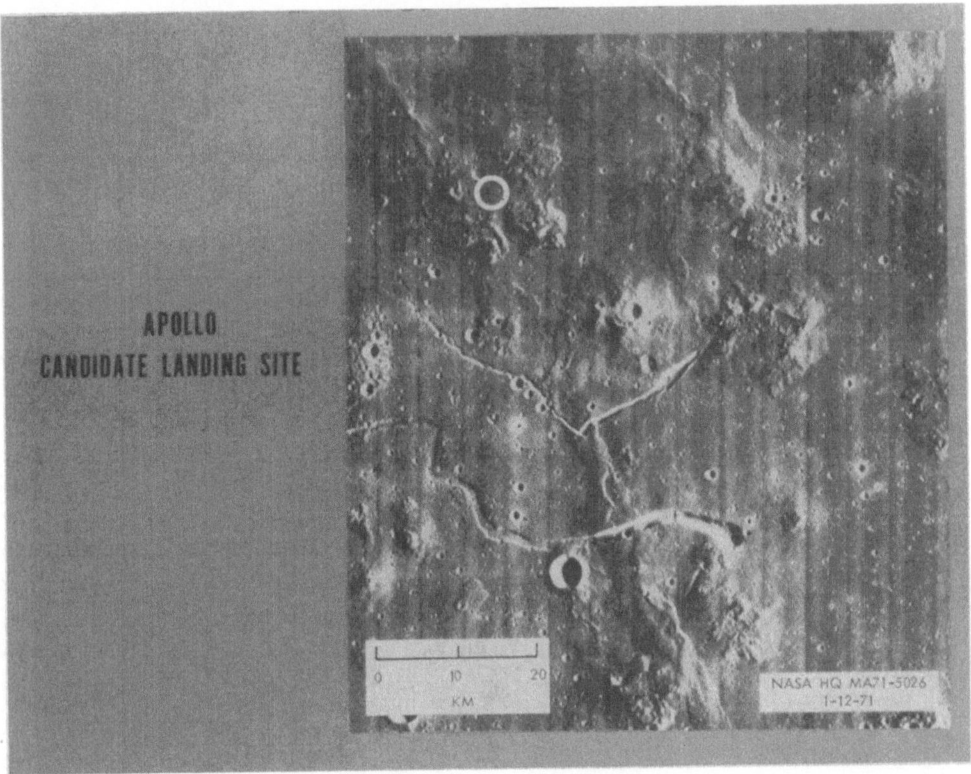

Fig. 13. Marius Hills.

In site selection, crew safety is always our primary consideration, of course. There are a number of other operational constraints that play a role; in particular, the availability of sufficient photography. Within the constraints we have, we want to take every feasible step to ensure the sites selected offer the highest potential for scientific gain. Any suggestions from qualified groups or individuals such as those present are most welcome.

ENGINEERING POTENTIAL FOR LUNAR MISSIONS AFTER APOLLO

JAMES D. BURKE

Jet Propulsion Laboratory, Pasadena, California, U.S.A.

Abstract. With results of past and present lunar exploration programs in hand, it is now possible to outline the next logical steps in the exploration and use of the Moon. This paper describes first, a set of mission objectives and engineering capabilities considered desirable and feasible for projects following soon after Apollo, and second, some objectives for the farther future.

1. Introduction

Assuming success in the remaining Apollo missions, what are the prospects for continuing lunar achievements? This paper will describe some desirable and feasible missions that could be executed at reasonable cost. There is, of course, no certainty now that the U.S. will carry on such a program, since the required national decisions are still in the future. However, by discussing the options now, we can help to assure that any future program will be efficient and scientifically productive. It is a privilege to present ideas developed at JPL to members of the IAU, for one of our recognized goals is to increase international participation, not only in our present projects but also in the early planning of possible future programs. The IAU, with its long tradition of free and prompt information exchange, can help not only in the scientific review and discussion of our plans, but also in cultivating support for these plans by scientists in other countries, whose efforts may be complementary to our own.

2. Rationale for Further Lunar Investigations

At JPL we have proceeded from the viewpoint of an evolutionary cosmogony, regarding the Moon as a source of information on the various stages of a postulated grand process that begins with nucleosynthesis and stellar evolution, continues

TABLE I

Uses of Moon for investigating evolution

Stage of evolution	Typical lunar investigations
Stellar	Wide-spectrum astronomy
Planetary	Selenodesy, geophysics, geology
Biological	Organic geochemistry, volatiles
Social	International programs
Interplanetary	Communications experiments

Urey and Runcorn (eds.), The Moon, 104–120.

through the formation of planets and life, and extends on into the unknown evolutionary future of life forms and their artifacts, here in the Earth-Moon system or elsewhere. Some of the more obvious uses of the Moon for investigating (or advancing) these stages of evolution are listed in Table I.

A. STELLAR EVOLUTION

The Moon is already in use as a site for observations relevant to astrophysics. The Apollo solar-wind experiments and the Lunokhod X-ray telescope are examples. However, to reach the goal of founding automated or manned lunar observatories, as long desired by every true amateur of the Moon, astronomers and engineers will have to work together to demonstrate the merits of such observatories in competition with Earth-orbital and Earth-based techniques. Because there are still so many useful observations that can be made from on or near the Earth [1], resources continue to be allocated mainly here, in preference to Moon-based efforts.

To understand this value tradeoff and to predict its future course remain as challenges to those of us who advocate and plan lunar programs; we would be most grateful for advice and assistance from members of the IAU.

B. PLANETARY EVOLUTION

The idea that the Moon is a source of information dating possibly from the time of planetary accretion has been a main force in past and present lunar exploration programs. During more than a decade at JPL [2] and [3], studies have been made to define lunar investigations pertinent to the problem of the origin and evolution of planets. In our more recent work [4]–[9], we have concentrated on investigations, particularly those involving remotely-controlled lunar surface traverses, that can build upon the results of the Apollo program and so extend our planetological knowledge of the Moon. The engineering prospects emerging from these studies are discussed in Section 3. The scientific requirements include imaging and sampling of lunar surface materials, elemental and mineralogical measurements, and geophysical measurements such as thermal, seismic, and gravity profiles, with the object of determining the source, pre-accretion history, mode of accumulation, and post-accretion history of the lunar material. In addition to their intrinsic interest, these ancient events demand understanding for the broader view: by developing a bulk compositional model and a compatible thermal history for the Moon, we may elucidate the early history of the Earth-Moon system (cf., for example, [10]–[13]) and thence learn more about the conditions required for the origin of life.

Assuming success in all currently-planned Apollo experiments and an adequate data analysis, we can expect to have a thorough understanding of lunar surface properties at six sites, a good passive seismic survey, the beginnings of a good Earth-Moon laser ranging network, and a variety of information on geologic processes in the lunar environment. The Apollo orbital experiments, coupled with the results of earlier missions [14] should yield metric photography over broad regions, improved knowledge of the Moon's equatorial geopotential and surface shapes, and a variety of remote-sensor data on surface physical and chemical properties from the equator to middle latitudes. Polar surveys and far-side landings are excluded; some engineering aspects of such missions are mentioned in Section 3.

A logical post-Apollo planetological program, then, should include polar lunar orbital surveys, long surface traverses, and continuing Earth-based experiments such as the precision tracking of optical and radio reflectors or beacons on the Moon. There are also applications in planetology for telescopes on the Moon, for example in the spectrometry of planetary atmospheres, but as mentioned earlier the planning of such missions is not now far advanced.

C. BIOLOGICAL EVOLUTION

The Apollo lunar surface samples from equatorial maria are anhydrous and practically devoid of organic or proto-organic materials [15]. These regions of the Moon may be representative of one kind of lifeless planetary environment and can therefore serve as controls in future exobiology investigations. In the continuing exploration of the Moon, it will be desirable to search for subsurface and polar volatiles, to determine the composition of highland rocks and soils, and to extend the Apollo results into an understanding of the biological potential of all lunar materials including, for example, carbonaceous meteorite residues. Automated surface traverse missions can be useful in these searches.

Experiments have been made using Apollo samples to test the effectiveness of lunar soil as a growth medium. With reasonable engineering measures such tests could also be made remotely in small environmental shelters on the Moon; however, we are not now pursuing this development.

It remains to be seen whether or not the Moon was a part of a pre-biotic Earth, or whether it had a major role in modifying the Earth's early environment, and so contributed to the local beginning of life. Insight into this problem may be gained through further study of the mechanics of the Earth-Moon system. If so, the precise radio and laser measurements mentioned earlier for planetology may be important also in a biological context.

D. SOCIAL EVOLUTION

There can be no doubt that the study of the Moon has played a significant role in the growth and change of human institutions. One has only to recall the development of optical astronomy, celestial mechanics, and navigation to see that this is so. Now, in the act of reaching toward the Moon, we may be increasing our own chance of success as a species [16]. At the very least, the energy and the ingenuity dedicated to lunar programs have been subtracted from the total available for weapons. In a more positive sense, we are now demonstrating the power of large, systematic problem-solving organizations on a scale known previously only in war. And above all, we have opened a new arena for collaboration and stimulating, non-destructive competition [17]. Members of the IAU, because of their long experience in international efforts, are especially well prepared to observe both the prospects and the limitations of such programs, and to assist in the practical planning of new lunar ventures to take advantage of this historic opportunity.

E. INTERPLANETARY EVOLUTION

As the concept of extraterrestrial intelligence passes from the realm of fantasy and speculation into the realm of sober inquiry [18], it is appropriate to consider the role of the Moon in a future stage of evolution characterized by exchanges of information across the cosmos. The most obvious possible use of the Moon is as a site for sensitive radio receivers with large antennas on the far side, shielded from the radio noise of Earth. Without guessing as to the likelihood of having such interplanetary communications at interstellar distances, we can note that, over the long term, the required engineering developments may move forward during radio astronomy research and spacecraft tracking from the Moon.

3. Engineering Prospects

A. EARTH-MOON TRANSPORTATION

One factor governing the design of a post-Apollo lunar exploration program is the available launch-vehicle capacity. For U.S. flights following shutdown of Apollo and Saturn 5 production, the candidate launching systems are (a) Atlas-Centaur, (b) various versions of the Titan 3, and eventually (c) a reusable manned orbital shuttle with added upper stages. Table II lists approximate lunar spacecraft masses [19]

TABLE II

Lunar vehicle performance

		Approximate spacecraft mass, kg		
No.	Launch vehicle	At translunar injection	In lunar orbit	On lunar surface
I	Atlas-Centaur	1500	1000 (200)[a]	500 (50)[a]
II	Titan 3 C	3000	2000 (500)[a]	1000 (150)[a]
III	Shuttle/Up. Stg. (or Titan 3/ Centaur)	6000	4000 (1000)[a]	2000 (300)[a]

[a] Instrument payload portion

and scientific payloads for representative members of this launch vehicle family. Figures 1 and 2 show present versions of the Atlas-Centaur and Titan 3, and Figures 3 and 4 show the approximate spacecraft size envelopes available within the Atlas-Centaur (OAO) and Titan-Centaur (Viking) nose fairings. Figure 5 shows the size of these vehicles relative to Saturn 5.

Fig. 1. Atlas-Centaur at Liftoff.

Fig. 2. Titan 3 vehicle shortly after launch.

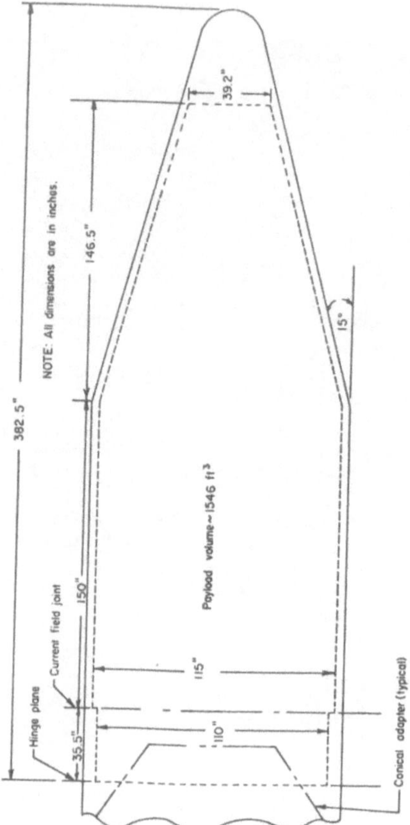

Fig. 3. Centaur nose fairing dimensions for Orbiting Astronomical Observatory.

B. SPACE VEHICLE AND PAYLOAD COST

Another factor governing program design is, of course, cost. With limited budgets, it may not be possible to use the most powerful available versions of the Atlas and Titan-based launch vehicles, simply because of the higher cost of the larger spacecraft and scientific payloads. For this reason our studies include missions of relatively small capability in comparison to Apollo–missions which may not require the greatest lifting capacity available at a given future date. Even with this restriction, program costs may amount to hundreds of millions of dollars over the several years when a project is active. However, our studies and experiments show that very important data can be obtained with these modest payloads.

C. LUNAR ORBITAL MISSIONS

A class of experiments important in planetology can be carried out in lunar orbit [14]. The planned Apollo CSM instrumentation will be very useful, but the Apollo mission constraints prohibit achievement of highly-inclined orbits. Therefore one candidate post-Apollo mission is a polar orbiter with instruments similar to or developed from

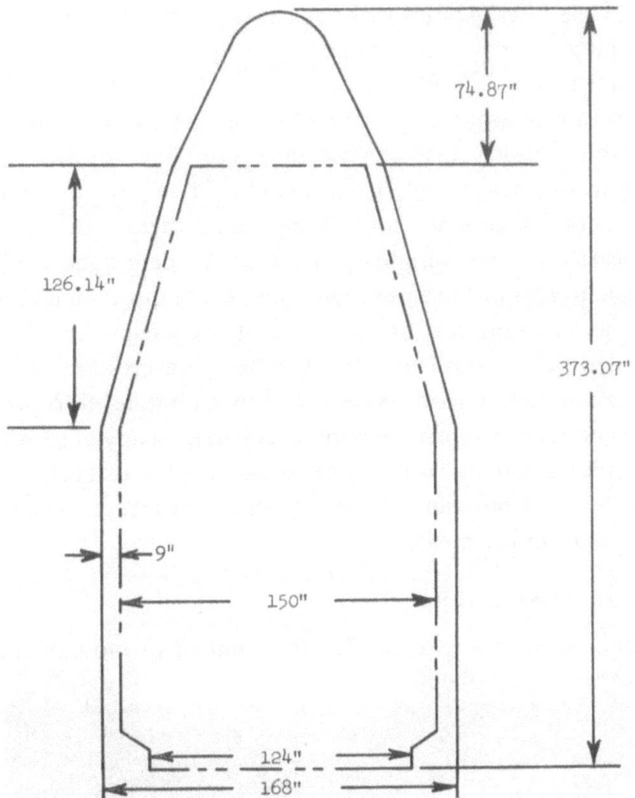

Fig. 4. Centaur nose fairing dimensions for Viking.

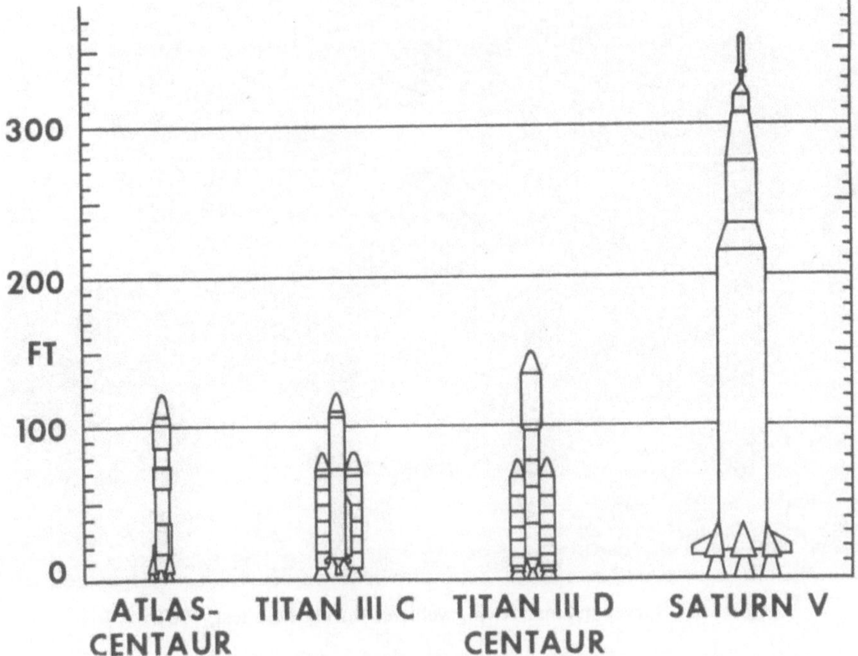

Fig. 5. Size comparison of launch vehicles.

those to be flown in the later Apollo CSM spacecraft. Even a simple polar orbiter, with just a radar altimeter, a gravity gradiometer, and a precision tracking transponder, could yield important planetological information.

The discussion of post-Apollo lunar orbital missions is complicated by the large number of practicable options. One can visualize useful results from a variety of missions extending from small, simple particle-and-field surveys to elaborate multi-purpose missions requiring complex and expensive spacecraft. A logical program would include at least those measurements necessary to define the meridional geo-potential and surface figures of the Moon, plus simple compositional remote-sensing measurements such as a gamma-ray survey. If a multi-frequency radio sounder could be carried, data might be obtained on subsurface temperatures and the presence or absence of permafrost in polar regions. Should a more elaborate effort prove possible, one or more metric and remote-sensing experiments based on any of several radiations and detector principles could be added. The main purpose of such measurements would be to extend, by a combination of observation and inference, verified point data from surface missions to larger areas.

D. LUNAR SURFACE TRANSPORTATION

Mobility on the lunar surface is essential for the conduct of most of the important

Fig. 6. Surveyor lunar roving vehicles during field test, 1964.

post-Apollo investigations. The technical foundation for such missions has been laid by developments such as the Surveyor Lunar Roving Vehicles (Figure 6), the Lunokhod, and the Apollo manned lunar rover. Post-Apollo scientific criteria require a long range (hundreds of km), and this raises several interrelated engineering problems which we have analyzed. The more important tradeoffs are summarized in Table III.

TABLE III

Engineering tradeoffs for automated long traverses

- Mobility and Hazard Avoidance
- Power, Speed, Range and Endurance
- Image Data Rate, Antenna Gain and Pointing Control
- Guidance and Navigation
- Automation and Human Intervention

1. *Mobility and hazard avoidance*

A basic requirement is that the machine must safely negotiate most of the surfaces encountered, and must avoid, with a certain margin of safety, surfaces on which it could overturn or get stuck. Obviously there is a tradeoff between high mobility, with attendant weight, power and complexity of chassis, wheels and drives, and high hazard-avoidance ability, with complex detection and reaction schemes. Our studies and tests, supported by those of General Motors and several other U.S. organizations, show that for automated missions high mobility (and consequently a more limited demand for hazard avoidance) is the preferred choice. Very good mobility (with acceptable weight and power) can be obtained through the use of a multi-wheel, articulated design with low footprint pressure. A similar choice was made by the Lunokhod designers. The manned Apollo rover, on the other hand, can and does have lower basic mobility since its operator is able to judge directly and avoid hazards.

2. *Power, speed, range and endurance*

Power supply is one of the main problems of the automated traverse. For a machine large enough to carry the desired payload (including at least sampling, imaging, and rudimentary sample analysis), the average power for travel and picture transmission may be several hundred watts, and the peak or emergency demand may exceed a kilowatt. There is a complicated tradeoff among speed, stop-go duty cycle, picture data rate, and other variables. In our studies this has been integrated with ground-control and hazard-avoidance considerations, to yield the conclusion that average speeds on a long traverse will be only one to a few km hr^{-1} – equivalent to a slow walk – and that therefore if it is to travel for hundreds of kilometers the machine must repeatedly survive the lunar night. Lunokhod 1, though it has not traveled so far, has now demonstrated the important night-survival function. The basic problems of equipment lifetime in the lunar environment now appear to be understood, as a result of Surveyor, Apollo, and Soviet experience. Analyses of Surveyor 3 parts returned by Apollo 12 [20] and [21] and ALSEP operating experience lend confidence to this conclusion. Since the roving mission must extend over several months, the prime power source must be either solar or nuclear, with a rechargeable battery to handle demand peaks.

3. *Image data rate, antenna gain and pointing control*

Another main problem is the delivery of pictures to the control site, with sufficient quality and frequency to permit human intervention when needed, either as a supplement to vehicle-borne hazard sensors or as an aid to steering, navigation, and scientific mission planning. This tradeoff works out in an interesting way. Tests by the Bendix Corporation have shown that there is an optimum picture frequency. With cameras mounted rigidly to the vehicle (i.e., not servo-stabilized), operator performance is actually better when still pictures are presented, at intervals of one to a few seconds, than it is when imaging is more nearly continuous, as in a cine film at tens of frames per sec. This helps, of course, with the power problem, since the rapid sending of television-quality images from the Moon requires a large radio power-gain product. Given a practical S-band transmitter of, say, 20 W output, and assuming reception by the 26-m antennas of the Deep Space Net or the Manned Space Flight Net, the imaging tradeoffs then require a vehicle antenna of modest gain, with a beamwidth of at least several degrees. Higher gain would reduce the power demand but would require a more complex system to keep the beam pointed at Earth despite vehicle motions. Therefore our lunar rover conceptual designs include S-band antennas with apertures of 30 cm or so, rather small in comparison, for example, to a Mariner antenna.

4. *Guidance and navigation*

Navigation studies and tests [24] show that the only practical scheme for small, long-range automated lunar rovers is visual landmark navigation, supplemented by simple celestial references such as Sun and Earth sensors, and by radio or laser ranging. Inertial systems, including the necessary stellar references, are too complex for this application.

5. *Automation and human intervention*

In all of these design tradeoffs, there is of course the question of how much the machine should know how to do, and how much its masters on Earth must do. With the lunar round-trip delay of a few seconds, plus the picture interval and human reaction times, it is clear that at a minimum the rover must be able to sense a hazard and stop before entering it. If it cannot do so, then it must travel slowly enough that the operator can intervene and stop or redirect its motion, based on his present (slightly obsolete) data. These relationships set limits on, for instance, the resolution of the TV system, the camera fields of view, the nominal cruising speed, the tolerable Sun angles, and the stop-command thresholds of devices such as tilt sensors. On the Moon, reasonable speed and safety can be achieved with quite limited autonomy, as demonstrated by the Lunokhod system. For roving on Mars, a much higher degree of autonomy would be required because of the greater communications time delay and the more limited data bandwidth.

E. INSTRUMENTATION

As pointed out in Section 2, the primary functions visualized for early lunar long-traverse instrument payloads are planetological: geologic reconnaissance, sampling and analysis of surface materials, and geophysical profiling. These experiments would be part of an integrated investigation consisting of (a) the existing diverse and detailed local observations from Apollo sites, (b) simpler observations over regional distances along the traverses, and (c) still less detailed, but broad and synoptic, observations from lunar orbit. Emphasis in the investigation as a whole would be placed on information useful in constructing bulk compositional models for the Moon, in deciphering the Moon's thermal or energy history, and in describing its early interactions with the Earth.

Fig. 7. Visual displays inside field-test truck. Top: panorama, simulating photofacsimile display. Center: TV close-up of rock sample. Bottom: two views of local scene.

Fig. 8. Rock sample viewed by TV camera. Sample and camera orientation, and zoom lens focus and magnification, are remotely controlled.

For these purposes the rover must have at least an imaging system designed for both general surveys of surroundings and close-up petrologic examination of samples, plus limited instrumentation (e.g., a scanning electron microprobe) for measuring the chemical content of lunar materials. Desirable additions would be instruments (e.g., an X-ray diffractometer/spectrometer) for discriminating among minerals, and instruments such as radio sounders, magnetometers, gravimeters, or seismometers for examining subsurface properties. A heat-flow experiment would be very desirable, but would be difficult on an automated rover, because of the precise down-hole temperature-gradient measurements required by present techniques.

In view of these desiderata, the JPL study team decided to concentrate on imaging and sampling as the most basic requirements, and we have made various experiments [7], [22] and [23] in this connection. Figures 7 and 8 show a recent field exercise on volcanic terrain in the California desert. A vehicle is set up with cameras and TV monitors viewing the local scene. The navigator and geologist are brought to the site blindfolded and placed inside, with maps and overhead photos simulating what would be available from lunar orbiters. Then, using only what they can see via television, they direct a sampling and mapping operation. The object is to determine various system design parameters, such as how much time is needed for the requisite

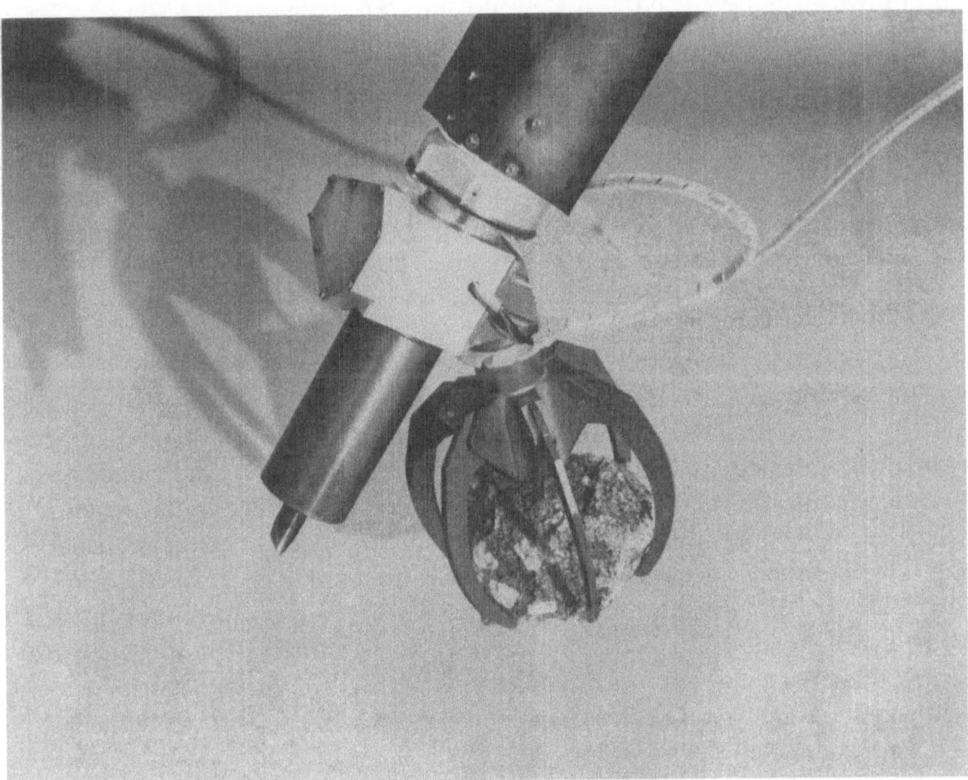

Fig. 9. Chipper-tong device for collecting and manipulating rock samples.

judgments, how good must the images be, which types of displays are preferred, and so forth. One result, for instance, has been the finding that mineral identification is quite good with a sample-viewing ability as shown in Figure 8, provided that the sample and camera can be manipulated to provide various magnifications and viewing and illumination angles. Samples must therefore be held so that they can be viewed, not enclosed as by the Surveyor scoop. Figure 9 shows a chipper-tong combination designed with this in mind. On the basis of these and other similar tests, we agree with our Soviet colleagues [24] that a lunar rover should have two types of cameras: a photofacsimile device – they call it a telephotometer – for panoramic, high-precision viewing, and a television camera or cameras for driving, where speed of display is more important than picture quality. Color imagery is not important in viewing the general lunar scene, but it is very desirable for the close-up examination of samples.

F. POLAR AND FAR SIDE MISSIONS

As mentioned earlier, Apollo exploration is limited to middle-latitude orbits and to surface sites on the Earthward side of the Moon. Ultimately we will want access to the surface in polar and far-side regions that do not have line-of-sight to Earth. To support such missions a communications relay will be needed. A promising concept is the 'halo orbit' [25] in which a relay satellite can be located near the Lagrangian libration point beyond the Moon and stabilized by small station-keeping impulses on a path that is continuously visible from Earth.

Landings in polar or far-side regions can be achieved via lunar orbit, as in Apollo and current Soviet automated missions. However, direct Earth-Moon transfers to these regions are also feasible, using slow outbound trajectories. There is even a class of lunar swing-by paths where the spacecraft passes near the Moon, then up out of the plane of the Moon's orbit and over its pole, to encounter the Moon again on the other side of Earth about two weeks later.

G. AUTOMATED SAMPLE RETURN

As demonstrated by Luna 16, lunar samples can be collected and returned to Earth by automated means. Return-path guidance is not needed, provided that the ascent velocity cutoff is accurate. There is a locus of sites in the Moon's eastern equatorial region where the required ascent vector is vertical; this removes the requirement for precise azimuth control of the launch direction. Luna 16 landed in this region. For launching from other sites, azimuth control appears to be no great problem; for example, it could be achieved at high latitudes by using a Sun-azimuth reference.

The concept of a rover delivering collected samples to a sample-return spacecraft is an attractive one and could indeed be realized by spacecraft in the Luna 16-17 series. There is a combined science and engineering tradeoff for roving missions with and without sample return. With sample return, the rover's instrumentation can be simpler, being mainly discriminatory rather than analytical, and the ultimate scientific yield may be great because of the power and variety of sample-analysis techniques available on Earth. However, for this value to be realized, the rover must make it all the way to the rendezvous (or must itself carry the sample-return rocket). Therefore the mission risk is increased. Also, the more limited the rover's instrumentation, the more restricted is its repertoire of actions in response to new findings along the traverse. Since automated missions are fundamentally less versatile than manned ones in this regard, this may prove to be a significant constraint. We therefore conclude that even if sample return is provided, it will be desirable to maximize the rover's analytical ability; this has the added advantage that less would be lost in the event of a failure somewhere along the way.

If the continued delivery of selected samples is assumed, the program plan tends to emphasize instrumentation for sample selection. If not, there must be increased emphasis on sample analysis by remote methods on the Moon. Clearly the latter method cannot provide the variety and precision of Earth-based analytical methods, so that more is dependent on inference. In either case, the suite of samples already gathered provides an essential point of departure. Looking ahead, it is probably desirable to develop at least some of the feasible remote-analysis techniques to prepare for exploring the surface of Mars, whence the return of samples is much more difficult than from the Moon.

4. Summary and Conclusion

We have seen that there are numerous opportunities for useful post-Apollo lunar activities. It is to be emphasized that our approach presumes the successful completion of the Apollo program. The kinds of remotely-controlled investigations that appear practical from an engineering and programmatic standpoint are not competitive with manned missions; rather, they should be designed to supplement and extend the results of present programs, making use of the Apollo and Soviet sample analyses and other data as a point of departure. As scientific knowledge and opinion evolve, changes in the program objectives described here are to be expected. But regardless of the detailed plan, lunar missions will continue to offer a fruitful opportunity for international efforts, a new means for observing the cosmos, and an understanding of the boundary conditions for the beginnings of life on Earth. Progress in all of these fields can be advanced by a suitable set of scientific missions to the Moon.

Acknowledgements

Much of the work described here was done by the JPL Lunar Studies Team under the direction of R. B. Coryell, and by members of the JPL Planetology Group including D. B. Nash, J. E. Conel, and F. P. Fanale. I thank L. D. Jaffe, R. A. Lyttleton, R. Choate, N. H. Horowitz, H. Bank, and G. K. Hornbrook for discussions, and T. Krupka for assistance in preparation. This paper presents the results of one phase of research carried out at the Jet Propulsion Laboratory, California Institute of Technology, under Contract No. NAS7-100, sponsored by the National Aeronautics and Space Administration.

References

[1] 'A Long-Range Program in Space Astronomy', Position Paper of the Astronomy Missions Board, NASA SP-213, Washington: July 1969.

[2] Hibbs, A. R.: 1959, 'Exploration of the Moon, the Planets, and Interplanetary Space', studies conducted for the National Aeronautics and Space Administration under Contract No. NASw-6, JPL TR 30-1, April 30.

[3] Adams, J. B., Conel, J. E., Dunne, J. A., Fanale, F. P., Holstrom, G. B., and Loomis, A. A.: 1969, *Rev. Geophys.* **7**, No. 3.

[4] Nash, D. B.: 1967, 'Sampling of Planetary Surface Solids for Unmanned in situ Geological and Biological Analysis: Strategy, Principles, and Instrument Requirements', TR 32-1225, Jet Propulsion Laboratory.

[5] Brereton, R. G.: 1968, 'Science Utility of Automated Roving Vehicles', SPS 37–51, Vol. III, Jet Propulsion Laboratory.

[6] Brereton, R. G.: 1968, 'The Objectives for Roving Vehicles in a Lunar Exploration Program', SPS 37–52, Vol. III, Jet Propulsion Laboratory.

[7] Brereton, R. G.: 1969, 'Comments on Geological Observations from an Automated Vehicle (Field Test)', SPS 37–55, Vol. III, Jet Propulsion Laboratory.

[8] Brereton, R. G.: 1969, 'Lunar Surface Gravity Investigations', SPS 37–57, Vol. III, Jet Propulsion Laboratory.

[9] Brereton, R. G.: 1969, 'Imaging and Sampling Requirements for an Automated Roving Vehicle', SPS 37–60, Vol. III, Jet Propulsion Laboratory.

[10] Cameron, A. G. W.: 1970, *Trans. AGU* **51**, 628.

[11] O'Keefe, J. A.: 1970, *Trans. AGU* **51**, 633.

[12] Singer, S. F.: 1970, *Trans. AGU* **51**, 637.

[13] Urey, H. C.: 1971, *Science* **172**, 403.

[14] Allenby, R. J.: 1970, *Space Sci. Rev.* **11**, 5.

[15] 'Apollo 12, Preliminary Science Report', Prepared by NASA Manned Spacecraft Center, NASA SP-235, 1970.

[16] Symington, J. W.: 1970, in '"For the Benefit of All Mankind." A Survey of the Practical Returns from Space Investment', Report of the Committee on Science and Astronautics, U.S. House of Representatives.

[17] 'Space Cooperation Between the United States and the Soviet Union', Hearing before the Committee on Aeronautical and Space Sciences, U.S. Senate, March 17, 1971.

[18] *Extraterrestrial Life: An Anthology and Bibliography*, compiled by Shneour, E. A., and Ottesen, E. A., Publication 1296A, National Academy of Sciences, National Research Council, Washington, D.C., 1966.

[19] 'Launch Vehicle Estimating Factors', for use in Advance Space Mission Planning, NHB 7100.5, NASA, U.S. Government Printing Office, January 1971.

[20] 'Test and Evaluation of the Surveyor III Television Camera Returned from the Moon by Apollo XII', Hughes Aircraft Company, December 31, 1970.

[21] Jaffe, L. D.: 1970, *Science* **170**, 1092.

[22] Strand, J. N.: 1970, 'Tests of Instruments Proposed for the Scientific Payload of a Lunar Rover', SPS 37-62, Vol. III, Jet Propulsion Laboratory, March 24, 1970.

[23] Burke, J. D., Choate, R., and Coryell, R. B.: 1971, *JPL. Quart. Tech. Rev.* **1**, 131.

[24] Brereton, R. G., Burke, J. D., Coryell, R. B., and Jaffe, L. D.: 1971, *JPL. Quart. Tech. Rev.* **1**, 125.

[25] 'The Utilization of Halo Orbits in Advanced Lunar Operations', Farquhar, R. W., GSFC Document No. X-551-70-449, Goddard Space Flight Center, December 1970.

D. PETROLOGICAL STUDIES OF THE MOON

EXPERIMENTAL PETROLOGY AND PETROGENESIS OF APOLLO 12 BASALTS

D. H. GREEN

Australian National University, Canberra, Australia

Abstract. Experimental studies at 1 bar and up to 30 kbar establish the crystallization sequences for basalts 12021, 12065, 12022, 12009 and 12040. Olivine is the liquidus phase at low pressures, with minor chromium spinel and pigeonitic clinopyroxene joining the olivine at lower temperatures or accompanying the olivine in the less magnesian basalts (12021, 12065). At higher pressures, subcalcic clinopyroxene becomes the liquidus phase except in the most magnesian basalt (12040) where orthopyroxene joins the olivine and becomes the liquidus phase at pressures of 25 kbar. Integration of experimental studies with observed mineralogy of natural rocks shows conclusively that the basalt compositions studied do not lie on a plagioclase + pyroxene + spinel ± olivine cotectic nor have these rocks been derived by accumulation of olivine or pyroxene into such a low temperature cotectic liquid. The Apollo 12 basalts provide clear evidence for the genesis of olivine-rich basalts in the lunar interior. The nature of the source rock is deduced to be pyroxenite or olivine-bearing pyroxenite in which orthopyroxene is probably the major phase with lesser sub-calcic or pigeonitic clinopyroxene. The 100 Mg/(Mg + Fe) ratio of the source region in the deep lunar interior is 75–80.

Urey and Runcorn (eds.), The Moon, 123.

ELECTRON MICROSCOPIC STUDIES OF SOME LUNAR MINERALS

P. E. CHAMPNESS
Dept. of Geology, University of Manchester, England

and

G. W. LORIMER
Dept. of Metallurgy, University of Manchester, England

Abstract. Ion-thinned samples of rock 12052 have been examined in the electron microscope at 100 kV. This rock contains numerous phenocrysts of pyroxene, $Ca_{1-P}(Mg, Fe)_{1+P}Si_2O_6$, and a few rounded olivine crystals $(Mg, Fe)_2SiO_4$ set in a finer-grained groundmass of pyroxene, plagioclase $(Na, Ca)(Al, Si)_4O_8$, ilmenite, $FeTiO_3$ and other minor minerals (Champness *et al.*, 1971). The cores of the pyroxene phenocrysts consist of a calcium-poor pyroxene, pigeonite, with an epitaxially grown rim of a calcium-rich pyroxene, augite. Both the pigeonite and augite are zoned, and indication of non-equilibrium crystallisation (Bence *et al.*, 1970).

The outer parts of the pigeonite cores of the phenocrysts show very finescale composition modulations on (001) (major) and (100) (minor) (Figure 1). This texture is interpreted as exsolution by a spinodal mechanism (Cahn, 1968).

The augite rim of the phenocryst exhibits a coarse exsolution of pigeonite on (001) on which is superimposed fine-scale composition modulations on (001) and (100) (Figure 2).

The two-stage exsolution process indicates that the cooling rate of the pyroxene increased suddenly, possibly when the magma was extruded on the Moon's surface.

Antiphase domains produced during the polymorphic transition from space group $C2/c$ to $P2_1/c$ have been found in all primary and precipitated pigeonites (Figure 3). This confirms Morimoto and Tokonami's (1969) suggestion that all clino-pyroxenes have the $C2/c$ space group at high temperatures.

Cristobalite, a high-temperature polymorph of silica, occurs as very small grains in the groundmass. It contains numerous microtwins on $\{100\}$ planes produced during the $\beta \rightarrow \alpha$ polymorphic transition which occurs at 268 °C (Figure 4a). Diffraction patterns show extensive diffuse streaks in various directions (Figure 4b) which correspond to intersections of planes in reciprocal space with the Ewald Sphere. These diffuse planes originate from the cooperative thermal vibration of Si-O chains parallel to $\{100\}$ and $\{111\}$.

(More detailed accounts of the results reported here can be found in Proceedings of the 25th Anniversary Meeting EMAG, Inst. of Physics, 324–327 (1971) *Contr. Mineral. and Petrol.* **33**, 171–183 (1971) and in the Proceedings of the 5th International Materials Symposium, Berkeley, California (1972).)

Urey and Runcorn (eds.), The Moon, 124–128.

References

Bence, A. E., Papike, J. J., and Prewitt, C. T.: 1970, *Earth Planetary Sci. Letters* **8**, 393.
Cahn, J. W.: 1968, *Trans. AIME* **242**, 166.
Champness, P. E., Dunham, A. C., Gibb, F. G. F., Giles, H. N., MacKenzie, W. S., Stumpfl, E. F., and Zussman, J.: 1971, *Proc. Second Lunar Sci. Conf.* **1**, 449.
Morimoto, N. and Tokonami, M.: 1969, *Amer. Mineral.* **54**, 725.

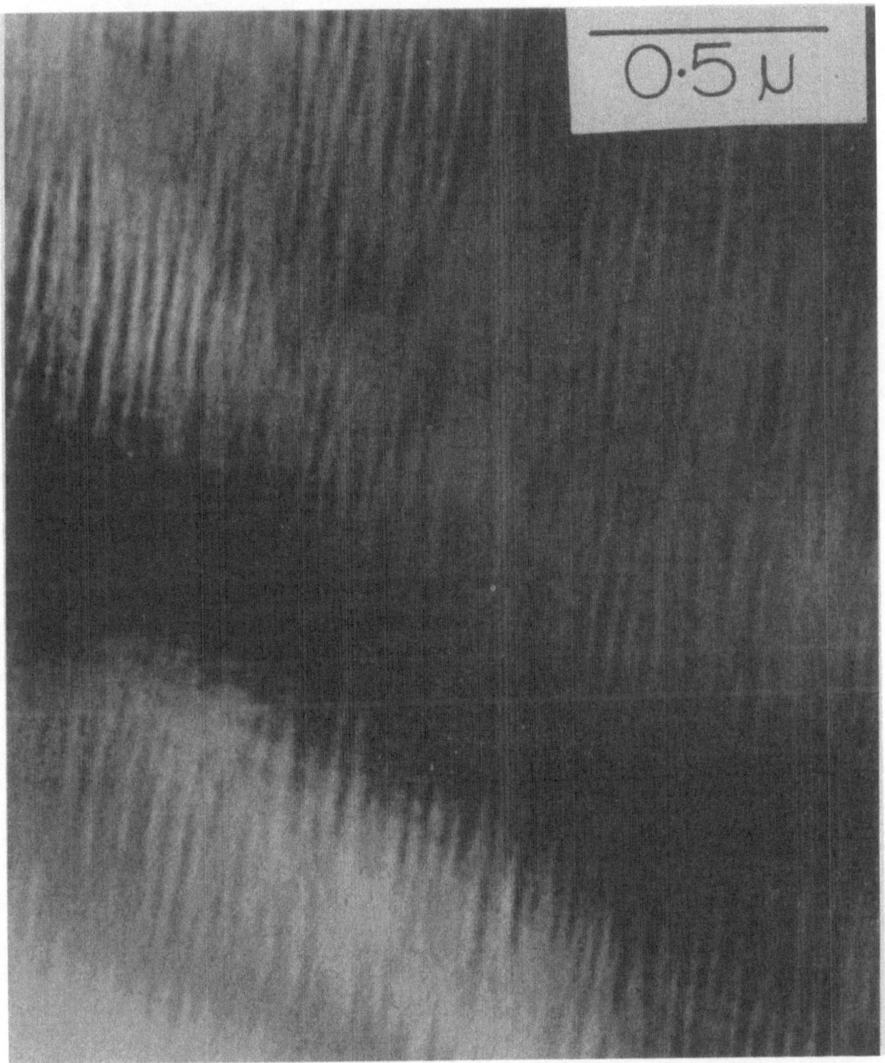

Fig. 1. Electron micrograph of a pigeonite grain in the groundness showing simultaneous composition modulations on (001) (running approximately N-S) and (100).

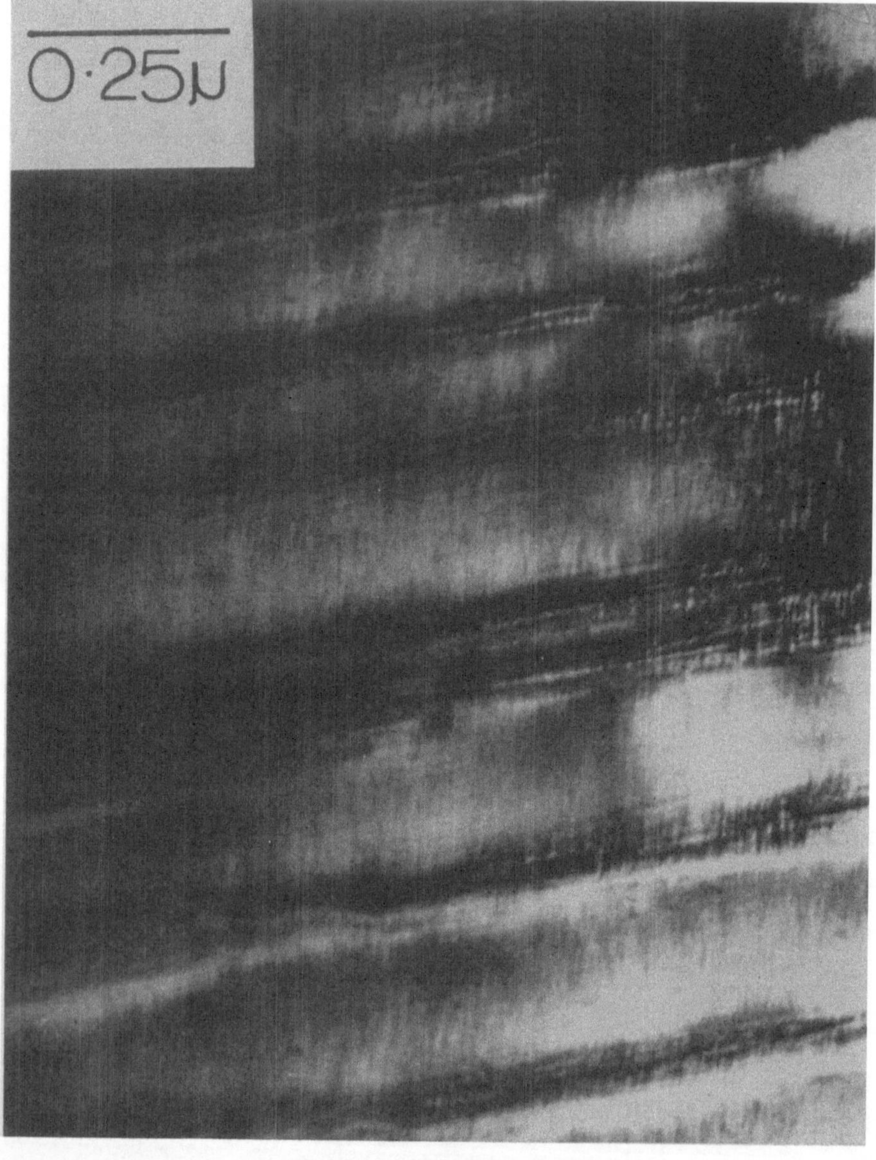

Fig. 2. Electron micrograph of the augite rim of a phenocryst pyroxene showing coarse exsolution of pigeonite on (001) and superimposed fine-scale composition modulations on (001) and (100).

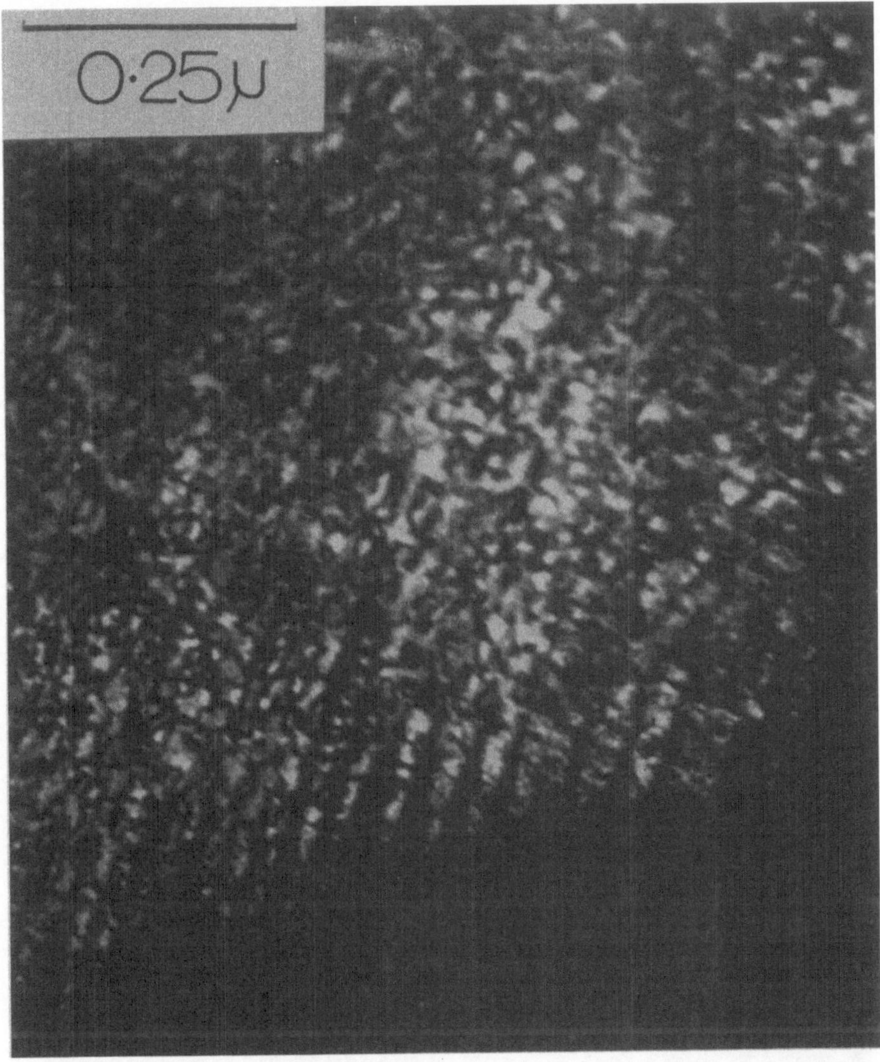

Fig. 3. The outer of the pigeonite core of a phenocryst pyroxene which has unmixed augite. The pigeonite shows antiphase domain boundaries.

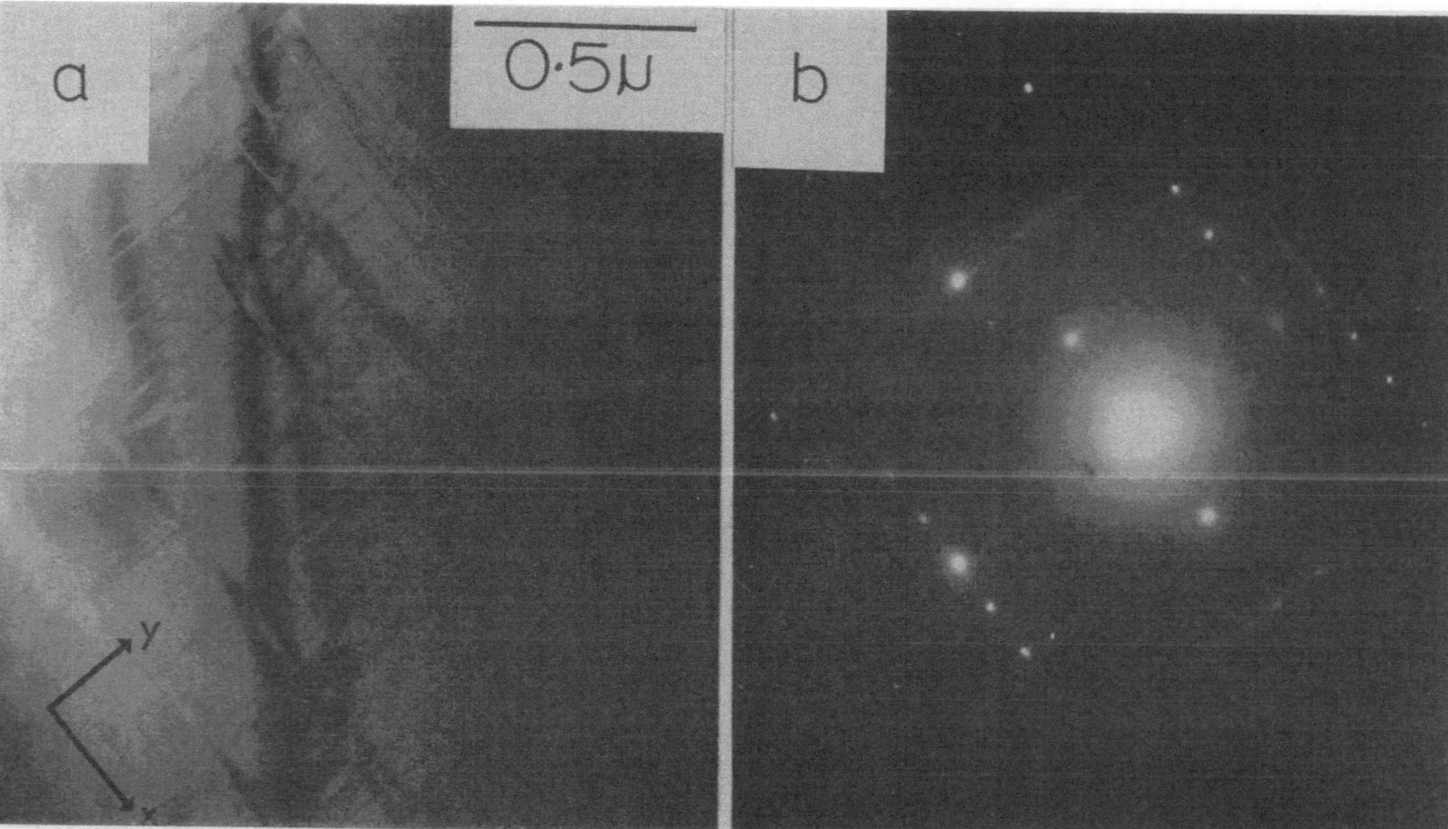

Fig. 4. (a) A cristobalite grain from the groundmass showing microtwins produced during the $\beta \rightarrow \alpha$ transition. (b) Diffraction pattern from the grain showing diffuse streaks resulting from thermal diffuse scattering.

MARIA LAVAS, MASCONS, LAYERED COMPLEXES, ACHONDRITES AND THE LUNAR MANTLE

G. M. BIGGAR, M. J. O'HARA and D. J. HUMPHRIES
Grant Institute of Geology, University of Edinburgh, Scotland

and

A. PECKETT
Department of Geology, University of Durham, England

Abstract. Experimental data show Apollo 11 and 12 lava compositions to be controlled by fractional crystallization close to the lunar surface, in a process which yields achondrite-like igneous rocks as underlying complementary crystal accumulates. Volatilization losses during eruption can account for most other chemical differences between lunar lavas and common terrestrial magmas. No specific hypotheses of the composition, mineralogy, or origin of lunar interior can be sustained until the extent of these processes is known. A terrestrial upper-mantle-type lunar interior cannot yet be excluded. The assumption that maria surface lavas are primary partial melts is unjustified and leads to a postulated lunar interior with too low Mg/Mg+Fe to serve as a source for Apollo 14 and other igneous liquids. Other workers' uncontrolled visual estimates of crystallinity in experimental charges, purporting to show that maria lavas were not modified by low pressure fractionation, are irreconcilable with the chemistry of the residual liquids developed in our 'reversed' equilibrium experiments. The undesirability of using glass as a starting material for this type of experiment is re-emphasized.

1. Introduction and Statement

The preferred interpretation (Biggar *et al.*, 1971) of available lunar samples involves at least the following sequence of events:

(1) Lunar accretion.

(2) Primordial differentiation to yield crust by partial melting of interior.

(3) Continued but rapidly declining rate of meteoritic impacts, including formation of maria basins.

(4) Triggered release of further partial melts from the interior to fill maria basins with *lakes* of lava. (see 8)

(5) Slow consolidation of the lava lakes with strong fractional crystallization in each liquid body. (see 12)

(6) Random or intermittent eruption of underlying residual differentiated liquid through the crusts of the lava lakes to form the surface lavas sampled by Apollo 11 and 12 missions.

The age of these events is ~4.5 b.y. Apparent younger Rb/Sr isochron ages date one or more metamorphic events. *

Significant facets of this interpretation are that:

(7) There is no evidence that the lunar interior, or the *primary* partial melts derived from it, differ significantly in composition from those present in the *outer* parts of the Earth: there is some evidence that the two groups of materials are similar.

(8) Fire fountaining during initial lava eruption (see 4) led to small droplet forma-

* See footnote, p. 137.

Urey and Runcorn (eds.), The Moon, 129–164.

tion and selective volatilization of elements, as well as fall in oxygen fugacity to the point where metallic iron began to separate. Two of the four major chemical peculiarities of maria lavas (great depletion in all the volatile elements, and most siderophile elements) result directly from the loss of vapour phase at stage (4) coupled with the settling out of a layer of metal phase at the floor of the lava lake after fire fountain droplets had fallen back into the lake.

(9) The extent of metal phase production determines the mass concentration effects.

(10) Prolonged fractional crystallization, just as in large terrestrial lava bodies, greatly reduced the residual liquid volume while concentrating the remaining FeO, the TiO_2, and rare-Earth elements (REE) in that reduced volume, thus producing the third chemical peculiarity of maria surface lavas.

(11) Volatilization losses of Na_2O (in particular) at stage (4) caused excessive precipitation of calcium-poor pyroxenes accompanied by very calcic plagioclase from the differentiating lava lake. Prolonged plagioclase extraction aided by volatilization produced the fourth chemical peculiarity, the marked europium deficiency in the late residual liquids, which occasionally spilled out over the crust as lava flows.

(12) The petrographic character of the crystal accumulates in the underlying maria structure is such that fragments of them observed to fall upon the Earth would be classified as achondrites, irons or stony irons. The essence of the interpretation is conveyed in Figure 1.

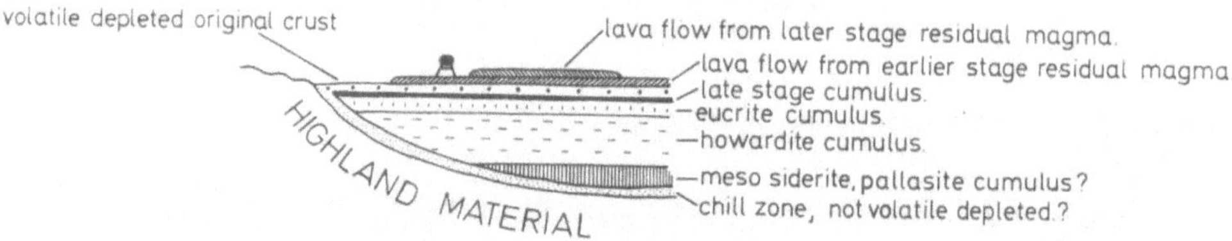

Fig. 1. Hypothetical cross section of lunar maria illustrating disposition of rock types consistent with experimental and petrological observations, topography, and Duke and Silver's (1967) interpretation of the achondrites (reproduced from Biggar *et al.*, 1971, Figure 10). Because both the chill zone and the first crust on the lake are formed from magma which may have fire-fountained on eruption, they will have suffered some volatilization losses, and iron precipitation, but not fractional crystallization. Impact melted glasses of howardite/eucrite-like composition are reported from soils at some sites (Marvin *et al.*, 1972).

This paper summarises the arguments demonstrating (i) that the maria are fractionated lava lakes, and not piles of successive individual flows. (ii) That their surface lavas represent strongly fractionated residual liquids from those lakes. (iii) That selective volatilization was a significant event. (iv) That the true age of the major igneous events is 4.5 b.y.* and (v) that the Mg/Fe+Mg of the lunar interior closely resembles that of the Earth's upper mantle.

The obscuring effect of the near-surface events which modify lunar rock compositions make it difficult to place any valid constraints on the nature of the lunar interior, beyond those already imposed by inference from astronomical observations. Firm

* See footnote p. 137.

choice between early fission, capture and conjugate birth as the origin of the Moon is impossible, as yet, on the strength of petrological or chemical data.

The interpretation presented above is founded upon certain observations of phase equilibria produced in our laboratory. Much of this paper is concerned with establishing the accuracy, sufficiency and relevance of these data. Experimental data from another laboratory have been alleged to invalidate our conclusions; some attention is given below to identifying serious discrepancies both within, and between those data and our own results, and to understanding the reasons for the discrepancies.

2. Maria Surface Lavas Are not Primary Partial Melts of Lunar Interior; Lunar Maria Are Sites of Extreme Fractional Crystallization

Strom (1971 and pers. comm.) reports stratigraphic sequences of very extensive thin flows emanating from wrinkle ridges in Mare Imbrium. The older lava flows are 'red', and the upper later flows are 'blue', indicating increase in TiO_2. Final effusions from the fractures appear to be small volumes of very viscous light coloured lava. The telescopic evidence for differentiation increasing with passage of time in a mare surface eruption sequence is strong.

Large lakes or chambers of magma undergo *extreme* differentiation by fractional crystallization during their solidification (Wager and Brown, 1968). At a late stage a very small fraction of the original liquid volume becomes grossly enriched in those rare elements selectively rejected by the crystals forming and sinking to the floor of the body to form cumulates. Periodic escape of such residual liquids onto the surface (due to foundering or impact fracturing of the lava-lake crust) will then give chemical sequences inverted with respect to those expected in a direct partial melting sequence, the latest lava flows being the most enriched in rare elements.

The elements which are selectively rejected by crystalline or immiscible liquid, phases separating at an early stage from basaltic magmas and which might, therefore, be expected to become enriched in the residual liquids developed during consolidation of the lava-lakes, may be subdivided as follows:

(i) *Volatile* e.g. H_2, O_2, H_2O, CO_2, F, Cl, Br, I, S, Se, Na, K, Rb, Cs, Zn, Cd, Hg, B, In, Tl, N, P, As, Sb, Te, Bi, Pb. These, however, will not reach high concentrations in residual liquids if a gaseous phase has separated at low pressure.

(ii) *Chalcophile* These include some of the volatile group, and will not reach high concentrations in residual liquids if a sulphide phase (crystal or liquid) has separated.

(iii) *Siderophile* e.g. Co, Ni, Pd, Pt, Os, Rh, Ag, Au. These will not reach high concentrations in residual liquids if oxygen fugacity has been low enough to cause the separation of an iron-rich metal phase.

(iv) *Lithophile* Any Fe remaining oxidised in the liquid after filling of the lava-lake will thereafter be strongly concentrated by fractional crystallization into the residual liquids together with Li, Sr, U, Th, Ce, Nb, Ba, Y, R.E.E., Ti, Zr, V, Cr. These will be progressively concentrated in the residual liquids regardless of volatilization, sulphide or metal phase precipitation, subject to the following provisos:

(a) Fe, Cr and V will remain high or be progressively concentrated provided there has not been precipitation of spinel (chromite, magnetite) or abundant Ca-rich pyroxene (diopside, augite).

(b) The light and heavy REE elements will be comparably concentrated provided that large amounts of Ca-rich pyroxene have not precipitated.

(c) Europium will not be concentrated if abundant plagioclase feldspar has precipitated, and *may have been selectively volatilized.*

The chemistry of maria surface lavas is marked by high concentration of Fe, Ba, Y, REE, Ti, Zr, V, Cr, without marked relative fractionation of the REE, but with a spectacular relative depletion of Eu. Extreme fractional crystallization and separation of calcic plagioclase and calcium-poor pyroxene, with little spinel or Ca-rich pyroxene is indicated.

In maria lavas great depletion of siderophile elements indicates extraction of a metal phase. Allowing for the initial Fe depletion caused by this, the present Fe/Mg ratio and high concentrations of REE, Ti, Zr, are such as might be generated in the final 1–2% of liquid in a consolidating lava lake.

The fractionated lava lake hypothesis *predicts* that the very high concentrations of elements such as Ti will be found in less than 2% of the total mass of the maria, and that these concentrations will increase in the surface lava flows from any one lava lake as they become younger, consistent with Strom *et al.*'s observation of increasing 'blueness' with passage of time, at least until a late stage when Ti-rich mineral begins to crystallize. Essene *et al.*'s (1970) hypothesis of successive flows of primary partial melts, each representing a very small fraction (less than 1%) of the source lunar mantle requires on the contrary (i) that concentrations of titanium should decrease, rather than increase with passage of time, and (ii) that huge volumes of the Moon's interior should have been 'worked over' and depleted of a minute fraction of partial melt, the mechanics of whose flow and collection into a few lava reservoirs presents severe problems. The mass balance problem is a principal objection to the successive lava flow hypothesis – it does not arise in the fractionated lava lake hypothesis because only 1–2% as much lunar mantle needs to be processed to yield the observed element concentrations in the last extreme differentiate of the lava lake.

One of the very minor but characteristic late stage rocks produced by extreme crystal fractionation in large terrestrial lava bodies is a hedenbergite-granophyre (Wager and Brown, 1968). Mason *et al.* (1971) now report a fragment of hedenbergite-granophyre from regolith at the Apollo 12 site.

3. Lunar interior; High Mg/Mg+Fe

Phase equilibria data from Apollo 11 samples did not exclude the possibility of a lunar mantle of terrestrial upper mantle composition (O'Hara *et al.*, 1970). Figure 2 summarises available relevant information on the partition of Mg and Fe between coexisting olivines and liquids, and the inferences which may be drawn from them about the Mg/Mg+Fe of the source regions of the liquids. The solid curve represents

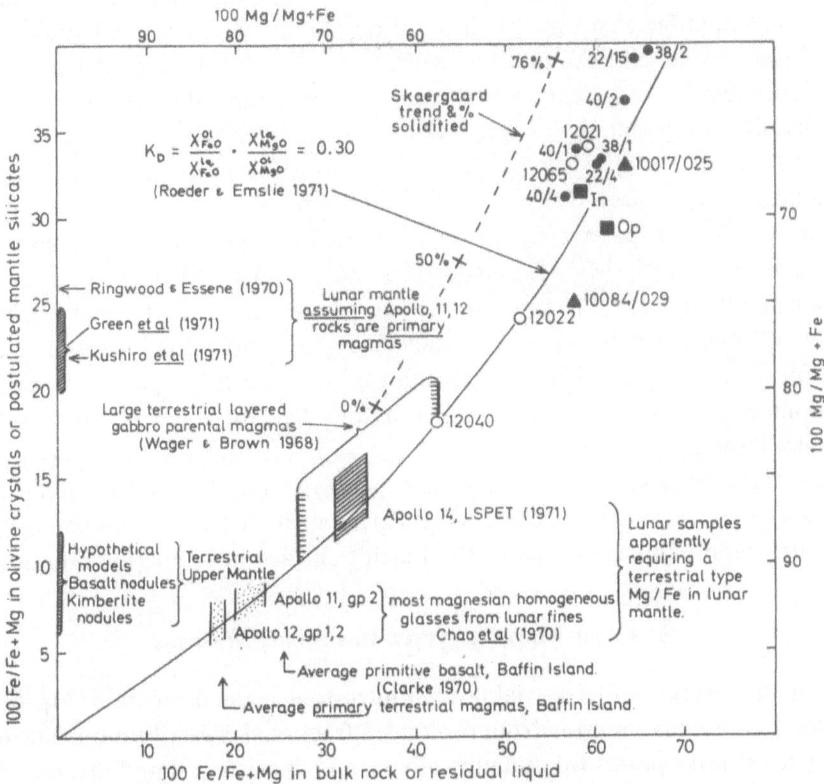

Fig. 2. Fe/Mg distribution between olivine crystals, or postulated lunar mantle and liquid, for Apollo 11, 12 and 14 samples; small filled circles and triangles from experimental data of Biggar *et al.* (1971); open circles from liquidus data of Green *et al.* (1971), filled squares from petrographic observation (James and Jackson, 1970) of the first olivines in the Apollo 11 intersertal (In) and ophitic (Op) group lavas.

an empirical partition function found to fit most terrestrial data. It is seen to fit data from quenching experiments on Apollo 11 samples (Biggar *et al.*, 1971, triangles), Apollo 12 samples (Biggar *et al.*, 1971, filled circles), liquidus olivine compositions quoted for four rocks (Green *et al.*, 1971, open circles) and phenocryst core compositions (James and Jackson, 1970 filled squares) in two groups (Op, ophitic; In, intersertal) of Apollo 11 lavas, plotted against average compositions of these groups (Compston *et al.*, 1970).

On the *assumptions* that Apollo 11 and Apollo 12 lavas represented primary partial melts, Ringwood (1970) and three other groups of workers have deduced relatively low Mg/Mg+Fe (0.75–0.80) in the silicates of the lunar mantle, as noted on the left side of the figure.

However, Apollo 14 samples have much higher Mg/Mg+Fe than the maria surface lavas, implying equilibration against silicates with Mg/Mg+Fe~0.87. These lavas cannot be interpreted as more advanced partial melts of some more iron-rich mantle because they have higher concentrations, than do lavas collected from maria-surfaces, of some incompatible elements which partition almost wholly into whatever volume of liquid is available. Nor can they be interpreted as partial melts depleted in metallic iron because relative to lavas collected from maria surfaces, they are richer in nickel, which dissolves preferentially in, and would be eliminated with a metal phase. These Apollo 14 liquids are clearly more primitive in Mg/Mg+Fe than those erupted at the Apollo 11 and 12 sites, which are presumably derived by fractional crystallization from some primitive parent. Brown *et al.* (1971) and Brown (pers. comm. 1971) report olivines with Mg/Mg+Fe$\sim$0.88–0.90 and comment on their implications for the lunar mantle. It may transpire that Apollo 14 magmas have the characteristics of primary magmas developed on a large scale during the initial differentiation and crust formation of the Moon (LSPET 1971) in a period when incompatible element concentrations and Fe/Mg ratio were sharply reduced in the lunar mantle. Still more magnesian basic glasses are reported by Chao *et al.* (1970) from the lunar fines, requiring a source region whose silicates have Mg/Mg+Fe$\sim$0.91–0.94 similar to that demanded by the most primitive terrestrial partial melts (Clarke 1970) and evidenced in the ultramafic nodules blasted to the Earth's surface in kimberlite eruptions.

4. Lunar Maria, Layered Intrusions, Mascons

The distinctive feature of terrestrial layered intrusions is the decrease of Mg/Mg+Fe in the residual liquids resulting from prolonged fractional crystallization. The dashed line (Figure 2) links points for cumulus olivine and coexisting liquid deduced for the successive residual liquids of the Skaergaard intrusion at various percentages of solidification of the original parental liquid (from Wager and Brown, 1968, Figure 119).

O'Hara *et al.* (1970) and Biggar *et al.* (1971) have proposed that solidification of the lunar maria lava lakes proceeds in a manner analogous to that seen in large terrestrial layered intrusions, with the modification that early precipitation of metallic iron has depleted siderophile elements and inhibited equivalent decrease of Mg/Mg+Fe in the liquid until larger percentages of the parental liquid have solidified. Early precipitation of c. 3.7% of the initial magma as metallic iron would inhibit the evolution of a Skaergaard-type trend so that more than 90% crystallization would be necessary for the liquid to reach the Mg/Mg+Fe ratios of lunar maria surface lavas.

The extent of metallic iron precipitation, and the resulting excess mass in the early stages of maria filling will vary with volatilization losses and hence with the mechanics of the first eruption process, and will, therefore, vary from one mare to another. This is only one of many possible internal explanations of mascons (O'Hara *et al.*, 1971), but the one which is supported by the experimental and petrological data (Biggar *et al.*, 1971).

If circumstances were such that a constant hydrostatic head and constant *volume* of liquid were maintained in the maria basins (possible if the level of the surface is maintained by a 'water-table' effect reflecting *widespread* partial melting at some depth) the process of densification of the erupted magma by loss of O_2, Na_2O and K_2O with accompanying precipitation of metallic iron might lead to c. 5% extra mass being consolidated in the basin. An excess mass $\sim 10^5$ g cm^{-2}, is required beneath the maria to yield observed mascon effects, equivalent to an extra 6% mass present in what would otherwise have been a gabbro column c. 6 km deep (well within the depth range possible in the lunar maria, and evident in terrestrial layered bodies e.g. Stillwater complex 5.2 km; Bushveld complex 7.3 km, with lateral extent of 300 × 400 km). Thus far, therefore, we find no constraints which exclude the possibility of a lunar interior and derived liquids comparable with those present in the Earth's upper mantle, and some evidence which points to considerable similarities.

All other factors being equal, the above hypothesis predicts that surface lavas from maria with larger mascons should exhibit higher non-volatile incompatible element concentrations at the same Fe/Mg ratio. The information required to test this prediction will take time to accumulate. It also predicts that lunar volcanic rocks will show some positive correlation between higher alkali retention and lesser depletion of siderophile elements (e.g. Ni) – this is true of Apollo 14 igneous rocks compared with Apollo 11 and 12 materials.

5. Selective Volatilization from Lunar Lavas

The rate of mass loss of Na_2O from a basic magma which is necessary to extract 1% Na_2O by weight from the liquid is readily calculated.

During fire fountaining, droplet surfaces will not be rapidly cooled because they are surrounded by other equally hot droplets. Assuming an ejection velocity of $\sim 10^4$ cm s^{-1}, flight time $\sim 10^2$ s and a droplet size of $\sim 10^{-2}$ cm, a loss rate of $\sim 2.5 \times 10^{-6}$ g cm^{-2} s^{-1} at 1500 K would suffice. Loss rates of 10^{-4} to 10^{-5} g cm^{-2} s^{-1} are required for similar losses from thin sheets of liquid flowing up to a hundred kms across the lunar surface, or fire fountain droplets averaging up to 1 cm in size.

Experiments on a terrestrial iron-rich tholeiite magma near its liquidus temperature are summarised in Table I. The mass loss rate of Na_2O is $>10^{-5}$ g cm^{-2} s^{-1} at ~ 1450 K and $fO_2 \sim 10^{-11}$ to 10^{-15} bar and is appreciably higher than reported by O'Hara *et al.* (1970) when $fO_2 \sim 10^{-8}$ to 10^{-9} bar. This directly measured loss rate is similar to those inferred by application of the Langmuir equation (rate $= P_v \sqrt{(\text{mol } wt/2\pi RT)}$, vaporization coefficient of 1) to De Maria *et al.*'s (1971) vapour pressures, or by assuming a rate $\sim 10^2$ times greater than for pure SiO_2 or bulk tektite compositions (Chapman and Scheiber 1969). The lower loss rates for K than Na presumably reflect the lower activity of potassium in the melt.

Selective loss of Na_2O at least *must have occurred during eruption* of lunar lavas, and the gross depletion of these rocks in all volatile elements points to the general nature of this conclusion.

Restoration of alkalis and other volatiles including oxygen to the lunar lava compositions leads, among many other possible parents, to compositions familiar among terrestrial basalts. It is unsafe to assume that the present lava compositions are those

TABLE IA

Volatilization study on Scourie iron-rich tholeiite dike; $fO_2 \sim 10^{-12}$ bar

	Time h	Temp °C	Pressure (torr)	Comments	Na_2O wt %	K_2O wt %
Start	–	–	–		2.38 +[a]	0.92 +[a]
V.63	1	1150–60	10^{-4}	part devitrified; bubbled surface, black glass powder	1.93	0.88
V.62	5	1160	10^{-4}	part devitrified; bubbled surface, brown glass powder	1.51	0.97
V.64	24	1160	1.5×10^{-4}	trace devitrification; smooth surface, brown glass powder	0.60	0.91
V.57	60	1175	2×10^{-5}	trace devitrification; smooth surface, brown glass powder	0.16	0.25
V.58	240	1175	2×10^{-5}	trace devitrification; smooth surface, brown glass powder	0.13	0.12
V.59	430	1175	$\sim 5 \times 10^{-5}$	trace devitrification; smooth surface, brown glass powder	0.20	0.21

[a] Not adjusted for H_2O, Fe_2O_3 in rock sample. Alkalis determined by flame photometer. Surface area of sample exposed ~ 2.9 cm^2

TABLE IB

	I	II	III	IV
SiO_2	50.49	51.29	53.43	52.19
TiO_2	2.87	2.81	2.82	2.97
Al_2O_3	12.48	12.44	12.96	12.91
FeO	16.57	16.28	16.03	17.13
MnO	0.26	(0.26)	(0.26)	0.27
MgO	4.85	4.84	4.92	5.02
CaO	8.89	8.97	9.24	9.19
Na_2O	2.41	2.13	0.09	0.09
K_2O	0.93	0.98	0.25	0.25
P_2O_5	0.25	–	–	–
	100.00	100.00	100.00	100.02

I – Scourie dike, 10727 from thin vein at Geodh Eanruig (O'Hara, 1961; Table III, locality II, p. 856, plate XVIII), recalculated less H_2O and with all iron as FeO, from complete wet method analysis by M. R. Saunders.

II – Same material after 2 h at 1200 °C in molybdenum capsule at $pO_2 \sim 10^{-12.0}$ bar, recalculated to 100% less P_2O_5 from microprobe analyses of glass fragments.

III – Same material as I after 60 h at 1175 °C in molybdenum crucible at low pO_2 and total pressure of 2×10^{-5} torr. Electron microprobe analysis.

IV – Analysis I recalculated less P_2O_5 and appropriate alkalis to compare with III.

Note: alkali loss apparent between I and II; apparent loss of FeO from glass III (? reduced and reacted with crucible) and pick-up of SiO_2 (? from cracking of silicones oil of diffusion pump).

of the pre-eruption liquids, let alone the primary partial melts, and premature to ascribe their volatile-poor characteristics to the Moon as a whole on the strength of that assumption.

The feldspar phenocrysts in samples 14073, and 14310 record pre-eruption levels of Na_2O and K_2O in the liquids *at least* 2–3 times higher than the levels on final consolidation of these lavas (Brown and Peckett, 1971). Restoration of the missing alkalis and c. 0.25% O_2 to the present composition of 14310 leads to a composition quite familiar among *inter alia* the Kamchatka calcalkaline volcanics.

Volatilization which would have accompanied eruption and which selectively separates Rb from Sr, would have destroyed the orderly 4.5 b.y. model ages and the 4.5 b.y. isochron defined by the different bulk rock samples collected at the Apollo 11 (ophitic group) and Apollo 12 sites. Therefore, no igneous event can have affected these rocks since 4.5 b.y. (compare Urey *et al.*, 1971) and the internal mineral isochrons from individual rocks may date devitrification or other metamorphic events. 3.3–3.7 b.y. ago. (Liquids low in volatiles, alkalies and Fe^{3+} such as were produced in the experiments of Table I, were very slow to nucleate and some did not devitrify even in cooling times of 1 h from 1175 °C to 700 °C.)*

6. Accuracy Precision and Significance of the Experimental Data

Experiments performed on lunar samples and related compositions define, at various pressures, the (liquidus) crystalline phases stable with a liquid of the sample composition. This information can then be used to deduce the crystalline species present in the solid material with which the liquid last equilibrated *provided that* the pressure of that last equilibration is indicated by other information, or can be reliably estimated.

If that pressure is *assumed*, provisional conclusions only can be reached (O'Hara *et al.*, 1970); but addition of 10% of what you fancy to produce a more convenient liquidus phase is less than objective.

We have presented experimental evidence that maria surface lava compositions have low pressure cotectic character, controlled by crystal-liquid equilibria at low pressure, leading to the preceding interpretation of maria structure and origin. Others claim there is no evidence of significant low pressure control; go further in claiming that there has been no control of particular lunar lava compositions at any pressure other than that of genesis by partial melting of the lunar mantle; and assume the depth of that partial melting process. From this basis they deduce a model of a volatile depleted, strongly reduced pyroxenite or olivine-pyroxenite lunar interior. If our experimental observations are correct, maria surface lavas are not primary magmas and the model developed by Ringwood and Essene (1970), Ringwood (1970) and Green *et al.* (1971) is a chimera.

The controversy centres on the accuracy of our observation of close approach to cotectic character at low pressure, which has been criticized by Ringwood (1970) and

* Wasserburg and Papanastassiou (1971) suggest crustal or soil contamination of the lavas, vitiating this argument and most petrogenetic discussion based on minor element chemistry.

Green *et al.* (1971), on the strength of their visual estimates of crystallinity, in their products obtained from starting material long deprecated for its propensity to yield metastable products. Moreover, much information necessary for the evaluation of their statements was not presented by Green *et al.* (1971). The temperatures and durations of individual experiments were not given, nor is it clear which experiments used natural and which used synthetic materials (p. 602*). Without these data, claims to have produced the same result by different techniques are insubstantial. Details of the containers, atmospheres, temperatures and durations used in the initial glass preparations are needed to define the nature of the starting material. No calibration of the temperature or oxygen fugacity actually achieved in the charges was reported, and the pressure of some experiments is apparently unspecified (p. 610*). There were no data to establish the precision achieved in their visual estimates of percent crystallinity (Figure 2 caption*) which yield unrealistic residual liquid compositions.

There seem to be several inconsistencies in the data which were reported. For example, 12040 is given as showing plagioclase entry between 1120 and 1140°C (Figure 1*), near 1140°C (p. 603*), or between 1150°C and (presumably) 1140°C (p. 607*; but no 1150°C run is shown in their Figure 1). The percentage of crystals when plagioclase first appears in 12040 is reported variously as 70% (Figure 2*) or 50–60% (text on same page). There are no runs (Figure 1*) to substantiate the remark (p. 603*) that ilmenite enters at close to 1100°C in 12040. The percentage of crystals at plagioclase entry in 12021 and 12065 are given as 40% and 45% (Figure 2*), but as 20–30%, and 30–40%, respectively, in the text (p. 607*). It was claimed (p. 611*) that 12021 could be derived by extraction of olivine, spinel and minor pigeonitic pyroxene from 12009, yet the data (Figure 2*) indicate that a pyroxene olivine ratio of 3/1 would be required in the extract from 12009 to yield a residuum with the phase behaviour of 12021 at temperatures of c. 1160°C. Unlisted experiments of unknown bracket width and duration, which were apparently no more than syntheses from different starting materials *as loaded*, (see Figure 18, sequence 5 below) were presented (p. 608*) and claimed as reversals, and the reader is not told which compositions were used for these experiments.

Green *et al.* (1971) exploit small composition differences between experimental and natural minerals to stress the importance of 'direct supporting evidence from each sample'; but surely there is a prior need to place the experimental observations on a quantitative and objective basis.

Ringwood and Green (1972) have published a further criticism of our results which is disappointing for its misrepresentation of our statements, and for its reliance yet again on subjective visual estimates of crystallinity as a basis for criticism of our quantitative data. We have *never* stated that plagioclase was liquidus phase in natural Apollo 11 or 12 basalts**, nor that the rocks were *on* the plagioclase-precipitating cotectic under the conditions encountered on the lunar surface, nor have we 'qualified' our original position that the lavas have compositions *close to* the 1 atmosphere cotectic (too close

* All page and figure references are to Green *et al.* (1971a).

** 10084 was a soil sample.

for this to be a coincidence). Our position is adequately stated by Biggar *et al.* (1971), p. 634, para 2. We did not *claim* that other workers results were invalid because of the use of glass as a starting material; we have *suggested* that it may have had something to do with the discrepancies in observations between ourselves and Ringwood and Essene (1970) while clearly recognizing that the other two groups whom Ringwood and Green (1972) introduce into the argument worked on compositions substantially removed from those under discussion (Biggar *et al.*, 1971, p. 631, para 2). The reader must judge for himself, in the light of the following remarks, however, how much of the actual disagreement stems from the choice of starting material, how much from inadequate calibration of techniques, and how much from uncertainties in more subjective observations. The loss of alkalis was not "claimed to cause an expansion of the primary fields of crystallization of mafic minerals": no variation in an intensive variable can change the extent of primary phase fields, and, therefore, of cotectic liquid compositions in a system of fixed composition. The effect of the loss of alkalis and oxygen is to shift the composition of the remaining condensed system towards ferromagnesian minerals as claimed by Biggar *et al.* (1971, p. 634) who showed (ibid, p. 635, Equation (2)) a reaction in which the plagioclase declines from 64% to 55.5% consequent on alkali loss. The key question is whether the change in normative feldspar composition from An_{50} to An_{80}, combined with the increase in pyroxene at the expense of olivine will offset the reduction in total plagioclase; and problems raised by possible reduction of iron oxides further complicate the issue.

The reader may want more proof that addition of alkalis suppresses rather than enhances the initial appearance of plagioclase than the difference between the two approximate visual estimates of 40–50% (at 1125±5°C) and 35–50% (at 1115±5°C) crystallinity at plagioclase entry with or without alkalis respectively, given by Ringwood and Green (1972, Table II; what the alkali addition does to the *temperature* of plagioclase or other phase entries is, of course, irrelevant). We cannot yet provide a definitive answer for the iron rich composition of maria basalts; for the more magnesian compositions of Apollo 14 samples addition of alkalis *alone* undoubtedly lowers the temperature of plagioclase appearance and (visual estimates!) moves the composition towards the cotectic liquid composition *from* the plagioclase liquidus field. Raising or lowering fO_2 opposes this effect.

Contrary to Ringwood and Green's (1972) suggestion, nothing in any of the reported phase data, nor any conceivable data for effusive rocks could ever support the view that certain basalts were derived from regions below those at which plagioclase was a stable phase in the lunar interior. Such an interpretation depends upon totally different criteria. Their statement is only valid if 'residual' is inserted after 'stable' and the magmas are truly primary.

This disagreement is becoming long on immoderate wording and adrenalin, rather than on discussion of scientific data and techniques; we must, however, thank Ringwood and Green (1972) for pointing out our error in drawing of one of our figures from our data table. Biggar *et al.* (1971, p. 633) carries an error (9.1% TiO_2) in the text. 9.2% is correct; in Table IV, column 1, Cr_2O_3 = 0.46%, not 0.06%.

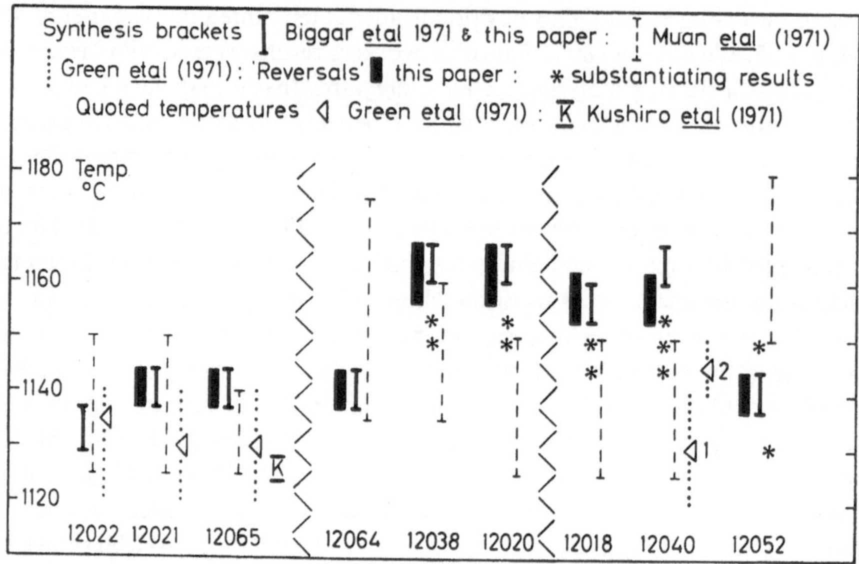

Fig. 3. Comparison of atmospheric pressure quenching data on the temperature of plagioclase appearance for 9 rocks reported on by four laboratories. Results substantiating one side of a bracket are shown where relevant by asterisks. Muan *et al.* (1971) finally reported plagioclase entry at 1125–1150 °C in 12052: Green *et al.* (1971b) now concede ± 5° control precision and ± 10 °C accuracy in their data. See Table IX, footnote (h).

TABLE II

Results of calculations of crystallinity contrasted with visual estimates[a]

100 % rock =	%	residual liquid + %	sp	ol	pig	plag	Estimated % crystals at plagioclase entry	% crystals visual estimates Green *et al.* (1971)
12040[b]	60	cotectic glass 1160	1	29	9	–	40	50 to 60 (text)
12040[b]	54	12038 rock	1	32	10	3	46	70 (Figure 2)
12052[c]	74	12064 cotectic glass at 1137 °C	1	3	17	6	<25	
12065[c]	77	Plagioclase already fairly common	1	4	12	6	<20	~ 40 (text)
12021[c]	80		tr	1	14	5	<15	~ 45 (Figure 2)
12064[c]	88		tr	1	7	4	~5	~ 30 to 40 (text); ~ 40 (Figure 2)
12022[c]	71	cotectic glass 1129 °C	1	23	0	5	<25	~45 (Figure 2)

[a] These and subsequent calculations are advanced to show a *major* discrepancy between estimates of crystallinity by different methods. Quibbling over the choice of mineral compositions for the calculations could be endless, but would not lead to large changes in the final results.

[b] Calculation in Table III.

[c] These calculations used chrome spinel (Table IV, note 5), plagioclase (An_{85}), and pigeonite and olivine from 12038/2 (Biggar *et al.*, 1971, Tables III, IV).

TABLE III

Solutions to calculations for 12040

	12040[a]		12038 +	12040	12040	
	(Kushiro and Haramura, 1971)	(Scoon, 1971)	46% crystals [b]	1160°C glass + 40% crystals [c]	liquid at plagioclase entry Calculated [d]	Actually [e]
SiO_2	43.59	43.82	43.88	44.02	37.93	46.86
TiO_2	2.48	2.74	1.96	2.70	8.30	4.17
Al_2O_3	7.34	7.40	7.71	7.22	27.40	11.50
Cr_2O_3	0.70	0.70	0.67	0.76	−0.05	0.17
FeO	20.87	20.81	21.37	21.19	11.15	17.58
MnO	0.26	0.26	0.28	0.24	tr	0.20
MgO	16.66	16.08	16.43	15.83	−8.48	7.10
CaO	7.90	7.86	7.31	7.73	22.79	11.93
Na_2O	0.16	0.20	0.39	0.31	0.67	0.51
K_2O	0.05	0.04	0.04	0.05	0.21	0.07
P_2O_5	0.02	0.07	0.02	–	0.08	–

[a] Normalised to 100%.

[b] 32.1% olivine (Fo_{66} 12040, 1160°C), 10.4% pigeonite (12040, 1160°C Biggar *et al.*, 1971) 1.0% spine, (12009, Green *et al.*, 1971) and 2.5% An_{85} calculated to fit Kushiro and Haramura analysis.

[c] 29.3% olivine (Fo_{66}), 9.2% pigeonite, 1.2% spinel; calculated to fit Scoon analysis.

[d] Extracting 30% olivine, 45% pigeonite (1136°C, Biggar *et al.*, 1971), 1% spinel (12009, 1150°C, Green *et al.*, 1971); mineral proportions measured from Green *et al.* (1971, Figure 2 first version issued at Houston conference, showing plagioclase entry at c. 1130°C as per their Figure 1. Later revised values dealt with in Table IV). Note the extraordinary values of CaO, Al_2O_3, very low SiO_2 and excessive depletion in ferromagnesian constituents.

[e] From Biggar *et al.* (1971, Table V).

[f] Biggar *et al.* (1971, p. 619) required at least three lava flows for the simplest admissible explanation of their experimental data. Precise calculations by James and Wright (unpublished) convince us that a more complicated model, involving a greater number of flows is required to explain the detailed geochemistry.

We emphasize that:

(1) The data of Biggar *et al.* (1971) establish significantly higher temperatures of plagioclase entry (Figure 3) than those claimed by Green *et al.* (1971). Muan *et al.* (1971), also using natural rock powder as starting material, could compromise with Biggar *et al.* (1971) on plagioclase entry at 1140–1160 °C in most cases. Green *et al.* (1971) and Kushiro *et al.* (1971), both using glass as a starting material, would select a temperature in the 1120–1140 °C range, but also agree with Muan *et al.* (1971).

(2) Calculations of percent crystallinity based on analyses of the coexisting phases in experimental products from Apollo 12 rocks (Tables II to IV) indicate much less crystallization at observed plagioclase entry (Figure 4) than claimed by Green *et al.* (1971) if levels of MgO, Al_2O_3, SiO_2 in the liquids are to be realistic.

(3) Products obtained from glass starting materials can be so fine grained (plate I) that percent crystallinity and crystalline-phase ratios are difficult to estimate, microprobe analysis of the residual glass is nearly impracticable, phases such as plagioclase are very difficult to detect in small amounts, and augite becomes difficult to distinguish optically from pigeonite.*

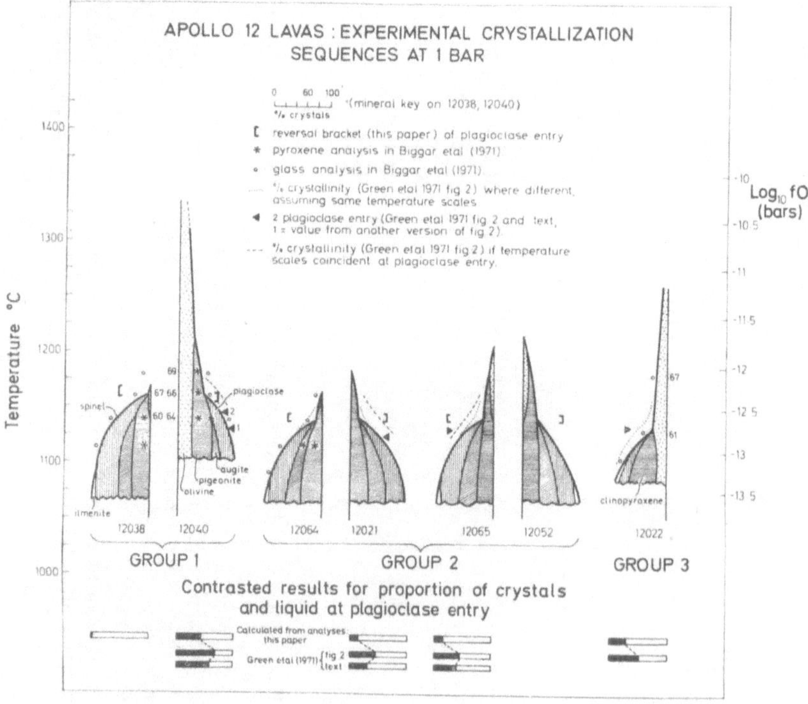

Fig. 4. Graphical representation of co-variation in fO_2, temperature, phase assemblage and percentage crystallinity in Apollo 12 rock compositions (after Green *et al.*, 1971, Figure 2, but with percent crystallinity based on calculations, and the calcium-rich and calcium-poor clino-pyroxenes distinguished except in 12022; note contrast between our calculated, and Green *et al.*'s estimated crystallinity at plagioclase entry, shown by proportional length of black bars).

* Green *et al.* (1971b) refer to this problem, but were confident they could distinguish pigeonite from inverted protohypersthene!

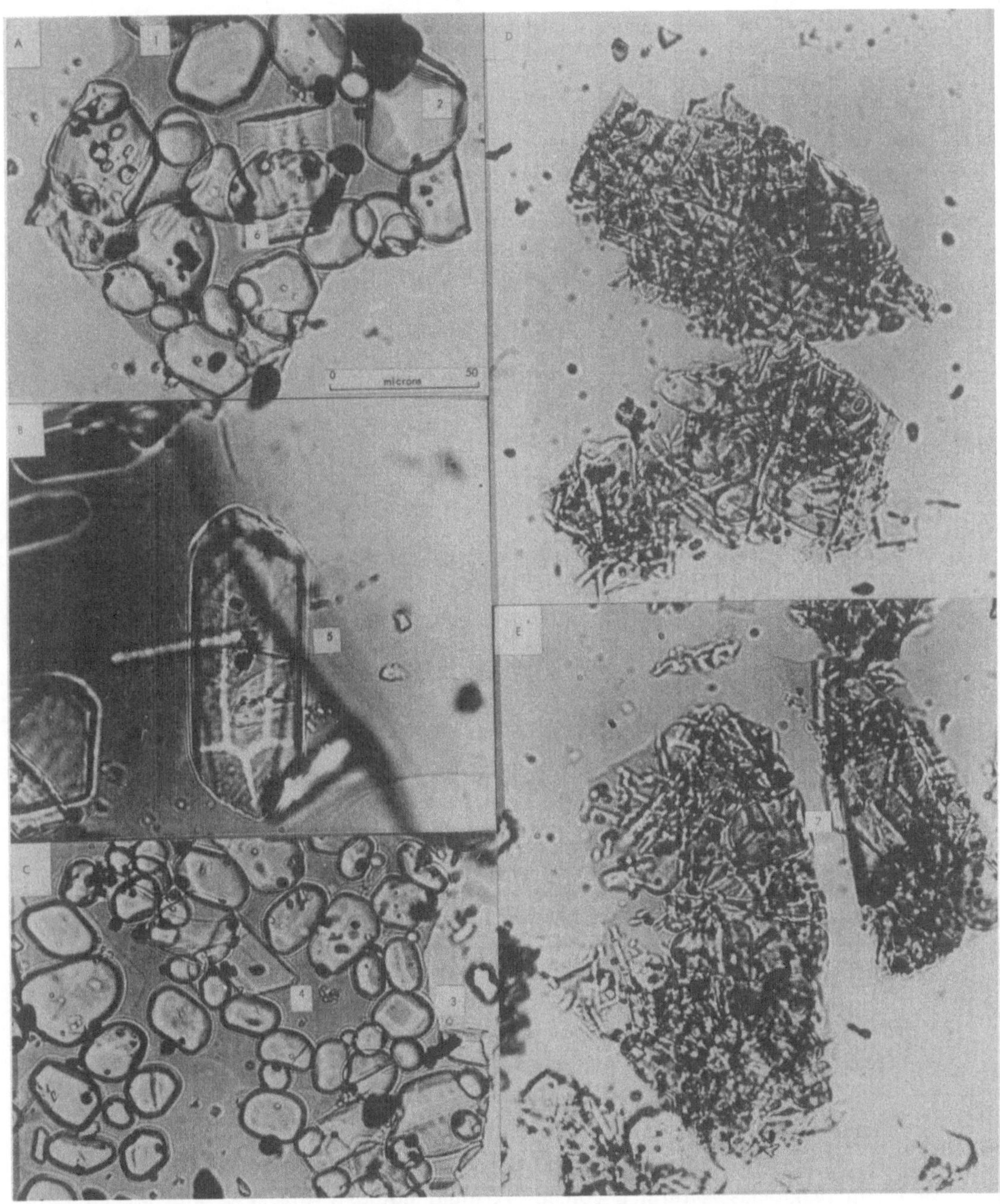

Plate I. Microphotographs (all same scale, see A) of products from reversal experiments on 12040 at 1162°/1153°C; A, B, C from rock powder starting material (experiment 035) showing euhedral olivines (1–3), pyroxene (4) and, again, with curving crack (5), plagioclase (6) and opaque spinel in glass. D, E from largely glass starting material (experiment 117) containing large olivines (7) as in A, C, but interlocking mesh of acicular pyroxenes whose minuscule width makes phase identification, estimation of crystallinity and microprobe analysis extremely difficult.

TABLE IV

Results of extract calculations, using Green *et al.*'s (1971, Figure 2) visual estimates of crystallinity, compared with the actual analysed cotectic liquid compositions

	12021 – 40% (A)	12065 – 45% (B)	12040 – 70% (C)	12040 – 50% (D)	12009 – 50% (E)	Observed real values 12038, 40, 64 cotectic liquids	12022 – 43% (F)	12022/15 observed 1129°C
SiO_2	43.16	41.77	37.68	41.06	42.33	46 ± 1	48.15	43.77
TiO_2	5.09	5.30	6.71	4.29	4.91	4 ± 0.5	–2.62	6.76
Al_2O_3	16.56	18.29	22.08	13.63	15.67	11.4 ± 1	15.20	9.92
Cr_2O_3	–0.56	–0.55	–0.08	0.37	–0.28		0.53	0.12
FeO	21.09	21.78	20.33	20.64	18.75	19 ± 1.5	17.20	20.53
MnO	0.23	0.22	0.37	0.21	0.25		0.14	0.30
MgO	–2.14	3.56	–6.69	6.68	4.64	6.1 ± 0.5	5.84	5.72
CaO	15.89	16.18	18.83	12.56	12.88	11.5 ± 0.4	14.51	11.36
Na_2O	0.54	0.40	0.53	0.32	0.46		0.51	0.73
K_2O	0.12	0.13	0.17	0.10	0.13		0.12	0.13
P_2O_5	0.02	0.04	0.07	0.04	0.14		0.43	

Extracts, all obtained from measurement of Green *et al.* (1971, Figure 2). Slightly lower figures are indicated in their text, and these reduce, but do not cure, the discrepancies..

A pigeonite[1] 39%, spinel[2] 1%; B pigeonite[1] 44%, spinel[2] 1%; C olivine[3] 26%, pigeonite[4] 43%, spinel[5] 1%; D olivine[3] 18.6%, pigeonite[4] 30.7%, spinel[5] 0.7%; E olivine[6] 7%, pigeonite[7] 42%, spinel[2] 1%; F olivine[8] 13%, pigeonite[9] 19%, ilmenite[10] 11%.

1 = 12021, 1150°C; 2 = 12009, 1150°C; from Green *et al.* (1971).
4 = 12040/2, 1137°C; 7 = 12064/5, 1114°C; 8 = 12022/15, 1129°C; 9 = 12040/2, 1137°C; from Biggar *et al.* (1971).
3 = Fo_{65} + 0.5 Cr_2O_3, 0.3 CaO; 5 = 12064, 6 (Haggerty and Meyer, 1970).
6 = Fo_{55}; 10 = 12022, 1103°C (unpublished).

Note that all solutions based on Green *et al.* (1971) visual estimates have attained *conspicuously* higher Al_2O_3, CaO concentrations than the analysed cotectic liquids which are already saturated with these components (as plagioclase). This results from gross overestimation of percentage of olivine, pyroxene. Solutions to 12021, 12040 and 12065 show excessive depletion in SiO_2, MgO because estimates of the amount of pyroxene present are excessive. In 12022, high SiO_2, low FeO and negative TiO_2 result from excessive ilmenite extract. Negative Cr_2O_3 values are not significant.

(4) Green *et al.*'s (1971) visual estimates of crystallinity yield values which are too high overall, and too high in pyroxene/olivine ratio, to apply to our experimental products. Residual liquid compositions, calculated by extracting their estimated percentages of crystals at plagioclase entry, frequently have negative components (Tables III and IV) have plagioclase-forming oxides well in excess of measured values in plagioclase-saturated cotectic residual liquids (Table IV), and are either grossly depleted, or are excessively rich in SiO_2 relative to the analysed residual liquids (Table IV). These discrepancies place the calculated liquids, when in positive composition space at all, in regions where plagioclase is likely to crystallize first and for some interval before the appearance of olivine or pyroxene (Figures 5 and 12).

(5) Similar criticisms apply to Ringwood's (1970) estimate of 30–50% crystallization

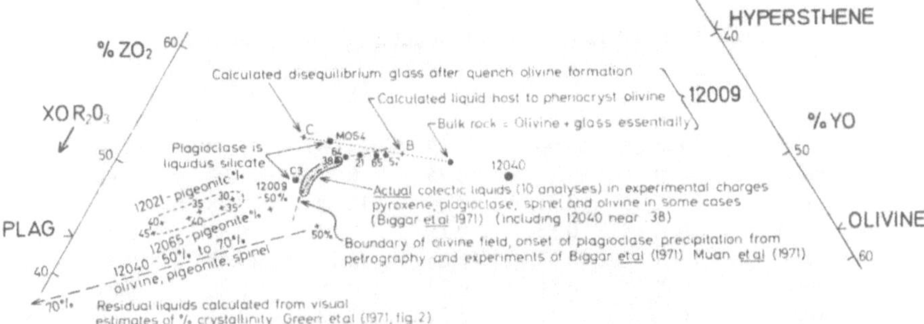

Fig. 5. Projection from diopside into the plagioclase-olivine-hypersthene-(silica) plane, of selected experimentally investigated natural and synthetic Apollo 12 materials, and analysed plagioclase-saturated liquids, to contrast with alleged plagioclase-saturated liquids, calculated by subtracting visual estimates of percentages of crystals in experimental charges. MOS 4 is from Muan *et al.* (1971). Calculated compositions with negative MgO plot within the figure provided there is sufficient FeO to compensate. The result would be unaffected by altering the Fe/Mg ratio in the pyroxene or olivine extracted *in this projection.*

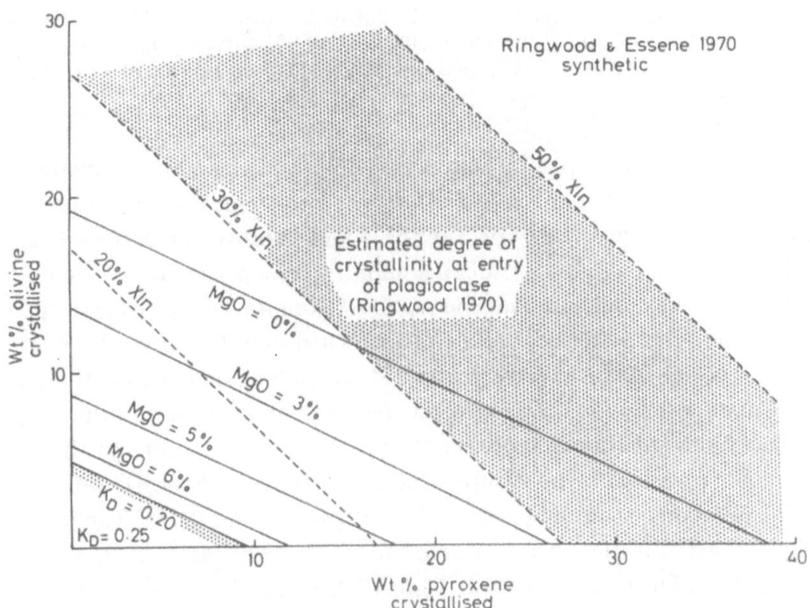

Fig. 6. Graphical presentation of results of extraction of 3% armalcolite (which reduces TiO_2 to cotectic levels at acceptable K_D and MgO figures) and variable combinations of olivine and pyroxene (Ringwood and Essene, 1970; Tables I(i), II(iii) and III, col. 1) from a synthetic Apollo 11 composition, contoured for resultant K_D olivine-liquid distribution coefficients, residual MgO in liquid, and total percentage of crystalline material present. High crystallinity is incompatible with acceptable K_D values and observed MgO contents of 6–7% (Biggar *et al.*, 1971; Table VI).

TABLE V

Apollo 11 lavas, cotectic liquid composition calculations

	I	II	III	IV	V	VI
SiO_2	42.60	43.00	44.89	44.09	40.69	40.76
TiO_2	8.86	8.76	8.25	8.31	11.92	11.98
Al_2O_3	11.01	10.31	9.95	10.09	7.78	7.84
Cr_2O_3	0.12	0.12	0.15	0.09	0.34	0.39
FeO	17.78	17.68	16.83	18.13	19.49	19.54
MnO	0.28	0.31	0.22	0.28	0.28	0.29
MgO	6.12	6.47	5.39	5.92	7.51	7.56
CaO	12.41	11.34	12.73	11.07	10.76	10.81
Na_2O	0.43	0.72	0.66	0.62	0.51	0.45
K_2O	0.09	0.37	0.39	0.59	0.30	0.43
P_2O_5	0.12		0.24		0.18	
S	0.18				0.23	

I – Apollo 11 ophitic group average (Compston *et al.*, 1970) minus 5% ilmenite, 4% olivine, 0.5% spinel.

II – Cotectic liquids, average from 10017, 10084 (Biggar *et al.*, 1971, Table VI); note CaO, Al_2O_3 *lower* than in I.

III – Apollo 11 intersertal group average, (Compston *et al.*, 1970) minus 10% ilmenite, 4% olivine, 4% pigeonite, 4% augite.

IV – 10017 residual liquid at 1133°C (Biggar *et al.*, 1971), note CaO, *lower* than in III.

V – Apollo 11, 10017 rock (Compston *et al.*, 1970).

VI – Best fit to V using 10017 liquid at 1133°C plus 0.8% olivine (Fo_{65}), 0.2% spinel (10020–40 Haggerty *et al.*, 1970), 10.1% ilmenite (10020–40), 19.8% augite (10024 – 23/grain 14 Kushiro and Nakamura 1970) and 2.6% pigeonite (10024 – 23/grain 14): K_D ol – lq = 0.26. Negative pigeonite shows that a more calcic augite should have been used.

of Apollo 11 lavas before plagioclase entry (Figure 6); excessive depletion, in MgO and, above all, in MgO relative to FeO result. Petrological observation (James and Jackson 1970; see Biggar *et al.*, 1971, p. 637) and various calculations (Table V and Figure 7) based on analysed residual liquids from experiments carried out on the natural rock samples, indicate that the Apollo 11 ophitic group (B) lavas precipitated plagioclase and 4 or 5 other crystalline phases when only 10% crystalline. The Apollo 11 intersertal group (A) lavas were nearer 25% crystalline on reaching the same condition. New data (Prinz *et al.*, 1971) for the average composition of lithic fragments at the Apollo 11 site yield more representative average compositions of the high potassium and low potassium lavas groups which are strikingly close in composition to the average of two cotectic liquids analysed by Biggar *et al.* (1971), Table VI. Selected oxide figures, listed in the above sample order, are: SiO_2 42.8, 42.9, 43.0; TiO_2 9.6, 8.7, 8.8; Al_2O_3 10.6, 11.8, 10.3; FeO 18.0, 17.0, 17.7; MgO 7.0, 7.2, 6.5; CaO 10.7, 11.4, 11.3. These compositions will reach the plagioclase-precipitation cotectic at still lower amounts of crystallization.

(6) Primary magmas produced by partial melting of unknown source rocks at an elevated pressure are randomly chosen compositions with respect to the low pressure phase equilibria. The probability of such compositions exhibiting the close approach to low pressure cotectic character actually observed in Apollo 11 and 12 lavas by

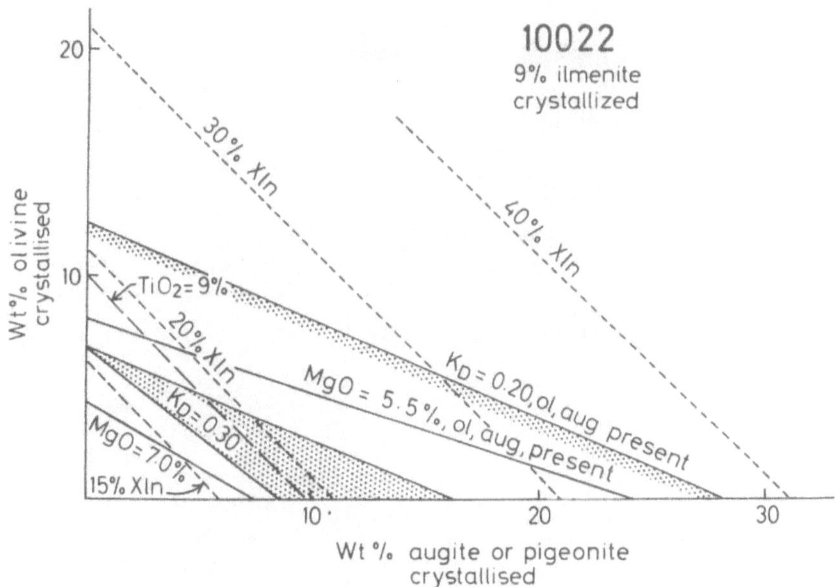

Fig. 7. Similar presentation to Figure 6 for rock 10022, using analysed mineral (core) compositions (Kushiro and Nakamura 1970) showing (shaded field) a range of acceptable MgO contents in the residual liquid at acceptable K_D values consistent with 15–25% crystallization. Results with $K_D < 0.20$ are deemed unacceptable. Alternative sets of K_D and MgO contours result from the choice of pigeonite or augite as the pyroxene crystallizing.

TABLE VI

Probabilities, ϕ, of observed approaches to cotectic behaviour being coincidence

Source	Laki 1783 Iceland			Kilauea Hawaii			Apollo 11				Apollo 12	
							Ophitic		Intersertal		12038	12064
P^*		3			3		6		6		4	5
X		0.10			0.05		0.10		0.25		0.05	0.10
C	4	6	10	4	6	10	7	9	7	9	8	8
ϕ	0.028	0.033	0.22	0.007	0.022	0.071	0.00005	0.0004	0.0046	0.027	0.0036	0.0027

Calculated from the relationship

$$\phi = \sum_{P=P^*}^{P=C} \frac{(C-1)!}{(P-1)!\,(C-P)!} \cdot X^{(P-1)}\,(1-X)^{(C-P)}$$

where C = number of significant components, P^* = number of crystalline phases present, X = weight fraction present as crystals. This relationship is approximately true provided liquid loci are not strongyl curved.

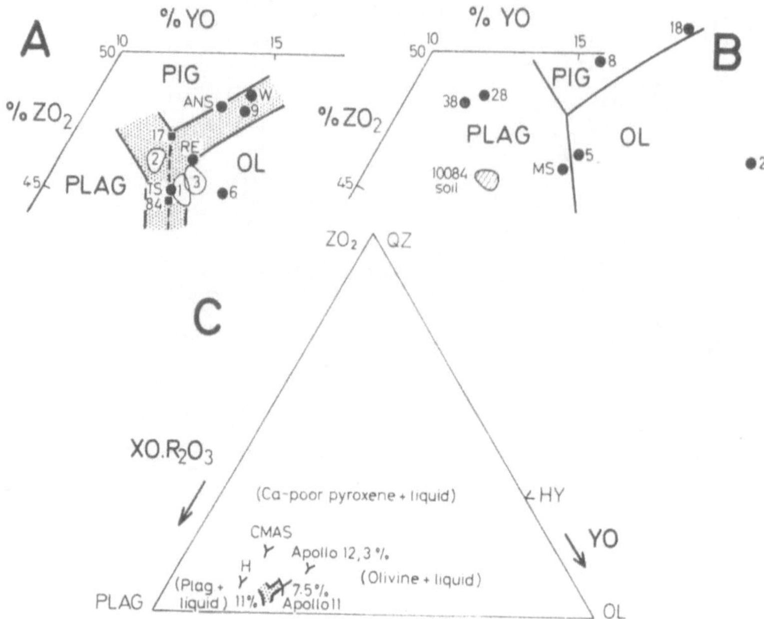

Fig. 8. Projection (O'Hara, 1968; Jamieson, 1970) from diopside into the plane plagioclase-olivine-hypersthene (quartz) showing the boundaries between the primary liquidus phase volumes of plagioclase, pigeonite and olivine as deduced from experimental and petrographic data for (A) synthetic and natural compositions of high TiO_2 content, ($\sim 11 \pm 1\%$) making no allowance for TiO_2-rich phase crystallization at liquidus, and for (B) intermediate TiO_2 content ($\sim 7.5 \pm 1.5\%$). Compositions 2, 5, 6, 8, 9, 18, 28, 38 from O'Hara *et al.* (1970); ANS from Anderson *et al.* (1970); MS from Muan and Schairer (1971); RE from Ringwood and Essene (1970); TS from Tuthill and Sato (1970); W from Weill *et al.* (1970). Numbered fields enclose analyses of lavas of ophitic groups 1, 2; intersertal group 3, and 10084 soil as in Biggar *et al.* (1971, Figure 8), while 17, 84 mark the analysed residual liquids of rocks 10017, 10084 at the $8.75 \pm .5$ TiO_2 level (Biggar *et al.*, 1971, Table VI). Figure 7c compares the deduced cotectic liquid compositions with the analogous cotectic liquids in the Apollo 12 rocks at $\sim 3\%$ TiO_2; in Hawaiian tholeiitic lavas (H); and in the iron, sodium and titanium free system CaO-MgO-Al_2O_3-SiO_2 (CMAS).

coincidence is measured in parts per ten thousand to parts per hundred at best (Table VI). When the new average compositions of Apollo 11 lavas (Prinz *et al.*, 1971), and the near coincidence of cotectic temperature and eruption temperature is taken into account these probabilities decline still further.

(7) Cotectic liquid compositions are best established by bracketing, i.e. by finding closely related compositions lying in different primary liquidus phase volumes. This has been done for Apollo 11 compositions (O'Hara *et al.*, 1970, Figures 2, 3; Biggar *et al.*, 1971, Figures 8, 9; and this paper Figures 8, 9 and 10). A bracket on plagioclase entry was established by two laboratories from synthetic compositions resembling Apollo 12 lavas (Figure 5). Cotectic character can also be established by analysing the cotectic liquids and other experimentally produced phases, and thence proving low percentage crystallinity by calculation, as done here (Tables II, III and V). Least

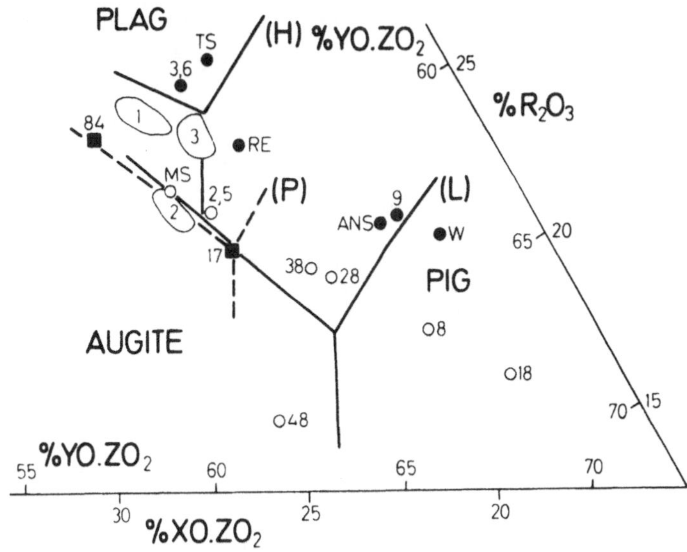

Fig. 9. Part of the projection from olivine into the plane hypersthene ($YO.ZO_2$)-wollastonite ($XO.ZO_2$)-R_2O_3, treating the same data as in Figure 8 to deduce the apparent boundaries between the plagioclase, augite and pigeonite crystallization fields at saturation with olivine, when TiO_2 contents are high ($\sim 11 \pm 1\%$; boundaries (H), solid circle data points and natural rock, fields 1,3; drawn without allowance for TiO_2-rich phase crystallization), or low ($\sim 7.5 \pm 1.5\%$, boundaries L, open circle data points), and the preferred position (P) at simultaneous saturation with TiO_2-rich oxides and the silicates, $TiO_2 \sim 8.75 \pm 0.5\%$ from probe analyses of experimentally produced residual liquids, and ophitic lavas group 2 (compare Biggar *et al.*, 1971, Figure 9). 10084 soil (not shown) falls in the plagioclase field of the (L) boundaries between MS, 84 and 1.

satisfactory is the indirect method of estimating, visually or otherwise, the percentage of crystals present on reaching cotectic crystallization, because it carries no internal check on consistency, and is susceptible to subjective and systematic errors in arriving at the estimates, particularly in fine grained products (plate I; D, E).

Close approach to cotectic composition is often, but not necessarily, accompanied by the appearance of many crystalline phases within a narrow temperature interval. However, neither the size of the temperature interval over which a number of crystalline phases appear, nor the size of the temperature interval between liquidus and solidus are criteria of closeness to cotectic composition. O'Hara *et al.* (1970) were incorrectly criticized (Ringwood, 1970, p. 6455) directly or by implication for the use of such criteria, but neither the abstract from which Ringwood drew his quotations, nor the full statement (O'Hara *et al.*, 1970, pp. 695, 704), is written in terms which excuse such a misrepresentation. The only authors who, to our knowledge, have attempted to utilize the size of the temperature interval as a criterion in this respect are Green *et al.* (1971, p. 608), in search of criticisms of Biggar *et al.* (1971), and Ringwood and Green (1971).

(8) The techniques used in the Edinburgh laboratory did not reveal a spuriously close approach to cotectic character in the random compositions represented by

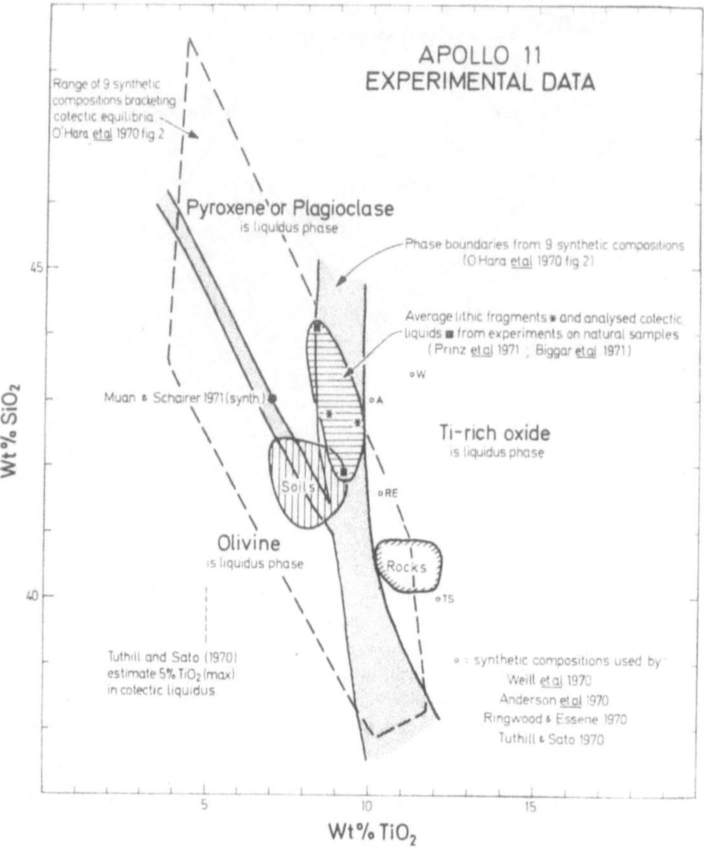

Fig. 10. Simplistic SiO_2 v. TiO_2 representation of liquidus fields, average rock compositions and cotectic liquid compositions encountered in Apollo 11 natural rocks and synthetic simulations. (N. B. Use of a higher oxygen fugacity caused plagioclase to appear as liquidus phase at the centre of the figure).

erroneous Apollo 12 analyses (Biggar *et al.*, 1971), nor has it revealed it in a suite of Reunion lavas known to depart somewhat from low pressure cotectic liquid compositions in their natural crystallisation (D. J. Humphries, in press). These techniques do, however, determine cotectic liquid compositions and temperatures identical (e.g. Figure 11) with those reported by other laboratories also engaged in long-term programs of atmospheric pressure experiments in systems of industrial and geological interest. With these techniques we also observe a close approach to cotectic character in the Laki fissure eruption (J. D. Bell and D. J. Humphries, in press) whose petrography demonstrates eruption with three phenocryst species when as little as 5% crystalline. Similar demonstrations of objectivity and applicability to natural circumstances are available for the technique used by Muan *et al.* (1971), these authors also remarking on the low degree of crystallinity at plagioclase entry.

(9) Available data from relevant synthetic systems (Figure 12) supports the bound-

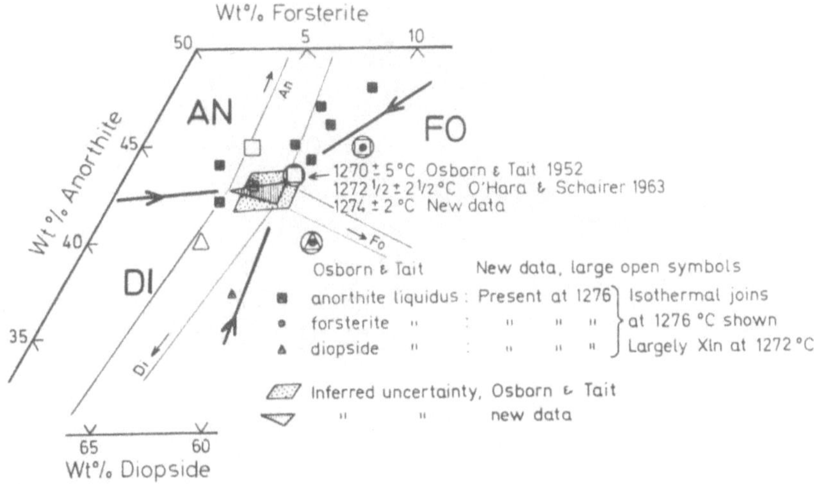

Fig. 11. Part of the plane anorthite-diopside-forsterite comparing atmospheric pressure phase boundaries (Osborn and Tait, 1952), and results for five compositions made and run in the Edinburgh laboratory illustrating close interlaboratory agreement on cotectic composition and temperature.

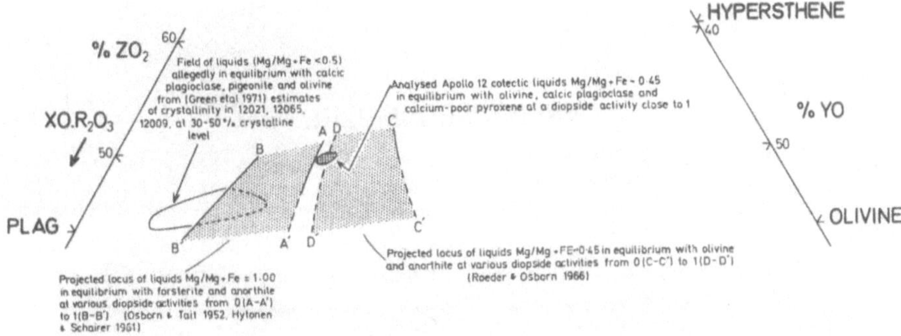

Fig. 12. Projection from diopside (as in Figure 6) into the plane plagioclase-olivine-hypersthene (silica) showing the loci of liquids in equilibrium in synthetic systems with olivine, plagioclase and one or two pyroxenes. Diopside activity has been set at zero in the olivine-plagioclase-silica plane, and at 1 when calcium-rich clinopyroxene coexists with olivine and plagioclase.

ary marking onset of plagioclase crystallisation found by Biggar *et al.* (1971) and Muan *et al.* (1971) in Apollo 12 compositions.

(10) Interpretation of the phase equilibria and lava chemistry turns on the *average* rock composition at the Apollo 12 site (Figures 13, and 14). The fundamental implications of this decision for lunar petrogenesis are illustrated by Figure 15. The hypothesis that 12009 or 12040 represent the parental magma at this site is untenable. The compositions of the average fines; average basaltic lithic fragments from the solls; impact melted glass fragments least affected by alkali depletion; and even the weighted average of the large rocks distributed for analysis (liable to bias towards extreme

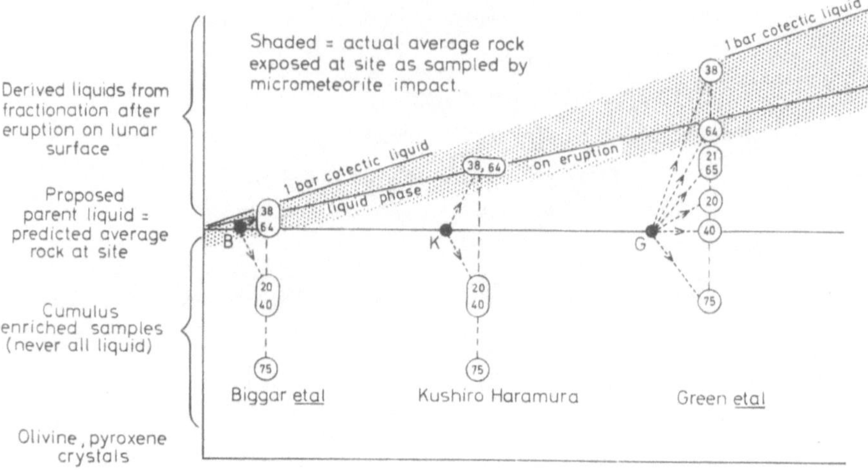

Fig. 13. Graphical summary of alternative interpretations of Apollo 12 site petrogenesis and their implications. The status of 12009, also regarded by Green *et al.* (1971) as possible parental liquid is discussed in the text. The alternative status of Apollo 12 rock samples (indicated by final two digits of sample numbers) in the three interpretations is apparent.

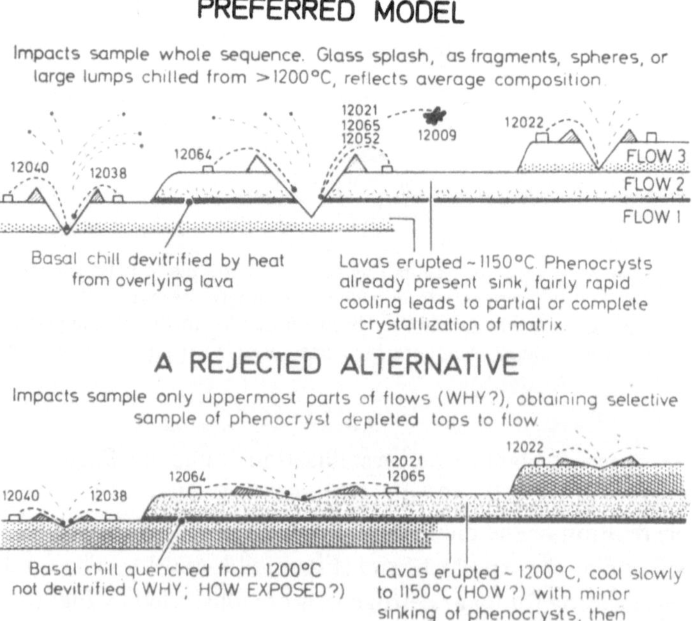

Fig. 14. Schematic cross-sections corresponding with two extreme alternatives from Figure 13, illustrating their implications for Apollo 12 site geology. Individual globs of impact melt depart from average rock composition, but on average short travelled material reflects rock average.

compositions) all fall much closer to the analysed low pressure cotectic liquids (Figure 16, Table VII) than to 12009 or 12040. The average composition of materials exposed near this site resembles that of sparsely porphyritic rocks such as 12038, and 12064, identified by Biggar *et al.* (1971) as the parental magmas, close to cotectic composition at atmospheric pressure and truly cotectic at a slightly higher pressure and volatile content that prevailed within the lava lake.

(11) The phenocrysts probably formed prior to eruption from the lava lake and precipitated at a pressure higher than atmospheric. Changes in alkali content and oxygen fugacity accompanied eruption, and atmospheric pressure experiments do not exactly reproduce the conditions of phenocryst formation. Too much should not be

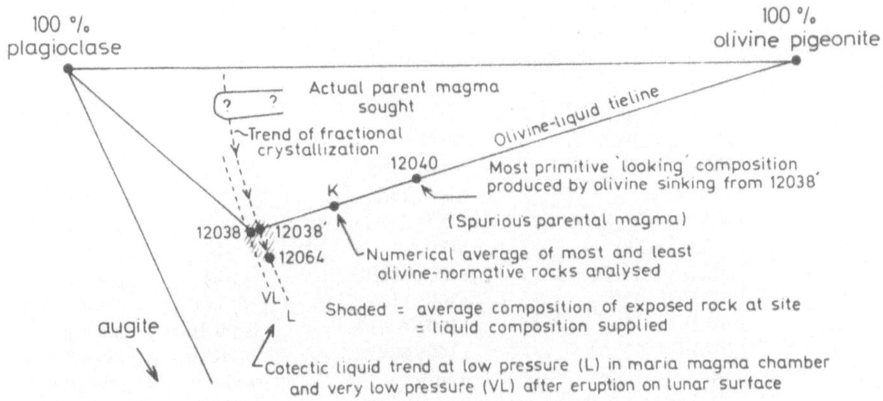

Fig. 15. Simplified phase diagram illustrating the experimental significance of the alternative interpretations of Figure 13, and their implications for studies of the lunar interior.

TABLE VII

	SiO_2	TiO_2	Al_2O_3	Cr_2O_3	FeO	MgO	CaO	Na_2O	K_2O
(1) Average basaltic lithic fragments[a]	47.2	3.0	14.0	tr	15.8	8.0	10.9	0.5	0.2
12038 rock[b]	47.1	3.3	13.0	–	17.7	6.6	11.4	0.7	0.1
(2) K 10 glass[a]	40.3	3.5	7.1	0.6	23.2	16.6	8.0	tr	tr
12040 rock[b]	43.9	2.7	7.4	0.7	20.8	16.1	7.9	0.2	tr
(3) K 9 glass[a]	43.1	4.4	8.5	0.3	23.0	11.0	9.4	0.4	0.2
12009? impact vitrophyre[c]	45.0	2.9	8.6	0.5	21.0	11.5	9.4	0.2	0.1

[a] Keil *et al.* (1971); [b] Kushiro and Haramura (1971); [c] Compston *et al.* (1971).

(1) Comparison of average of 28 basaltic lithic fragments in Apollo 12 soil with rock 12038 (suggested parental cotectic liquid to eruptive sequence) showing similarity; average fragments are *not* similar to 12040, 12009, below, in SiO_2, Al_2O_3, Cr_2O_3, MgO or CaO.

(2) Comparison of impact melted glass fragment in Apollo 12 soil, with rock 12040, a coarsely crystalline potential target rock formed by crystal accumulation.

(3) Comparison of impact melted glass fragment in Apollo 12 soil with rock 12009, a largely glassy sample with thermal history more akin to that of impact melts than to the great majority of the igneous rock samples.

TAI

Summary of experimental data relevant to entr

Run No.	Samples	Temperatures °C	Time (h)	Starting material as loaded
A. *Reversals*				
031	12021, 38, 52, 64, 65	1144; 1137	5; 5	rock powder
032	12018, 20, 38, 40	1167; 1156	5; 5	rock powder
033	12018, 20, 38, 40	1166; 1153	4; 12[a]	rock powder
035	12018, 20, 38, 40	1162; 1153	$4\frac{1}{2}$[g]; 16	rock powder
111	12020	1167; 1156	5; 5	glass (1340, $1\frac{1}{3}$ hr)[b,]
112	12020, 12038	1166; 1153	4; 12[a]	glass (1340, $1\frac{1}{3}$ hr)[b,]
117	12018, 20, 38, 40	1162; 1153	$4\frac{1}{2}$[g]; 16	glass (1330 $1\frac{1}{4}$ hr)[b,]
B. *Syntheses from 'different' starting materials (pseudoreversals)*				
030	12038	1167	5	rock powder
034	12018, 20, 38, 40	1153	$6\frac{1}{2}$	rock powder
036	12018, 20, 38, 40	1153	72	rock powder
113	12020, 12038	1300; 1153	24; $6\frac{1}{2}$	glass[b, c]
114	12020 only	1340; 1153	$1\frac{1}{3}$; $6\frac{1}{2}$	glass[b, c]
116	12018, 20, 38, 40	1153	72	glass[b, c]
015	12022, 38, 52	1129	$1\frac{1}{2}$	rock powder
016	12022, 38, 52	1300; 1129	1; $1\frac{1}{2}$	glass[b, c]
017	12022, 38, 52	1300; 1129	1; $1\frac{1}{2}$	glass + 5% rock
109	12052	1300; 1041	24; 0.16	glass[b]
110	12020, 38, 52, 64	1340; 1041	1; 0.16	glass[b]
?	n. d.	n. d.	n. d.	glass[c, d]
?	n. d.	n. d.	n. d.	glass[c, d]
?	n. d.	n. d.	n. d.	devitrified glass[e]
?	n. d.	n. d.	n. d.	devitrified glass[e]
C. *Charges converted to liquid, or mainly liquid then temperature lowered substantially*[f]				
006	18, 20, 21, 22, 38, 40, 52, 64, 65	1250; 1154	1; $4\frac{1}{2}$	rock powder
013	18, 20, 21, 22, 38, 40, 52, 64, 65	1306; 1110–1127	1; 1	rock powder
101	12022, 38, 64	1228; 1121	1,24	rock powder
102	12022, 38, 64	1225; 1119	1,1	rock powder
104	12022, 38, 64	1320; 1116	2,1	rock powder
Muan and Schairer (1971)	Apollo 11 synthetic	1200; 1090	n. d.	devitrified glass
Roedder and Weiblen (1970)	Inclusions in olivine	> 1200; 1125/ 1065	many hours	glass

All Edinburgh experiments carried out at fO_2 of Fe/FeO equilibrium.

[a] held 3 h each at 1159, 1156 during temperature reduction.

[b] glass prepared, quenched, then reloaded. Picritic samples still retain some olivine.

[c] glass may have devitrified while coming up to temperature (see 109, 110).

VIII

plagioclase and use of glass as starting material

Comments

plagioclase destroyed except in 12038; renucleated in all.
plagioclase destroyed; renucleated 20, 38 only.
plagioclase destroyed; renucleated 12038 only.
plagioclase destroyed; renucleated in all four samples (less than in 034)

plagioclase renucleated
plagioclase renucleated in both
plagioclase renucleated in all samples

trace olivine only
well distributed scarce plagioclase
well distributed scarce plagioclase
plagioclase present in 12038
plagioclase present
plagioclase present, pyroxene with multiple twinning
same phases present in all;
plagioclase less abundant in products from
glass bearing starting materials
variably devitrified on reaching 1041 °C;
charges spent 9 mins > 1000 °C; 17 mins > 700 °C;
plagioclase in some. See text.
Green *et al.* (1971) "confirmed temperature
of disappearance of
plagioclase": claimed as reversal.

No plagioclase; large pyroxene prisms regenerated, showing curving cracks, some polysynthetic twinning, in 12020, 40, 65
A few crystals free from cracks had straight extinction

No plagioclase; little pyroxene in 38, 64 (2–50 °C below
equilibrium entry).
No plagioclase in 12022 (contrast 016)
No plagioclase in 12022,64 (10–20 °C below equilibrium)
No plagioclase in any (10–45 °C below equilibrium)
No plagioclase (90 °C below equilibrium entry)

No plagioclase or pyroxene renucleation.

[d] glass prepared by few minutes heating at high temperature (Green, pers. comm).
[e] conditions of devitrification not stated.
[f] runs were kept in the furnace and not quenched, merely lowered to the second temperature quoted, and finally quenched.
[g] 2½ h at 1160, then 2 h at 1162, lowered in 1½ h to 1153.

made out of small differences in composition between experimentally produced liquidus phases and natural phenocryst cores, least of all when the phase in question is in reaction relationship with the liquid and effects leading to excessively magnesian residual compositions can come into play (O'Hara, 1963 p. 42). Those effects are likely to be most marked in rocks such as 12065 where the ratio of olivine resorbed to olivine precipitated is highest, and the occurrence of the most magnesian olivines within the reaction product (pigeonite; Green *et al.*, 1971, p. 605) underlines the need for caution.

(12) A liquid such as 12064 *can* be derived by pigeonite + plagioclase dominated fractionation from liquid such as 12038 (Biggar *et al.*, 1971, Table VII). Green *et al.* (1971, p. 611) have not examined the data sufficiently closely in reaching their conclusions.

(13) 12009 existed as a liquid plus c. 9% olivine crystals at the lunar surface, and even that olivine may be a quenching product if Green *et al.* (1971 p. 601) are correct in identifying it as skeletal. This vitrophyre was very rapidly cooled from temperatures of c. 1200 °C or possibly considerably higher, whereas most other rocks at the site were cooled more slowly, and from temperatures of c. 1140–1170 °C. The distinctive thermal history of 12009 * also characterises impact melted glasses, and impact melted glass spatter is prominent in samples 12017, 12030 and 12054 (LSPET 1970). Impact-melted glasses approaching 12009 and 12040 in composition have been reported (Figure 16, Table VII) as required by our interpretation of 12009 as a fragment of

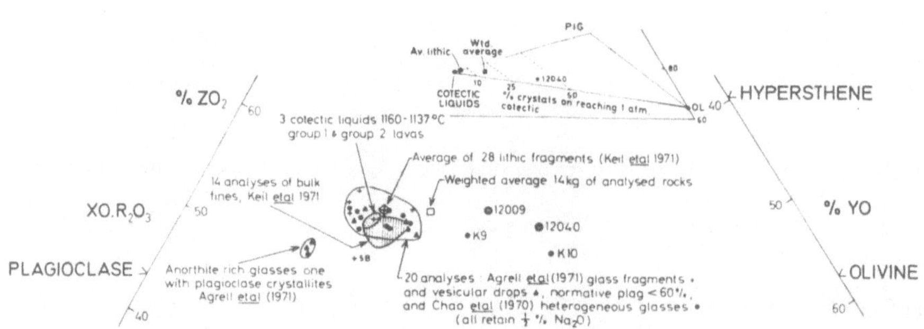

Fig. 16. Projection from diopside into the plane plagioclase-olivine-hypersthene (quartz) (analogous to Figures 6, 12) contrasting Green *et al.*'s (1971) suggested parental magmas 12009, 12040 with the analysed experimental cotectic liquids (Biggar *et al.*, 1971); the compositions of glass fragments and particles little affected by alkali, or silica depletion, taken to approximate to the average rock composition exposed near the site; and the weighted average composition of 14 kg. of 16 rock samples analysed from this site. The average of 28 lithic basaltic fragments in the soil from this site is also shown.

Two impact melted glass spheres conspicuously poorer in SiO_2 or richer in potential ferromagnesian minerals than the majority are indicated, to support the contention that vitrophyres such as 12009 may represent larger globs of impact melted phenocryst-rich target rock.

* The rapid heating and cooling make preservation of pre-existing chemistry, even of alkalis, unremarkable.

TABLE IX

Details of furnace assembly calibration

Carried out between run and run		Calibration point	Width of bracket solid-liquid °C	Date carried out 1970/1971	Correction to be applied temperature from e.m.f. (a) (b)
(initial calibrations)		Au	5.1	13 Oct. 19 Oct.	+0.8
		L2S	3.2	21 Oct. 22 Oct.	−2.1
003	004	L2S	3.5	5 Nov. 5 Nov.	−1.0 (c)
including	009	L2S	2.0 (d)	25 Nov.	−1.6
including	010	Au	Solid (e)	26 Nov.	(+0.8)
019	020	L2S	2.0 (d)	17 Dec.	−1.5
020	021	L2S	Solid (f)	24 Dec.	(−1.5)
after 024		Diopside	2.7	4, 5 Feb.	+1.5
Furnace used for other calibrated work at higher temperatures					
initial calibration, then					
101	102	L2S	(d)	22 Apr. 1971	+3.6
102	103	Au	2.3	30 Apr. 1 May	+9.5
104	105	L2S	4.4	10 May 11 May	+9.6
used for 106, 107, 108 (glass production) and the replaced by new thermocouple junction.					
initial calibrations		L2S	2.0 (d)	19 May	+4.0
		L2S	(g)	20 May	
then used for 030, 031, 109, 110, 032, 111, 033, 112		Au	3.0	24 May 25 May	+6.6
033	034	L2S	(g)	4 June	+4.0

(a) These are the corrections applied to the measured e.m.f. to calculate the temperatures quoted by O'Hara *et al.* (1970), Biggar *et al.* (1971), and this paper.

(b) A further linear interpolation between 0°C at 1064.4°C and 4.9°C at 1208°C must be added to these quoted temperatures to convert them to the International Practical Temperature Scale of 1968. This further correction mainly results from the redetermination of the melting point of lithium metasilicate as 1208°C on IPTS 1968 (Biggar, in press).

(c) Based on the wide bracket (3.5°C). The next line of the table involved a smaller bracket and for greater accuracy the value −1.0 should be replaced by the value between −1.6 (the later and better result) and −2.1 (the value used earlier).

(d) In melting interval known to be less than 2° (Biggar and O'Hara, 1969).

(e) Thermocouple drift cannot have been greater than 0.8°.

(f) Confirmed solid at 3.1° below the melting interval result of line 4. Therefore downward drift of thermocouple emf has not been greater than 3.1° in fact previous correction of −1.5 continued in use.

(g) Confirmed solid at 2.0°C below result of 19th May.

(h) Control precision is better than ±0.5°C (Biggar and O'Hara, 1969); accuracy relative to stated scales is ±2.5°C. Contrast Green *et al.* (1971b) who give ±5.0°C, ±10°C respectively.

impact-melted target material from the phenocryst-enriched base of one of the flows (Figure 14).

(14) Results presented here (Table VIII) and by Biggar *et al.* (1971) were obtained from close temperature brackets in thermally explored and properly calibrated furnaces

(Biggar and O'Hara, 1969; and Table IX, Figure 17 this paper), at controlled, known, and calibrated oxygen fugacities (Figure 5; Biggar *et al.*, 1971, Table II), in containers which do not react with the charges (Table X). All temperatures quoted in this and our previous work must be *raised* 2–5 °C to correspond with the International Practical Temperature Scale (1968) and our revised melting point for Li_2SiO_3 (1208 °C; IPTS, 1968) N.B. The emf of a thermocouple indicates nothing more than the temperature of the thermocouple tip (Figure 17). Sample preparation techniques do not

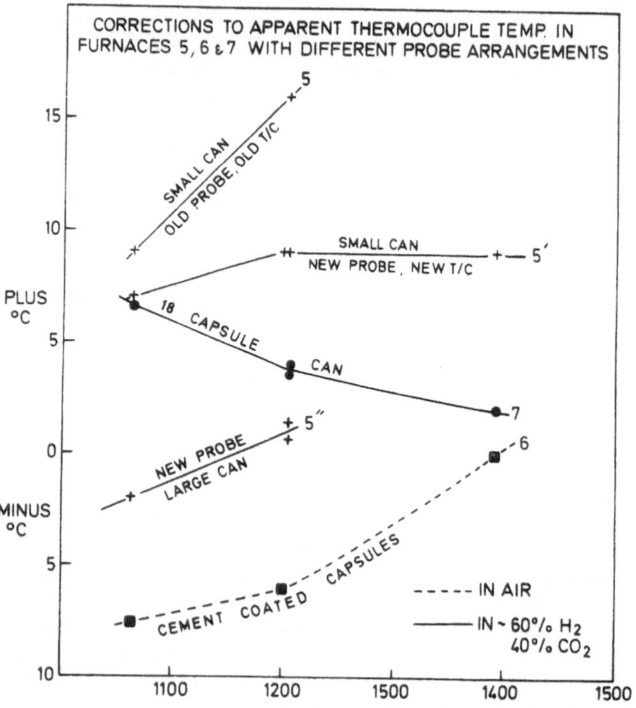

Fig. 17. Plot of corrections to be added to apparent thermocouple temperature as a function of temperature in furnaces of the type used in the studies at Edinburgh.

TABLE X

	SiO_2	TiO_2	Al_2O_3	FeO	MgO	CaO
12038 Rock	47.1	3.3	13.0	17.7	6.6	11.4
12038 1179 °C, 5 h 0% crystals	46.9	3.4	12.4	17.6	6.8	11.2
12038 1160 °C, 5 h <5% crystals[a]	46.8	3.5	12.3	17.7	6.6	11.2
12064 Rock	46.4	4.1	10.5	19.9	6.4	11.7
12064 1160 °C, 5 h <5% crystals	46.4	4.0	10.6	19.4	6.6	11.6

[a] cotectic liquid, plagioclase, olivine, spinel and two pyroxenes.

Table X, comparing glass analyses from all liquid, or nearly all liquid experiments with bulk rock composition, to illustrate composition control, absence of sampling errors conducive to premature plagioclase crystallization, and absence of iron loss. Data from Biggar *et al.* (1971) Table V.

bias the charge compositions, least of all towards excessive potential plagioclase (Table X).

The quenching and microprobe data refer to assemblages of homogeneous equilibrium phases, the apparent variation (< ±2% Fo in olivine; < ±3% Wo, 2–4% En in pyroxene; and ±1–4% major oxides in the glasses, equivalent to ± c. 1% olivine in the norm) not greatly exceeding the precision of the method, and arising principally from the difficulty of ensuring that the volume sampled is located exclusively within one phase (c.f. Gibb, 1971) even in the relatively coarse-grained products from rock powder (plate I). No such check could be carried out on the much finer grained products from glass. Total reconstruction of the loaded natural minerals is of course

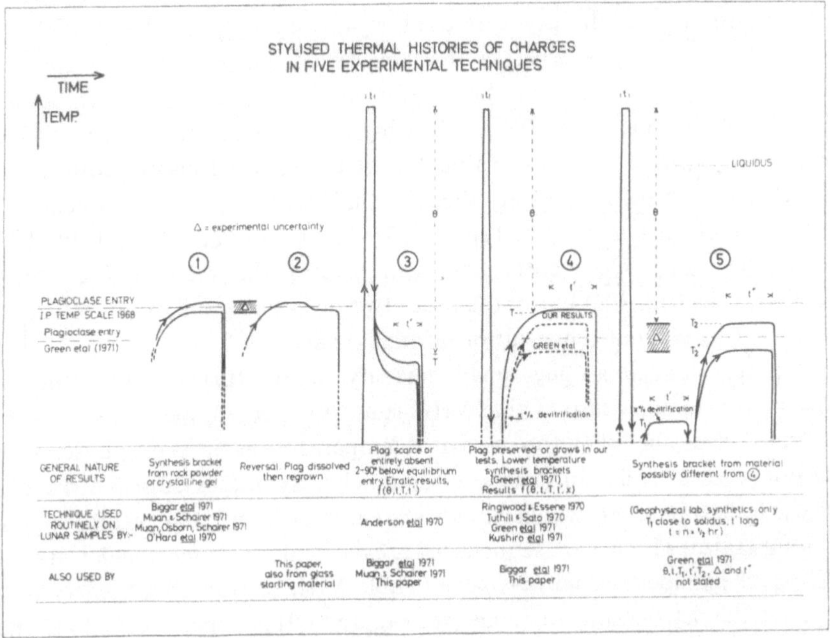

Fig. 18. Diagrammatic representation of thermal sequences used in various experiments.

TABLE XI

Microprobe analysis of cotectic plagioclase-bearing liquids in reversal experiments 1162°C/1153°C showing similarity of results from two starting materials

	SiO_2	TiO_2	Al_2O_3	Cr_2O_3	FeO	MnO	MgO	CaO	Na_2O	K_2O	Total
12038/035 (powder)	47.27	3.49	12.12	0.25	17.35	0.31	6.31	11.39	0.62	0.13	99.24
12038/117 (glass)	47.10	3.51	12.47	0.19	16.68	0.28	6.41	11.70	0.62	0.13	99.09
12040/035 (powder)	46.53	4.64	11.33	0.19	16.94	0.22	6.95	12.11	0.46	0.13	99.50
12040/117 (glass)[a]	47.21	4.06	10.46	0.28	17.17	0.25	7.94	11.42	0.33	0.13	99.25

[a] Minute crystal size may mean that some pyroxene lay within sample area.

proved by the reduced Al_2O_3 content of the pyroxenes, and the destruction of the zoning.

(15) The only effects likely to arise from the presence of 0.25 bar H_2O in our furnace atmospheres are accelerated reaction rates, and a slight movement of the total liquid composition towards the primary phase volumes of olivine and pyroxene, which would diminish, not enhance, the appearance of cotectic character.

(16) Thermal sequences used by various workers on Apollo samples or synthetic simulations in different types of experiment are summarised in Figure 18. The entry of plagioclase has been 'reversed'* (Figure 18, sequence 2) in group 2 rocks 12021, 12052, 12064 and 12065 between 1144 and 1137 °C (Table VIII, run 031). Plagioclase has been demonstrated to be present in products of group 1 rocks 12018, 12020, 12038 and 12040 at 1153 °C (Figure 18, sequences 1 and 4), irrespective of whether rock powder or largely glass charges were used as starting material (Table VIII, runs 034, 036, 113, 114, and 116) or whether the runs lasted 6 h or 72 h. The appearance of plagioclase in these four rocks has been reversed (Figure 18, sequence 2) between 1162 and 1153 °C (Table VIII, runs 035, 117) using rock powder and largely glass starting materials. Microprobe analyses (Table XI) of the residual glasses conform, in each case, to the trend previously established by analyses of liquids developed in 5 h syntheses (Figure 18, sequence 1) at 1160, 1137 °C (Biggar *et al.*, 1971, Table V). In two of the samples plagioclase entry has been reversed between 1167 °C and 1156 °C (Table VIII, runs 032, 033) but in the other two even this small extent of superheating was sufficient to prevent renucleation of plagioclase.

(17) Charges loaded as glass may devitrify in our furnace atmospheres before arriving at run temperature (Table VIII, runs 109, 110). Experiments loading glass or devitrified glass may, therefore, arrive at temperature as the same phase assemblages and do not necessarily demonstrate even so much as syntheses from contrasted starting materials. Experiments in which liquid is created at high temperature, then lowered to run temperature (Figure 18, sequence 3) are particularly prone to delayed or deferred nucleation of plagioclase and pyroxene (Table VIII, section C), an effect beautifully displayed in the petrography of 12009 (see Figure 5). It is imperative to limit the superheating used in the first stage of glass preparation for 'reversal' experiments. Other experiments (Table VIII, runs 015, 016, 017) indicate that charges loaded as glass may yield less plagioclase than is obtained from rock powder under the same conditions. The two effects discussed last might be enhanced when dry atmospheres are used.

(18) Nucleation from glass does not always lead immediately, nor even rapidly, to stable phase assemblages (Ostwald, 1897; Volmer, 1939). Products are sensitive to the previous thermal history of the glass (Table VIII, sections B, C). When glasses have been prepared by the several (half-hour) fusions at very high temperature necessary to guarantee homogeneity (Schairer, 1959), and are then used without prior devitrification or annealing, difficulties must be expected. The accumulated experience of the past is worth repeating lest it be altogether overlooked,

* An absolutely valid reversal of an equilibrium involving a phase (e.g. glass) capable of inheriting a metastable internal structural state from the previous thermal treatment is very difficult to devise.

Nearly all silicate glasses readily undercool. Unless crystals are present in the glass before attempts are made to study phase changes with temperature erratic results and metastable equilibrium may be obtained.... In most silicate systems the determination of the appearance of a crystalline phase at equilibrium practically requires that that phase be present in the starting material. (Schairer 1959).

In glasses which devitrify to olivine, pyroxenes and plagioclase, metastable assemblages may form in minutes and persist for weeks unless devitrification is carried out close to the solidus (O'Hara and Schairer, 1963); these metastable crystals may subsequently enter into metastable crystal-liquid equilibria.

Summarising, then, we consider the assertions and criticisms by Ringwood (1970) and Green *et al.* (1971) concerning the atmospheric pressure phase equilibria to be unsubstantiated and somewhat authoritarian. We welcome any discussion of atmospheric pressure data on lunar or terrestrial materials which is based on fully reported, quantitative data obtained by well described techniques in calibrated equipment, preferably not relying solely on glass as a starting material.

Sparsely porphyritic rocks like 12038, 12064 *are* close in composition to low pressure cotectic liquids precipitating plagioclase, spinel, two pyroxenes, and olivine. The plausibility of the choice of these compositions as the parental magmas to local post-eruption differentiation at the Apollo 12 site must be judged by the reader in the light of two facts: The average rock exposed to meteorite bombardment resembles the sparsely porphyritic lavas (Figure 16): When erupted these lavas were in that early stage of crystallization which permits the creation of a small proportion of compositions such as 12040, 12052 by sinking of the dense crystals through the low viscosity liquid.

Whether vitrophyres, like 12009, with its extreme composition and distinctive thermal history which allies it directly with impact melted materials, can safely be interpreted as anything but impact 'splash' must also be judged by the reader.

Even neglecting the *certainty* of alkali and other element losses by volatilization during eruption, which must be restored before any lunar lava can be treated as a primary magma, it is our opinion that *none* of the rocks taken from the Apollo 11 or 12 sites represents even a volatile depleted primary magma from the deep lunar interior; the only identifiable parental magmas bear the stigmata of composition control by low pressure fractional crystallisation.

7. Summary of Conclusions

Protohypersthene*-crystallizing basic lavas from the Ocean of Storms were erupted as low pressure cotectic residual liquids from an underlying magma body, already as much as 80–98% crystallized. The lunar maria are differentiated lava lakes, filled by rock types familiar among the achondrites, floored by early metal precipitates which form mass concentrations, and capped by lava flows representing second-stage extrusion of the late residual liquids of the fractionation sequence. The distinctive geochemical and petrographic characteristics of this magmatic kindred relative to terrestrial ex-

* See addendum.

perience were controlled by volatilization losses during eruption and crystallization. The igneous events leading to maria formation may have been very early and triggered by major impacts.

Acknowledgements

The Natural Environment Research Council (U.K.) supported this research. We thank G. M. Brown, J. E. Dixon, P. L. Roeder and W. C. Storey for their critical reading of this paper.

References

Agrell, S. O., Long, J. V. P., and Reed, S. J. B.: 1971, 'A comparison of Glasses from the Apollo 11 and 12 Missions' (unpublished).

Anderson, A. T., Crewe, A. V., Goldsmith, J. R., Moore, P. B., Newton, J. C., Olsen, E. J., Smith, J. V., and Wyllie, P. J.: 1970, *Science* **167**, 587.

Biggar, G. M. and O'Hara, M. J.: 1969, *Mineral. Mag.* **37**, 1.

Biggar, G. M., O'Hara, M. J., Peckett, A., and Humphries, D. J.: 1971, *Proc. Second Lunar Sci. Conf., Geochim. Cosmochim. Acta Suppl. 2*, Vol. 1, 617.

Brown, G. M., Emeleus, C. H., Holland, J. G., Peckett, A., and Phillips, R.: 1971, *Proc. Second Lunar Sci. Conf. Geochim. Cosmochim. Acta Suppl.* 2, Vol. 1, 583.

Brown, G. M. and Peckett, A.: 1971, *Nature* **234**, 262.

Chapman, D. R. and Scheiber, L. C.: 1969, *J. Geophys. Res.* **74**, 6737.

Chao, E. C. T., Boreman, J. A., Minkin, J. A., James, O.B., and Desborough, G. A.: 1970, *J. Geophys. Res.* **75**, 7445.

Clarke, D. B.: 1970, *Contrib. Mineral. Petrol.* **25**, 203.

Compston, W., Chappell, B. W., Arriens, P. A., and Vernon, M. J.: 1970, *Proc. Apollo 11 Lunar Sci. Conf., Geochim. Cosmochim. Acta Suppl. 1*, Vol. 2, 1007.

Compston, W., Berry, H., Vernon, M. J., Chappell, B. W., and Kaye, M. J.: 1971, *Proc. Second Lunar Sci. Conf., Geochim. Cosmochim. Acta Suppl. 2*, Vol. 2, 1471.

De Maria, G., Balducci, G., Guido, M., and Piacente, V.: 1971, 'Mass Spectrometric Investigation of the Vaporization Process of Lunar Samples', *Second Lunar Sci. Conf., Geochim. Cosmochim. Acta Suppl. 2*, Vol. 2, 1367.

Duke, M. B. and Silver, L. T.: 1967, *Geochim. Cosmochim. Acta* **31**, 1637.

Essene, E. J , Ringwood, A. E., and Ware, N. G.: 1970, *Proc. Apollo 11 Lunar Sci. Conf., Geochim. Cosmochim. Acta Suppl. 1*, Vol. 1, 385.

Gibb, F.: 1971, *Contr. Mineral. Petrol.* **30**, 103.

Green, D. H., Ringwood, A. E., Ware, N. G., Hibberson, W. O., Major, A., and Kiss, E.: 1971a, *Proc. Second Lunar Sci. Conf., Geochim. Cosmochim. Acta Suppl. 2*, Vol. 1, 601.

Green, D. H., Ware, N. G., Hibberson, W. O., and Major, A.: 1971b, *Earth Planetary Sci. Letters* **13**, 85.

Haggerty, S. E. and Meyer, H. A. O.: 1970, *Earth Planetary Sci. Letters* **9**, 379.

Hytonen, K. and Schairer, J. F.: 1961, *Yearb. Carnegie Inst. Wash.* **60**, 125.

James, O. B. and Jackson, E. D.: 1970, *J. Geophys. Res.* **75**, 5793.

Jamieson, B. G.: 1970, *Mineral. Mag.* **37**, 537.

Keil, K., Prinz, M., and Bunch, T. E.: 1971, *Proc. Second Lunar Sci. Conf., Geochim. Cosmochim. Acta Suppl. 2*, Vol. 1, 319.

Kushiro, I. and Haramura, H.: 1971, *Science* **171**, 1235.

Kushiro, I. and Nakamura, Y.: 1970, *Proc. Apollo 11 Lunar Sci. Conf. Geochim. Cosmochim. Acta Suppl. 1*, Vol. 1, 607.

Kushiro, I., Nakamura, Y., Kitayama, K., and Akimoto, S.: 1971, *Proc. Second Lunar Sci. Conf., Geochim. Cosmochim. Acta Suppl. 2*, Vol. 1, 481.

LSPET (Lunar Sample Preliminary Examination Team): 1970, *Apollo 12 Preliminary Science Report*, NASA, Washington, D.C.

LSPET (Lunar Sample Preliminary Examination Team): 1971, *Apollo 14 Preliminary Science Report*, NASA, Washington, D.C.

Marvin, U. B., Reid, J. B. Jr., Taylor, G. J., and Wood, J. A.: 1972, *Abstracts 3rd Lunar Sci. Conf.*, NASA Houston.

Muan, A., Hauck, J., Osborn, E. F., and Schairer, J. F.: 1971, *Proc. Second Lunar Sci. Conf., Geochim. Cosmochim. Acta Suppl. 2*, Vol. 1, 497.

Muan, A. and Schairer, J. F.: 1971, *Yearb. Carnegie Inst. Wash.* **69**, 243.

O'Hara, M. J.: 1961, *Mineral. Mag.* **32**, 848.

O'Hara, M. J.: 1963, *Amer. J. Sci.* **261**, 32.

O'Hara, M. J.: 1968, *Earth Sci. Rev.* **4**, 69.

O'Hara, M. J., Biggar, G. M., Richardson, S. W., Jamieson, B. G., and Ford, C. E.: 1970, *Proc. Apollo 11 Lunar Sci. Conf., Geochim. Cosmochim. Acta Suppl. 1*, Vol. 1, 695.

O'Hara, M. J., Biggar, G. M., Richardson, S. W., Ford, C. E., and Jamieson, B. G.: 1971, *Phys. Earth Planetary Int.* **4**, 181.

O'Hara, M. J. and Schairer, J. F.: 1963, *Yearb. Carnegie Inst. Wash.* **62**, 107.

Osborn, E. F. and Tait, D. B.: 1952, *Amer. J. Sci.* **250A** (Bowen vol.) 413.

Ostwald, W.: 1897, *Z. Phys. Chem.* **22**, 289.

Prinz, M., Bunch, T. E., and Keil, K.: 1971, *Contrib. Mineral. Petrol.* **32**, 211.

Ringwood, A. E.: 1970, *J. Geophys. Res.* **75**, 6453.

Ringwood, A. E. and Essene, E.: 1970, *Proc. Apollo 11 Lunar Sci. Conf., Geochim. Cosmochim. Acta Suppl. 1*, Vol. 1, 769.

Ringwood, A. E. and Green, D. H.: 1972, 'Crystallization of Plagioclase in Lunar Basalts and Its Significance' (in press).

Roedder, E. and Weiblen, P. W.: 1970, *Proc. Apollo 11 Lunar Sci. Conf., Geochim. Cosmochim. Acta Suppl. 1*, Vol. 1, 801.

Roeder, P. L. and Emslie, R. F.: 1971, *Contr. Mineral. Petrol.* **29**, 275.

Roeder, P. L. and Osborn, E. F.: 1966, *Amer. J. Sci.* **264**, 428.

Schairer, J. F.: 1959, in J. O'M. Bockris, J. L. White, and J. D. Mackenzie (eds.), *Physico-Chemical Measurements at High Temperatures*, Butterworths, 117.

Scoon, J.: 1971, *Proc. Second Lunar Sci. Conf., Geochim. Cosmochim. Acta Suppl. 2*, Vol. 2, 1259.

Strom, R. G.: 1971, *Mod. Geol.* **2**, 133.

Tuthill, R. L. and Sato, M.: 1970, *Geochim. Cosmochim. Acta* **34**, 1293.

Urey, H. C., Marti, K., Mei-Kao, L., and Hawkins, J. W.: 1971, *Proc. Second Lunar Sci. Conf., Geochim. Cosmochim. Acta Suppl. 2*, Vol. 2, 987.

Volmer, M.: 1939, *Kinetik der Phasenbildung*, Steinkopff, Dresden.

Wager, L. R. and Brown, G. M.: 1968, *Layered Igneous Rocks*, Oliver and Boyd, Edinburgh.

Wasserburg, G. J. and Papanastassiou, D. A.: 1971, *Earth Planetary Sci. Letters* **13**, 97.

Weill, D. F., McCallum, I. S., Bottinga, Y., Drake, M. J., and McKay, G. A.: 1970, *Proc. Apollo 11 Lunar Sci. Conf., Geochim. Cosmochim. Acta Suppl. 1*, Vol. 1, 937.

Addendum. Green *et al.* (1971b) doubt our finding of protohypersthene on the strength of their experiments carried out under appreciably different conditions. The facts to bear in mind are: (i) Biggar *et al.* (1971) analysed inverted protohypersthene from 12040 ($Ca_7Mg_{67}Fe_{26}$) and observed inverted orthorhombic phase present in the range 1160–1200 °C, and 1135–1160 °C in 12021. Eruption temperatures lay within these ranges, hence inverted protohypersthene is to be expected in the rocks, all other factors being equal. (ii) Ross *et al.* (1971, Abstract, *2nd Lunar Science Conf.*) reported inversion of calcium-poor pigeonite cores ($Ca_9Mg_{63}Fe_{28}$) from 12021 to orthorhombic structures (? space group) at 1160–1215 °C, in their heating experiments. (iii) Hollister *et al.* (1971, *Proc. 2nd Lunar Sci. Conf.*, p. 529–557) report orthorhombic cores in pigeonite from 12065. Aspects of their analysis 9-1-6 ($Ca_6Mg_{66}Fe_{28}$) are compared with our inverted protohypersthene (12040/4) TiO_2 0.43, 0.43; Al_2O_3 1.01, 1.08; Cr_2O_3 0.77, 0.46 (misprint in earlier text); CaO 2.90, 3.39; FeO 17.96, 16.72; MgO 23.50, 23.81. (iv) Brown *et al.* (1971), Gay *et al.* (1971, *Proc. 2nd Lunar Sci. Conf.*, p. 377–392) suggest that curving cracks in pigeonite phenocryst cores of 12040, 12065 may reflect the $C2/c \rightarrow P2_1/c$ inversion in pigeonite; Ross *et al.* (1971) did not report $C2/c$ structure formation but Prewitt *et al.* (1971, *Proc. 2nd Lunar Sci. Conf.*, p. 59) did for material of this general composition at 1000 °C. A few cracks are present in the larger prisms of pigeonite in our runs, and the $C2/c \rightarrow P2_1/c$ inversion may also have occurred. Gay *et al.* (1971) comment on unusual pyroxene trends in 12065, matched by a terrestrial protohypersthene precipitating rock. (v) Yoder, Tilley and Schairer (1964, *Yearb. Carnegie Instn. Wash.* **63**, 121–129) reported an orthorhombic calcium-poor pyroxene stable at these relevant compositions and temperatures, identified as protohypersthene by the same authors (1963, *Yearb. Carnegie Instn. Wash.* **62**, 84–95) with detailed discussion (1963, p. 90). (vi) Low-calcium-pyroxene cores comparable with Biggar *et al.*'s (1971) analysed protohypersthene form a smaller part of the rock powder as loaded, than do the large crystals with curving cracks which we interpret as inverted protohypersthene. Furthermore, had our protohypersthene been undissolved residual phenocryst cores it should have been more abundant in, not absent from, the lower temperature experiments. (vii) Finally we refer to results in Table VIII, run 006 which seem to eliminate worries about unmelted residual material.

PETROCHEMISTRY AND CHEMICAL FEATURES OF LUNAR GLASSY SPHERULES

R. TRIGILA

University of Rome, Rome, Italy

Abstract. The major elements of glassy spherules from samples 12001.73, 12057.60 and 12070.37 of lunar fines have been determined by electron probe analysis. The bulk chemistry of these spherules (which are between 200 μm and 62 μm in size) falls in most cases within the range of gabbro-peridotite and gabbro-anorthosite.

In individual spherules, the grade of chemical homogeneity is highly variable and can be related to the possible presence of small inclusions or sometimes to slight concentration gradients evident near the peripheries of the particles. These features, together with the presence of vesicles of different size in association with strong alkali enrichment along the boundaries of such cavities can account for relationships between the chemistry of the particles and their genetic history.

1. Introduction

The origin and composition of the glasses from samples of the fines collected in the Apollo missions have already been described by several investigators.

Comprehensive studies on the morphological and chemical characteristics of these materials have been published by Agrell *et al.* (1971), Chao *et al.* (1970) and McKay *et al.* (1970). Among others, Isard (1971) and Carusi *et al.* (1972) made models to explain on a theoretical basis the formation of glassy particles by impact of meteoritic bodies on the lunar surface.

It is generally accepted that the great majority of lunar glasses are impactoclastic in origin rather than representative of liquids of magmatic origin. Basic considerations on the relationships between abundance, size, shape and chemical composition of the glasses found in the lunar regolith seem to support this view. In particular, the widespread occurrence of shock structures, of glass corresponding in composition to single mineral phases and of a fine powdery coating on the particles investigated can be well explained by impact events. The occasional presence of flow structures and of iron droplets inside the glasses are further confirmation of very rapid melting and subsequent quenching over a range of physical parameters far from those characteristic of magmatic processes. Researches on pit-craters have led to the conclusion that the chemistry of the glassy particles is much more affected by the size of the impacting body in relation to the grain size of the target rock than by a selective distribution of major elements as a consequence of the impact event.

On the other hand it is also possible that some homogeneous glasses have originated through volcanic activity. Unfortunately no investigations have as yet been possible on lunar glass spheres of undoubted volcanic origin. Some indications may however be obtained from a comparative study of terrestrial glasses and lavas from the same eruption. Such work is now in progress on material collected during the latest eruption of Etna and the chemistry is in some ways similar to that of the lunar rocks.

Urey and Runcorn (eds.), The Moon, 165–179.

TABL

Electron Probe Analys

Sample	40	81	12	7	12070.37 10	37	16	41	102
Shape	lent.	sphere	sphere bowl	sphere	sphere	fragm.	sphere	sphere	half sphere
Opt. char.	isotr.	isotr.	isotr.	semi opaque	isotr.	isotr.	isotr.	isotr.	anisotr.
Colour	light brown	brown	greenish	dark grey	color-less	light green	brown	light green	light green
n	1.65	n.d.	1.65	1.61	1.60	1.62	1.67	1.61	n.d.
SiO_2	39.1	42.6	43.0	44.7	47.1	47.3	48.5	49.4	40.7
TiO_2	3.2	2.8	3.0	2.1	0.3	2.0	2.9	2.2	0.7
Al_2O_3	15.3	12.5	13.6	16.1	23.9	15.7	14.5	13.8	31.0
FeO	16.0	17.7	16.3	11.7	4.0	10.4	12.0	10.4	4.2
MgO	12.6	12.4	10.6	10.2	8.8	9.8	10.7	9.9	8.6
CaO	12.3	11.1	10.9	14.1	15.5	13.0	11.4	9.4	14.9
K_2O	0.2	0.2	0.4	0.2	0.1	0.4	0.1	0.8	0.2
Na_2O	0.2	0.2	0.4	0.3	0.7	0.4	0.1	0.9	0.3
Total	98.9	99.5	98.2	99.4	100.4	99.0	100.2	96.8	100.6

2. Petrochemical Considerations

The major elements of 19 glassy particles extracted from samples 12001.73, 12057.60, 12070.37 of lunar fines have been determined by electron microprobe to supply data on bulk compositions and the relationships between such compositions and those of some elements of unique distribution possibly as a consequence of the glass formation processes.

Table I gives values for the oxides of Si, Ti, Al, Fe, Mg, Ca, K and Na, together with some physical characteristics. Silica ranges almost regularly from 39 to 49% with the exception of the ellipsoid 145 in which it is 54.2%. Regular variations can be observed for MgO (8.5–15.2%), for CaO (9–15.5%) and to a minor extent for the alkalies, which normally range from 0.1 to 0.5%, but in sphere 41 K_2O is 0.8% and Na_2O is 0.9%. Most of the values for alumina lie between 8.6 and 16.1%, the few remaining, between 20.7 and 31.0%. This kind of distribution can be observed for FeO and TiO_2, their values being bimodal: a major group for the high values and a minor one for the low values. In Figure 1 where FeO has been plotted against Al_2O_3 a negative correlation is clearly shown.

Niggli basis bonds and standard katanorms are plotted in Tables II and III respectively. Most of the glasses are characterized by the predominance of the normative paragenesis feldspar-pyroxene-olivine and lie within the field of gabbro composition. A few glasses (102, 10, 133) have a very high content of normative plagioclase and can be equated with gabbro-anorthosite to anorthosite types while others (40, 81,

I

of Apollo 12 glasses

		12001.73					12057.60		
107	105	104	103	172	138	133	142	144	145
tear drop	sphere	ellip.	sphere	tear drop	half ell.	tear drop	ellip.	ellip.	half ell.
semi opaque	semi opaque	isotr.	isotr.	isotr.	isotr.	anisotr.	isotr.	opaque	isotr.
dark brown	brown	brownish	light brown	brown	brown	dark grey	reddish oran.	grey	brown
n.d.	n.d.	1.67	1.67	n.d.	n.d.	n.d.	n.d.	n.d.	n.d.
40.9	40.9	41.1	42.4	44.0	44.9	47.4	47.8	49.4	54.2
3.4	2.8	2.9	3.0	2.8	2.6	2.1	4.0	2.9	2.3
12.0	14.0	15.4	12.2	12.8	10.5	20.7	8.6	11.2	13.0
16.7	14.2	15.5	16.8	15.6	17.4	8.0	17.9	15.2	12.5
11.1	15.2	12.9	12.0	11.9	10.9	8.5	11.3	10.3	8.8
12.5	9.5	11.0	10.0	12.7	12.7	13.0	9.0	10.6	10.1
0.5	0.2	0.2	0.2	0.3	0.1	0.2	0.2	0.4	0.4
0.5	0.1	0.1	0.2	0.5	0.1	0.4	0.3	0.4	0.6
97.6	97.0	99.1	97.0	100.6	99.2	100.3	99.1	100.4	101.9

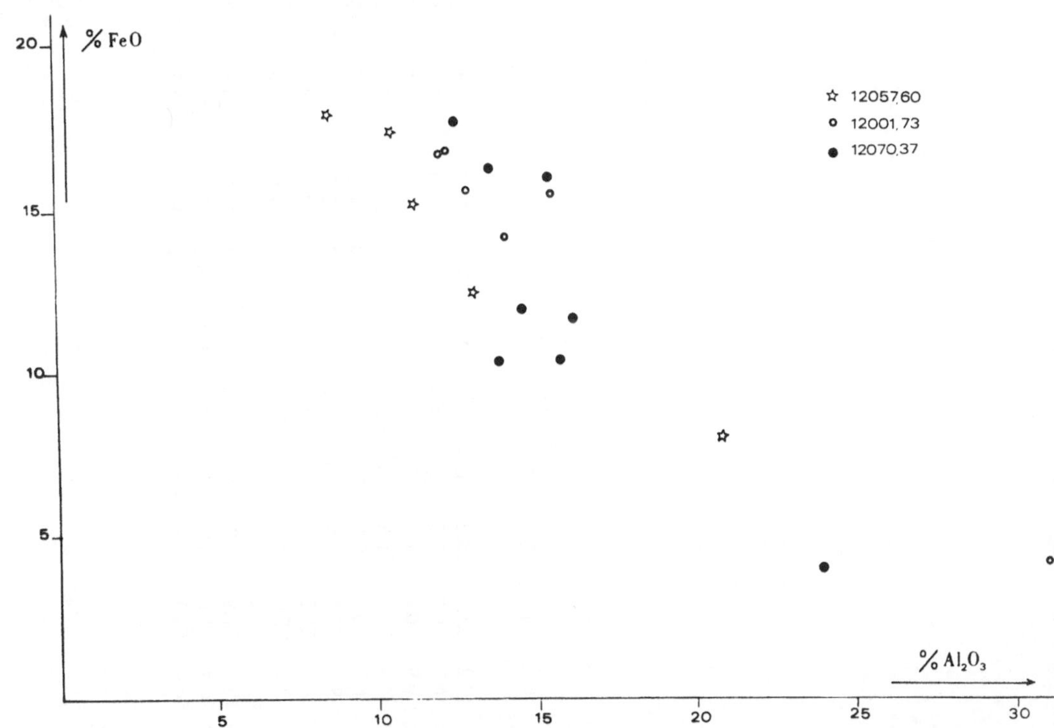

Fig. 1. Diagram showing the inverse correlation of Al_2O_3 with FeO. Glasses analyses from different samples plot with the same distribution.

TABLE II

Basis bonds and QLM values of the analyzed glasses

	12070.37								12001.73						12057.60				
Sample	40	81	12	7	10	37	16	41	102	107	105	104	103	172	138	133	142	144	145
Q	19.2	21.9	24.2	26.8	32.5	29.9	31.4	32.7	29.6	20.2	21.2	21.9	23.5	22.7	24.5	33.2	27.9	30.0	36.5
Kp	0.7	0.7	1.5	0.9	0.4	1.6	0.4	3.0	0.7	1.9	0.7	0.7	0.7	1.0	0.3	0.9	0.8	1.4	1.5
Ne	0.9	0.9	1.7	1.4	2.8	1.9	0.4	3.9	1.2	2.2	0.4	0.4	0.9	2.0	0.3	1.7	1.3	1.7	2.5
Cal	24.9	20.2	21.6	25.7	36.9	24.4	23.7	20.2	43.2	18.6	23.1	25.2	20.3	19.7	17.7	32.8	13.7	17.2	19.5
Cs	6.3	6.9	6.1	8.5	4.7	7.6	5.4	4.5	5.3	10.2	3.1	4.1	5.6	9.4	10.8	3.0	7.2	7.6	5.4
Fa	19.0	21.1	19.8	13.8	4.5	12.3	14.2	12.7	4.8	20.3	17.0	18.4	20.6	18.3	21.0	9.4	21.7	18.1	14.6
Fo	26.7	26.3	22.9	21.4	18.0	20.8	22.5	21.4	14.7	24.1	32.5	27.2	26.2	24.9	23.5	17.6	24.5	21.9	18.4
Ru	2.3	2.0	2.2	1.5	0.2	1.5	2.0	1.6	0.5	2.5	2.0	2.1	2.2	2.0	1.9	1.4	2.9	2.1	1.6
Q	19.2	21.9	24.2	26.8	32.5	29.9	31.4	32.7	29.6	20.2	21.2	21.9	23.5	22.7	24.5	33.2	27.9	30.0	36.5
L	26.4	21.8	24.9	28.0	40.1	27.9	24.5	27.1	45.2	22.6	24.2	26.3	21.9	22.7	18.3	35.4	15.8	20.4	23.4
M	54.4	56.3	50.9	45.2	27.4	42.2	44.1	40.2	25.2	57.2	54.6	51.8	54.6	54.6	57.2	31.4	56.3	49.6	40.1

TABLE III

Standard katanorm of the analyzed glasses

	12070.37								12001.73						12057.60				
Sample	40	81	12	7	10	37	16	41	102	107	105	104	103	172	138	133	142	144	145
Q	0	0	0	0	0	0	1.0	1.8	0	0	0	0	0	0	0	0	0	0.6	8.1
Or	1.2	1.2	2.5	1.4	0.7	2.6	0.6	4.9	1.2	3.1	1.2	1.2	1.2	1.6	0.5	1.6	1.3	2.4	2.4
Ab	0.1	1.4	2.9	2.4	4.7	3.2	0.7	6.5	0.9	3.7	0.7	0.7	1.5	3.3	0.5	2.8	2.2	2.9	4.2
Ne	0.8	0	0	0	0	0	0	0	0.7	0	0	0	0	0	0	0	0	0	0
An	41.4	33.7	36.0	42.8	61.6	40.7	39.6	33.6	72.0	31.0	38.4	41.9	33.8	32.9	29.4	54.6	22.8	28.7	32.5
Wo	8.5	9.1	8.1	11.4	6.2	10.1	7.2	6.0	0	13.6	4.1	5.5	7.4	12.5	14.4	4.0	9.6	10.1	7.2
Hy	0	20.2	22.0	18.4	6.0	26.4	18.9	16.9	0	6.7	16.1	10.9	27.5	18.1	28.0	12.5	29.0	4.1	19.5
En	0	0	0	2.7	10.6	18.7	30.0	28.6	0	0	0	0	0.6	0	7.1	22.1	30.9	29.1	24.5
Fa	19.3	6.0	3.4	0	0	0	0	0	4.7	15.3	5.0	10.4	0	4.7	0	0	0	0	0
Fo	26.7	26.4	22.9	19.4	10.0	6.8	0	0	14.7	24.1	32.5	27.3	25.8	24.9	18.2	1.0	1.3	0	0
Sp	0	0	0	0	0	0	0	0	5.3	0	0	0	0	0	0	0	0	0	0
Ru	2.3	2.0	2.2	1.5	0.2	1.5	2.0	1.6	0.5	2.5	2.0	2.1	2.2	2.0	1.9	1.4	2.9	2.1	1.6

107, 105, 104) show olivine predominating over pyroxene and a shift to gabbro-peridotite composition.

A comprehensive picture of the bulk chemistry of the analysed glasses comes from the representation of the Niggli QLM triangle where Q is the basis quartz and L and M represent the leucocratic and melanocratic basis values respectively (Figure 2). If we now consider an impactoclastic origin for most of the particles which Mason and Melson (1970) reported as mainly produced by melting of various proportions of pyroxene, plagioclase and olivine or of preexisting glasses, one would expect to find some trend towards these mineral compositions. According to Chao *et al.* (1970) two different trends can be recognized in the Apollo glasses: a major one connecting the gabbro compositions with the anorthositic and a minor one in the direction of increasing or decreasing olivine. This situation may be recognized for the group of glasses analyzed here: almost all the points are distributed within the field PFR below the saturation line PF. The occurrence of the oversaturated composition in glass 145 could also be explained by a melting of a high silica parent rock.

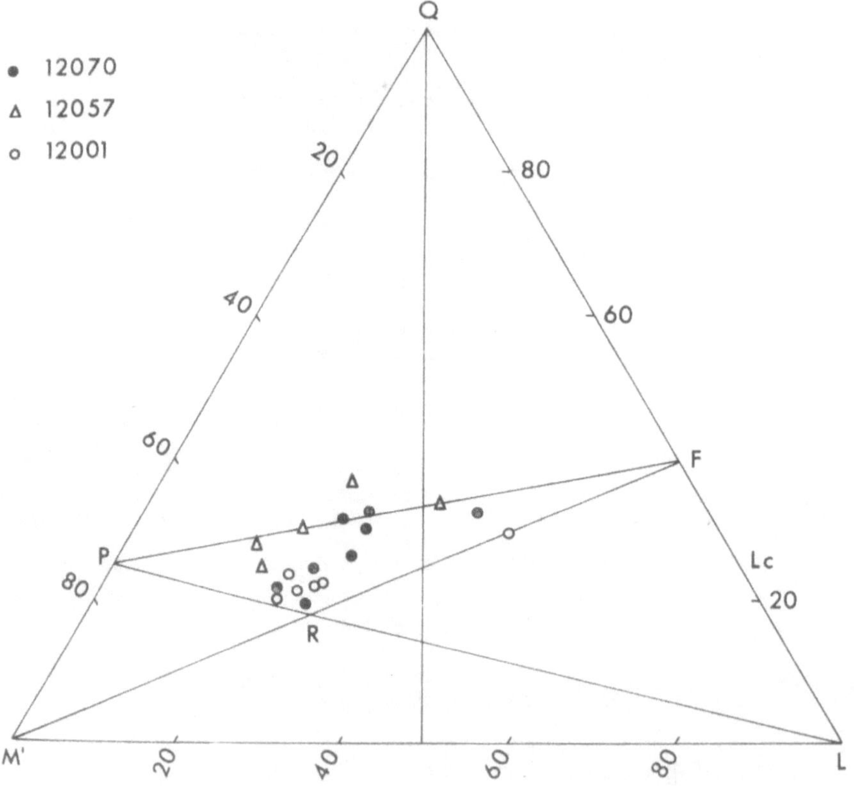

Fig. 2. The Q-L-M diagram. Almost all the glasses analyses plot in the field P-F-R which corresponds to the major paragenesis pyroxene-feldspar-olivine. The line Q-50%L divides the gabbro rich compositions from the anorthosite rich compositions.

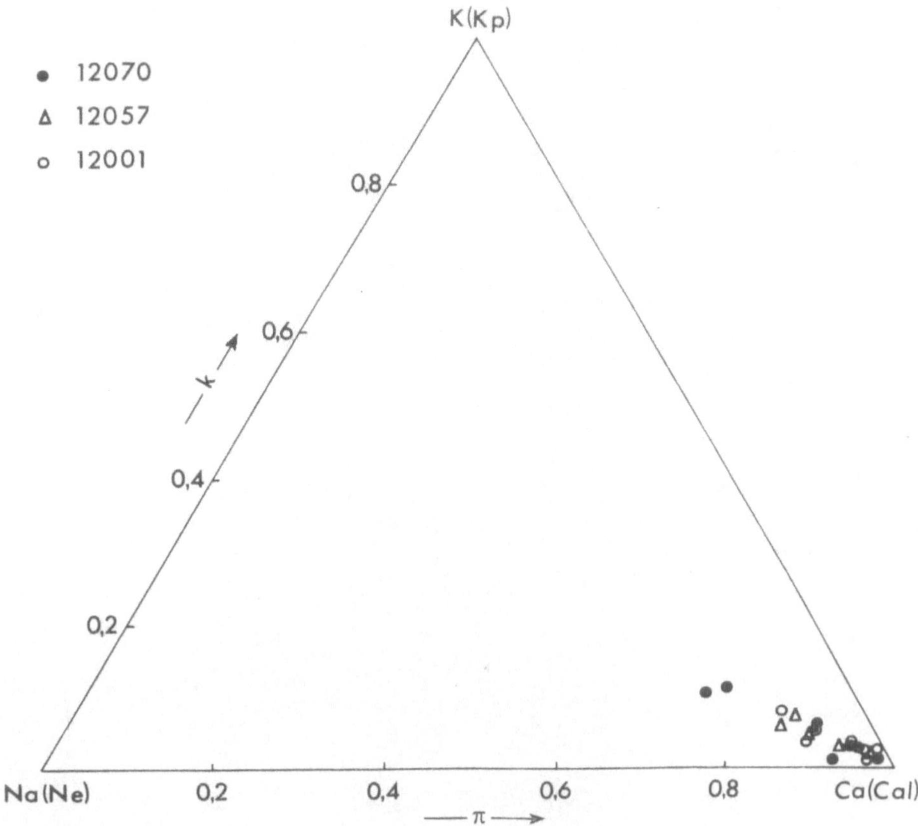

Fig. 3. The feldspar diagram shows the balance between normative orthoclase and albite along the line 0.4k.

Figure 3 gives the composition of the normative feldspar. Most of the values are in the range 85–95% An, in good agreement with the feldspar compositions already observed for Apollo 12 rocks and lithic fragments. In several cases, some sort of balance can also be observed between the relative amounts of the Na and K normative feldspars. This does not seem to fit in with the general prevalence of normative albite over normative orthoclase in lunar rocks or lithic fragments (Keil *et al.*, 1971). As depletion in alkalies has been inferred for a certain group of glasses as a consequence of their formative processes (Agrell *et al.*, 1971) a differential volatilization of alkalies through the impact mechanism may account for the modification of the K/Na ratio in the normative feldspars.

The relationships between Mg, Fe and Ca in the mafic components is shown in Figure 4 where the Fo, Fa and Cs values of the Niggli basis bonds have been plotted. As already noticed by Chao *et al.* (1970), there is a balance of the Mg and Fe silicates for most of the glasses.

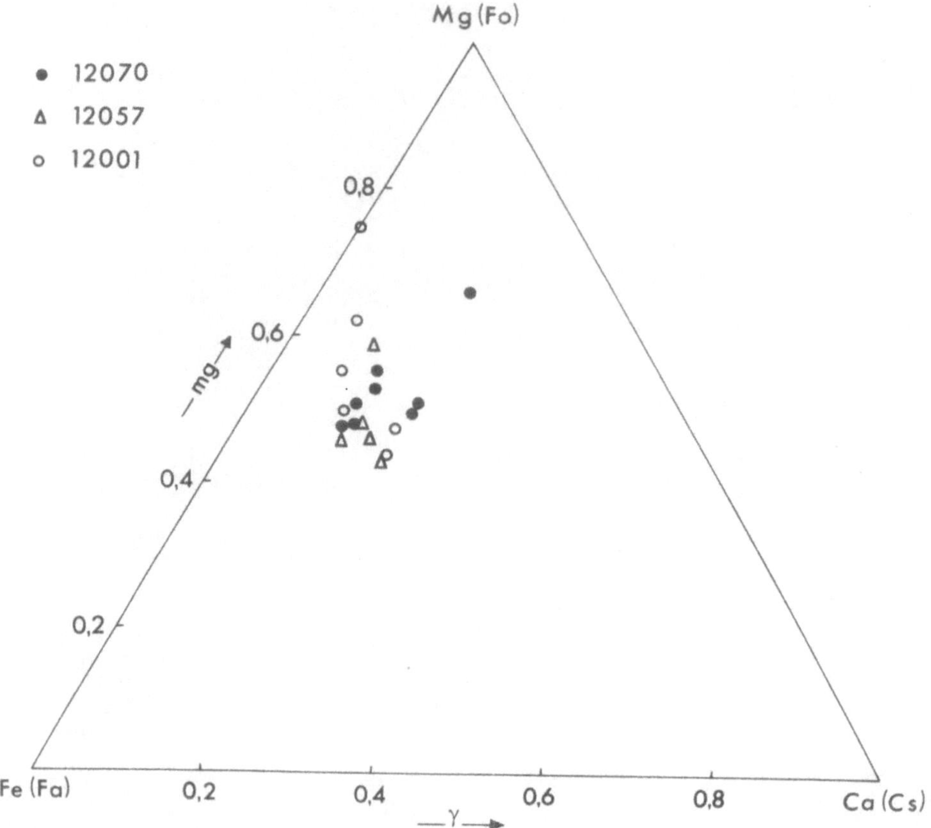

Fig. 4. The Mg-Fe-Ca triangle illustrates the quantitative relationships of Fe, Mg, and Ca silicates in the mafic minerals.

3. Homogeneity

The grade of homogeneity of lunar glassy particles is highly variable and can be related to the presence of crystalline or glassy inclusions or, on a smaller scale, to slight concentration gradients sometimes evident near the boundary of a particle.

An approximate estimate of the range of variability in homogeneity of the analyzed glasses is shown in Table IV, where all the coefficients of variation of Si, Al, Fe and Ca X-ray intensities are given. Counting statistics are less than 2% for Si, less than 3% for Al, 1–3% for Fe and 1–3% for Ca. On these grounds, about half of the analyzed glasses are shown to be definitely inhomogeneous.

Let us consider for instance the profile given in Figure 5, referring to a traverse on a pseudo-equatorial section of an ellipsoidal glass fragment. The sharp increase in the silica content near the left edge is associated with the almost total depletion in Al_2O_3, FeO, CaO, MgO and is evidence of a nearly pure silica inclusion. Some other glasses, such as sphere 10 (Figure 6) show only smaller scale inhomogeneities, but

TABLE IV

Coefficients of variation of Si, Al, Fe, Ca X-ray intensities

	12070.37								12001.73						12057.60				
Sample	40	81	12	7	10	37	16	41	102	107	105	104	103	172	138	133	142	144	145
Si	1.8	14.6	3.9	8.0	1.2	6.1	1.9	2.2	3.7	6.8	2.9	3.1	2.6	3.5	2.5	11.3	2.7	4.2	33.6
Al	7.4	0.8	6.1	8.3	6.1	5.0	3.4	25.9	6.8	15.5	4.7	9.3	3.8	16.0	5.9	39.9	5.3	11.6	11.8
Fe	4.7	5.7	6.4	4.3	19.9	14.5	5.0	10.6	18.2	37.8	3.3	2.6	4.1	8.2	2.7	29.9	2.5	18.6	31.9
Ca	3.6	10.2	4.1	2.6	5.3	3.7	4.1	7.7	6.1	15.6	4.5	33.4	3.6	4.6	4.1	17.3	2.5	14.4	26.5

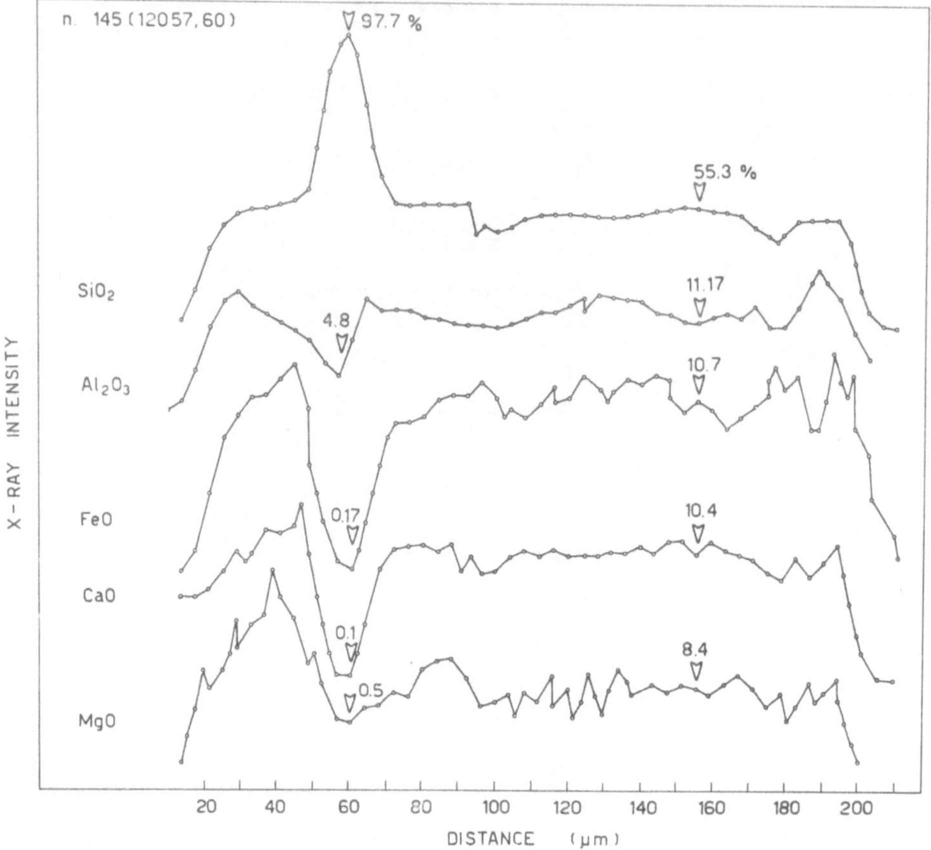

Fig. 5. Slow scan profiles of the ellipsoidic glass No. 145. The profiles are from count rates at different scales and the figures on each profile indicate concentration variations.

at the margin, the alkalies reach levels greater than the reported value for the bulk composition. Other anomalies are evident for Fe, Ti and probably Mg, which are also enriched at the boundary. According to Cooper *et al.* (1971) such inhomogeneities can represent a diffusion zone between two adjoining mineral compositions. The symmetrical enrichment at both sides of the traverse may suggest another possible explanation for alkalies involving condensation in the impact cloud.

Some of the analysed glasses show a more or less pronounced vesiculation (Figure 7). Drops and other less regular shapes are in general more vesicular than perfect spheres and this has been explained by Agrell *et al.* (1970) by perfect melting, degassing and homogenization for the latter, while the former have undergone nearly complete melting, partial degassing and homogenization.

Glass 172 is ovoid in shape, inhomogeneous and vesiculated. A partial traverse on a pseudo-equatorial section shows a marked inhomogeneity near the margin on the left side and a large vesicle on the right side (between 50 and 70 μm in distance)

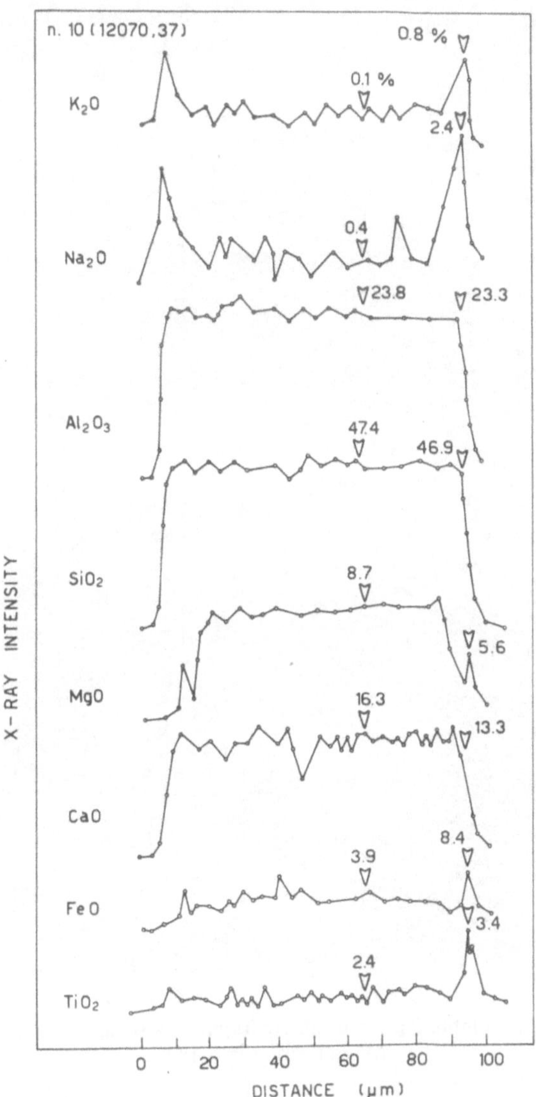

Fig. 6. Slow scan profile of the glassy sphere No. 10. The particle has a diameter of about 100 μm, is fairly homogeneous and free from vesiculation. The profiles are from count rates at different scales; the figures on each profile indicate concentration variations.

associated with a strong increase in the alkali content (Figure 8). A similar pattern can be seen in Figure 9. The glass consists of a sphere of gabbroic composition and appears hollow, the inside being occupied by a single vesicle. In this glass the increase in alkali along the vesicle wall appears very high: something like 25 times (K_2O) and 40 times (Na_2O) the recorded values over the whole section of the glassy particle. The experimental studies of Blonder *et al.* (1970) on heating basalts with a laser source

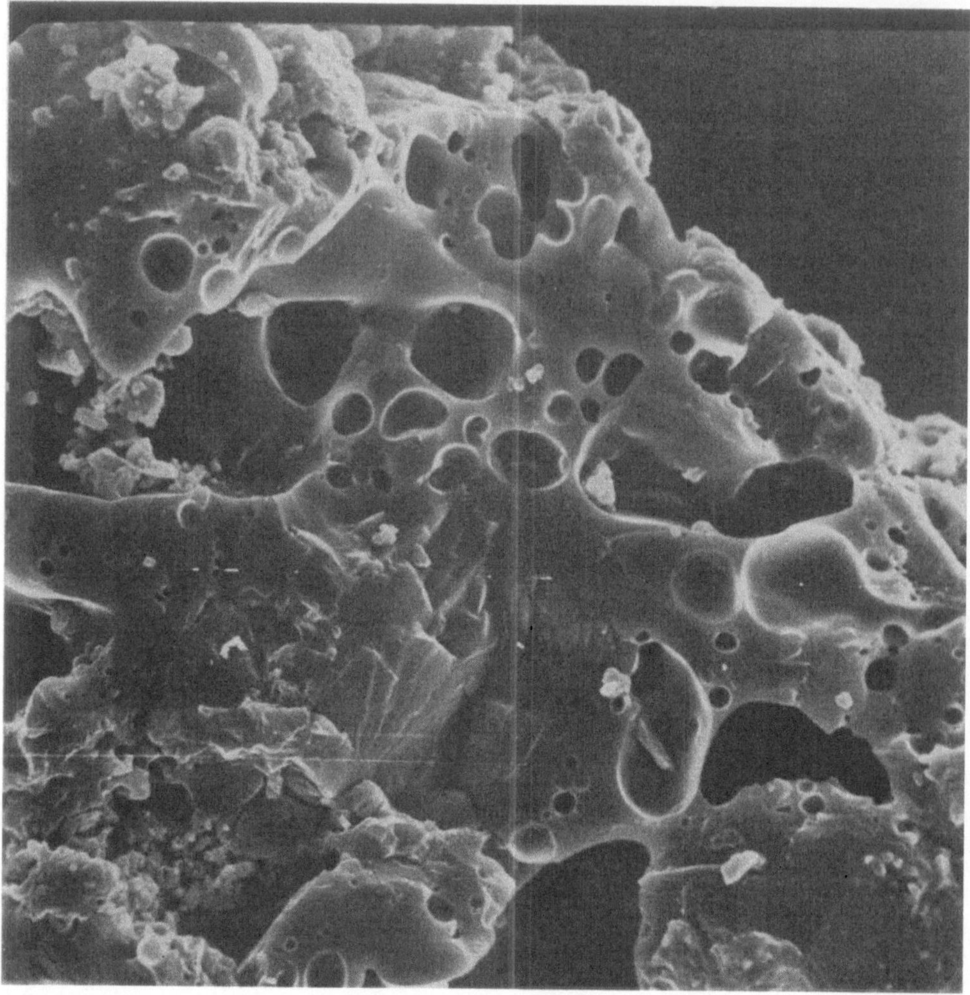

Fig. 7. SEM picture of an inhomogeneous glassy fragment also showing an extensive vesiculation. Magnification is $\times 1800$.

show there is strong alkali enrichment in the vapour phase formed by the volatilization of the material at the point of impact. Condensation of this vapour phase along the walls of vesicles could explain the observed alkali enrichment.

4. Conclusions

The chemistry of the glasses analysed falls within the range of types already observed by other investigators; most of the particles can be plotted in the field of gabbro composition and show a major trend towards the light compositions termed 'anorthosites' in established lunar nomenclature. The occurrence of an oversaturated composition

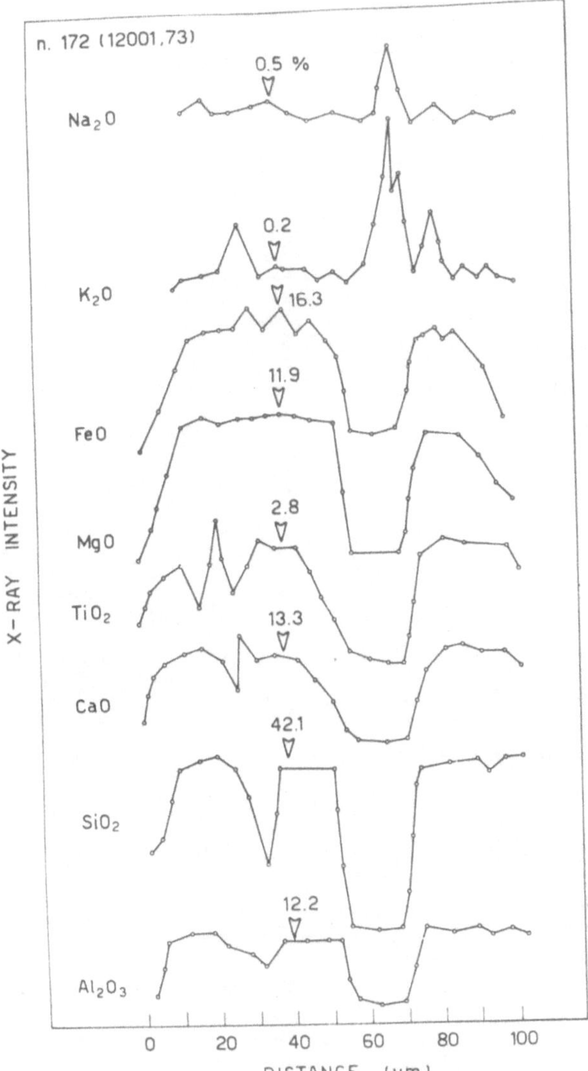

Fig. 8. Slow scan profiles of the glassy particle No. 172. The reported partial traverse has been taken from a pseudo-equatorial section of the glass 200 μm in size. The profiles are from count rates at different scales and eventually are not exactly from the same line.

requires a high silica parent material or is more unlikely due to some kind of fractionation during the glass formation process.

The origins reported for most of the glassy shapes have been ascribed to the melting in various proportions of mineral phases such as pyroxene, olivine, or of other glasses as the result of the impact of meteoroids on the lunar surface.

The low alkali values often observed in the glasses analyzed and the ratios of the

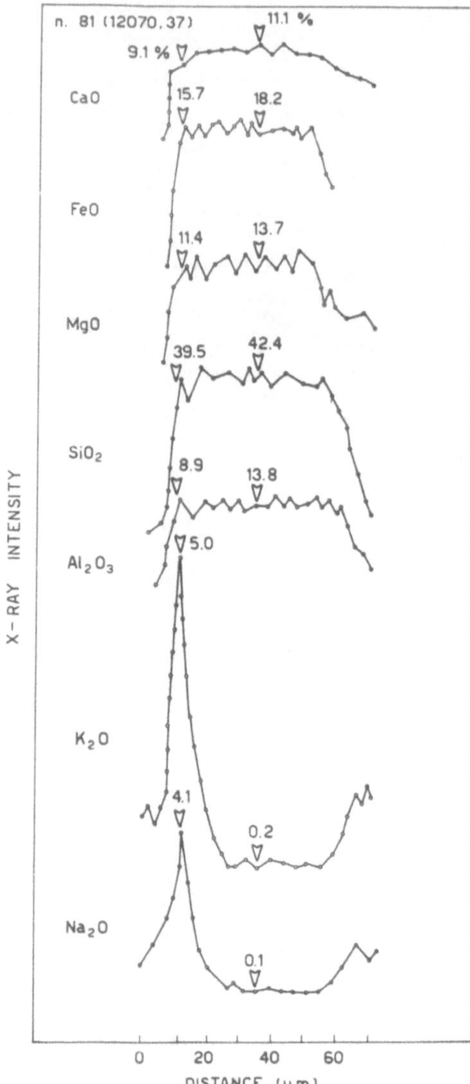

Fig. 9. Slow scan profiles of the glassy sphere No. 81. Traverse is from a pseudo-equatorial section of the particle with diameter about 160 μm. The inside is hollow, being occupied by a single vesicle. The profiles are from the solid part of the particle (the outside is on the right while the inside is on the left) and show the strong alkali enrichment along the boundaries of the vesicle. The figures on each profile indicate concentration variations.

Na and K values that sometimes differ from those previously observed for rocks and lithic fragments are probably related to the peculiar behaviour of the alkalies in the impact event. On the other hand, the microprobe examinations of vesiculated glasses lead to the conclusion that if there is a strong alkali enrichment at the boundary between glass and vesicle, this can be explained as the result of condensation of these elements after volatilization in the impact cloud and of the condensed material being trapped within the vesicles by immediate cooling.

Acknowledgements

The writer wishes to express his gratitude to Dr and Mrs S. O. Agrell who very kindly reviewed the manuscript. Many thanks are due to Professor B. Accordi and Professor C. Lauro for making this work possible and to the colleagues of CNR research group on the lunar dust for helpful discussions. The research has been supported by a CNR grant.

References

Agrell, S. O., Long, J. V. P., and Reed, S. J. B.: 1971, 'Glasses from Apollo 11 and 12 Soils and Microbreccias', Second Lunar Science Conference (unpublished proceedings).

Blander, M., Keil, K., Nelson, L. S., and Skaggs, S. R.: 1970, *Science* **170**, 435.

Burri, C.: 1964, *Petrochemical Calculations*, Sivan Press.

Carusi, A., Coradini, A., Fulchignoni, M., and Magni, M.: 1972, this volume, p. 180.

Chao, E. C. T., Boreman, J. A., Minkin, J. A., James, O. B., and Desborough, G. A.: 1970, *J. Geophys. Res.* **75**, 7445.

Cooper, A. R., Varshneya, A. K., Swift, J., and Yen, F.: 1971, 'Properties of Lunar Glasses', Second Lunar Science Conference (unpublished proceedings).

Derby, J. V., Lewis, V. A., Hale, D., Legrone, H., and Naughton, J. J.: 1971, 'Investigation of Lunar Erosion by Volatilized Alkalis', Second Lunar Science Conference (unpublished proceedings).

Isard, J. O.: 1971, 'The Formation of Spherical Glass Particles on the Lunar Surface', Second Lunar Science Conference (unpublished proceedings).

Keil, K., Prinz, M., and Bunch, T. E.: 1971, 'Mineralogical and Petrological Aspects of Apollo 12 Rocks', Second Lunar Science Conference (unpublished proceedings).

Mason, B. and Melson, W. G.: 1970, *The Lunar Rocks*, Wiley-Interscience.

McKay, D. S., Greenwood, W. R., and Morrison, D. A.: 1970, *Geochim. Cosmochim. Acta Suppl. I* **1**, 673.

FORMATION OF LUNAR GLASSY SPHERULES: A DYNAMICAL MODEL

A. CARUSI and A. CORADINI

Istituto di Geologia e Paleontologia
University of Rome, Italy

and

M. FULCHIGNONI and G. MAGNI

Laboratory of Astrophysics, Frascati, Italy

Abstract. Glassy spherules ranging from 200 μ to 62 μ in size have been separated from lunar dust samples No. 12001.73, 12057.60, 12070.37. Most of them are regular in size (spherical, ellipsoidic, dumbbell, teardrop, etc.); some are irregularly shaped.

A tentative dynamical model of the evolution of a rotating melted spherical drop of homogeneous glassy material has been built in order to explain the observed forms. We suppose such fluid to be originated from the impact of meteoroids on the lunar surface. The energy balance between the projectile (meteoroid) and the target (lunar surface) has been calculated supposing that the impact gives rise to strong shock waves in both bodies.

Equations of the model have been solved numerically and a good agreement between these results and the experimental data regarding small spherules has been obtained.

1. Introduction

The samples of lunar dust which have been used for the experimental researches are No. 12001.73, 12057.60, and 12070.37 and consist of about half a gram each of lunar fines. The purpose of our studies is to determine some correspondence between the shapes of glassy spherules observed in the sample of lunar dust and the one we obtained from a theoretical model.

We assume the hypothesis of meteoritic origin of lunar spherules; then we studied the impact processes of a meteorite on the lunar surface, which allowed us to formulate a dynamical model correlating the various forms of lunar spherules with the characteristics of a homogeneous fluid generated by impact.

2. Meteoroids Impact Phenomena

The primary meteoritic hypothesis of lunar glassy spherules origin has to be excluded. In fact the chemical composition of the lunar spherules is usually missing among the cosmic dust collected by satellites and rockets. In Table I the average chemical composition of the spherules and of the corresponding lunar soil is reported. The chemical composition of the two is quite similar.

Besides, the probability of finding such a figure ($=1000$ spherules mm^{-3}) of micrometeorites is really low if their spatial distribution follows the mass law [Whipple, 1968]

$$f(m)\,\mathrm{d}m = Am^{-\alpha}\,\mathrm{d}m$$

Urey and Runcorn (eds.), The Moon, 180–184. *All Rights Reserved.*

TABLE I

	Average lunar spherules chemical composition (%)	Average lunar soil chemical composition (%)
SiO_2	0.45	0.42
TiO_2	0.05	0.03
Al_2O_3	0.15	0.14
FeO Fe_2O_3	0.15	0.17
CaO	0.10	0.10
MgO	0.09	0.11
MnO	0.02	0.02
Na_2O K_2O	0.01	0.05

where A and α are constants experimentally determined. Considering the micrometeorites impact as a random event, we will therefore suppose that the spherules are the secondary products of a meteoritic impact on the lunar surface.

It is possible to describe this phenomenon in the following way: at the beginning (compression phase) the meteoroid (projectile) impacts the lunar surface (target) and generates a strong shock wave system which causes both in the projectile and in the target an increase of pressure of up to some megabars.

During this phase the target material is stressed much more than its rupture point, than we can consider the process as a hydrodynamic one.

When the wave front reaches the free surface of the involved materials a rarefaction waves system is generated which precedes the ejection of high velocity material.

Let us now examine the energetic aspects of the phenomenon (Gault and Heitowit, 1963). On first estimate we can describe the phenomenon in the following way. We suppose that the surface of the projectile and the target are plane and semi-infinite.

In a reference system solidale with the unaltered medium the equations by Rankine-Hugoniot are still valid

$$\mathrm{d}_0 W = \mathrm{d}(W - w) \tag{1}$$

$$p = \mathrm{d}_0 W w \tag{2}$$

$$E - E_0 = \tfrac{1}{2} p (1/\mathrm{d}_0 - 1/\mathrm{d}) = \tfrac{1}{2} w^2 \tag{3}$$

The system is completed by the experimental equation

$$W = a + bw \tag{4}$$

W is front of shock wave velocity and w is the material velocity, a and b are constants experimentally determinated (Al'tshuler *et al.*, 1961; Lombard, 1961; Rice *et al.*, 1958; Kormer *et al.*, 1962).

Adding to preceding equations the condition

$$V_i = w_p + w_t \tag{5}$$

where w_p, w_t are the velocities of the material in the projectile and in the target, it is possible to find the values of the pressure and the energy which belong to the two bodies.

In Table II for an impact velocity of 13 km s^{-1} the projectile and the target material

TABLE II

Impact velocity $V = 13$ km s^{-1}

	Cu	Zn	Ag	Cd	Au	Pb	Bi	Fe
Projectile material velocity (km s^{-1})	8.60	8.08	8.83	8.34	9.70	8.45	8.31	8.44
Target material velocity (km s^{-1})	4.40	4.42	4.17	4.66	3.30	4.46	4.69	4.56
Compression								
Projectile pressure (dyne $cm^{-2} 10^{-14}$)	0.12	0.08	0.15	0.11	0.32	0.13	0.11	0.11
Target pressure (dyne $cm^{-2} 10^{-13}$)	0.12	0.15	0.11	0.13	0.75	0.12	0.14	0.13
% energy trapped in the projectile	0.55	0.53	0.56	0.54	0.62	0.55	0.54	0.54
% energy trapped in the target	0.45	0.47	0.44	0.46	0.38	0.45	0.46	0.46
Rarefaction								
% irreversibly trapped energy in the projectile	0.62	0.54	0.49	0.43	0.44	0.42	0.36	0.57
% irreversibly trapped energy in the target	0.73	0.65	0.77	0.69	0.97	0.72	0.68	0.70

velocity, the projectile and the target pressures, and the percent of energy irreversibly trapped in the compression and rarefaction phases have been reported for some pure element projectile and for basaltic target.

3. Dynamical Model

The energy trapped by the target is enough to provoke the volatilization, the melting and the ejection of prevailingly molten material. The fluid is in dynamic condition; consequently it has a large kinetic energy per unit of volume and therefore has the tendency to divide itself in small fluid particles in rotation too.

Owing to the surface tension those particles get a spherical form. It has been observed experimentally that most of the lunar dust particles have a regular enough form.

About 40% of the glassy particles are almost perfectly spherical forms.

Non-spherical particles can be fully irregular or may show some elements of symmetry; the last ones can be grouped in various classes: ellipsoidic forms, cigars, dumbbells, teardrops (Fulchignoni *et al.*, 1971).

To explain the observed morphologies we have built a dynamical model of the evolution of a rotating melted material spherical drop governed by simultaneous actions of surface tension and centrifugal force.

Owing to the high available energy it will be possible for a large range of angular velocities to distribute themselves among the fluid particles in a statistical way. The possible situation of stable or unstable equilibrium will depend on the dimensions of the particles and their kinetic energy.

In the case of instability it is possible to calculate the particle breaking-time; these times are comparable to the cooling times obtained by black-body radiation law.

Considering the opacity (Isard, 1971), the solidification time became much larger, therefore it is reasonable to assume that the experimentally obtained forms are equilibrium forms.

Considering for the surface tension potential

$$V_{ts} = T \int_{\sigma} \mathrm{d}\sigma \tag{6}$$

where T stands for surface tension coefficient, and σ is the particle surface and for the centrifugal force potential

$$V_c = \tfrac{1}{2}\omega^2 d \int r^2 d\tau \tag{7}$$

where d is the density of the particle, ω is the angular velocity, r is the radius and τ the volume, the total potential of the system will be

$$V = V_c + V_{ts}.$$

Putting now r function of ϑ, φ all space surfaces can be represented; choosing for r an adequate parametric expression, V_c and V_{ts} will be functions of the adopted parameters $c_1 \ldots c_n$.

To determine the equilibrium conditions of the drop under the action of the above-mentioned forces, let us minimize their total potential with the restricting condition of the volume constant.

This problem has been solved numerically.

In applying the above-mentioned method we have calculated the equilibrium surfaces of a fluid supposed to be homogeneous, obtained varying only one free parameter starting from the spherical configuration. In this particularly simple case the equilibrium surfaces can be nothing but rotation ellipsoids.

On Figure 1 values of the parameter (eccentricity) at the equilibrium in function of ω for some values of the volume have been reported.

Increasing the parameter number and choosing for $r(\vartheta, \varphi)$ an expression

$$r^2(\vartheta, \varphi) = c^2 \cos 2\vartheta + \sqrt{k^4(\vartheta, \varphi) - c^4 \sin^2 2\vartheta}\,.$$

We can see that increasing ω we obtain some shapes which approximate the entire range of experimentally found shapes.

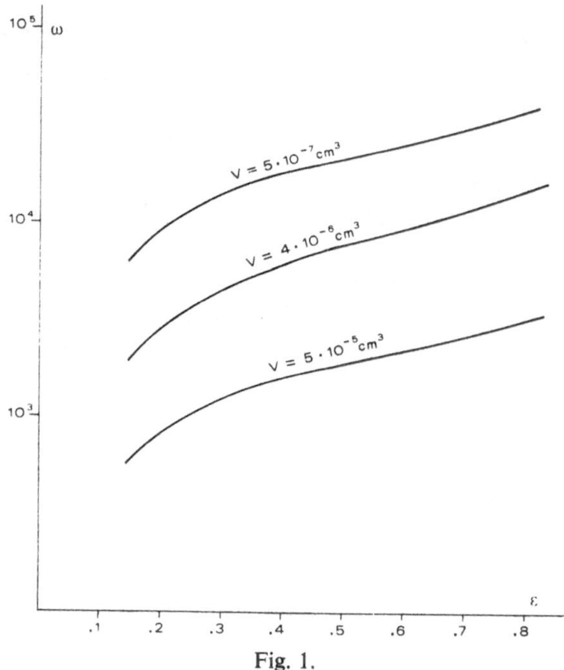

Fig. 1.

The results numerically obtained simulating the meteorite impact of pure metal on the basaltic target give a representation of the phenomenon in agreement with the experimental results.

In fact a large number of particles with a chemical composition similar to soil have been found. The balance between the internal and kinetic energy distributed between the projectile and the target, gives values for the kinetic energy of the ejected melt fluid that are compatible with the minimum threshold required by the dynamic model to obtain the particle morphologies.

The obtained results may be improved by refining the hypothesis. But the lack of complete experimental data (i.e. the coefficients of the equations describing the evolution of the shock wave in the solid) may render useless a more sophisticated approach.

References

Al'tshuler, L. V. Bakanova, A. A., and Trunin, R. F.: 1961, *Soviet Phys. JEPT* **15**, 65

Fulchignoni, M., Funiciello, R., Taddeucci, A., and Trigila, R.: 1971, 'Apollo 12 Lunar Science Conference' (unpublished proceedings).

Gault, D. E. and Heitowit, E. D.: 1963, *Proceedings of Sixth Hyper Velocity Impact Symposium, Cleveland Ohio, April 30, May 1–2* **2**, 419.

Kormer, S. B., Urlin, V. D., and Popova, L. T.: 1962, *Soviet Phys.* **3**, 1547.

Isard, J. O.: 1971, 'Apollo 12 Lunar Science Conference' (unpublished proceedings).

Lombard, D. B.: 1961, 'The Hugoniot Equation of State of Rocks', URCL 6311, University of California, Lawrence radiation laboratory.

Rice, M. H., Mac Queen, R. G., and Walsh, J. M.: 1958, in P. Turnbull (ed.), *Solid State Physics* **4**, 63, Academic Press.

Whipple, F. L.: 1968, in Ľ. Kresák and P. M. Millman (eds.) 'Physics and Dynamics of Meteors', *IAU Symp.* **33**, 481.

E. LUNAR TECTONICS

LUNAR MARE RIDGES, RINGS AND VOLCANIC RING COMPLEXES

ROBERT G. STROM

University of Arizona, Tucson, Arizona, U.S.A.

Abstract. Mare ridges often consist of two separate but related features: (1) a broad gentle arch overlain by (2) a sharper, more contorted ridge. Often these sharper, secondary ridges have flowed into craters in the adjacent terrain indicating they are extrusions. Major flows in Mare Imbrium appear to have issued from several prominent mare ridges. These flows show color differences which may be related to their abundance of titanium. The association of flows with mare ridges, the broad arching linked with many ridges and the coincidence of linear ridges with the directions of major fracture patterns in the highlands indicates that arched mare ridges are dike, sill or laccolithic-type intrusions along major fractures. In many cases these intrusions appear to have broken through the surface to form short flows and bulbous lava extrusions. Mare ridges unassociated with arching are probably lava extrusions only.

Mare ridges often take the form of rings indicating that they developed along ring fractures. Several linear mare ridges and mare ridge rings have bright hills situated along them and are considered to be post-mare volcanic hills of more siliceous composition than the maria. Evidence suggests that many of the ring structures are post-mare volcanic ring complexes formed over large igneous masses.

1. Introduction

Mare ridges are ubiquitous features of the maria and their origin has a strong bearing on the nature of the processes that have shaped the maria. Both a tectonic and magmatic origin has been suggested for the formation of these ridges, but the evidence for either origin has been inconclusive until the recent acquisition of the Lunar Orbiter photography. New evidence from this photography strongly implies that both tectonism and magmatism have played a role in their formation, but magmatic activity seems to have been the principal process in forming the ridges as they appear today.

Mare rings (sometimes referred to as 'Ghost Rings') are nearly circular structures which occur exclusively in the maria and consist of mare ridges or intermittent groups of relatively low, bright hills situated along the ridges. These structures have been interpreted as the rims of craters which were almost completely buried by mare lavas and, therefore, of pre-mare age. This interpretation was challenged by Fielder (1965) who suggested that many of these features (which he termed elementary rings) are young, post-mare extrusions of volcanic material along ring fractures. Recently, Fulmer and Roberts (1967) and Guest and Fielder (1968) have found evidence on the Lunar Orbiter photography which supports this theory. The evidence consists primarily of fine, preferentially oriented lineaments on the hills, and the general structure and geological relations of the ring structures.

The present paper is primarily concerned with the origin of mare ridges and their association with certain ring structures. There appears to be a complete transition from linear mare ridges, through mare ridge rings, to rings made up almost entirely of

Urey and Runcorn (eds.), The Moon, 187–215.

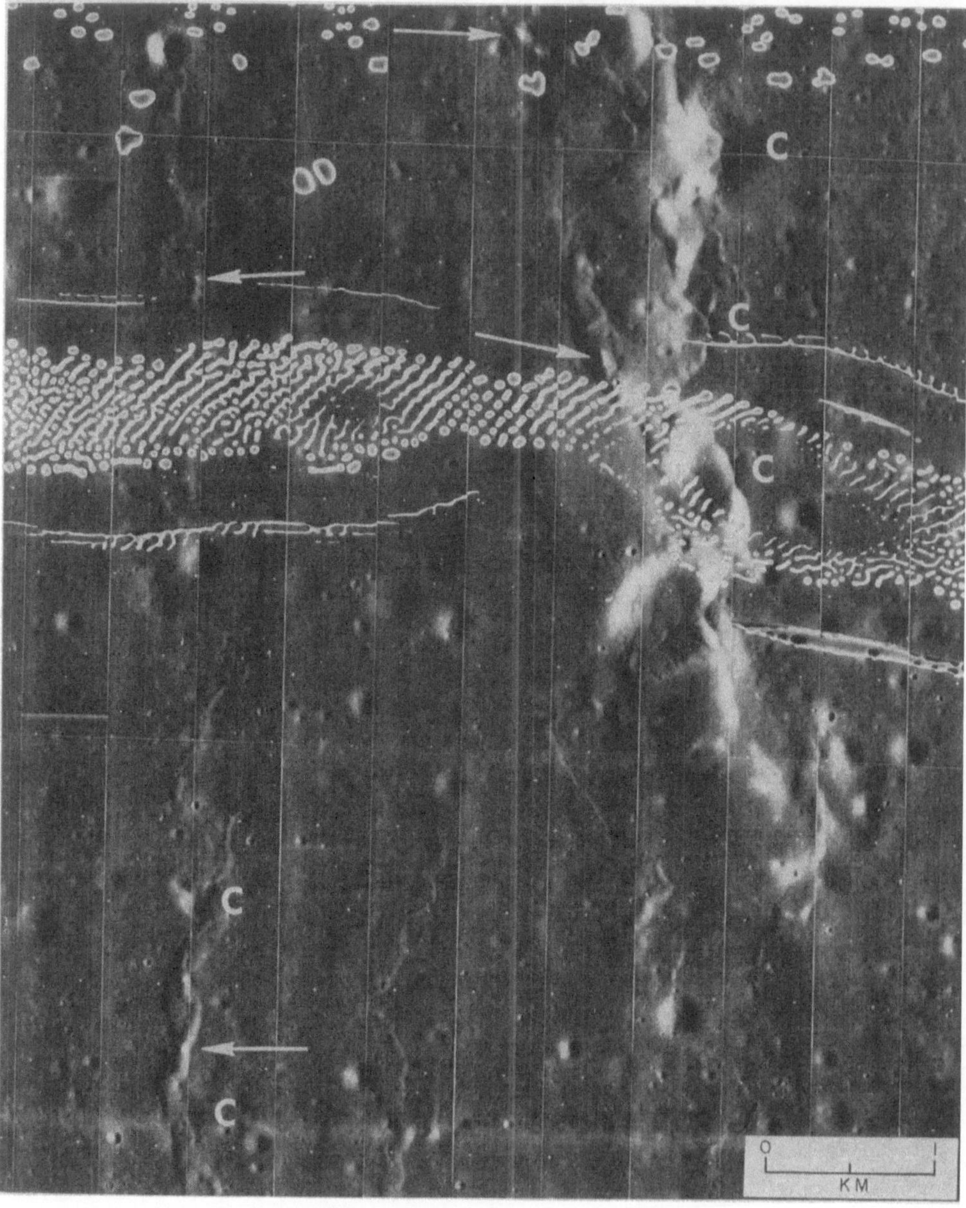

Fig. 1. Mare ridges in Mare Serenitatis near the Littrow Rilles. The ridges have flowed into and partially filled craters located at *c*. Small bright incipient hills similar to those in Figure 15 and discussed in the text are indicated by arrows. The horizontal white patterns are bimatt processing defects. (*NASA Lunar Orbiter 5)*

bright hills situated on mare ridges. These associations indicate that certain ring structures are post-mare volcanic ring complexes. Several bright hills seem to be superimposed on mare ridges which indicates they are post-mare volcanic hills rather than flooded portions of the highlands.

2. Mare Ridges

Mare or 'wrinkle' ridges are long, low ridges which are concentrated in, but not limited to, the maria. They range from less than 1 km to over 20 km wide and may reach lengths of several hundred kilometres. Their height varies from a few metres to about 200 or 300 metres. Many of the ridges consist of a broad arch capped by a sharper, more irregular secondary ridge. The mare ridges seen on the Lunar Orbiter photography are generally the sharper portions of the structures. These photographs primarily were taken at a Sun angle of about 20° but the broad arching is generally visible only at Sun angles less than 15°. Sometimes these gentle arches are asymmetrical and, if a secondary ridge occurs on them, it is usually situated on the crest of the arch.

Fig. 2. Mare ridges in Mare Tranquillitatis that have flowed into craters located at *A*. The arrow represents the direction and extent of slope of the arch upon which the righthand ridge lies. For the location of this area see Figure 3. (*NASA Lunar Orbiter 1*)

Usually the secondary ridges comprise about 25–60% of the total width of the structures. Although the association of sharp ridges with broad arches is relatively frequent, there are numerous examples of sharp ridges unaccompanied by arching.

The general geometric arrangement of the ridges may be sinuous, linear or concentrically oriented with respect to the borders of the circular maria. However, the Orbiter photography has shown that even the sinuous and arcuate ridges often consist of short, linear segments. These linear segments, as well as the linear mare ridges, often coincide with the directions of the major lineament systems (grid system) found in the lunar highlands (Fielder and Kiang, 1962; Strom, 1964). Recently Fryer (personal communication) has found a striking agreement in orientation between the mare ridges in the western part of Oceanus Procellarum and the lineaments in the adjacent highlands. This coincidence in direction of linear mare ridges and the fracture pattern in the highlands strongly suggests that these ridges are situated along major fractures in the lunar crust.

It has been suggested that the mare ridges are anticlinal or fault ridges formed by compression, or ridges resulting from the intrusion and/or extrusion of magma along

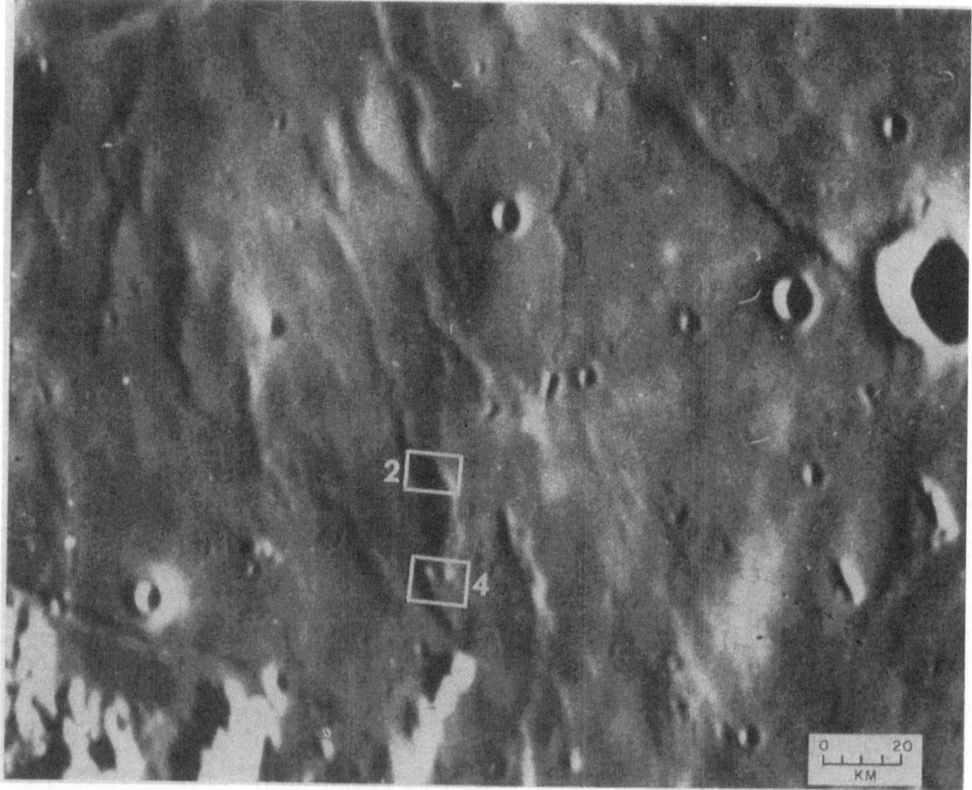

Fig. 3. Earth-based photograph of the mare ridges shown in Figures 2 and 4. Notice the extent of the arch in this low-illumination (7°) photograph which is not readily discernible in Figure 2 taken at a Sun angle of 20°. *(LPL Catalina Obs. photo)*

fractures. The Orbiter photography shows rather conclusively that the second explanation is basically correct. Figure 1 is a portion of a mare ridge at the eastern border of Mare Serenitatis near the Littrow rilles. The ridge has a very ropy texture and has flowed into and partially filled three craters. The texture and lobate character of the flow front where it overlaps the most northerly crater is characteristic of a relatively viscous lava flow. A smaller ridge to the west also has been extruded into several shallow craters. Figure 2 shows similar extrusions which form ridges in Mare Tranquillitatis and also partially fill or overlap several craters (A). The larger structure to the east is a secondary ridge that caps a gentle asymmetrical arch visible on low-illumination Earth-based photographs (Figure 3). This secondary ridge is located at the crest of the arch. The arrow in Figure 2 represents the slope direction and western extent of the arch. The eastern extent of the arch coincides with the position of the sharp ridge. Figure 4 shows a short. hummocky flow (F) about 1.8 km long which forms part of a ridge in the same vicinity (see Figure 3). This flow is clearly superimposed on the adjacent terrain showing the extrusive origin of this ridge. Notice the

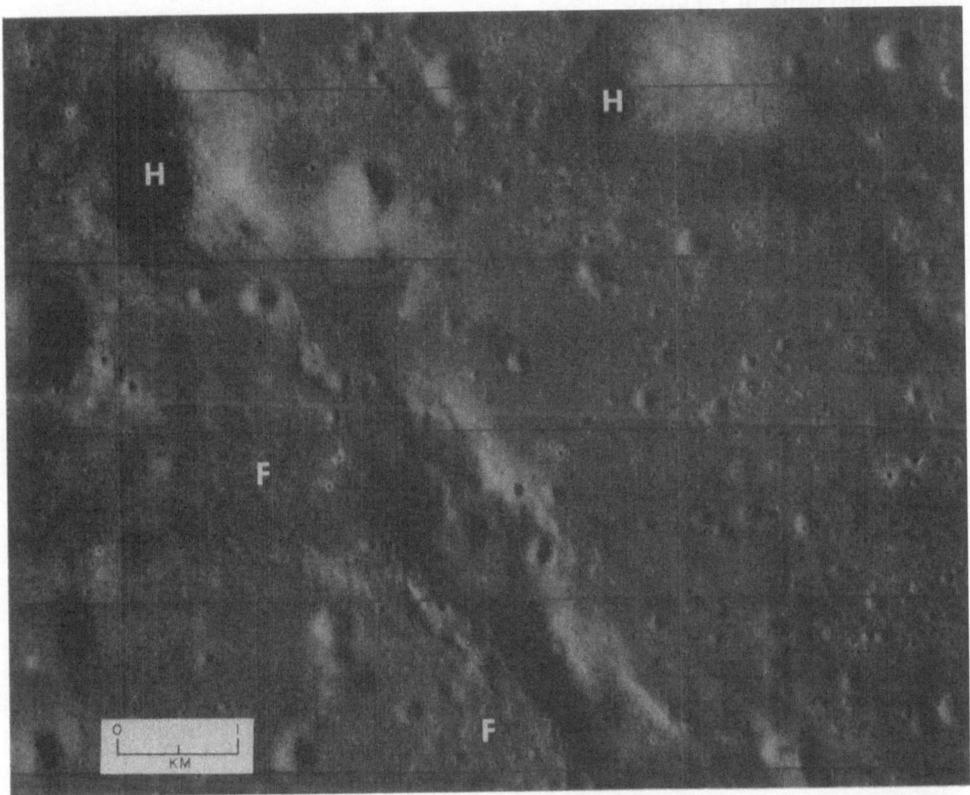

Fig. 4. A mare ridge in Mare Tranquillitatis (see Figure 3) showing an associated short hummocky flow (*F*). Two bright hills (*H*), discussed in the text, are situated along the ridges. *(NASA Lunar Orbiter 1)*

small bright hills (H) which appear to be integral parts of the ridges and are similar to the bright hills discussed later.

In several instances mare ridges traverse portions of the highlands where they also form short flows with well-defined fronts which have flowed into craters. This indicates that mare ridges are deep-seated structures which follow zones of weakness not necessarily restricted to the maria.

The examples mentioned above are only several of many ridges which show similar phenomena. The broad, gentle arching associated with many ridge systems implies that uplift has been a major factor in shaping these ridges. The fact that many of the sharper ridges which cap broad arches have partially filled pre-existing craters strongly indicates that the secondary ridges are extrusions of post-mare age. Their positions on the uplifted arches suggests that this uplifting was the result of magmatic intrusions, which may have migrated up major fractures and been injected between planes of lava flows in the form of sills or laccoliths. The position of many extrusive ridges on the crests of the arches – the area of maximum curvature – indicates that in these areas the surface was fractured by the uplift allowing the magma to reach the surface in the form of short flows and bulbous lava extrusions. The proposed relationships are shown diagrammatically in Figure 5.

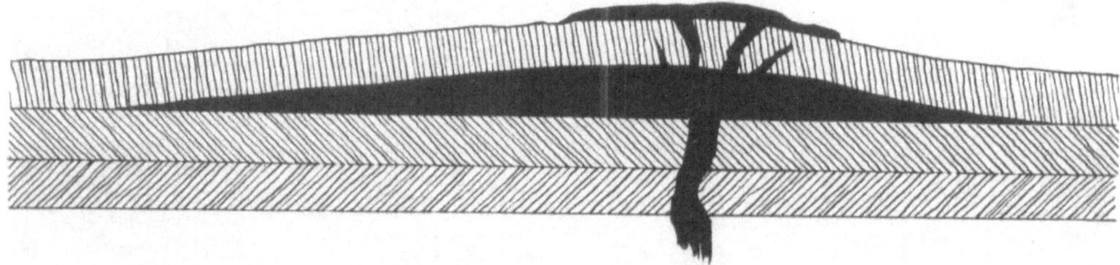

Fig. 5. Diagrammatic representation of the proposed intrusive/extrusive relationship of mare ridges exhibiting uparching and extrusions. See text for explanation.

Mare ridges are very complicated structures, and the explanation of their origin expressed here should be considered a generalization applying to the broad arches with associated sharp secondary ridges. There are numerous ridges unassociated with arches which are probably extrusions along fractures or faults. Rarely, on the other hand, ridges consist of a broad gentle arch only, and may represent intrusive action unaccompanied by extrusion. However, the basic premise that the ridges are tectonically controlled features of intrusive and/or extrusive origin seems well supported by the evidence at hand.

The mare ridges concentrically oriented with respect to the borders of the circular maria probably have a similar origin but have been formed along concentric fractures related to the circular basins rather than linear fractures related to the grid system.

Near the central and west central portion of Mare Imbrium occur an extensive series of flows, several of which are associated with, and appear to have issued from, prominent mare ridges. Figure 6 is a detailed map of the flows based on Lunar Orbiter

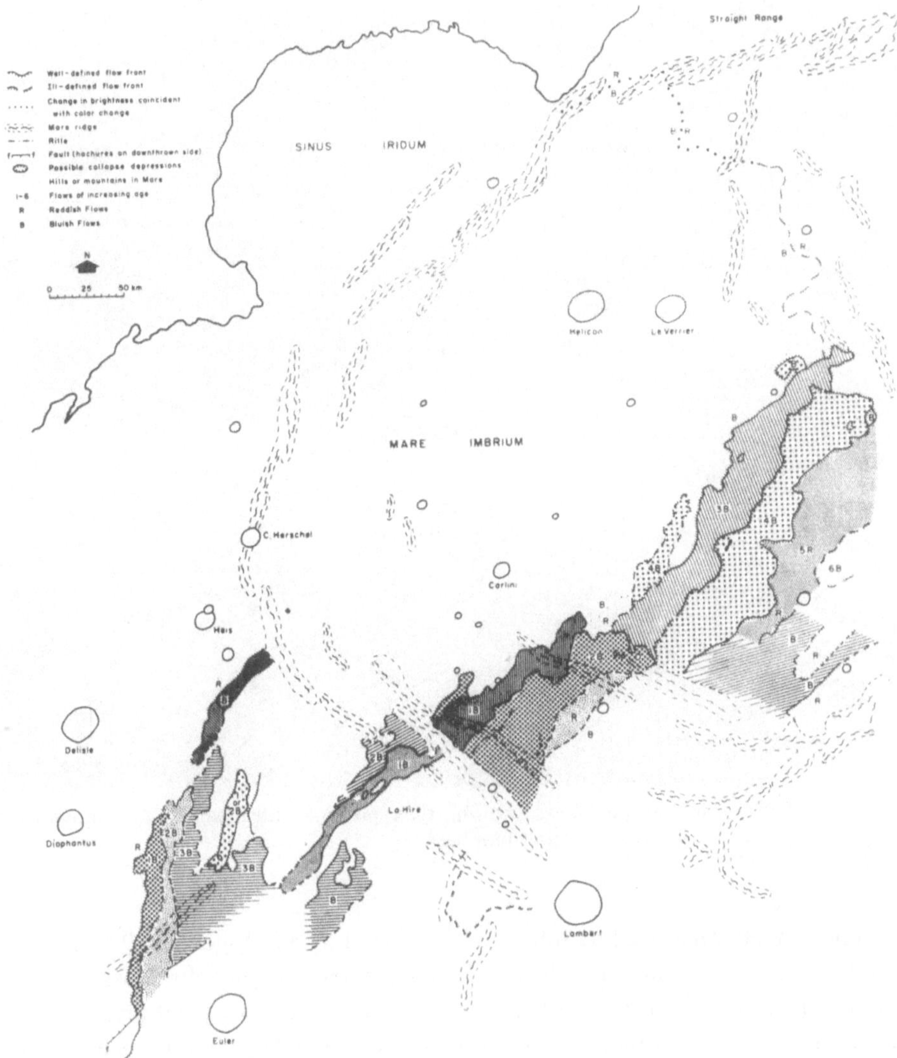

Fig. 6. Map of the flows in Mare Imbrium.

and low sun angle Earth-based photography, whereas Figure 7 is a map of the main flows superposed on a photograph of Mare Imbrium in order to show the position and areal extent of the flows relative to the Imbrium basin. Near the center of the mare the flows trend in a northeasterly direction, but in the west central portion their trend is in a more north-south direction. The flows are long plateau-like features which range from 65–235 km in length to 10–60 km in width, averaging about 130 km long and 28 km wide. Shadow measurements indicate they vary in thickness from about 10–40 m. The main flows cover a minimum area of 3×10^4 km^2. Their total volume may exceed 1000 km^3.

Fig. 7. Map of the main Imbrium flows showing their position and areal extent relative to the Imbrium basin.

The flows are characterized by lobate outlines as shown in Figures 8, 9 and 10, and the distal end of the flow photographed by Lunar Orbiter 5 (Figure 10) shows secondary flow units which have issued from the main flow terminus. The great areal extent of the flows, their lobate outlines, and the presence of secondary flow units are characteristic of basaltic lava flows. This interpretation if supported by the basaltic composition and extrusive nature of the rocks returned from Mare Tranquillitatis and Oceanus Procellarum by the Apollo 11 and 12 astronauts, and the Surveyor 5 and 6 chemical analyses in Mare Tranquillitatis and Sinus Medii.

Furthermore, the boundaries of the flows often coincide with color boundaries in the mare. Figure 11 is a photograph showing color boundaries in Mare Imbrium, Mare Frigoris and the northern part of Oceanus Procellarum. The darker the tone, the redder the lunar surface. This photograph was prepared by Ewen Whitaker from a composite consisting of an original negative ultraviolet plate and a positive copy of an original negative infrared plate. The color differences were verified by photoelectric calibration by T. Gehrels. The color and relative age of the individual Imbrium flows are indicated on the map (Figure 6). The younger uppermost flows are usually the

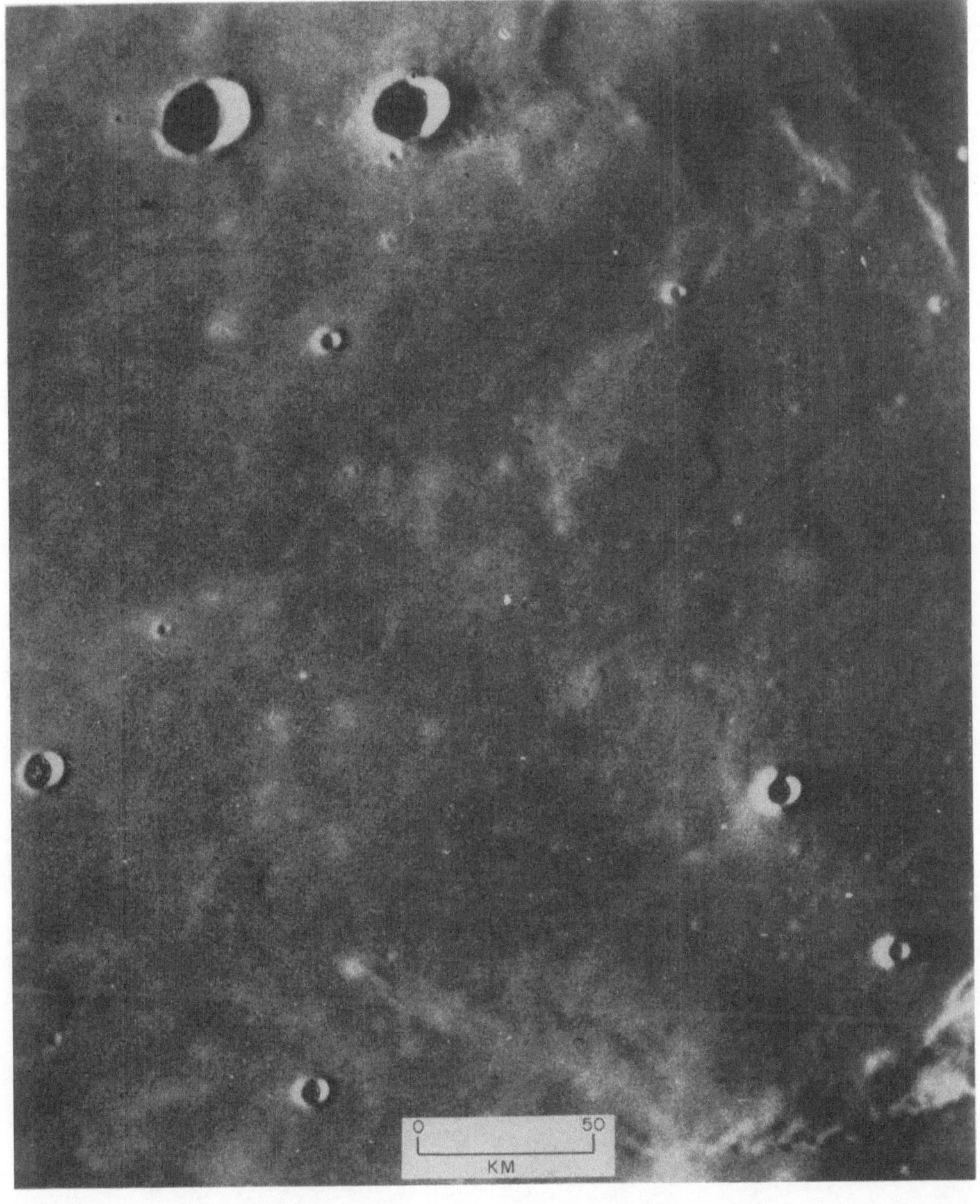

Fig. 8. Earth-based sunset photograph of several Imbrium flows. *(Mt. Wilson Obs. photograph)*

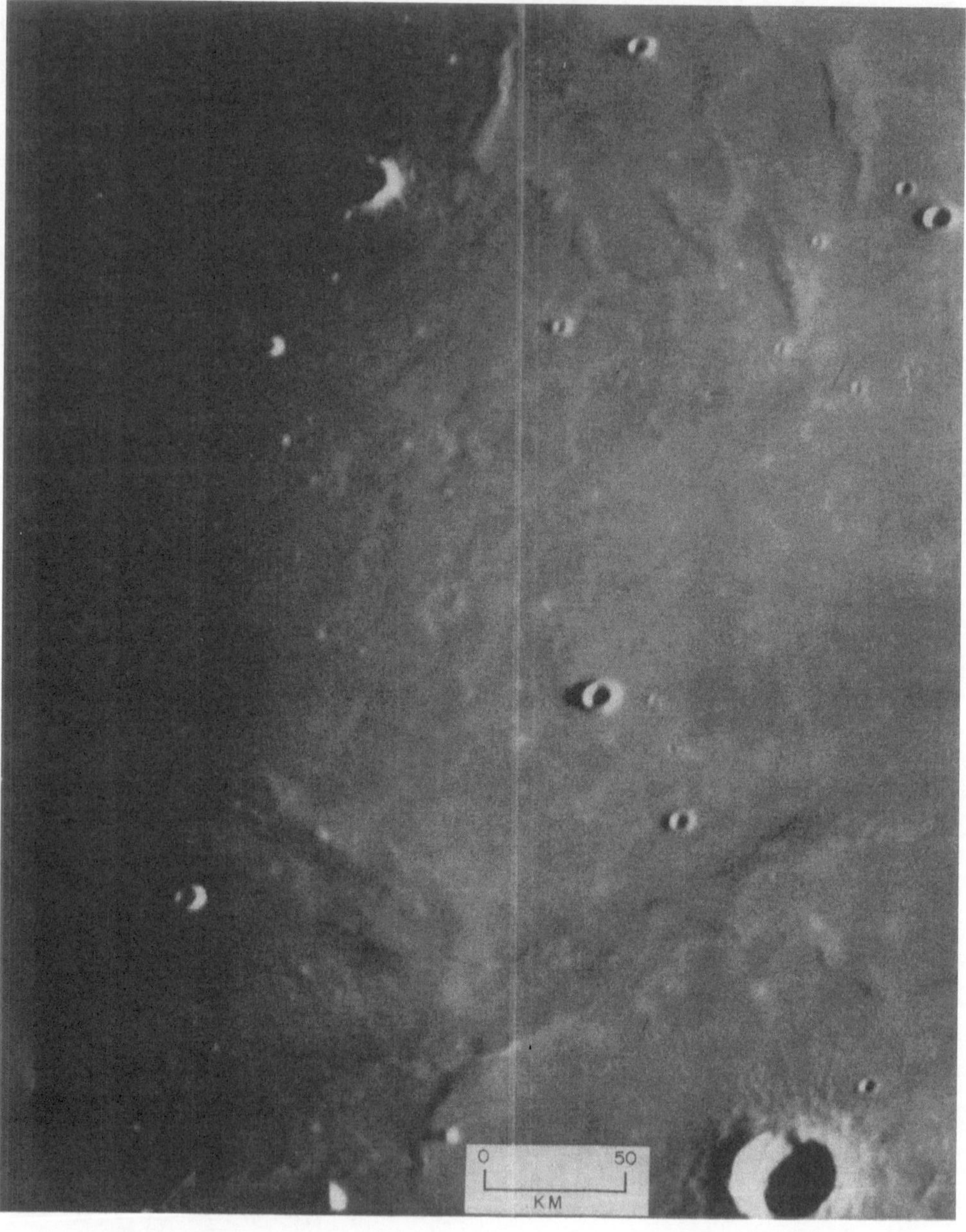

Fig. 9. Earth-based sunrise photograph of the flows shown in Figure 8. *(LPL Catalina Obs. photograph)*

Fig. 10. Lunar Orbiter 5 photograph of one of the Imbrium flows. The arrows point to secondary flow units which have issued from the terminus of the main flow.

bluest; the older deeper-lying flows, which are partly covered, are usually (but not always) the reddest. These color differences may be due to slight compositional differences in the basalts. Figure 12 is a preliminary generalized map of the color differences in the lunar maria compiled from Figure 11 and similar color photographs constructed by Ewen Whitaker. Also shown are the locations of the Surveyor and Apollo sites where chemical data or rock samples have been obtained. Both Apollo 11 and Surveyor 5 are located in Mare Tranquillitatis which is a bluish mare; Surveyor 6 and Apollo 12 landed in Sinus Medii and Oceanus Procellarum respectively at sites which are intermediate in color but toward the reddish side. Table I lists the chemical composition at the Surveyor and Apollo 11 and 12 sites. A comparison of the abundance of the various major rock forming elements at the sites in the bluish mare (Mare Tranquillitatis) and the intermediate reddish maria (Sinus Medii and Oceanus Procellarum) indicates that the abundance of titanium may be responsible for the color differences. Both sites in the bluish mare have from 2–3 times more titanium than those in the intermediate reddish areas. Furthermore, the highlands are redder than the most reddish mare areas, and the Surveyor 7 site in the highlands north of

Fig. 11. Earth-based composite of UV and IR photographs prepared by E. Whitaker showing color differences in the maria. See text for explanation.

TABLE I

Chemical composition (% by weight) of Apollo and Surveyor sites

Oxide	Surveyor 5	Apollo 11 basalts (aver.)	Surveyor 6	Apollo 12 basalts (aver.)	Surveyor 7	Apollo 11 anorthosites	Apollo 12 sample 12013
SiO_2	46.4	40.8	49.1	40.0	46.1	45.7	61.0
TiO_2	7.6	10.5	3.5	3.7	–	0.3	1.2
Al_2O_3	14.4	10.0	14.7	11.2	22.3	30.5	12.0
FeO	12.1	18.8	12.4	21.3	5.5	4.5	10.0
MgO	4.4	7.5	6.6	11.7	7.0	4.8	6.0
CaO	14.6	10.9	12.9	10.7	18.3	15.8	6.3
Na_2O	0.6	0.48	0.8	0.45	0.7	0.35	0.69
K_2O		0.21		0.06		–	2.0
MnO		0.22		0.26		0.1	0.12
H_2O		0.005					

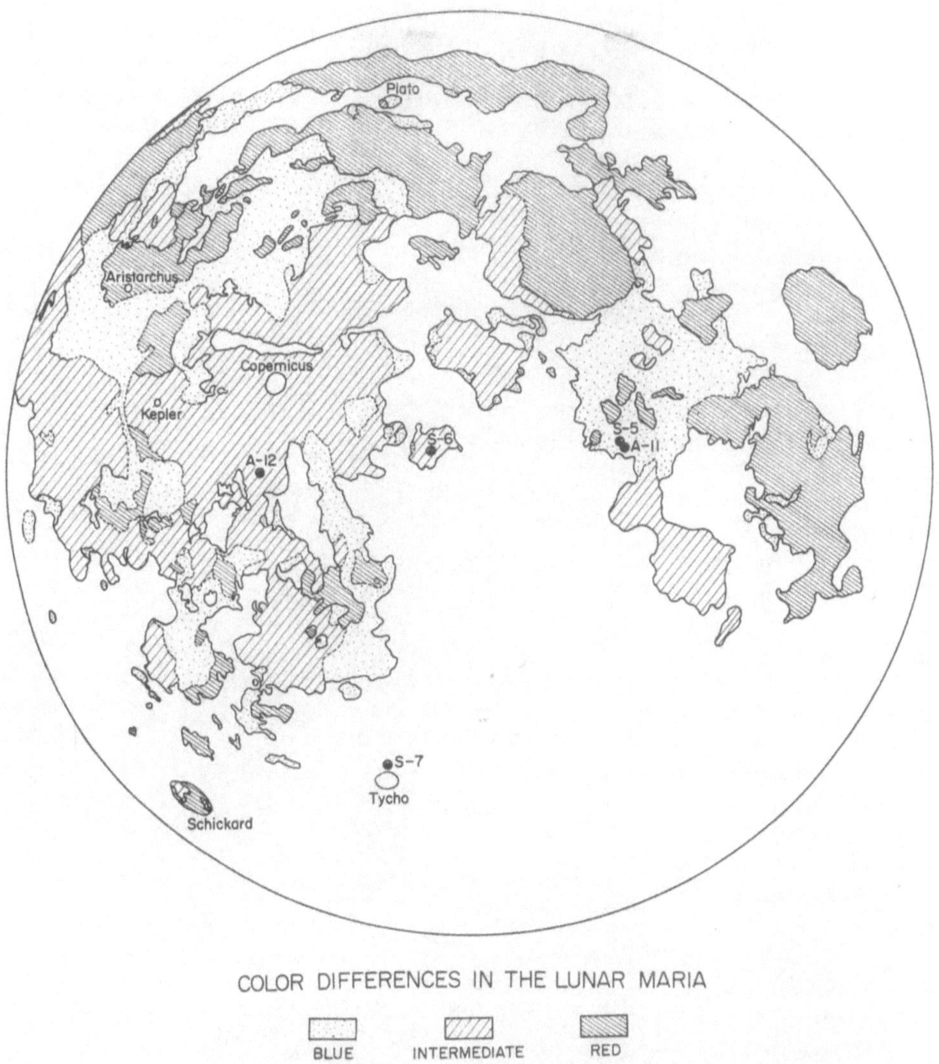

Fig. 12. Preliminary generalized map of the color differences in the lunar maria.

Tycho and the Apollo 11 and 12 anorthosite fragments thought to originate in the highlands have very low titanium abundances. Therefore, the color differences in the mare basaltic lavas may represent differences in the titanium content; the bluer the mare, the higher the titanium abundance.

3. Bright Hills Associated with Mare Ridges

Numerous bright hills in the maria are intimately associated with mare ridges. The hills have a higher albedo than the maria (confirmed on Earth-based full Moon

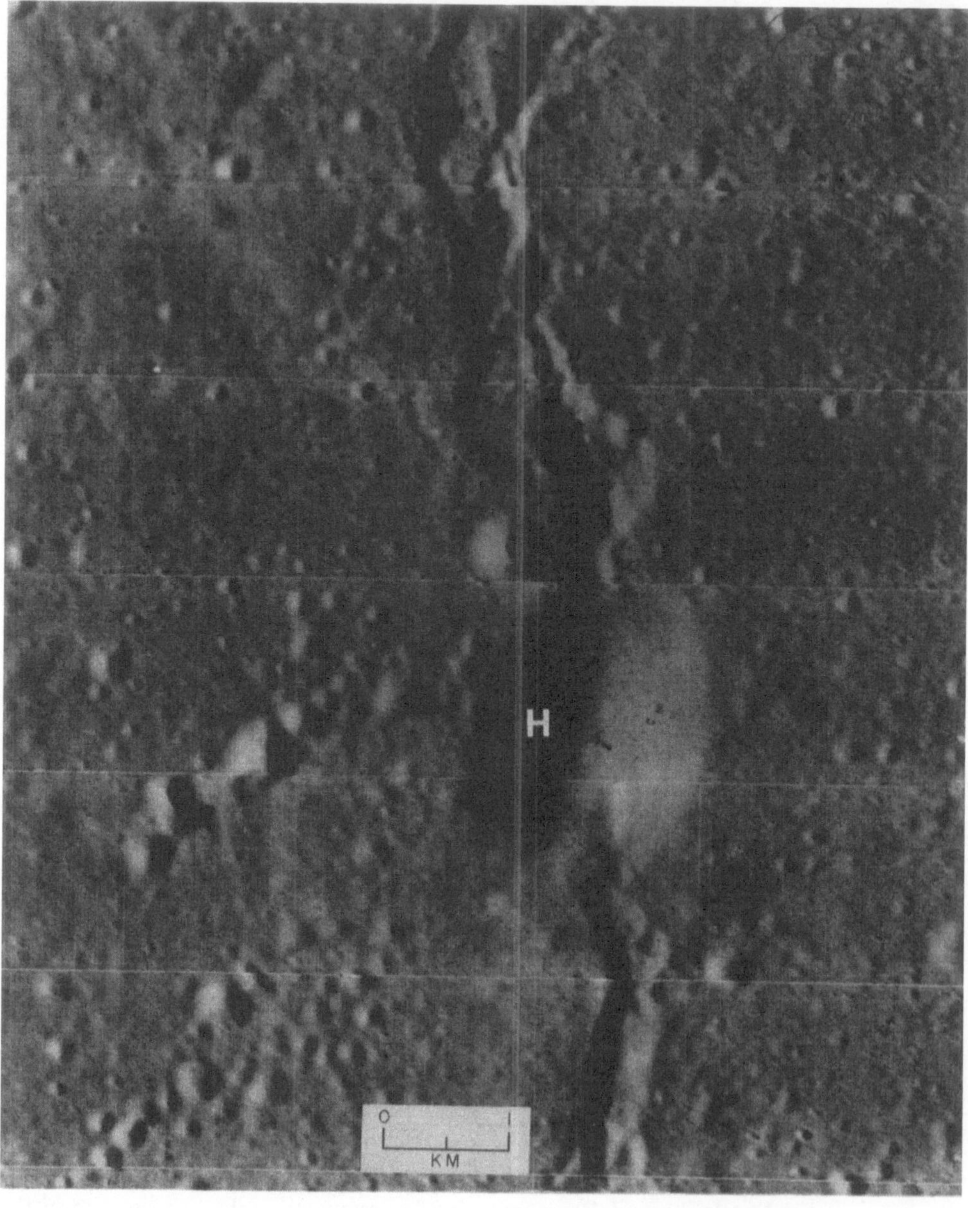

Fig. 13. Conical bright hill (*H*) overlying mare ridge in Oceanus Procellarum. (*NASA Lunar Orbiter 1*)

photographs), and exhibit similar morphological and reflectivity characteristics as the highlands. They range in height from a few tens of metres to over 0.5 km.

Figure 13 shows a bright, conical hill about 2 km in diameter and 300 m high situated on a prominent mare ridge in Oceanus Procellarum. The ridge is very sharp and fresh appearing but has had no visible effect on the hill. If the hill were older than the ridge one would expect the hill to have been affected by the ridge, particularly since the hill is relatively small compared with the ridge. Since this is not the case, it seems clear that the hill overlies the ridge and is therefore younger than both the ridge and the surrounding mare. A previous reference was made to the bright hills situated on the ridge in Mare Tranquillitatis and shown in Figure 4. These hills are identical in

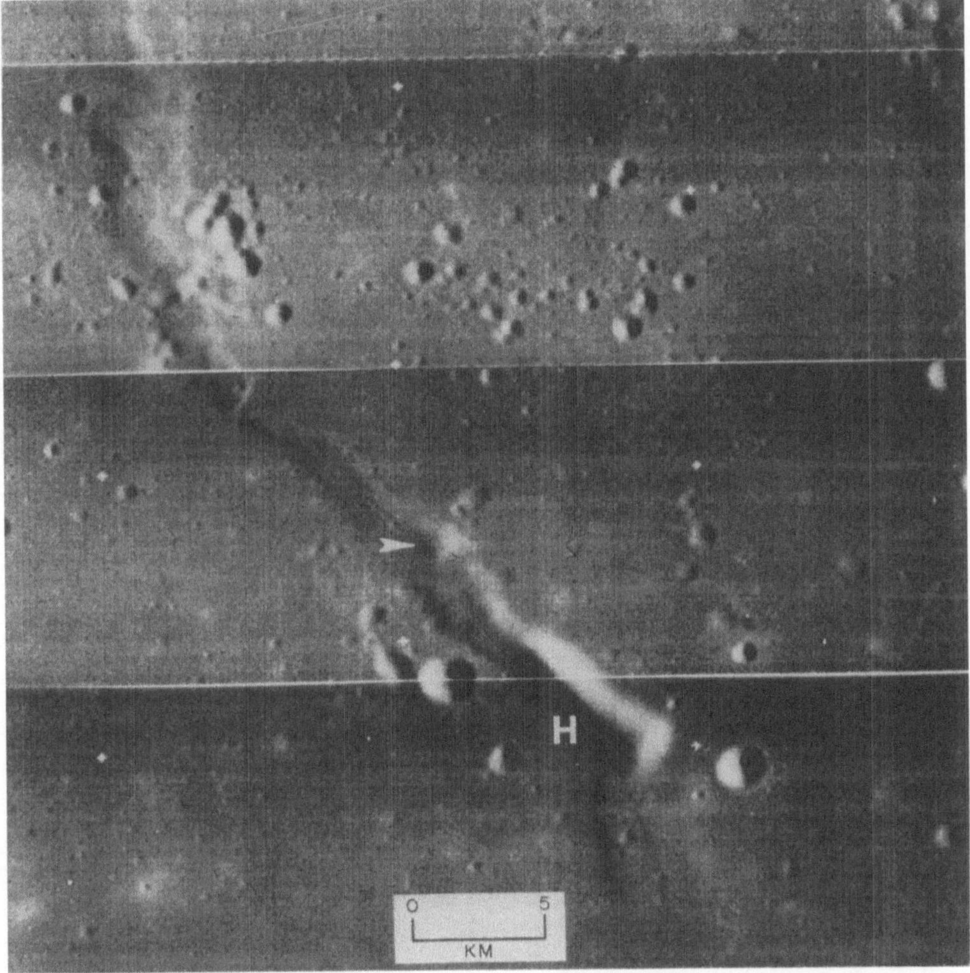

Fig. 14. Bright linear hill (*H*) forming the terminus of a mare ridge in Oceanus Procellarum. A small conical hill situated on the ridge is indicated by the arrow. *(NASA Lunar Orbiter 4)*

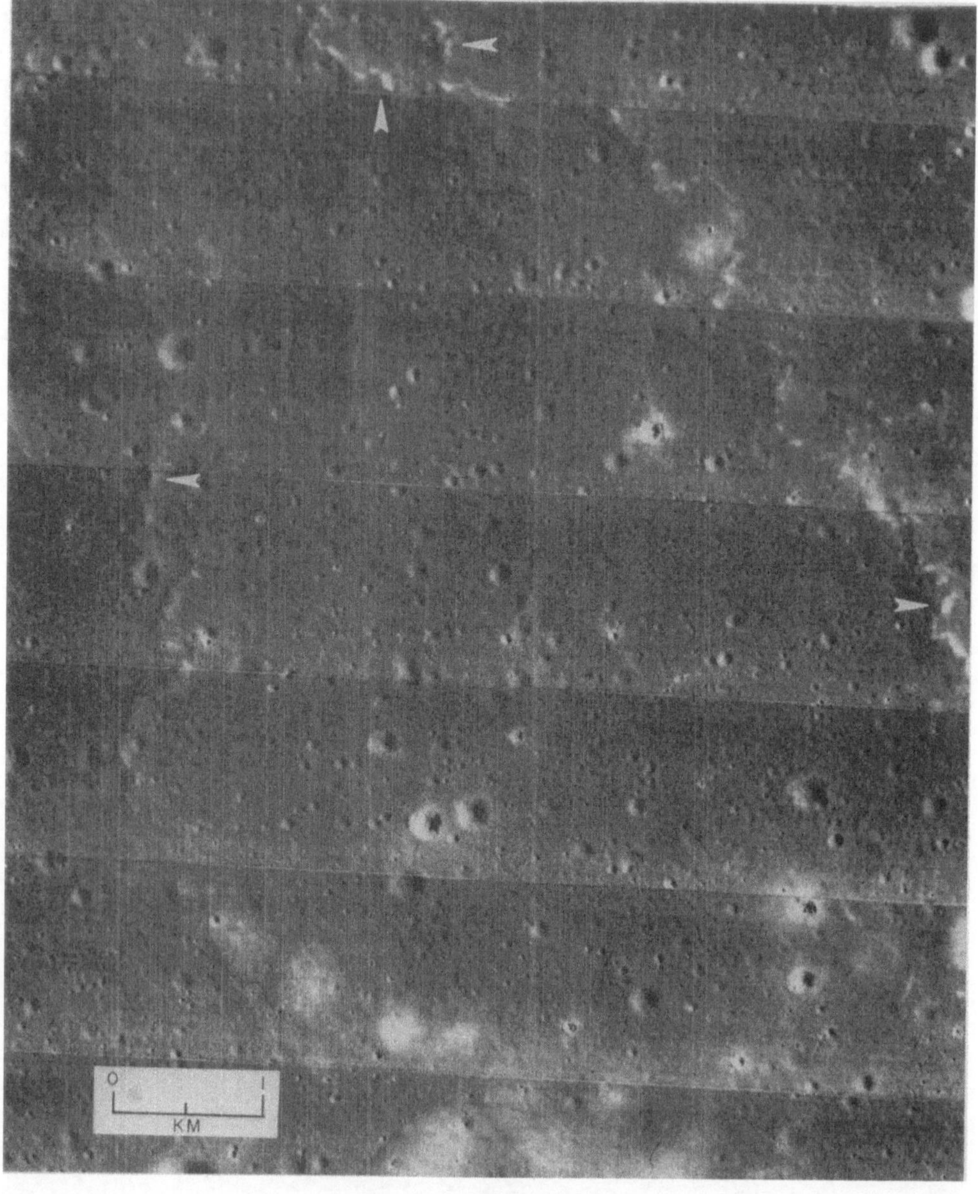

Fig. 15. Small bright incipient hills (arrows) located on mare ridges in Oceanus Procellarum northeast of Flamsteed P. See Figure 23 for location. *(NASA Lunar Orbiter 1)*

morphology and albedo to the hill shown in Figure 13 and are clearly related to the ridges on which they lie. Again the ridges have had no visible effect on the hills indicating the hills formed after the ridges.

Figure 14 shows a bright hill 7 km long, 1.7 km wide and about 400 m high which is located near the terminus of a mare ridge in Oceanus Procellarum. The long axis of the hill follows the direction of the ridge suggesting a genetic relation to the ridge. Since the axis of both the hill and ridge coincide with one of the main directions of the global grid system (NW-SE), they are probably controlled by a major fracture related to the system. About 4 km northwest of the elongated hill is another bright conical hill which lies on the ridge.

Occasionally ridges contain small, bright incipient hills or knobs which are similar to the above mentioned hills but appear not to have developed to the same extent as the other hills. Figure 15 shows a series of such hills located on two ridges northeast of Flamsteed P in Oceanus Procellarum (see Figure 23). On the examples shown, the hills are about 70 to 140 m in diameter and about 1–20 metres high. Other possible incipient hills occur on the ridges shown in Figure 1. These small hills may represent a transitional phase in the development of the larger, bright hills.

The association of mare ridges with bright hills clearly indicates a genetic relation between them. The examples mentioned above, and numerous others, are not chance occurrences on the ridges because they are isolated hills or groups of hills which are relatively distant from the highlands. Hills which are very near the highlands and appear to be flooded portions of the highlands usually do not occur on ridges although there are numerous mare ridges in the area. Furthermore, the hills show no visible effect from the ridges upon which they are situated and therefore, they probably are younger than the ridges. Certainly the small, incipient hills are post-mare since they usually occur only on, and are smaller than, the sharpest part of the ridge and clearly form an integral part of these ridges.

Since the ridges are most likely intrusive/extrusive phenomena located along fractures, it seems probable that the hills associated with these ridges are of volcanic origin, possibly consisting of pyroclastic and/or viscous extrusions. The fact that the hills have a higher albedo than the mare lavas and are of different morphology suggests they differ in composition. Although the extrusive portions of mare ridges have the same albedo as the adjacent mare terrain, they appear to have been more viscous than the lavas which flooded the basins; their texture is more ropy and they have flowed only a short distance. This higher viscosity may reflect a generally lower iron and higher silica content relative to the Apollo 11 and 12 basalts. The hills may consist of more highly differentiated material similar to the siliceous rock of intermediate composition (No. 12013) returned by the Apollo 12 astronauts (*Science*, 1970b) or, less likely, anorthosites (*Science*, 1970a).

4. Mare Ridges in the Form of Rings

Locally, mare ridges take the form of rings which vary in size from 1 to 2 km to over

80 km in diameter. Many of the rings are circular but a large number are elliptical to irregular in shape. The character of the ridges that form rings is identical in most instances to normal mare ridges. One exception is a peculiar ring, R, which appears to consist of flows arranged in a circular pattern (Figure 16). This ring is situated southwest of Flamsteed near a highly contorted mare ridge and is about 8 km in diameter. Figures 17 through 20 are Lunar Orbiter photographs of several rings ranging in size from 3 to 55 km in diameter. In most cases the rings are situated on linear ridges and these ridges may terminate at the ring boundary, deviate in an arcuate pattern to form one side of the ring (Figure 19) or bifurcate to form both sides of the ring. The centers of several rings appear to be slightly depressed with respect to the exterior plain, while

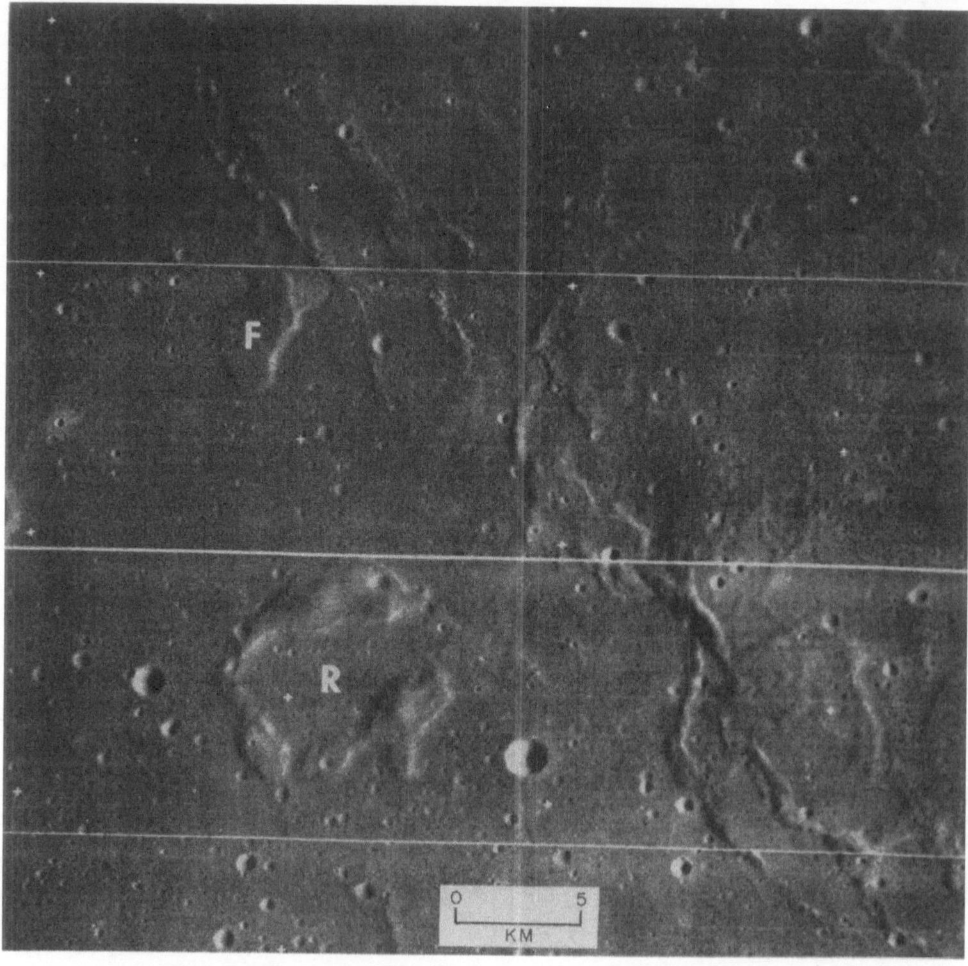

Fig. 16. A ring (*R*) in Oceanus Procellarum apparently composed, at least in part, of lava flows. Notice the flow (*F*) associated with a mare ridge, and of similar morphology as the ring. *(NASA Lunar Orbiter 4)*

Fig. 17. Irregular ring structure (*R*) 3 km in diameter associated with a mare ridge in Oceanus Procellarum. To the left of this small ring is a larger ring (*L*) formed by the NW trending mare ridge. *(NASA Lunar Orbiter 4)*

in others the ring forms a plateau covering most of their centers (Figure 18). There appears to be gradational stages in the development of the rings; from partial semi-circular rings to completely formed rings.

The best example of a mare ridge ring is the Lamont structure in Mare Tranquillitatis (Figure 20). This structure is a double ring about 70 km in diameter with subradial ridges. The subradial ridges are equally as prominent as the ridges that comprise the rings. Lamont has been studied in detail by Guest and Fielder (1968) and, therefore it will be mentioned only briefly here. They conclude, from a study of the peripheral arcuate rilles associated with the margins of a broad depression which surrounds the western portion of the structure, that the formation of Lamont was accompanied by subsidence, and that the ridges are extrusions. The present author agrees with this interpretation. Evidence that the sharper parts of the Lamont ridges are extrusions is shown in Figure 21. This is an Apollo 10 photograph of part of the

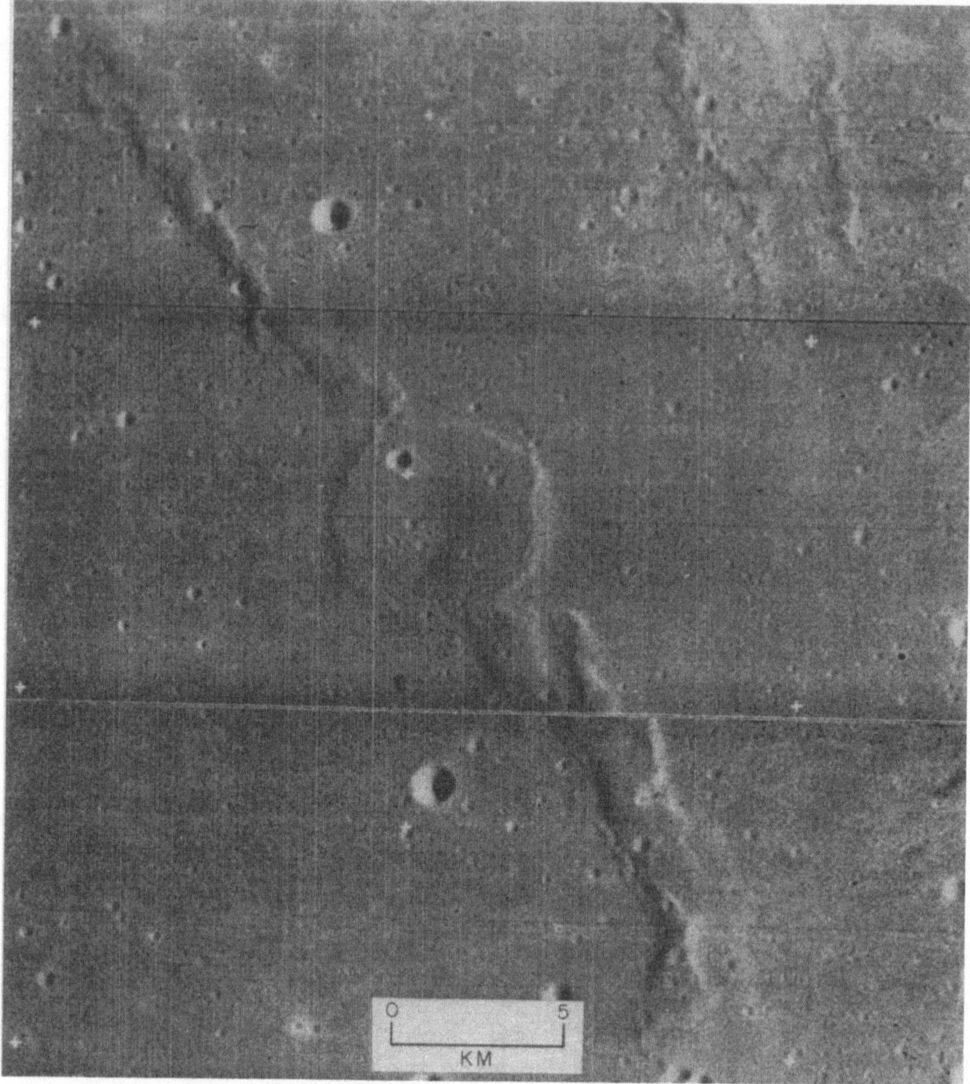

Fig. 18. Ring structure 5 km in diameter formed along a NW trending mare ridge. Note that most of the interior of the ring, except the extreme centre, is an elevated plateau, and that the linear portion of the mare ridge does not traverse the ring. *(NASA Lunar Orbiter 4)*

eastern outer ring clearly showing that the ridge has been extruded into a chain of craters indicated by the arrows in Figures 20 and 21. However, the ridge shown in Figure 21 is only the sharper part of a much broader gentle arch visible on the low-illumination Earth-based photographs (Figure 20). Therefore, at least this part of the Lamont ring appears to have formed by both intrusive and extrusive processes. The crater chain in question is radial to Theophilus, and, if it represents a secondary

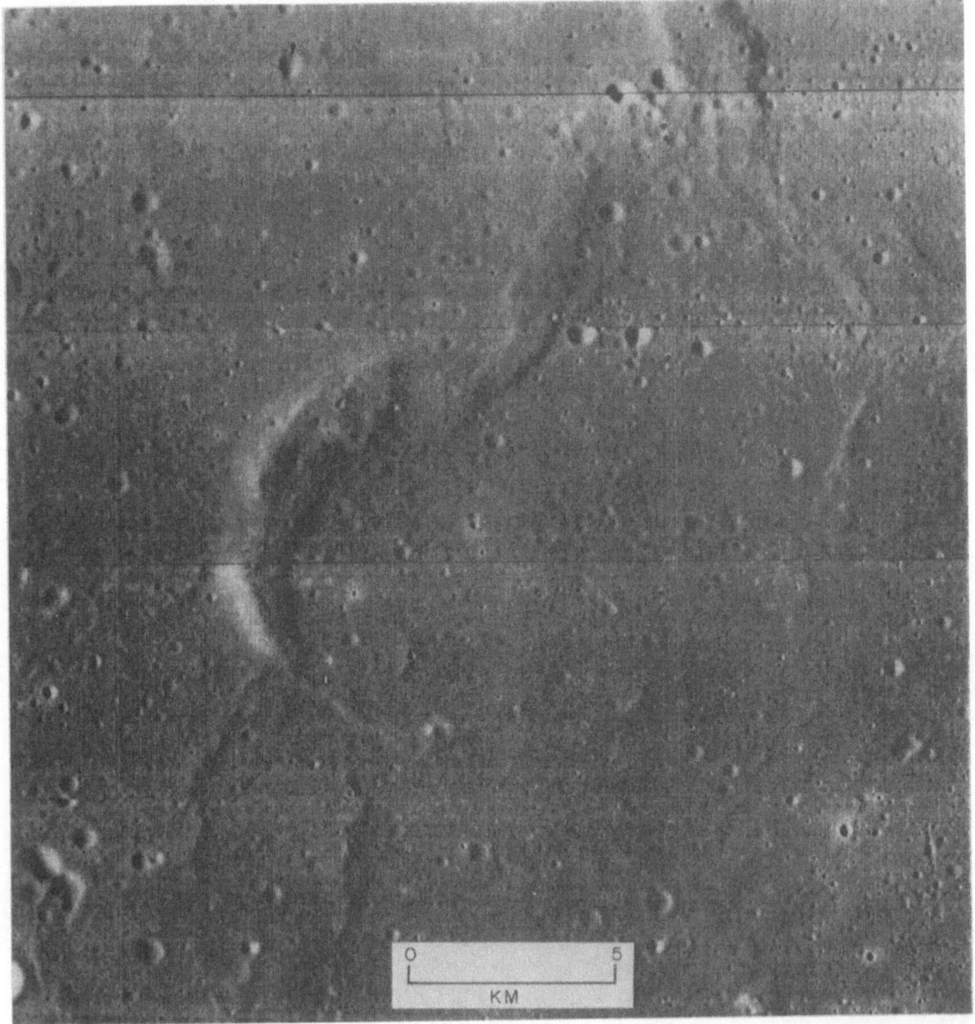

Fig. 19. Partial ring structure in Mare Fecunditatis. Notice that the mare ridge broadens where it forms the ring. *(NASA Lunar Orbiter 1)*

impact from Theophilus, then this ridge, and possibly the entire Lamont structure, is younger than Theophilus. This again indicates that mare ridges and mare ridge rings are relatively young features of post-mare age.

The mare ridges which comprise the rings are in almost every case identical to normal ridges found in linear or arcuate arrangements. Since the mare ridges are almost surely intrusion/extrusion phenomena, the ring structures must be of the same origin. The circular nature of these structures indicates they have formed along ring fractures somewhat similar to terrestrial ring dike complexes. The possible origins of these ring fractures will be discussed later.

Fig. 20. The Lamont ring structure in Mare Tranquillitatis. The arrow indicates the location of the crater chain shown in Figure 21. *(LPL Catalina Obs. photo.)*

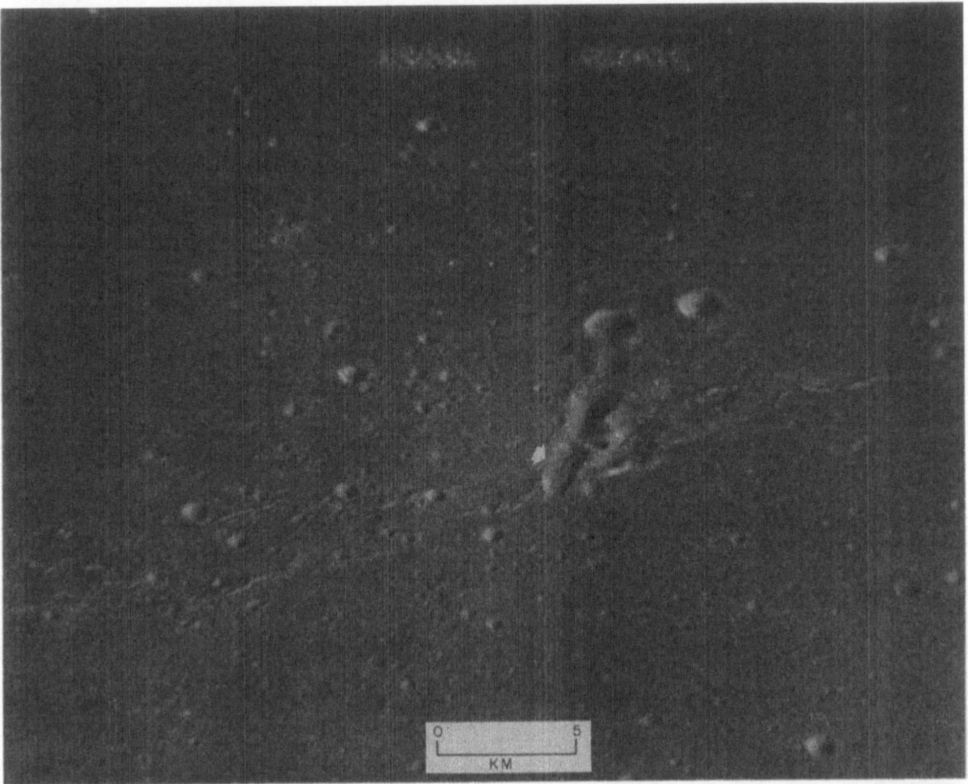

Fig. 21. Mare ridge, forming part of the outer ring of Lamont, which has been extruded into a crater chain. *(NASA Apollo 10 photo)*

5. Composite Ring Structures

Composite rings are circular mare ridges along which are situated bright hills. In almost no case do the hills form a continuous ring; they are usually discontinuous, their strongest development generally occurring along the east and west portions of the ring. Often the northern and southern portions of the ring are subdued or entirely missing. The hills comprising the rings have the same morphology and albedo as the previously mentioned bright hills on linear ridges and, therefore, may have a similar origin and composition.

Figures 22 and 23 are examples of mare ridge rings with associated bright hills. One of the most well-developed ring structures is Flamsteed P in the southern region of Oceanus Procellarum shown in Figure 23 and recently discussed by Guest and Fielder (1968). This structure, 100 km in diameter, is composed of a double ring, the outer ring consisting of a mare ridge or broad gentle arch with bright hills situated along it, and the inner ring consisting only of a mare ridge. It is somewhat similar to the Lamont structure, but contains bright hills and lacks subradial ridges. The mare ridge portions of the rings are related to a prominent linear ridge to the north which

Fig. 22. Composite ring in Oceanus Procellarum south of Reiner. The southern portion of the ring is composed of two bright hills (*H*) situated along and elongated in the same direction as the mare ridges comprising the ring. *(NASA Lunar Orbiter 4)*

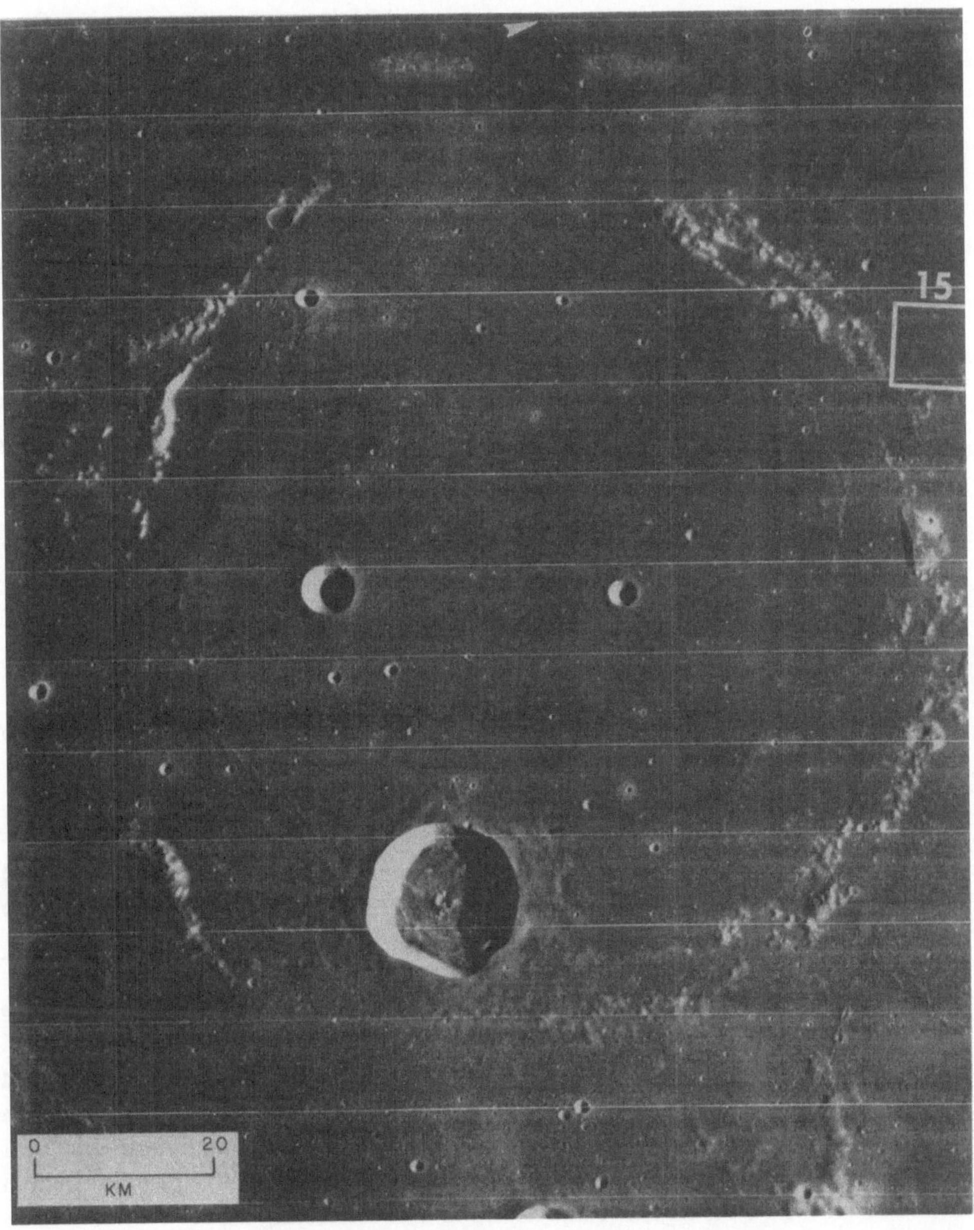

Fig. 23. The Flamsteed *P* composite ring structure in Oceanus Procellarum. The outlined area represents the location of Figure 15, while the arrowhead points to several small hills situated on the linear mare ridge which, further south, forms part of the eastern ring of Flamsteed *P*. Notice that the ejecta blanket of the Flamsteed crater is faintly visible on the outside of the ring but is completely missing from the inside. *(NASA Lunar Orbiter 4)*

has bifurcated to form the outer, and also possibly the inner, ring structure. On a portion of this linear ridge (arrowed in Figure 23) are three or four small, linear knobs which appear to be incipient hills. Both the northern and southern portions of the ring and about 32 km of the western outer ring lack bright hills. Mare ridges tend to connect individual hills on the ring.

The fact that the ejecta blanket surrounding the crater Flamsteed has been partly covered on the interior of the ring but is intact on the outside, as can be seen in Figure 23, shows that the interior is composed of more recent lavas which must be younger than the Flamsteed crater. This is substantiated by crater counts on the outside and inside of the ring by Fielder *et al.* (1968) and Fryer and Titulaer (1969) which show that the crater density on the interior is appreciably lower than on the exterior. The group of low hills forming the southern portion of the crater ring appears to have been partly covered by the ejecta blanket of Flamsteed indicating they are older than this crater. These facts strongly suggest that the Flamsteed ring has undergone a protracted development.

If the hills of the outer ring are extrusive, the sequence of events for the formation of Flamsteed P would be as follows:

(1) General flooding of the mare with highly fluid basaltic lava.

(2) Formation of the outer mare ridge ring and bright hills.

(3) Formation of Flamsteed crater.

(4) Flooding of the interior of Flamsteed P with highly fluid basaltic lava, possibly erupted from ring fractures later utilized in stage 5.

(5) Formation of the inner ring ridge.

If the bright hills are remnants of a buried crater rim it could be argued that the distribution of the ridges was controlled by the underlying crater. However, it is very difficult to explain the observation that none of the interconnecting ridges, so closely associated with the hills, cut the hills despite the fact that the ridge would be younger. No matter what the origin of the hills there is clearly late flooding and ridge formation within the ring indicating that this structure was the locus of igneous activity over a protracted period of time.

Figure 24 is a 36 km ring in the northern part of Mare Fecunditatis photographed by the Apollo 8 astronauts. The western side of the ring consists of a mare ridge with an associated short flow (F). This ridge continues to the north as a more-or-less linear mare ridge. The eastern side consists of a discontinous, arcuate chain of hills of about the same albedo as the adjacent mare terrain. The interesting aspect of these hills is that they consist of individual low conical or dome-shaped hills which resemble in many respects several of the volcanic cones and domes in the Marius Hills, and the chain of very dark volcanic hills in the southern part of Fra Mauro. On the summits of at least three hills there appear to be small pits which may be volcanic vents.

The association of bright hills with mare ridge rings and the similarity of these hills to those superimposed on the linear ridges suggests that the mare ridge rings with bright hills are post-mare volcanic ring complexes. The rings are no more than variations of the linear mare ridges which have apparently formed along ring fractures.

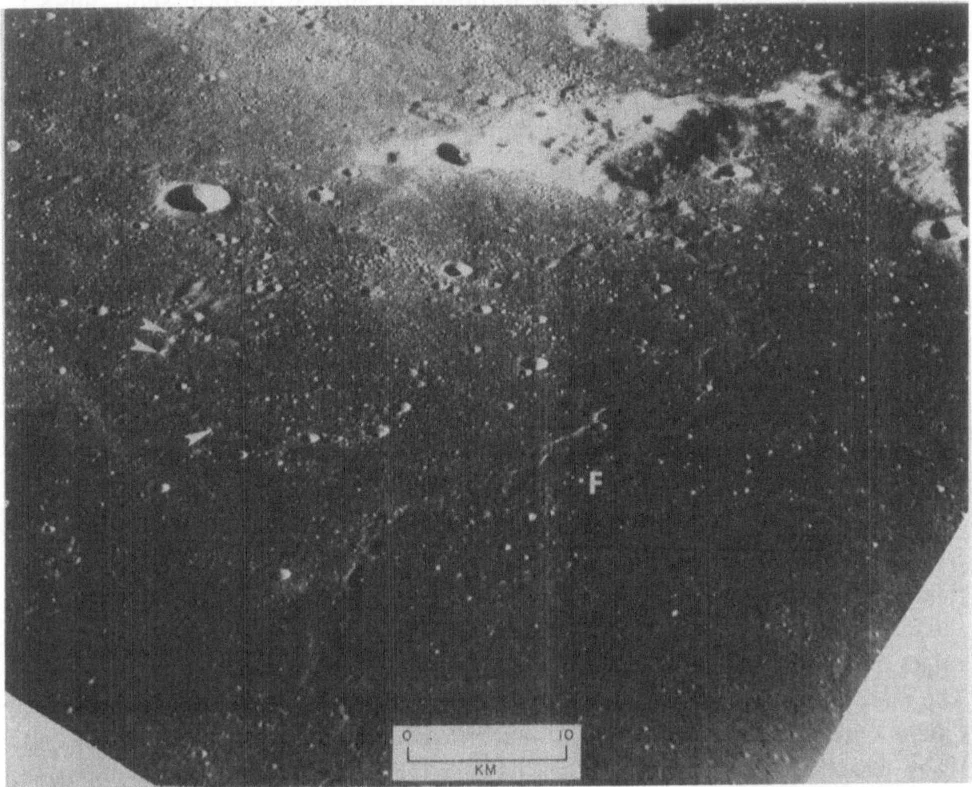

Fig. 24. Composite ring 38 km in diameter in Mare Fecunditatis. The righthand portion of the ring consists of a typical mare ridge with an associated flow (*F*), while the other side of the ring is composed of low hills or domes. Several of the hills may contain summit pits indicated by the arrows. *(NASA Apollo 8 photo.)*

The low cones or bright hills seem to be later eruptions which formed over the ridges. The high albedo and highland-type morphology of the hills suggest that they are of different composition from the mare ridges and may represent a late magmatic differentiate which evolved from igneous bodies associated with ridges.

6. Summary and Conclusions

It has been shown that mare ridges often consist of two separate but related features: (1) a broad gentle arch overlain by (2) a sharper more contorted ridge. The sharper, secondary ridge is usually situated on the crest of the arch and, in many instances, has clearly flowed into and partially filled pre-existing craters in the adjacent terrain. This relationship of broad arching with extrusive activity indicates that the mare ridges which show this relation consist of intrusions – probably in the form of sills or laccoliths intruded along planes between successive lava flows – that have uparched the surface and subsequently broken through the surface to form short flows or bulbous lava extrusions. The uparching apparently caused fracturing of the uplifted surfaces providing egress for the magma to reach the surface. Other ridges consist only of extrusions, while still others seem to represent gentle uplifts possibly caused by sill or laccolithic intrusions unaccompanied by an extrusive phase.

The fact that many mare ridges coincide with the directions of the global fracture pattern found in the highlands indicates that they are located along major fractures in the lunar crust. It is possible that the fractures along which the mare ridges formed were a major source of the low viscosity lavas that comprise the bulk of the maria. This view is strengthened by the fact that very extensive flows in Mare Imbrium seem to have issued from several prominent mare ridges. The ridges as they appear today probably represent the last intrusive/extrusive phase of this flooding.

There is at least one compelling and several probable examples of isolated bright hills overlying linear mare ridges. There is little doubt that these hills, at least, are post-mare volcanic cones or domes. The high albedo and highland-type morphology of these hills implies a composition different from that of the maria. Similarly, the contorted nature, ropy texture and short flow distances of the extrusive parts of the ridges suggest that they were more viscous than the bulk of the lavas which comprise the maria. Therefore, it seems possible that the igneous masses which gave rise to the ridges underwent magmatic differentiation prior to being extruded onto the surface. The mare ridge extrusions of the same albedo as the maria may represent an early differentiate with a higher silica and lower iron content relative to the Apollo basaltic rocks, whereas the bright hills may be a late differentiate of more siliceous composition similar to the siliceous rock of intermediate composition (No. 12013) returned by the Apollo 12 astronauts (*Science*, 1970b).

The fact that ordinary mare ridges take the form of rings must mean that they have formed along ring fractures in the same manner as the other ridges. This would be true for both mare ridge rings with and without bright hills. There are two possible explanations for the origin of the lunar ring fractures: (1) they are the result of very large cylindrical intrusions which have been intruded along major faults and fractured the surface in an annular pattern, or (2) they represent ring fractures or zones of weakness associated with the rims of buried craters which have been reactivated by later major fracturing. The former origin is somewhat similar to the formation of terrestrial ring dike complexes, whereas the latter has no known terrestrial counterpart. In several cases, such as Sinus Iridum and Respold R, linear mare ridges are diverted in an arcuate path around partially flooded highland craters. The arcuate part of the ridge is at the location where the crater rim is missing. This suggests that the formation of the mare ridges in these areas was influenced by pre-existing crater rims and were formed along reactivated ring fractures associated with the buried part of the rims, or that the major fractures were diverted around the buried crater rims because they represent a pre-existing zone of weakness. In either case the buried crater rims would be the indirect cause of the arcuate ridges. This may be the explanation for some of the rings in the maria. However, it is difficult to explain how the circular shape of the smaller rings, many of which are smaller than the ridge, could be so perfectly preserved by this mechanism. One would expect that the intrusive and extrusive activity along the fracture would completely obliterate such small, buried craters. Furthermore, many of the larger rings are quite irregular in shape, being very far from perfect circles. If pre-existing craters influenced the shapes of these ridges one would expect them to be more circular. The 8 km diameter ring comprised of flows near Flamsteed (Figure 16) is separated from, but possibly related to, the nearby mare ridge, and could only have been extruded along a separate ring fracture. This separate ring fracture is probably best explained by the intrusion of an independent igneous body rather than reactivation of fractures related to a buried crater. If the latter explanation applied there should be many other examples of this unique structure.

Also the Flamsteed P structure has undergone a late stage of flooding of its interior after the main period of mare deposition, indicating a protracted period of development. Although the bright hills comprising the Flamsteed ring may be remnants of a buried crater rim, it is difficult to understand the late flooding of its interior unless this structure was underlain by an independent magma source capable of expelling large volumes of lava. (The volume of lava necessary to fill Flamsteed P to a depth of 10 m is 80 km^3).

For these reasons I favour the hypothesis that many of the mare ridge rings are post-mare volcanic ring complexes associated with large, igneous bodies rather than reactivated fractures associated with buried crater rims; and that most of the bright hills associated with mare ridges are post-mare volcanic domes or cones of more siliceous composition.

It should be emphasized that I do not consider all incomplete rings in the maria to be of post-mare age. There are numerous examples of partially buried rims which are clearly pre-mare, flooded craters.

References

Fielder, G.: 1965, *Lunar Geology*, Lutterworth, London.

Fielder, G. and Kiang, T.: 1962, *Observatory* **82**, 8.

Fryer, R. J.: 1970, 'The Origin of Lunar Craters', in *Geology and Physics of the Moon*, Elsevier Pub. Co., Amsterdam (in press).

Fryer, R. J. and Titulaer, C.: 1969, *Comm. Lunar Planetary Lab.* **8**, 51.

Fulmer, C. V. and Roberts, W. A.: 1967, *Icarus* **7**, 394.

Guest, J. E. and Fielder, G.: 1968, *Planetary Space Sci.* **16**, 665.

Science, 1970a, **167**, 447–781 (The Moon Issue: Analyses by several authors.)

Science, 1970b, 'Preliminary Examination of Lunar Samples from Apollo 12', by Lunar Sample Preliminary Examination Team, 167, 1325, Table II.

Strom, R. G.: 1964, *Comm. Lunar Planetary Lab.* **205**, 216.

A POSSIBLE MECHANISM OF THE GENERATING OF THE UNUSUALLY LONG LUNAR SEISMIC OSCILLATIONS

I. P. PASSECHNIK and D. D. SULTANOV

Physics of Earth Institute, U.S.S.R. Academy of Sciences, Moscow, U.S.S.R.

Abstract. One of the possible reasons of the long duration of lunar seismograms are multiple reflections in the upper part of lunar cross-section, similar to those generating under Earth conditions. While registration of seismic oscillations in a region with stratified low-velocity sediment rocks of 6–7 km thickness the authors observed long seismic low-decaying oscillations. The observed decay was about 40 db during 30 min.

The long duration records are obtained only on the distances where the intense overcritical reflections exist from the M-boundary situated at a depth of 50 km.

These waves generated numerous multiple reflections, longitudinal, shear and converted waves in the upper part of the cross-section. These body waves are responsible for the generation of the surface waves, recorded in the subsequent parts of the seismograms. Periods of body waves are about 1–1.5 s., those of surface waves – about 1.5–3 s.

Extrapolating similar mechanism of the wave generation to lunar media, one should assume the presence of a layered upper part of the cross-section and the possibility of the sufficiently intense wave incidence on its lower boundary.

It is possible if either reflecting boundaries or a vertical velocity gradient are present in the underlying media. The first variant is accompanied with overcritical reflections, the second one – with intense refracted waves.

Lunar seismograms and their description are presented in numerous papers of American authors (Latham *et al.*). The main peculiarity of these records, received in the lunar maria from sources of slight intensity, is the long duration of oscillations and the slow time decay. The mechanism of generation of such long records is not at present clearly understood. The source intensities at the points of impact were equivalent to only one and ten tons of TNT respectively. The registration of oscillations was at distances of 75 and 146 km. In terrestrial conditions the durations of records from sources of such low intensity at the same distances do not exceed some dozens of seconds, the decay of the oscillations being about 100 dB. The absolute values of the maximal displacement amplitudes on the terrestrial and lunar records are approximately equal in magnitude and are limited to the range from a tenth part to tens of millimicrons, their period T being about 1–2 s.

In the problem discussed the most interesting question is: what is the mechanism of generation and spreading of the lunar seismic oscillations so as to cause such poor time decay and specific form of envelope? The question has been studied in some papers (Latham, Gold *et al.* [1–3], U.S.A., Muchametdzanov *et al.* [4] U.S.S.R.). We think that a satisfactory explanation has not yet been obtained. The different mechanisms of generation of long seismic lunar oscillations discussed in [1–3] are disputable and more convincing proofs are needed. Some hypotheses, as, for example, the mechanism of multicascade re-descent of ejected rocks accompanying high speed impact in lunar conditions, proposed by Muchametdzanov [4], can be rejected because the proof of the endogenic character of a part of the moonquakes is accepted (Ewing *et al.* [6]).

Urey and Runcorn (eds.), The Moon, 216–219. *All Rights Reserved.*

The authors of this paper tried to find records, received in terrestrial conditions, analogous to the lunar ones. If this were possible one could study in terrestrial conditions the mechanism of formation of the long records and extrapolate it to lunar records. For this purpose we studied the seismic records, received in terrestrial conditions at epicentral distances from 20–40 up to 300 km in regions with different seismo-geological structure of the upper part of the cros-section. While registering seismic oscillations in a region with stratified low-velocity sedimentary rocks of 6–7 km thickness the authors observed long, slowly decaying seismic oscillations. The observed decay was about 40 dB during 30 min.

The long duration records are obtained only at distances more than 90 km where intense overcritical reflections exist from the *M* boundary situated at a depth of 50 km. These waves generated numerous multiple reflections, longitudinal, shear and converted waves in the upper part of the cross-section. These body waves are responsible for generating the surface waves, recorded in the subsequent parts of the seismograms. Periods of body waves are about 1–1.5 s. Those of surface waves about 1.5–3 s.

Extrapolating a similar mechanism of wave generation to lunar media, one should assume the presence of a layered upper part of the cross-section and the possibility of sufficiently intense wave incidence on its *lower* boundary. Either reflecting boundaries or a vertical velocity gradient must be present in the underlying media. The first possibility is accompanied by overcritical reflections, the second by intense refracted waves.

Two parts of the long record are shown in Figure 1. In regions where low-velocity sedimentary rocks are absent, long oscillations are also absent. Latham *et al.*, using an averaged curve of the time dependence of the oscillation amplitude decay, made

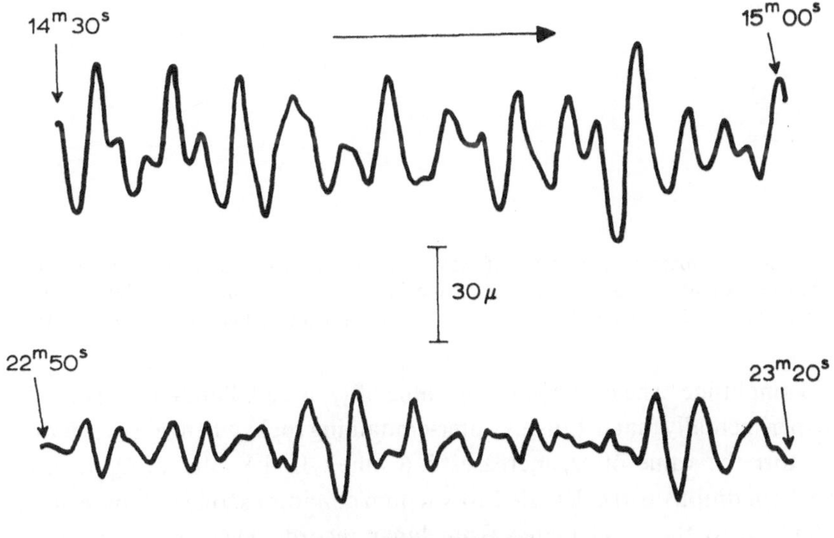

Fig. 1. Two fragments of long seismic terrestrial record, duration is over 23 min. Vertical component.

estimates of Q_t, the quality factor parameter, characterising the diminution of the oscillation intensity during a period. The record received from the impact of the lunar module gave a value $Q_t = 3600$. It is necessary to note that this parameter Q_t is not the analogue of the quality factor Q, characterising oscillation intensity decay per wavelength because of the non-ideal elasticity of the media. The calculated value of Q_t is some effective parameter, caused by the joint influence of the layered character and the non-ideal elasticity of the media.

Using the same method we determined Q_t from the records received in terrestrial conditions in the region of existence of overcritical reflections from the M-boundary. Figure 2 represents the averaged curves of the time dependence of the ratio current to

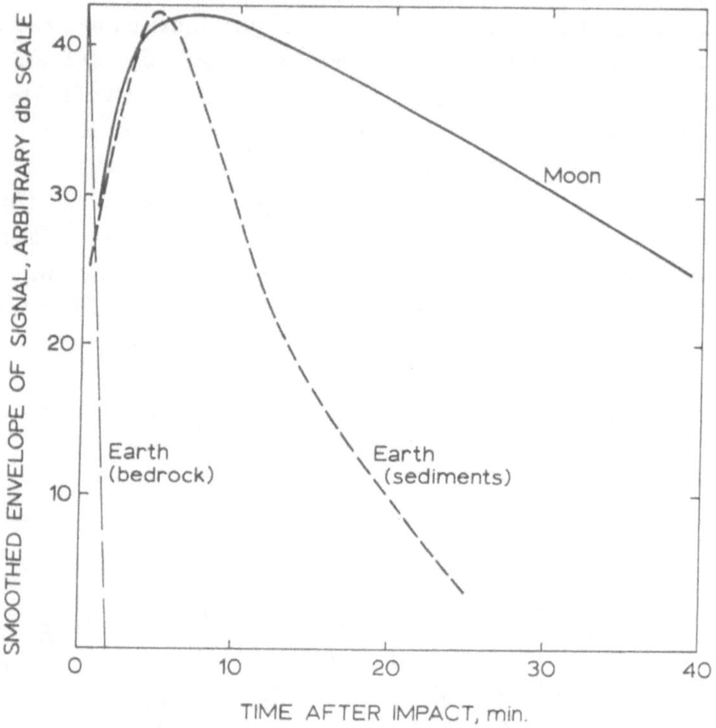

Fig. 2. Smoothed observed envelopes of wave train (vertical component) – the ratio of current amplitude to maximal one. Solid line – the same for the lunar record caused by the impact of the lunar module [1,2]. Dashed lines – the records, obtained in different terrestrial conditions.

maximal amplitude. The corresponding value of Q_t is 550. Pandit and Tozer [5] established experimentally that in porous water-containing rocks when pressure is diminished to 10^{-5} torr the value of Q_t increases 5–6 times. If the value of Q_t determined for terrestrial conditions is recalculated to vacuum conditions, one obtains a value practically identical to those calculated from lunar records, even neglecting lower lunar gravity forces. This also may be evidence in support of the proposed mechanism of generating long lunar seismic oscillations.

References

[1] Latham, G. V., Ewing, M., Press, F., Sutton, G., Dorman, J., Nakamura, Y., Toksöz, N., Wiggins, R., Deer, J., and Duennebier, F.: 1970, *Science* **167**, 455.
[2] Latham, G. V., MacDonnald, W. G., and Moore, H.: 1970, *Science* **168**, 242.
[3] Gold, T. and Soter, S.: 1970, *Science* **172**.
[4] Muchamedzanov, A. K. and Davidova, G. V.: 1970, *Kosmich. Isled.* **8**, 795.
[5] Pandit, B. I. and Tozer, D. C.: 1970, *Nature* **226**, 335.
[6] Ewing, M., Latham, G., Press, F., Sutton, G., Dorman, J., Nakamura, Y., Meissner, R., Duennebier, F., and Kovach, R.: 1971, 'Origin and Evolution', in *Highlights of Astronomy* (ed. by C. de Jager), Reidel-Dordrecht, p. 155.

ON THE INTERACTION BETWEEN TECTONIC PROCESSES OF THE EARTH AND THE MOON

N. A. KOZYREV
The Pulkovo Observatory, Leningrad, U.S.S.R.

Abstract. At present seismographs are operating on the Moon as well, installed there owing to the successful Apollo missions. However these data are insufficient for detailed statistic investigations. That is why in case of the Moon we are to use indirect indications of its activity, such as the data on transient light phenomena from the catalogues by Miss B. Middlehurst. Among the great number of earthquakes there were chosen only the strong earthquakes (magnitude 6.5) with focuses deeper than 70 km. According to these characteristics 630 earthquakes were selected from 1904 to 1967. In the Middlehurst catalogue during the same period about 370 transient events on the Moon are registrated. A distribution of lunar events on the days of an anomalistic month gives evidence of the influence of the Earth's tidal forces (the Middlehurst effect). It appears that the distribution of earthquakes gives a similar curve. Thus the tidal interaction of the Earth and the Moon establishes certain synchronism in tectonic activity of these planets. The further statistic analysis reveals some more causal relation between the processes of the Earth and the Moon. Strongly pronounced maximum of lunar events is observed with the interval of 2–3 days after the earthquakes and the maximum of earthquakes – with quite the same interval after the lunar events. The peaks of these maxima exceed the mean number of events by a factor 3. The Moon Earth system is the astronomical example of a direct interaction of the processes in the neighbouring celestial bodies.

The corresponding experiments, made at the Pulkovo Observatory, confirm the possibility of immediate interactions of irreversible processes due to the change of physical properties of time. Thus we can form a chronology of orogenesis on the Moon judging from the data on the history of the Earth. Tectonic processes of the Earth and the Moon seem to be in such a close interaction as if the Moon were in direct contact with the Earth, i.e. in other words, were its seventh continent. These conclusions give evidence of the extreme importance of regular seismic observations on the Moon.

Seismic phenomena can serve as the direct, quantitative indication of tectonic activity of the planets. On the Moon such phenomena are being registered at present by seismographs installed there. However the data obtained embrace too short a period and are insufficient for statistic investigations. Therefore we are to use such an indirect indication of lunar activity as the transient luminous events on the Moon surface are observable from time to time. The vast catalogue of such events was compiled by Middlehurst *et al.* (1968). A certain number of erroneous information which could penetrate in the catalogue are not dangerous for the results of statistical investigations. However it is very important to account for extra-selectivity of the data in this catalogue, which can be explained by the specific conditions of lunar observations such as the Moon phases, height over the horizon, weather and even the heightened interest of astronomers towards the studies of the Moon surface. That is why the quantity of lunar events may not be compared directly with the quantity of simultaneous earthquakes.

Unlike to the lunar events, observed selectively, the total registration of earthquakes all over the Earth are being made regularly since 1904. It was necessary to choose from the great number of registered earthquakes those ones which gave evidence of tectonic processes of planetary scale. Therefore from the Gutenberg and Richter catalogue

Urey and Runcorn (eds.), The Moon, 220–225.

(Gutenberg and Richter, 1949) containing the list of earthquakes with the magnitude $\geqslant 7$ for the period from 1904 to 1946 the earthquakes with deep focus ($h > 300$ km) and intermediate ones ($70 < h < 300$ km) were selected. The data for the latest years, including 1967, and identical types of earthquakes (with the magnitude $\geqslant 6.5$) were selected from the International Seismic Bulletins. As a result the list of 630 earthquakes was compiled. In the Middlehurst catalogue about 370 transient events on the Moon were registered during the same period (from 1904 to 1967).

Gravitational interaction of the Earth and the Moon gives rise to the tides which can turn into a trigger mechanism responsible for a certain synchronousity in tectonic processes of these bodies. A potential of tidal forces on the Moon is 5 times as large as that on the Earth whereas the gravitation on the Moon surface is 6 times as small. Therefore deformations of the Moon surface can be larger than those of the Earth surface by a factor 30. In the same relations will be the variations of tidal deformations occurred due to a considerable eccentricity of the lunar orbit. A significance of these variations for processes on the Moon gives evidence of the dependence of the number of transient luminous phenomena on the days of an anomalistic month, i.e. a position of the Moon on the orbit. For the first time this dependence was discovered by Middlehurst (1967) on the base of her first catalogue (Middlehurst and Burley, 1966). The curve of distribution of the earthquakes shows a certain similarity of this dependence. The upper curve (Figure 1) gives the dependence of the number of selected earthquakes upon a position of the Moon on the orbit. The lower curve gives the number of lunar events taken from the second Middlehurst's catalogue for the same time. Unlike to the results of Miss Middlehurst the maxima were detected not only in perigee and apogee, but also near the moments of the highest speed of changing the geocentric distances. The upper curve of the number of earthquakes has also the similar peculiarities. Thus really the tides increase somewhat the probability of simultaneous events on the Moon and the Earth. This increase of probability can be

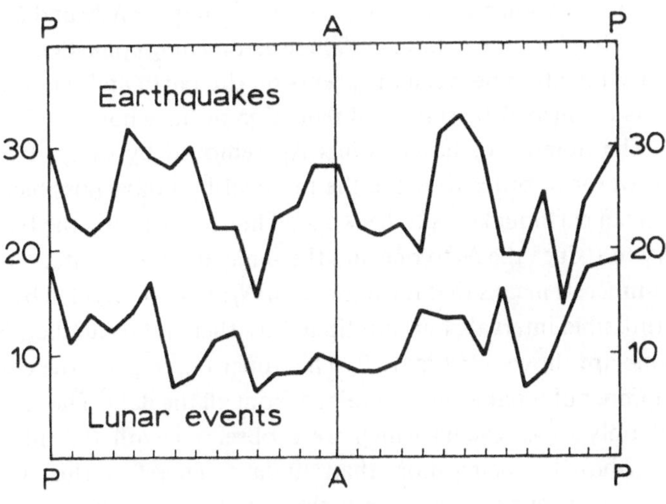

Fig. 1.

TABLE I

1897

⊕	☾ I	☾ II
12 VI	–	14 VI
5 VIII	–	–
20 IX	–	–
21 IX	21 IX	21 IX
–	–	(8–15 X)
–	–	9 XII

revealed only by superposition of the great number of anomalistic periods, while individual events occur often without any connection with tides. And indeed it is seen from the given curves that the number of events in the minima is no less than $\frac{1}{2}$ of those in maxima. Let us consider only these extremum values instead of the curves and assume them to agree precisely in time. The probability of agreement between lunar and terrestrial events will be proportional to $(1 + \frac{1}{2}\,\frac{1}{2})$ while a disagreement to $(\frac{1}{2} + \frac{1}{2})$. Thus according to the obtained curves the probability of coincidence of the events might increase not more than by 25% (rather – less). Therefore some occasional events must occur independently in practice. At the same time the direct comparison of the days, when the transient events were observed on the Moon, with the days of deep earthquakes shows remarkable coincidence. On April 1, 1969 the author obtained the spectrum of a transient red spot inside Aristarchus. The day before, on March 31 in the morning a deep earthquake took place in Egypt, and in the evening – at the seashore of Japan. In 1897 four deepest earthquakes (magnitude 8.5) occurred. In the table a comparison of the dates of these earthquakes with those of lunar events according to the first and second catalogues by B. Middlehurst is given. The first earthquake of the table is the great Indian earthquake. Via two days after it the volcanism in the Shröter Valley on the Moon was observed. The earthquakes on 20 and 21 of September were followed by the glow in Aristarchus. These astonishing coincidences give evidence of the direct causal relation between the events on the Earth and the Moon. Existence of this relation is confirmed by statistical reduction of the whole obtained data. There was calculated the number of lunar events N_i, removed by i days from the closest earthquake. However in order to make the removal by i days possible, the length of an interval between earthquakes is to be $d \geqslant 2i$. Thus for deriving the true distribution n_i it is necessary to reduce the N_i to one and the same number of intervals, for instance to their total number. It means that the number of N_i are to be divided by the calculated number of permissible intervals and multiplied by their total number. The results of such calculations are shown on Figure 2. The upper curve gives the corrected distribution of the number of lunar events, obtained from all the data. The lower curve gives the number of only those events which were observed with the intervals between neighbouring earthquakes being more than 30 days. Therefore this curve within the limit of 15 days can be constructed without any reduction, which is very important for

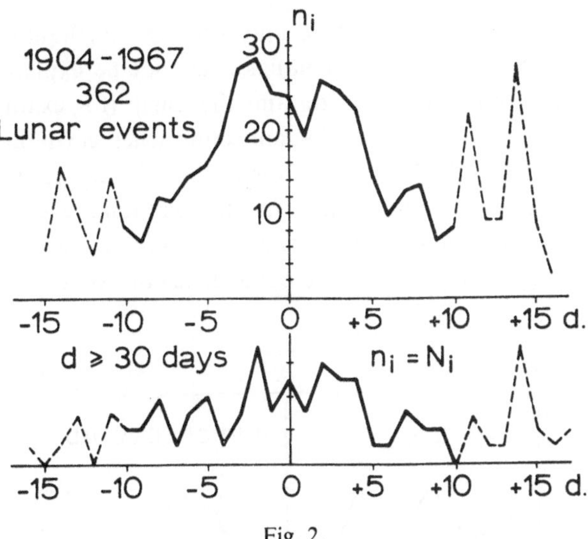

Fig. 2.

checking the results. The lunar events observed after the earthquakes are plotted to the right of the origin of coordinates whereas those observed before the earthquakes – to the left. The conditions of the Moon observations are independent on the earthquakes, unless the earthquake occurs just under the telescope. Therefore with the exception of such an unreal situation, incompleteness of the catalogue can only diminish the total number of data but cannot change their distribution. A lunar event being removed by more than 10 days from the nearest earthquakes, the reductional factor is more than 4. Then the obtained data becomes inaccurate and cannot be taken into account, that is why on the upper graph they are drawn by dotted lines. The spread of points on the lower graph is due to lack of data. Nevertheless this graph confirms that near to the earthquakes the number of lunar events increases by about a factor 3 as compared with the common background and that two peaks of maximum exist really. Any of the epochs detected from the whole data have also two peaks of maximum, which are nearly symmetric, at a distance of 2–3 days from the earthquakes. Even the small number group of events, taken for the short period (from 1948 to 1953) shows the analogous result (Figure 3).

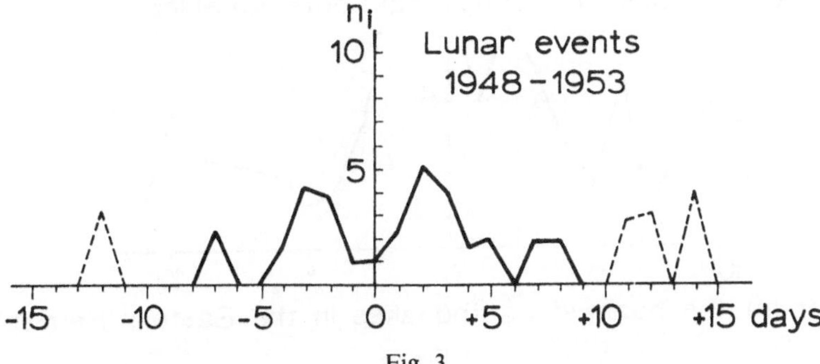

Fig. 3.

The existence of two maxima (before and after the earthquake) shows that the grouping of lunar events near the earthquakes cannot be explained by a common reason which influences upon the Moon and the Earth (for example the high solar activity). Rather there is a direct double-connection when earthquakes cause a lunar event with a delay of 2–3 days and, vice versa, lunar events cause terrestrial events with the same delay. It is interesting that coincidence of lunar events with earthquakes takes place only when the lunar and terrestrial crust are weakened by tides. Indeed the middle graph in Figure 4 shows that the coincidence of events occurs only near the

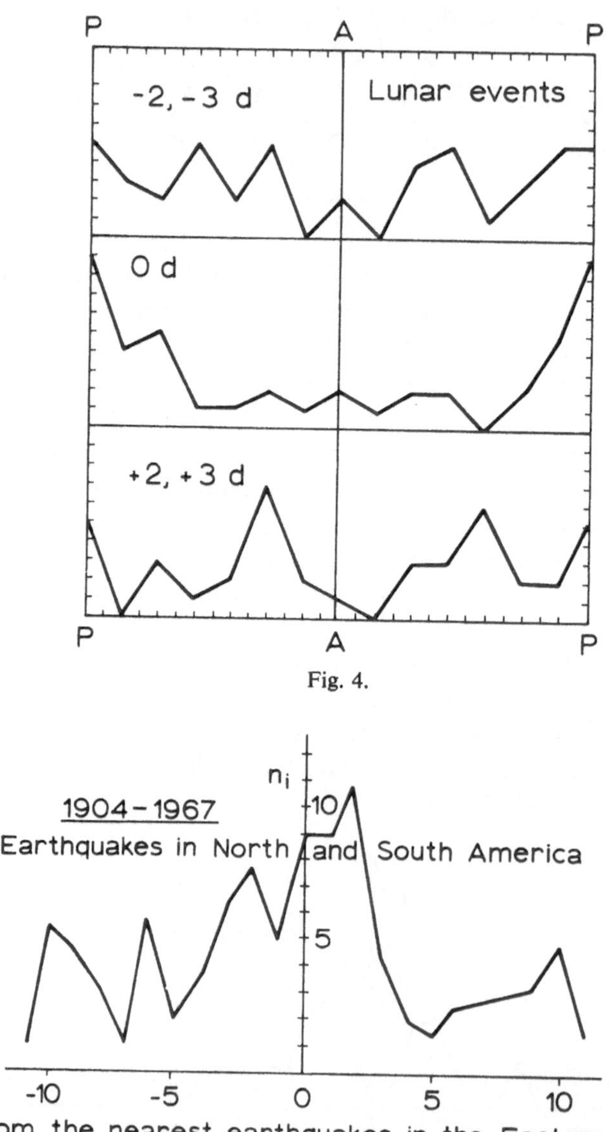

Fig. 4.

Fig. 5.

perigee. The preceding and following maxima, as it is seen from the lower and upper graphs, are quite independent on tidal effects. This result shows once more that the obtained interaction between earthquakes and lunar events cannot be explained by synchronism of tides. Most probably we have received an astronomical example of direct interactions between the processes occurred in the neighbouring systems. The indications of such interaction are present not only in the Earth-Moon system but also the double star systems (Kozyrev, 1967). Connections between processes could be established by current time, if its course is the physical property existing objectively. The corresponding experiments, made at Pulkovo, give evidence of the physical properties of time, varying in the vicinity of any irreversible process. These variations of time properties can be influenced by the other system. Thus the connections of systems in nature can be realized not only through space by means of fields but also through time by means of its physical properties. It is probable that by the same way the interaction between tectonic processes on the remote continents of the Earth is realized. Figure 5 gives an example of such interactions. It represents the distribution in time of the earthquakes in the America with respect to those in the eastern hemisphere (Eurasia). Most earthquakes of the 630 ones, considered by the author, occurred in the Eurasia and only 96 – in the America (Northern and Southern). This graph was plotted by the same method as that used in the analysis of the lunar events. The similarity of the obtained relations appears to be remarkable: the reaction of the Moon and the America on the earthquakes in the Eurasia is near the same. In both cases there was a marked delay of the maxima by two days and their heights were about 3 times as large as the background. Thus tectonic processes on the Earth and the Moon are in such interaction as if the Moon were the seventh continent of the Earth. Hence it means that the cycles of the Moon orogenesis must be synchronous with those of the Earth orogenesis. In this case we can construct the chronology of tectonic processes on the Moon from the data on the Earth history. For example it is quite possible that the recent Kopernic period of the Moon history coincides chronologically with the period of the Alpine orogenesis on the Earth.

The above investigations show that regular seismic observations on the Moon is very important for the study of its physical properties and also for solving the problems which are of a great scientific interest in principle. It is necessary to make every effort for making the permanent network of seismic stations on the Moon.

References

Gutenberg, B. and Richter, C. F.: 1949, *Seismicity of the Earth and Associated Phenomena*, Princeton.

Kozyrev, N. A.: 1967, *Colloquium on the Evolution of Double Stars*, Uccle.

Middlehurst, B.: 1967, An analysis of Lunar Events, *Rev. Geophys.*

Middlehurst, B. and Burley, J.: 1966, *Chronological Listing of Lunar Events*, NASA, Maryland.

Middlehurst, B., Barley, J., Moore, Patrick, and Welther, B.: 1968, 'Chronological Catalogue of Reported Lunar Events', NASA.

ON THE ORIGIN OF CENTRAL PEAKS IN THE CRATER FORMATIONS FILLED WITH MELT AFTER IMPACTS

L. KŘIVSKÝ

Astronomical Institute, Ondřejov, Czechoslovakia

Abstract. An attempt is made at explaining the successive origin of the central peaks and terrace ring-wall in the crater formations filled with melt after impacts.

1

In the course of experiments with freezing water containing an organic fine and coarse substance in a vessel at temperatures of -10 to $-15\,°C$ and freezing distilled and undistilled water in a vessel at temperatures of -3 to $-10\,°C$ the forming of the shape of the surface and the structure of the internal content were investigated. Description of one of the experiments: Freezing of undistilled water in an open glass vessel (height 12 cm, width 7 cm, water level at a height of 9 cm). The initial temperature of the water was $\sim 30\,°C$, the ambient temperature -5 to $-10\,°C$, and the temperature of the base on which the vessel was standing was $-5\,°C$. Conditions after about 10hr of exposure: the water froze in a thicker layer at the bottom of the vessel, the walls but less at the surface and remained clear; in the central part of the water the space with the unfrozen water had the shape of a smaller egg filled in its upper part with an air bubble. Surface profile: central peak of a height of about 0.5 cm with a broader base. Conditions after about 20hr of exposure: the water in the vessel froze completely; around the vertical axis of the vessel inside the ice there are regions of fine frozen-in bubbles pointing towards the axis and slightly upwards, and they are also ordered along the upper part of the axis under the surface where they point towards the central surface peak; the surface kept its previous shape.

The central peak is clearly due to the initial freezing of the boundary regions of the water which has the shape of a cylinder, and at his time gaseous components are released and pressed out into the central liquid regions. The latter freeze a little later. The volume pressure from the sides causes the unfrozen water in the region of the vessel axis to be compressed vertically, and the liquid with gaseous cavities is pushed up through the narrow central region to the weakly frozen surface were it also freezes. The central peak is formed by liquid eruptions of longer duration within a narrow space in the centre of the freezing surface, where the gaseous components are pressed out and released. These are in fact trivial processes occurring during the gradual non-uniform freezing of water in vessels of cylindrical or bowl-like shapes, when volume and structural differentiation occurs. These processes can be compared with the processes of solidification of melted rocks in craters, when the melted rocks after the impact and explosion of large meteors, or after an effusion of subsurface origin as a result of perturbation by impact. The mechanism described, due to non-uniform

Urey and Runcorn (eds.), The Moon, 226–230.

solidification and volume differentiation with the forming of the central peak would of course be applicable if water were present on the Moon in larger volumes in the solidifying and crystallizing melt and provided the melt would go down to a sufficient depth. The water could even be at the lower horizons, at the bottom of craters. This possibility has not been excluded yet. Analogous formations based on this 'water mechanism'' are also known from the Earth's surface. The cooling and solidifying connected with crystallization progressing from the edges to the centre of lunar craters may form an analogous mechanism of pressing out escaping gaseous and liquid constituents into the central regions, where extrutions would occur in the centre of the surface above the slowly solidifying surrounding surface; a cone with a 'funnel' would be formed through which the gases from the lower levels would escape at irregular intervals.

The size of the conical peak above the surface would depend on a number of parameters, e.g., on the depth of the region of the molten rocks, on the ratio of the diameter of the melt pool to its depth, on the physical and chemical properties of the melt (on the structural differentiation during the process of solidification), on the ambient temperature of the rocks around the crater, and on the temperature gradient along the radius.

It is also possible that the volume of the solidified melt itself would not have to be larger in comparison to the volume of the original liquid melt in order to form structural differentiation leading upto the formation of the central regions with escaping gases as is the case in the water experiments.

2

Should the melt be nearly waterless (and should there be no water even in the subsurface horizons), the process of structural differentiation will be applicable together with the dynamics within the spatially non-uniformly solidifying melt as a result of gravitational differentiation, which is well known, e.g., from the solidifying of molten basalts under laboratory conditions. The volume of a solid crystallized basalt is 3–10% smaller than the volume of the original melt. In the solidifying volumes of spherical shapes large cavities are formed in the centre (Kopecký, 1971).

Processes would occur on the Moon which are illustrated in sequence on the vertical section of the crater in the diagrams attached.

Figure 1a: (a) – regions of initial solidification of a basalt melt which was created by heating on impact or by partial effusion of the still liquid materials from below; heat losses by radiation and transfer after impact are the largest in these regions.

Figure 1b: (a_1) – the heavier sinking blocks of solidified and half-solid basalt melt; (b) – the hottest central regions where the gaseous constituents are concentrated. This lighter region has the tendency to ascend. The gaseous bubbles join up and larger and larger regions filled with gas are formed with a tendency of a more rapid ascension towards the surface. The gases escape from the bubbles mostly in the central regions on the surface (c). The gaseous bubbles captured temporarily or permanently by the top layer may form domes (d). Part of ascending melt flows out to the sides where it partly substitutes the sinking solid blocks. The sinking of the blocks (a_1) is the largest on the inside of the crater, and faults occur on the outer sides of the blocks. The outer edges of these slanting blocks form new lower inner edges of boundary ringwalls (e). This effect may be repeated several times as a result of the irregular decreases and turning of the solid boundary blocks around horizontal axes. The overall height of the surface gradually decreases (as a result of the total diminution of the volume of the solidifying melt). Cracks appear on the surface crust (f) and this is where the melt extrudes and the gas bubbles escape from the lower regions (b) above the sinking parts of the solid blocks.

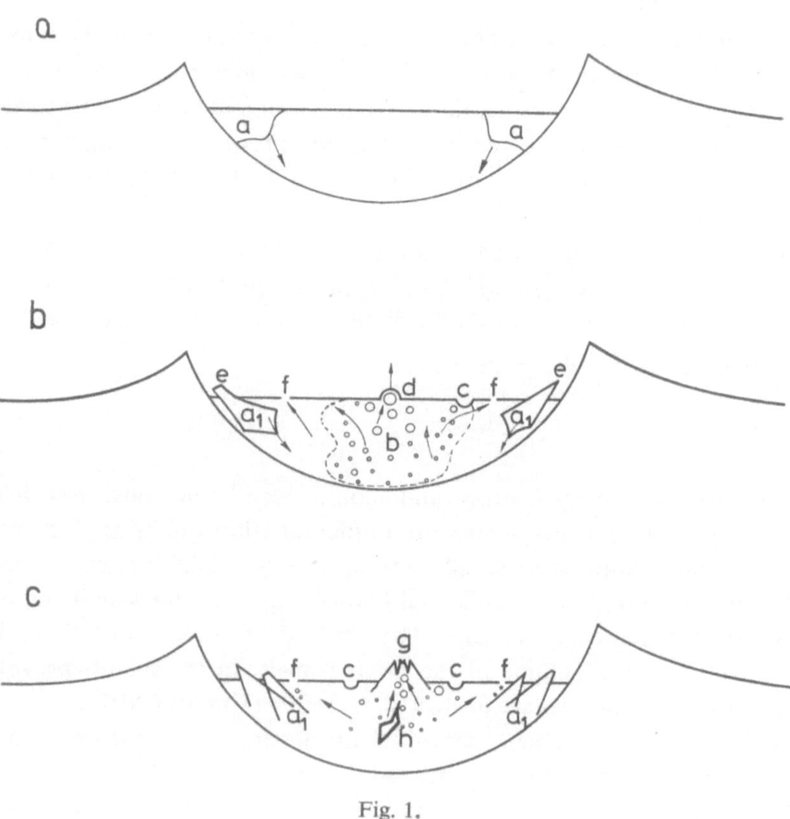

Fig. 1.

Figure 1c: If comparatively large bubbles escape from the central part of the melt towards the surface the surface crust are sometimes broken up and lifted (g). The central regions of this formation can then be filled by the melt coming from below. This process may be repeated several times, if several hot gaseous regions were formed and if the depth of the bowl-shaped melt region was sufficiently deep in comparison to the diameter of the crater. Impulses from neighbouring impacts causing moonquakes could be responsible for the formation of the central peaks (by rupturing the surface crust and causing the melt to flow out and gases to escape).

In the last phase also the central part solidifies. At this time cracks and cavities with little gas filling are formed in the internal parts (h), the cracks may interconnect bubbles filled with gases earlier captured by the semi-liquid melt. These processes may take thousands and tens of thousands of years, and sporadic extrusions and escapes of gases through the cracks from the cavities may go on for hundreds of thousands and millions of years. This need not be volcanism proper, but the processes could be called 'quasivolcanic', created in a melt within the region of crater after impact.

These process of differentiation and dynamics of a solidifying melt after large impacts may be used to explain the origin of the terrace-like inner boundaries of the crater ringwalls of some craters, the central regions with the domes or smaller craters (after gas escape), or of the central peaks (including the peak crater).

Should this analogy prove to be correct the central peaks in the crater formations would be due to the process of successive solidification of melts of limited volumes in the surrounding environment with temperature well below the point of solidification, and they would not have to represent an argument for the volcanic theory of the origin of some of the craters; some 'quasivolcanic' processes occurring later would thus evolve in the regions of the melt of bowl-like shape after impact.

These resultant effects are nearly identical with the effects which are caused by the solidification of melts which contain a high percentage of water, when the solidified parts could have a larger volume than the original melt. The physical dynamic process is however different in both cases considered; the measure of interconnection of both processes (1 and 2) may differ under lunar conditions.

Possible relations between meteorite impact and igneous petrogenesis were discussed also by French (1970).

3

If the volume of the solidified melt is neither larger nor smaller than the volume of the original melt, larger central peaks nor multiple terrace-like edges on the inner sides of the ringwalls would not be created with the exception of structural differentiation. Neither could gravitational differentiation and dynamics evolve properly in these cases when the crater deep is shallow (as compared to the crater diameter at melt surface level). The differentiation will also be conditioned by a certain sufficient temperature difference between the initial temperature of the melt lake and the ambient lower temperature of the surface rocks in the vicinity of the crater.

Acknowledgements

The author is indebted to Mrs B. Křivská (Ondřejov) for pointing out the phenomena connected with the solidification of liquids in limited volumes, to Prof. V. Guth

(Astronomical Institute Ondřejov) for his great interest and help and Ing. J. Valenta (Basalt Melting Works Stará Voda) for consultations and some of the experiments.

References

French, B. M.: 1970, *Bull. Vulcanologique* **34**, 466.
Kopecký, L.: 1971, private communication.

GEOLOGIC INTERPRETATION OF THE STUDY OF LUNAR ROCKS

LUBOMÍR KOPECKÝ

Geological Survey, Prague, Czechoslovakia

Abstract. The hypothesis on an internal origin of lunar rocks of gabbroic type is supported by experimental petrologic data obtained by the author. The temperature interval of their crystallization from a dry silicate melt is given.

Lunar gabbroic rocks of the Tranquillity Base are compared with paleobasalts of north-eastern Bohemia, eucrites and associated rocks of the British Isles and meteoritic eucrites of poikilophitic texture.

The metasomatic origin of lunar anorthosites displaying a mosaic texture in deeper parts of the volcanic apparatus is postulated.

An analogy is thought to exist between the Tertiary volcanic areas of Mull, Ardnamurchan and Rum in Scotland and the Ptolemaeus, Alphonsus and Arzachel regions of the Moon in the overall tectonic structure and the development of tectono-volcanic cirques. By analogy the possibility of the presence of more basic rocks in the rim and on the floor of the Moon subsidence calderas and of more acidic rocks of anorthositic to trachyandesitic chemistry in the central peak is presumed.

The mode of origin of double-craters of Cyrillus-Theophilus type is suggested to be similar to the origin of two consecutive volcanic rings of Mull.

The ring- and radial complexes of wrinkle ridges of Lamont type are compared with cone sheets and radial dykes of the central type tectono-volcanic complex of Mull.

1. Introduction

Lunar scientists are divided into two groups. Some are engaged in the direct study of the properties of the Moon and its rocks; the others are concerned with the published material on this subject and its interpretation. The latter group also includes a number of geologists who apply geological information about the Earth to analogous problems bearing on the Moon.

This contribution may be placed to the latter group and is based on published results of the studies of rocks from the Mare Tranquillitatis and Oceanus Procellarum.

2. Crystallization of Lunar Magma

Gabbroic rocks from the Mare Tranquillitatis, and especially from Oceanus Procellarum are strongly porous, with a frequent fluidal arrangement of the vugs, which is regarded justifiably as evidence of magma crystallization during surficial effusion [1]. The vug walls (0.1 mm to 4 cm across) consist of crystals of olivine, pyroxene, plagioclase and other minerals composing the host rock (Figure 1). This suggests crystallization of so-called dry magma, a process which can be expected to occur in the Moon. The dry lunar magma is a melt that is devoid of most of the crystallizers under the conditions of low gravity and, at the Moon's surface, of high vacuum. If present, the crystallizers make possible the separation of crystals even at low temperatures during cooling of the magma. This process, however, enables the crystallization of

Urey and Runcorn (eds.), The Moon, 231–245.

Fig. 1. Vugs in crystallized dry silicate melts. (a) Sample 12052, a typical equigranular crystalline Moon rock; vugs contain large euhedral crystals of pyroxene and olivine. NASA photo S-70-21320. (From *Science* **167**, 1970). (b) Laboratory melted and crystallized alkali olivine basalt. The photograph shows a wall of a huge vug, which is covered by skeletal crystals of olivine and pyroxene. Actual size 6 × 4 cm. *(Photo Geological Survey, Praha, Czechoslovakia.)*

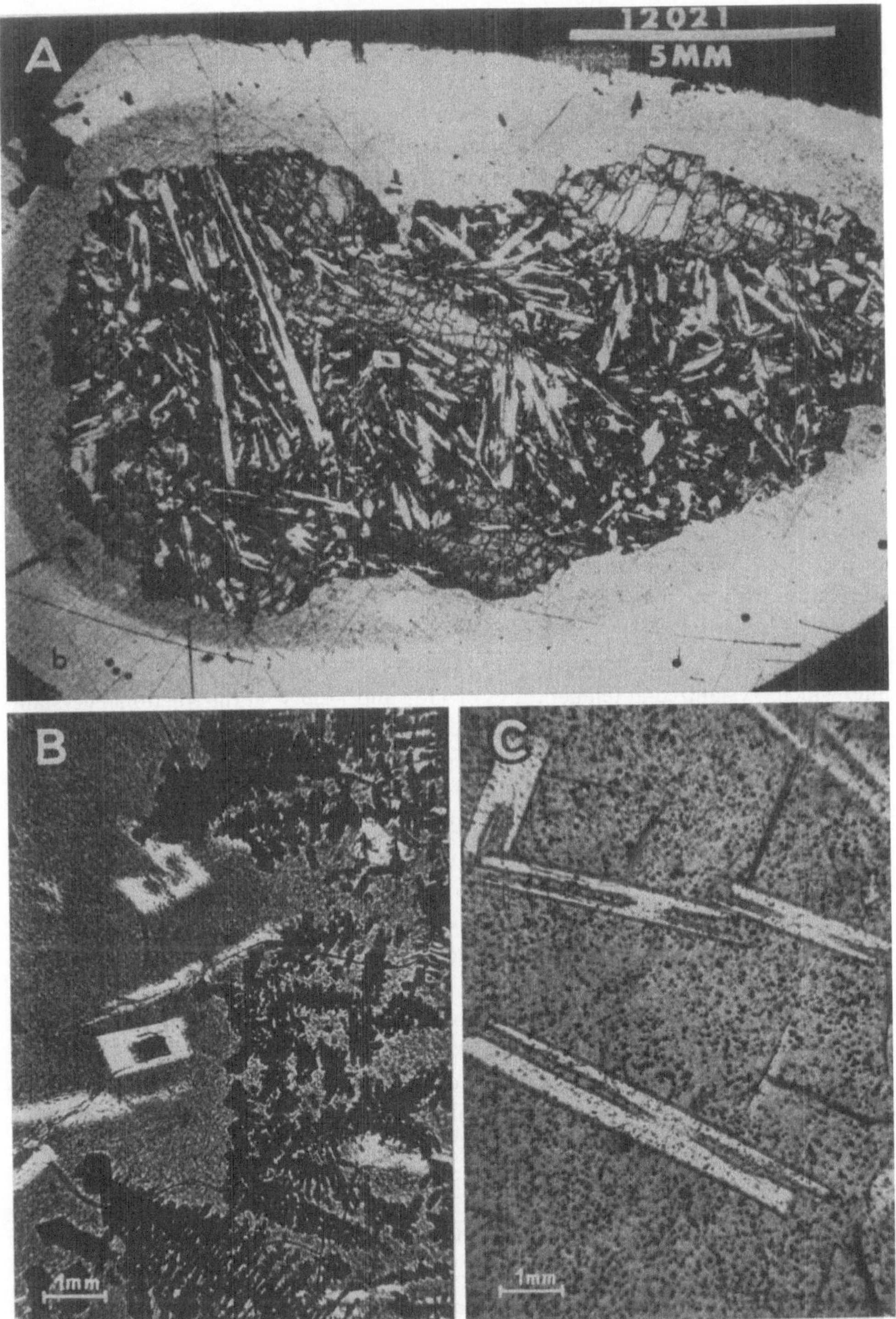

(For caption see p. 234.)

hydatogenetic minerals such as amphiboles, micas, alkali feldspars, although these are rare in the lunar rocks examined. The conditions under which lava congealed on the Moon are similar to those of laboratory rock melts prepared under normal terrestrial atmospheric conditions, which enable the crystallizers to escape. During their solidification pyrogenetic minerals are produced, such as magnetite, olivine, pyroxene and plagioclase [2]. They form walls of vugs as they do in lunar basic rocks (Figure 1b).

The experimental crystallization of dry silicate melts from fused terrestrial rocks which had the SiO_2 content of 38–45%, i.e. near to that of crystalline rocks of A, B types from Tranquillity Base, indicates that in this part of the lunar surface the rocks crystallized from magmatic effusions within the temperature interval of 1350–900 °C under an abrupt withdrawal of crystallizers and drop of temperature [3]. This conclusion agrees with the results obtained by other methods as described elsewhere [4].

The consistent results of experiments conducted by different methods in different laboratories show the crystallization of dry silicate melts to be applicable for the study of crystallization temperatures of the Moon rocks. The temperatures deduced together with the structural history point to demonstrably volcanic origin of gabbroic rocks of the Moon.

Also some of photomicrographs of lunar rocks provide interesting data for correlation with the crystallization of dry silicate melts. Various shapes of plagioclase crystals seen on photomicrograph of spec. 10046 [5] and of similar volcanic rock spec 12021 [6] (Figure 2a) are identical with skeletal crystal shapes of plagioclase (bytownite) which crystallized at temperatures of 1155 and 1150 °C at the fractional crystallization of tholeiite and amphibolite glass having 48.44 and 45.72% SiO_2 respectively (Figures 2b, c).

It can be concluded that the slightly skeletal forms of plagioclase from the samples taken at Tranquillity Base and Oceanus Procellarum could have separated from the rock melts with 38–45% SiO_2 at temperatures below 1150 °C. From the presence of skeletal forms a more rapid undercooling of the melt after effusion is inferable, probably close under the surface of the effusive body (to a depth of several decimetres). The influence of cooling of the lunar surface during the lunar night may also be taken into consideration.

In addition, the experimental petrographic observations exclude the impact origin of gabbroic rocks of lunar maria. It is well known that such extremely high temperatures as presupposed by high-velocity impact exclude any recrystallization of glass originated because of total loss of crystallizers and of liquidation of so-called crystallization seeds in the melt. But also the melting of Moon surface rocks by high pressure of low-

Fig. 2. Skeletal crystals as indicators of crystallization temperature of a silicate melt. (a) Sample 12021, a porphyritic gabbro with variolitic texture and with skeletal crystals of plagioclase. NASA, photo S-70-20749 (From *Science* **167**, 1970). (b) Cross section of plagioclase crystals separated from the laboratory melt of paleobasalt from Stará Paka (Bohemia, Czechoslovakia) at the temperature of 1155°C. Plane polarised. Photo L. Kopecký. (c) Longitudinal sections of skeletal plagioclase crystals separated from the amphibolite laboratory melt at the temperature of 1150°C. Plane polarised. *(Photo L. Kopecký.)*

velocity impact cannot produce glass with good crystallization properties. Theoretically, this glass is highly undercooled at the time of its origin and its high viscosity does not enable sufficient movement of ions for crystallization.

Contrary to the opinions of Metcalfe and Barricelli [7], the crystallization history of lunar rocks clearly shows that their origin is internal and are hence not derived by impact of meteorites.

From structural development and mineral composition of the gabbroic to peridotite rocks from Mare Tranquillitatis and Oceanus Procellarum, it is possible to conclude, in accordance with experimental petrographical data, that the rocks are of igneous nature and crystallized from lava flows of normal volcanic origin.

3. Some Terrestrial and Extra-Terrestrial Equivalents of the Main Lunar Rock Types from Mare Tranquillitatis and Oceanus Procellarum

There are a number of comparisons in literature between the lunar gabbros and basalts from Mare Tranquillitatis, on the one hand, and the terrestrial rocks of basaltic composition, on the other. In this context basalts of Permian age have been recorded from New Zealand [5], basaltic rocks crystallized from residual solutions rich in Fe and Ti [8], or basalts building up ocean floors [9].

As far as structural history is concerned, analogous rocks also occur among the Permo-Carboniferous basalts (so-called melaphyres) of the Bohemian Massif [4]. The rock types composed of calcic plagioclase, Ti-bearing augite, olivine and magnetite fully agree with some A-type crystalline rocks from Mare Tranquillitatis in the poikilophitic texture and grain-size (Figures 3a, b). They differ from the lunar basalts only in the absence of cristobalite, the lower basicity of plagioclase, and the content of magnetite instead of ilmenite.

The rocks of the group of quartz gabbros and euerites from the Scottish islands show a still closer similarity to lunar gabbros and basalts (Figure 3c). Unlike normal terrestrial gabbroic rocks, the eucrites of Scotland are close to ilmenite melabasalts and dolerites from Mare Tranquillitatis as well as gabbros, diabases and basalts from Oceanus Procellarum especially in an extremely high basicity of plagioclases; a feature which is decisive in establishing the relationship of igneous rocks.

Plagioclases of the basic rocks from Mare Tranquillitatis correspond to An_{64}–An_{90} [10], An_{71}–An_{87} [8], An_{60}–An_{100} [11], An_{78}–An_{93} [12], An_{73}–An_{81} [13], [14]. Labradorite is reported to predominate. Plagioclases of the basic rocks from Oceanus Procellarum correspond on the average to An_{50}–An_{90}, their average value also corresponding to labradorite (An_{80}) [6]. Plagioclase coming from the quartz gabbros to eucrites from Ardnamurchan Peninsula is bytownite to anorthite [15]. According to Richey, the quartz gabbro forming a marginal facies of the Great Eucrite of Ardnamurchan displays an ophitic texture, contains olivine and a high amount of iron ore. In essence, it differs from the lunar basalts only by the presence of quartz instead of cristobalite and by the content of magnetite instead of ilmenite. It has been established that the high ilmenite content is a specific feature of the rocks coming from Mare

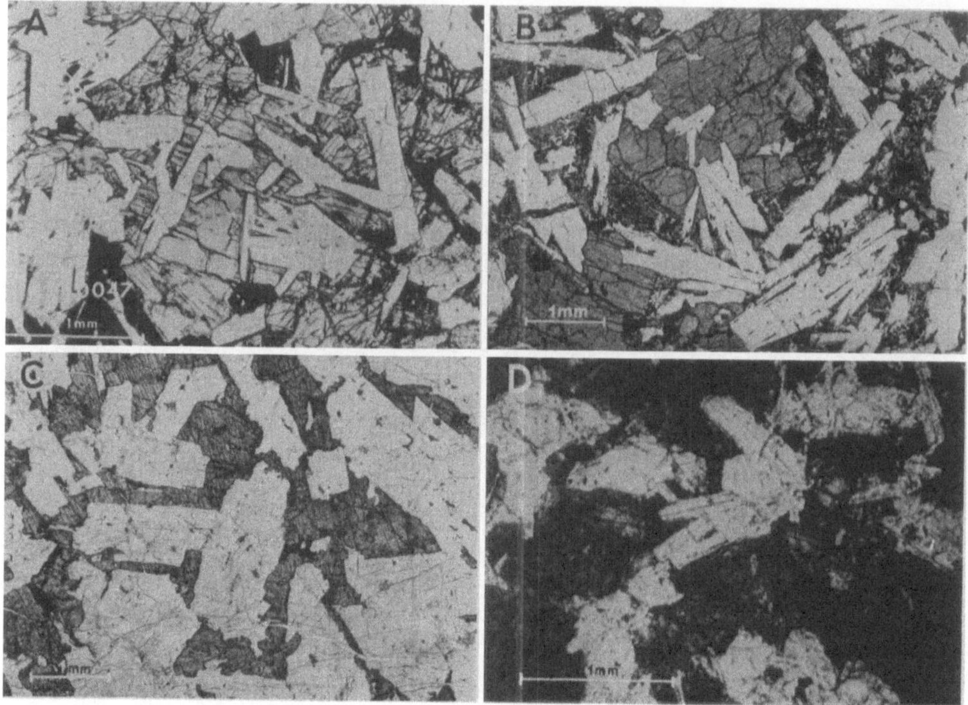

Fig. 3. Textures of the Moon, terrestrial and meteorite basaltic and eucritic rocks. (a) Poikilophitic texture of crystalline rock 10047 of Tranquility Base; the rock consists of Ca-rich plagioclase, clinopyroxene, ilmenite, minor olivine and cristobalite, and accessory minerals. Plane polarised. NASA, photo S/69-47907. (From Geotimes, October 1969.) (b) Similar texture and major mineralogy of terrestrial Upper Paleozoic basalt from NE-Bohemia (Czechoslovakia) – as shown in Figure 3a. Plane polarised. Geological Survey, Praha, photo 24336. (c) Poikilophitic texture of olivine eucrite from Ardnamurchan (near Lighthouse), Scotland. Plane polarised. Geological Survey, Prague, photo 26924. (d) Poikilophitic texture of eucrite material composed of pigeonite and bytownite and included in the Mt. Padbury meteorite. Plane polarised. *(From G. J. H. McCall, 1966.)*

Tranquillitatis, whereas the rocks from Oceanus Procellarum are poorer in ilmenite. Cristobalite in these lunar rocks resulted from crystallization of a dry rock melt, intrusive rocks of the Scottish circular complexes crystallized with a sufficient amount of water for the origin of quartz.

Of extra-terrestrial objects, the stony meteorites of eucrite-type composed essentially of pyroxene and anorthite may best be compared with lunar gabbros and basalts. As was pointed out by Shand [16], the meteorites with less than 10% of non-silicates highly resemble terrestrial igneous rocks. As was stated by Shand (as referred to by C. Rammelsberg), the frequently cited eucritic inclusion in the metallic meteorite from Juvinas (Figure 3d) has the same mineral composition as the anorthite gabbro. Furthermore, it displays a similar poikilophitic texture which suggets analogy between the meteoritic eucrites, lunar gabbros and terrestrial eucrites, which were named after the former (G. Rose in 1835).

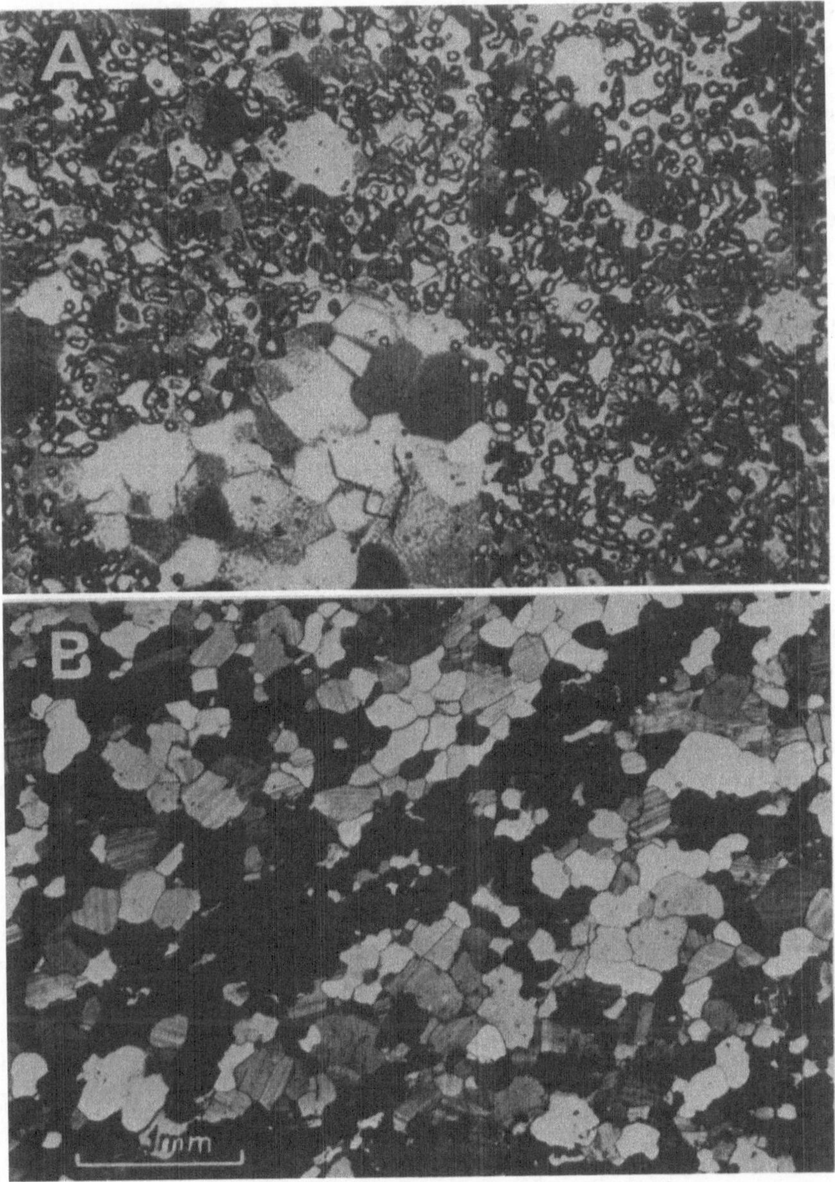

Fig. 4. Texture of the Moon anorthosite. (a) Gabbroic anorthosite with clots of true anorthosite. The polygonal (mosaic) texture of anorthosite may have formed by recrystallization. Oblique nicols. (From J. A. Wood *et al.*, 1970). (b) The same texture of alkaline metasomatic rock (fenite) from the České středohoří Mts. in Czechoslovakia; the polygons are made up of alkalic feldspars. Oblique nicols. *(Photo L. Kopecký.)*

According to Tschermak's view [17], which was later supplemented by von Wolff [18], stony and also metallic meteorites showing frequent brecciated structure of tuffaceous nature may represent ejecta of volcanic explosions produced by small cosmic bodies. Again, there seems to exist an analogy with the basic intrusive rocks of the Scottish islands, where repeated intrusions and explosions gave rise to brecciated eucrites.

The comparison of mineral composition and structural history of the lunar and terrestrial rocks of the gabbroic to eucritic types with meteoritic eucrites indicates that all these rocks originated in a similar manner. If the terrestrial eucrites and basalts are the crystallization products of magma intruded under or extruded on the Earth's surface during volcanic activity, then the above analogies prove unequivocally the magmatic origin of the lunar rocks discussed.

Lunar gabbros resemble lunar anorthosites in the composition of feldspars. Transitional types have been differentiated up to the anorthosite gabbro. The igneous origin of gabbroic to peridotite lunar rocks is accepted generally but the genesis of lunar anorthosites is not unambiguous. It is well known, however, that neither terrestrial anorthosites are regarded as igneous rocks by all geologists. Polygonal (mosaic) texture of lunar anorthosite (Figure 4a), uncommon in magmatic rocks, is explained by the accumulation of the early crystallizing plagioclase, although the possibility of its origin by recrystallization has also been suggested [19]. This view is supported by the comparison with polygonal textures of the terrestrial metasomatic rocks (Figure 4b). Finally there exists no reason for excluding the possibility that metasomatic rocks on the Moon are the result of thermal activity under the supply of material in the subsurface parts of volcanic apparatuses of large subsidence calderas.

Wood [19] compared chemical composition of the anorthosites from the breccias in the regolith of Mare Tranquillitatis with an analysis of ejectamenta of the crater Tycho, carried out by Surveyor 7. He believed the analysis to be clearly related. Anorthosites and more acid rocks may thus genetically be referred to large circular tectono-volcanic structures and volcanoes of the central type of terrae, including the crater Tycho. They can make up their central peaks often of very light colour, as is documented by the central peak of the crater Tsiolkovsky. This is in agreement with the principle that the rocks become more acid towards the centre of circular volcanic complexes of the Earth. There is also an analogy with central anorthosite elevation of the crater Manicouagan lying in the Canadian Shield (Figure 5) [20]. A high albedo suggests that some isolated mounts as well as domes in the maria or at the margins of the terrae may also belong to anorthosites. The presence of anorthosites in the terrae also corresponds to their low density (2.9 g cm^{-3}). Gabbroic rocks fill tectono-volcanic depressions of the maria together with peridotitic rocks of a higher density. The average density of gabbroic rocks (3.4 g cm^{-3}) which together with still heavier peridotitic rocks fill the tectono-volcanic depressions of maria, must be compensated for by lighter masses of the lunar crust within the area of terrae, so that average density value (3.34 g cm^{-3}) would be preserved.

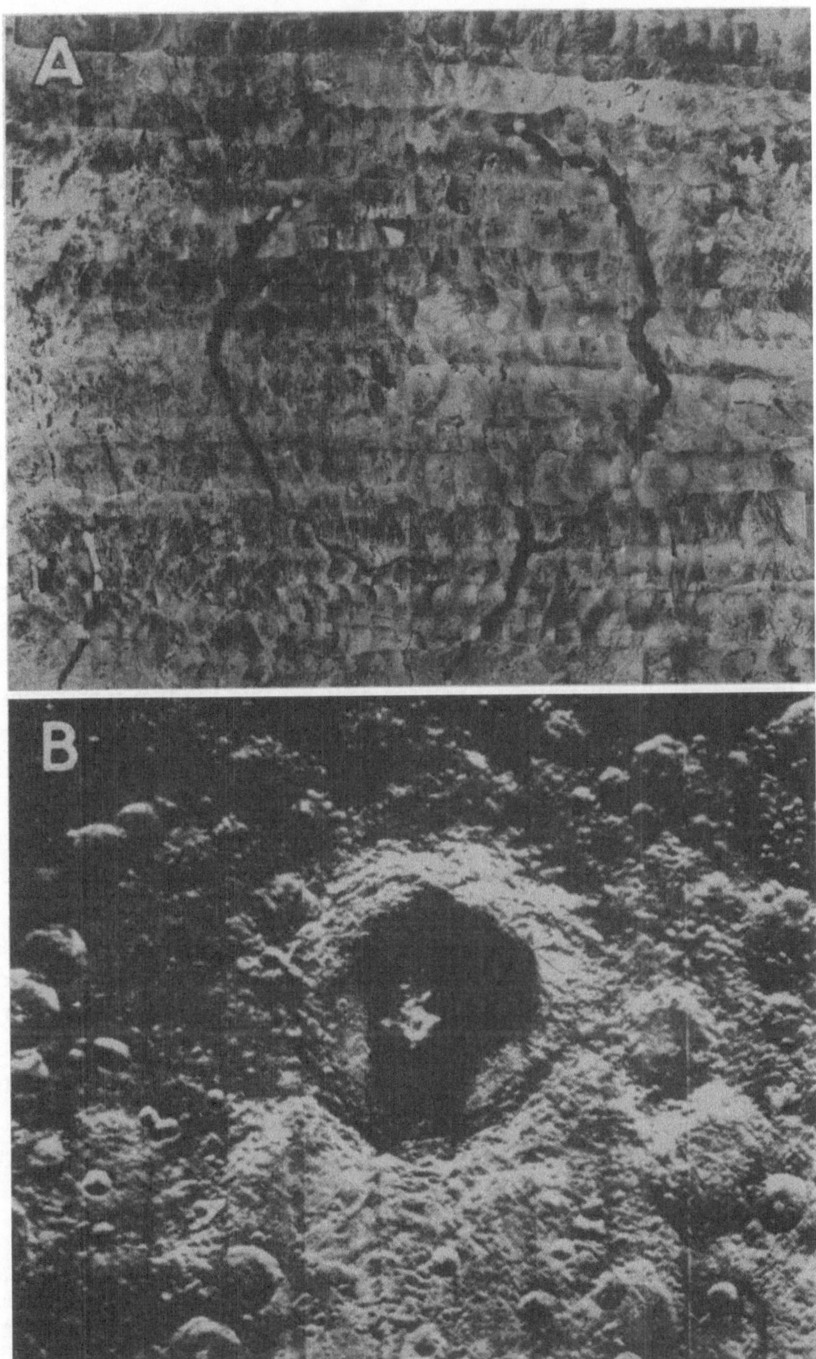

(For caption see p. 240.)

From the aforesaid conditions and assumptions it can be deduced that the accumulation of heavy masses in the maria (causing strong gravimetric anomalies) may also be responsible for the present-day passive rotation of the Moon [4].

4. Analogy of Tectono-Volcanic Structure of the Scottish Islands and Some Important Structures of the Lunar Crust

It is believed that the existence of basaltic volcanism in the maria is well evidenced, but it is not yet known with certainty (according to the Apollo 11 and Apollo 12 samples investigation results) whether it does occur in the terrae. This concerns, e.g., the dark spots on the bottom of Alphonsus crater [21] or the dark material composing bottoms of the large circular structures.

Assuming that the anorthosite material of the lunar breccia from the regolith of Mare Tranquillitatis is derived from the terrae [19], it is necessary to assume that anorthosite fragments were transported from the terrae to mare regolith together with a certain amount of the rocks of gabbroic to peridotite composition. It must then be adopted that a certain number of the fragments of basalts, gabbros and peridotites coming from a mare regolith, which were brought by Apollo 11 and 12, represent volcanic material that builds up (together with anorthosites) circular and polygonal continental formations. Furthermore, it may be deduced that the two types of lunar basic rocks, those of the maria and terrae (continents), do not differ essentially from one another. Hence, the basalts from the terrae are comparable with the eucrites (eucritic rocks) of the Earth.

Taking into account all what has been said above, comparison can be made between rock associations of the regolith of maria and analogous rock associations composing the volcanic circular complexes on the Earth. Such comparison is possible, for example, between the rocks from Mare Tranquillitatis and Ardnamurchan in Scotland (basalts, ilmenite gabbro, anorthosite gabbro, gabbroic anorthosite from the Apollo 11 collection, and gabbro, quartz gabbro and eucrites from Ardnamurchan Peninsula), or the rocks from Oceanus Procellarum and Rum in Scotland (pyroxene peridotite, olivine gabbro, gabbro and troctolite from the Apollo 12 collection and peridotite, anorthosite peridotite and eucrite of intrusive complexes of Rum).

If we adopt Spurr's application of Richey's principle of the cauldron subsidence

---

Fig. 5. Possible analogy between central anorthosite elevation of big cratonic craters on the Earth and central peaks of high albedo in big ring structures on the Moon. (a) The cratonic crater of Manicouagan in Canada is a tectono-volcanic ring of cauldron-subsidence origin. According to K. L. Currie (1965) it is a polygonal depression of a diameter of 60 km and with a central elevation. The depression is surrounded by a tectonic fracture at the surface filled with water (dark rim), and is covered by breccias and partly by andesite and basalt lavas. The central elevation is made up of anorthosite which is marked by its light colour on the photograph. Photo credit to the Geological Survey of Canada. (b) The ring structure Tsiolkovsky on the Moon displaying a light central elevation and dark bottom, possible flooded by a basic lava. Diameter about 240 km (150 miles). North to the left. *(From NASA SP-168: Exploring space with a camera, by E. M. Gortright, 1968.)*

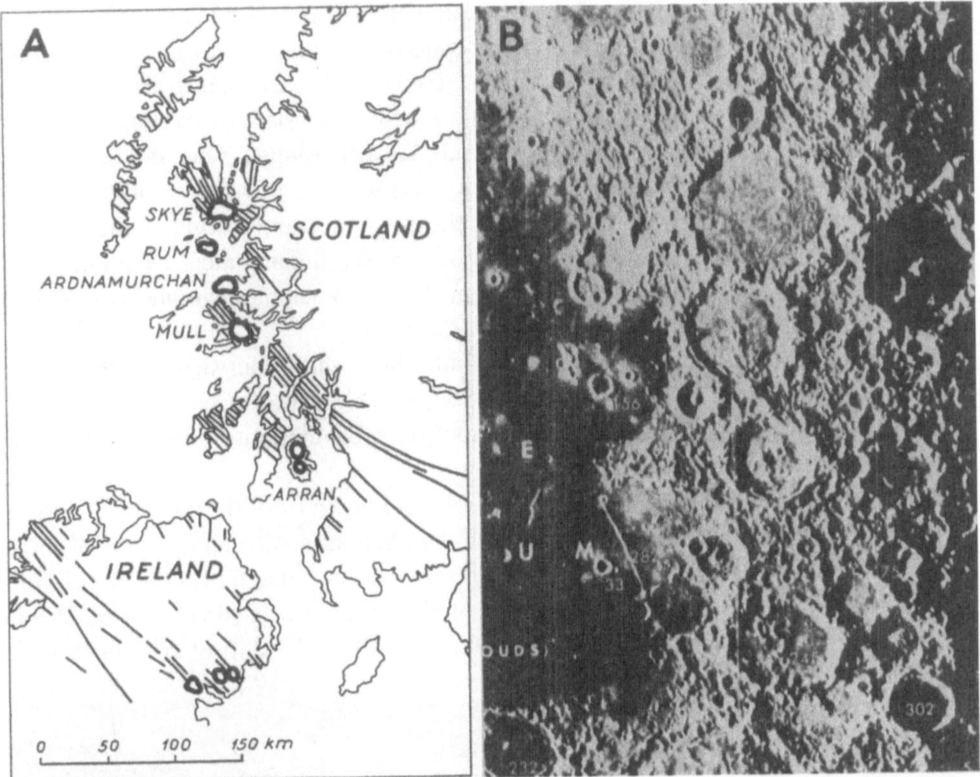

Fig. 6. Analogy of geological surface structure between the Moon and Earth. (a) The terra-mare boundary and the area of tectono-volcanic ring complexes Ptolemaeus, Alphonsus and Arzachel on the Moon, with a pronounced NW lineation marked by parallel ridges and valleys. (b) The area of central eucrite intrusion complexes of Scottish Islands display the same relation of tectonic lineation to the volcanic rings. The lineation is marked by basic dykes.

for the explanation of the origin of large lunar craters (circular structures), it is evident that there is an analogy between the lunar circular and polygonal subsidence calderas and central intrusive complexes of Scotland, not only in rock-type association but also in a detailed geological structure – except for the different degree of denudation (Figure 8).

Another striking analogy has also been observed in the geological structure of a wider area between the circular complexes of Skye, Ardnamurchan, Mull and Rum and the polygonal subsidence calderas of Ptolemaeus, Alphonsus and Arzachel (Figure 6). This analogy is documented by the following: (i) both regions were strongly affected by planetary tectonics trending north-northwest; (ii) the distribution of the individual subsidence calderas and their areal relationship to the zone of strong tectonic disturbance; (iii) and the detailed interaction of tectonic and volcanic activities within each of the calderas. In the Scottish Islands planetary tectonics effects are stressed by swarms of linear dykes of dolerite, basalt and augite andesite following the course of the so-called second volcanic zonal. In the Ptolemaeus Region and in other parts of the continents too, the panlunar grid system consists of conspicuous tectono-volcanic linear septa which pass into ridge sections of crater walls with apical craterlets [22].

Such a comparison is fully justified despite much larger sizes of the lunar calderas than attain their terrestrial equivalents of the Scottish islands. One should bear in mind that it is yet insufficiently known why tectono-volcanic forms of the Moon generally attain larger sizes than the genetically equivalent forms of the Earth*. Furthermore, subsidence calderas equalling in size the lunar craters have also been observed on the Earth, such as the Mogollon Plateau, a rhyolite circular complex in southern New Mexico, USA, as described by Elston [23].

The central intrusions of the Scottish islands show additional significant volcanic phenomena which may be regarded as analogous to the forms of circular lunar volcanism. This relates particularly to the shifting of the volcanic centres and the existence of systems of radial and circular dykes.

The classic form of shifting of volcanic centres is known from the volcanic intrusive circular complexes of Mull and Ardnamurchan in Scotland. Volcanic centres of these two islands moved along an axis [24] – apparently a deep-seated fault – the orientation of which is defined by the main lineament of the area. The existing early caldera is partly obliterated and buried by a late caldera in result of the shift of volcanic centre.

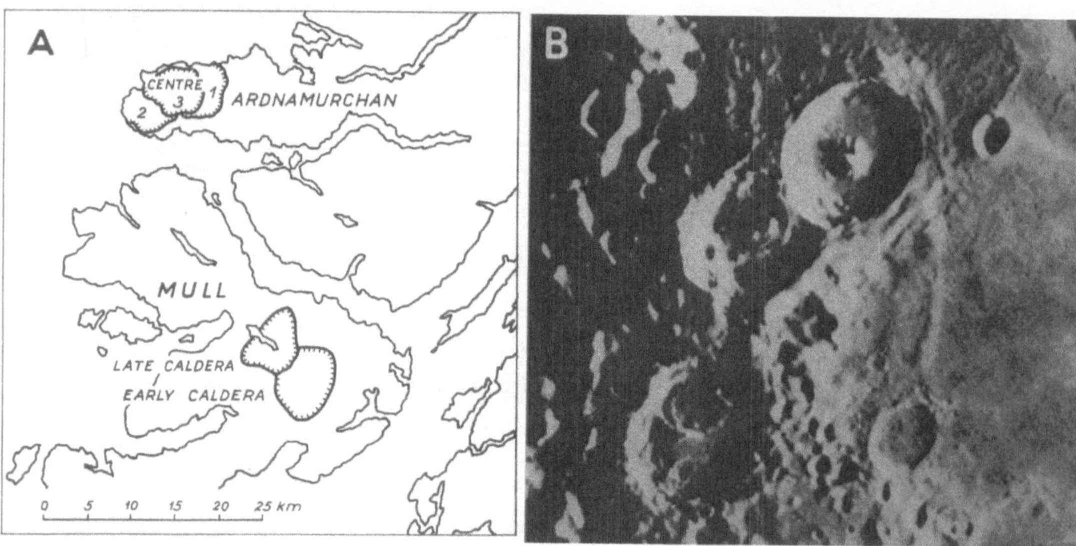

Fig. 7. Caldera doublets on the Earth and Moon. (a) The shifting of eruption centres of Scottish central intrusion complexes of the Mull and Ardnamurchan. The surface morphology of calderas has been effaced by denudation. In the sketch the margins of early and late calderas of Mull and three subsequent plutonic-vent complexes (centre 1–3) of Ardnamurchan are drawn. *(From J. E. Richey, 1961.)* (b) The doublet of subsequent equal tectono-volcanic ring complexes of Cyrillus (the older eruption centre) and Theophilus (the younger one) on the Moon. Surface volcanic morphology is preserved. *(From the Photographic Atlas of the Moon by Z. Kopal, 1965.)*

* A possibly higher thickness of the Moon's crust or its different mechanical properties as compared with the Earth's crust.

Fig. 8. The analogy of intrusive ring complexes of Scottish Islands and tectono-volcanic rings on the Moon. (a) Panorama of Great Eucrite and interior complex of ring-dykes of centre 3, Ardnamurchan. Outer ring of hills and dark foreground mark the outcrops of Great Eucrite. Low inner ring surrounding central knob of quartz-monzonite is the fluxion biotite-gabbro of Gledrian. Diameter 4.8 km *(From J. E. Richey, 1961)*. (b) Apollo 8 view of the ring structure of Goclenius, nearly 64 km from rim to rim. Photo CP 34430/26, Project Apollo 8 *(Original NASA photograph kindly offered by the U.S.A. Embassy in Czechoslovakia)*. Origin of both these types of ring structures is due to the cauldron subsidence phenomena. Also the deep structure and chemism may be analogous, only the degree of erosion differs. Note younger tectonic fractures crossing both, the terrestrial and the lunar circs.

Such displacement may be simple (as is the case with Mull) or complex (as with three volcanic centres in Ardnamurchan, Figure 7a).

The youngest centre of Ardnamurchan Peninsula morphologically resembles lunar circs with a central peak and an outer wall (Figure 8). It is 5 by 6.5 km in diameter, and consists of a set of circular dykes. Proceeding from the margin toward the centre the dykes are formed by gradually younger and more acid intrusions of gabbro, eucrite, quartz gabbro, eucrite, quartz dolerite, quartz gabbro, tonalite, and quartz monzonite of plutonic dimensions. As already mentioned above, this rock association is very similar to that of the regolith from Mare Tranquillitatis. Judging by the presence of anorthosite fragments, it must be assumed that in Mare Tranquillitatis is an abundance of rock material derived from the tectono-volcanic circular depressions of the terrae.

The origin of the double-craters on the Moon (and on the Mars), including the well-known pair of the craters Cyrillus and Theophillus and others (Figure 7B), can readily be explained by the migration of volcanic centres. It is very likely that a phenomenon of similar nature is also the origin of such paired circular structures as Craft and Cardanus, connected by a tectonic furrow and other similar doublets and triplets of lunar circs that lie on their common axis.

Other important forms of the volcanism of central type, the lunar analogues of which may be seen especially in wrinkle ridges of mare regions are the systems of cone sheets and radial dykes. Island Rum, a classical region of such central rock intrusions of rocks such as eucrite, allivalite, peridotite and gabbro, provides a good analogy in its rock associations to the rocks coming from the regolith of Oceanus Procellarum. A younger complex of cone sheets of Rum consists of gabbro and eucrite, and a youngest system of radial dykes comprises various basic rocks (Figure 9a). The system of these young dykes may also be regarded as possibly analogous, both in geological structure and mineral composition, to some typical lunar mare regions. Lamont, a system of circular and radial wrinkle ridges, north of the Apollo 11 landing site (Figure 9b), highly resembles the dyke complex of Rum. It has been established that systems of the wrinkle ridges generally show a similar orientation in all circular and polygonal maria. The wrinkle ridges may rightfully be interpreted as manifestations of the latest volcanism of central type of the maria [25] the largest lunar subsidence calderas – as they were called by von Bülow [26].

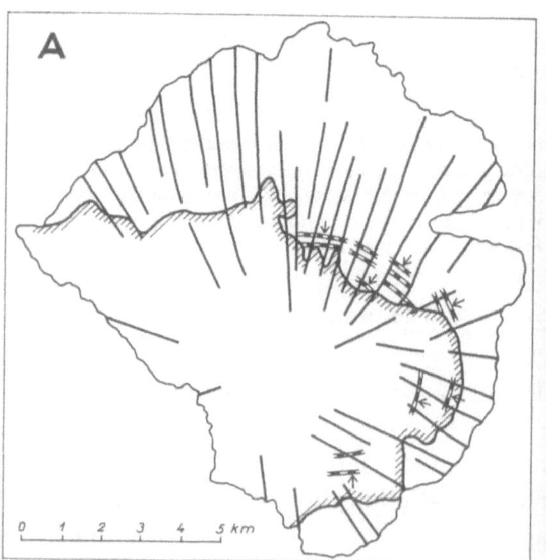

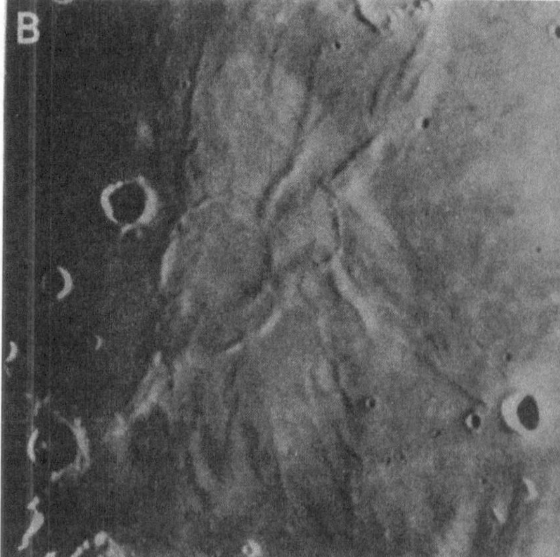

Fig. 9. Radial dykes and cone-sheets intrusion complexes on the Earth and on the Moon (a). Trends of an eroded complex of radial dykes and cone-sheets of the Scottish island of Rum *(From J. E. Richey, 1965)*. (b) Analogous structure of radially oriented and circularly shaped wrinkle ridges of the complex of Lamont in the Mare Tranquillitatis suggest that wrinkle ridges are linear volcanic features of deep-seated phoci.

Acknowledgments

Thanks are due to the officers of the American Embassy in Prague, especially to Mr. A. T. Falkiewicz, the former Press and Cultural Attaché, Mr. S. Vaňáč and Mr. J. Zábranský, farther to Dr. K. L. Currie, Department of Energy, Mines and Resources, Canada, Dr. R. E. Corcoran, Director, Department of Geology and Mineral Industries, Oregon, Dr. H. E. Newel, Associate Administrator and Dr. D. R. Morris, Deputy Assistent Administrator for International Affairs, NASA, and to Mr. H. Arnett, Pacific Power and Light Company, Wyoming, for the papers and photographs on this subject, and to Dr. H. J. Moore for the set of the Geologic Map of the Moon, which all would have otherwise been hardly accessible to the author.

Special thanks are expressed to Dr. J. Green, President of the IAP, for his revision of the manuscript. Professor S. K. Runcorn, Editor of the Symposium Proceedings, is highly appreciated for his personal assistance in preparing this paper for print.

Finally the author is highly grateful to Mr. A. Mackensie Smith, the Cultural Attaché, British Embassy in Prague. for his help in maintaining the author's correspondence with the Editor in early spring 1971.

References

[1] Lunar Sample Preliminary Examination Team: 1969, *Science* **165**, 1211.
[2] Kopecký, L. and Voldán, J.: 1965, *Ann. N.Y. Acad. Sci.* **123**, 2.
[3] Kopecký, L.: 1970, *Geologický průzkum* **XII**, 6.
[4] Summary of Apollo 11 Lunar Science Conference: 1970, *Science* **167**, 449.
[5] King, E. A. Jr., Carman, M. F., and Butler, J. C.: 1970, *Science* **167**, 650.
[6] The Lunar Sample Preliminary Examination Team: 1970, *Science* **167**, 1325.
[7] Metcalfe, R. and Barricelli, N. A.: 1970, *Moon* **1**, 2.
[8] Anderson, A. T., Jr., Grewe, A. V., Goldsmith, J. R., Moore, P. B., Newton, J. C., Olsen, E. J., Smith, J. V., and Wyllie, P. J.: 1970, *Science* **167**, 587.
[9] Katterfeld, G. and Shulz, S, Jr.: 1971. Porody lunnygh moreiy. Naukaizizn 2, Moskva.
[10] Keil, K., Prinz, M., and Bunch, T. E.: 1970, *Science* **167**, 597.
[11] Douglas, J. A. V., Dence, M. R., Plant, A. G., and Traill, R. J.: 1970, *Science* **167**, 594.
[12] Brown, G. M., Emeleus, C. H., Holland, J. G., and Phillips, R.: 1970, *Science* **167**, 599.
[13] Kushiro, I., Nakamura, Y., Haramura, H., and Akimoto, S. I.: 1970, *Science* **167**, 610.
[14] Weill, D. F., McCallum, I. S., Bottinga, Y., Drake, M. J., and McKay, G. A.: 1970, *Science* **167**, 635.
[15] Richey, J. E.: 1933, *Geol. Assoc.* **XLIV**, 1.
[16] Shand, J.: 1950, *Eruptive Rocks*, London, New York.
[17] Tschermak, G.: 1875, *Sitzber. Math.-Naturw. Kl. K. Akad. Wiss.* **71**, II, Wien.
[18] von Wolff, F.: 1914, *Der Vulkanismus*, I, Stuttgart.
[19] Wood, J. A., Dickey, J. S., Jr., Marvin, U. B., and Powell, B. N.: 1970, *Science* **167**, 602.
[20] Currie, K. L.: 1965, *Ann. N. Y. Acad. Sci.* **123**, 2.
[21] Green, J.: 1966, 'Lunar Exploration and Survival', Advanced Research Laboratories, Douglas.
[22] McCall, G. J. H.: 1965, *Ann. N.Y. Acad. Sci.* **123**, 2.
[23] Elston, W.: 1965, *Ann. N. Y. Acad. Sci.* **123**, 2.
[24] Richey, J. E.: 1961, British Regional Geology, Scotland', The Tertiary Volcanic District, Edinburgh.
[25] Cattermole, P.: 1967, 'The Nature and Origin of the Lunar Maria', Sborník věd. prací Vys. školy báňské v Ostravě, Řada horn.-geol., zvl. číslo, Ostrava.
[26] von Bülow, K.: 1965, *Ann. N. Y. Acad. Sci.* **123**, 2.

F. PHYSICAL PROPERTIES OF LUNAR SAMPLES

INTERFEROMETRIC STUDIES ON APOLLO 11 AND APOLLO 12 LUNAR GLASS OBJECTS

S. TOLANSKY

Department of Physics, Royal Holloway College, Egham, Surrey, England

Abstract. A review of results obtained by an examination with optical interferometry of a variety of glassy objects found in lunar fines from Apollo 11 and 12. Glassy spheres, cylinders, chips and fragments have been examined. It is established that many spherules show

(a) evidence of shock;
(b) evidence of impact with a small particle;
(c) very high specularity.

Some small pieces of lunar glass reveal the characteristic step pattern shown by glass shattered in the laboratory.

1. General Introduction

This report is devoted to 5 gm of Lunar fines from the Apollo 11 Expedition and three 0.5 gm samples of fines from Apollo 12. The 5 gm Apollo 11 sample will be discussed first. As received the fines had been pre-sieved through a 1 mm mesh sieve. By means of tweezers 207 glassy spherules and round-ended glassy cylinders were extracted. The majority of these objects (size range 1 mm to 0.1 mm) were reddish-brown glass and many exhibited a very high specularity. Some of the larger objects were of a greyish metallic lustre. A few were highly coloured, some green, one a deep blue. Although a considerable proportion of the fines consists of fragmental glass, the extracted objects did not exceed in amount some 0.01% of the total. If this fines sample is representative, there exist more than 40000 such objects per kg of dust. This figure refers only to objects big enough to handle and extract. When smears of the fines are examined with higher microscope powers (×500 or more) then larger numbers of such objects appear, even up to 100 per mg, i.e. the formidable number of 10^8 per kg of Apollo 11 Lunar dust. Even this is well exceeded in the Apollo 12 samples (see later).

These specular objects and numerous small irregular glassy fragments also found in the dust have been examined at magnifications mostly up to ×1000 and more, using 3 mm and 4 mm microscope objectives, with a specially designed interference microscope, whereby the chosen object is mounted on an inverted microscope and is matched against an interferometrically selected very thin glass cover slip. Some of the denser objects were matched against titanium-oxide coated glass cover slips to enhance interference fringe contrast. By using, variously, mercury green light, an unfiltered mercury light and white light, the object can be seen covered with interference fringes. Such fringes are in effect a micro-contour map of the surface wherein from fringe to fringe the height has changed by half a light wave (2730 Å for green mercury). Using simple two-beam systems one has no difficulty in recognising height changes down to as small, perhaps, as 100 Å or so. (With multiple-beam interference

Urey and Runcorn (eds.), The Moon, 249–263.

systems, one can resolve 5 Å features, but there is as yet no justification for pursuing interferometry to such fine limits with these particular Lunar objects.)

Characteristic features of typical objects are worthy of consideration. Figure 1a shows typical spherules and Figure 1b shows typical cylinders. (In each figure the largest object is $\frac{3}{4}$ mm.) Many spherules have impacted crusty tails (Figure 1c) which suggest that the particle has struck the Lunar dust at high velocity. An appreciable fraction show micro impact craters (Figure 1d) which indicate having suffered collision with a small solid object. Tail formation and crater production have clearly different origins. If it be postulated that many of the spherules have been high-speed

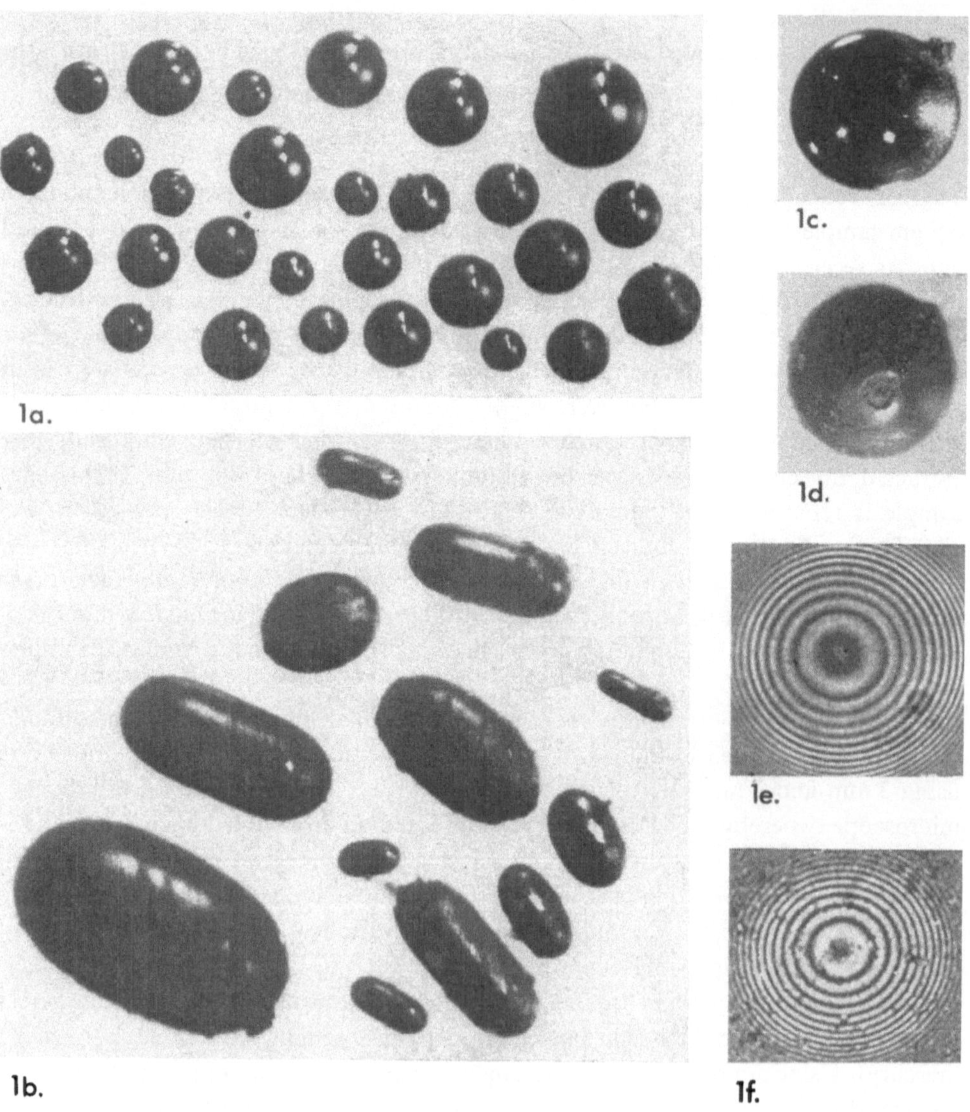

Fig. 1a–f.

projectiles, the tails can originate as a sintered mass formed by landing on dust. The craters suggest that such spherules have been hurled through a cloud of small solid particles.

2. Interferograms. Apollo 11, Sample No. 10084.15

A. SPHERULES

There are optical difficulties associated with photographing interference fringe systems from spheres of diameter 0.1 mm and less, due to small working distances of lenses, but largely due to difficulties created by focal localization of fringes produced by highly curved objects. Collimation is critical and it is necessary to image a point source precisely at the back focal plane of the objective in order to illuminate the object with parallel light at normal incidence and at the same time the object must be viewed critically in focus. With the small field of view, small working distance and small depth of focus, some delicate adjustment is involved.

(1) Some spherules (a minority) give very perfect regular Newton's ring interference fringe patterns, e.g. Figure 1e and Figure 1f ($\times$ 800). The remarkable perfection of circularity of these, which are micro-topographic contour fringes, is still maintained when the object is rolled to other regions of contact on the reference flat. This establishes that such objects are near perfect spheres. Clearly they must have been formed from small fluid glass droplets, subject only to surface tension forces, i.e. from droplets in free flight in a vacuum.

The notable smooth conformity of the fringes is a feature which shows that the specularity is today still very high, whatever the ages of the objects, which might even be many many millions of years, since ages between $3–5 \times 10^9$ yr are reported for various Lunar rocks. Clearly objects of such smooth specularity could not have suffered any erosion or leeching effects either from atmosphere, or liquid nor indeed through any solid friction occasioned by rolling.

The fringe contrast permits assessment of a rough value of the refractive index. It is probably that for many of the objects the refractivity lies in the range 1.50–1.55.

(2) Several spherules (Figure 2a and 2b $\times$ 800) show decisive evidence from the fringe pattern of having suffered collision damage, two being selected for illustration. Figure 2a reveals an impact crack, some half a light wave deep, whilst Figure 2b shows clear evidence of cracking and chipping. It is reasonable to argue that these effects arose through impact on landing, thus adding further evidence that many spherules have been high-speed projectiles.

Not all spherules show smooth surface finish. Figure 2c shows small scale micro-wrinkles. Figure 2d shows the region surrounding an impact crater.

(3) The glassy cylinders found, even if the long sides appear microscopically straight yet all show micro-waists when examined interferometrically. Figure 2e ($\times$ 1000) suggests that this cylindrical object was probably a segment of a thin jet or filament of molten glass, a filament which originally broke up into both droplets and round-ended cylinders. Cylinders of fluid would, through surface tension, try to pull themselves

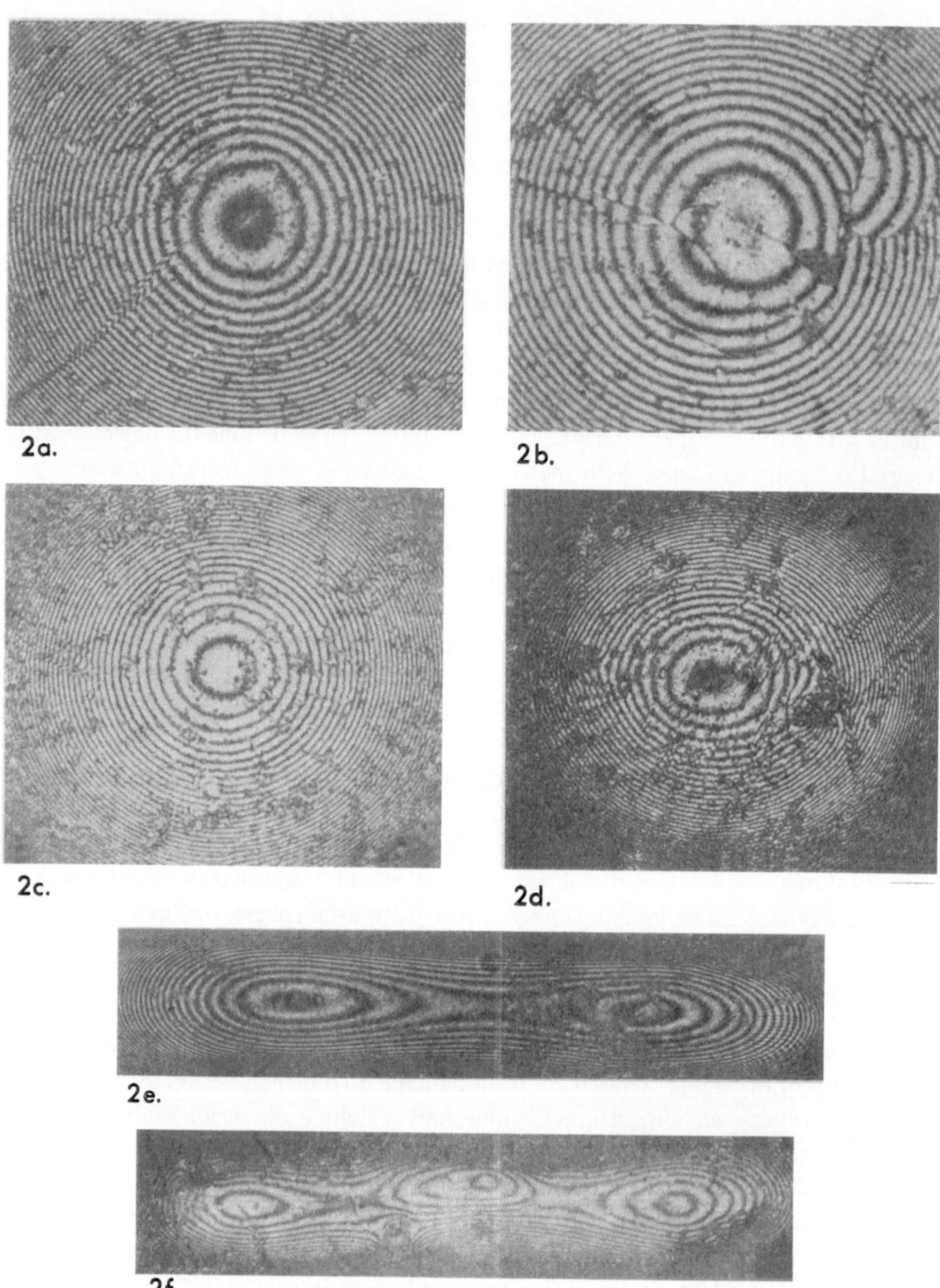

Figs. 2a–f.

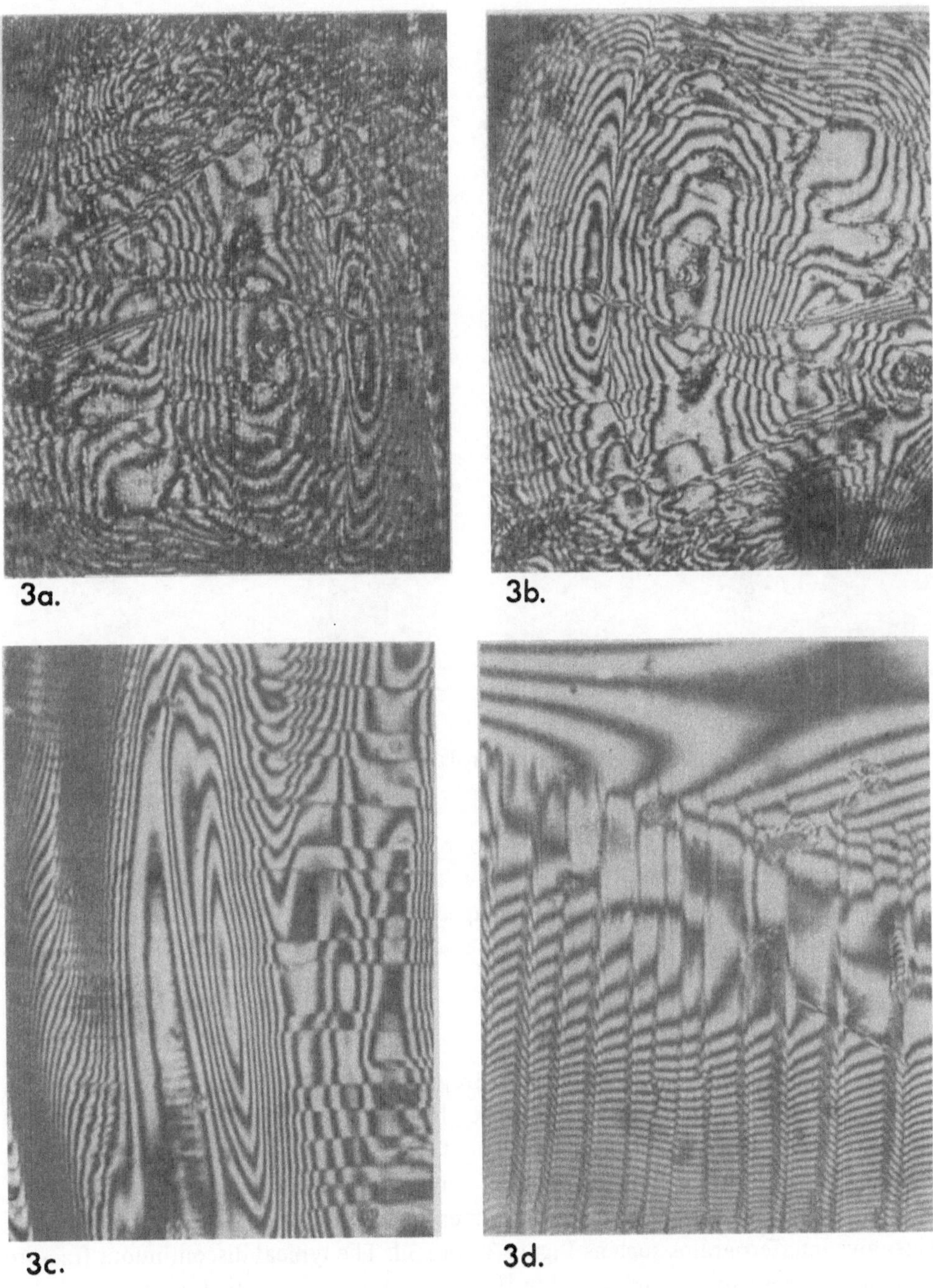

Figs. 3a–d.

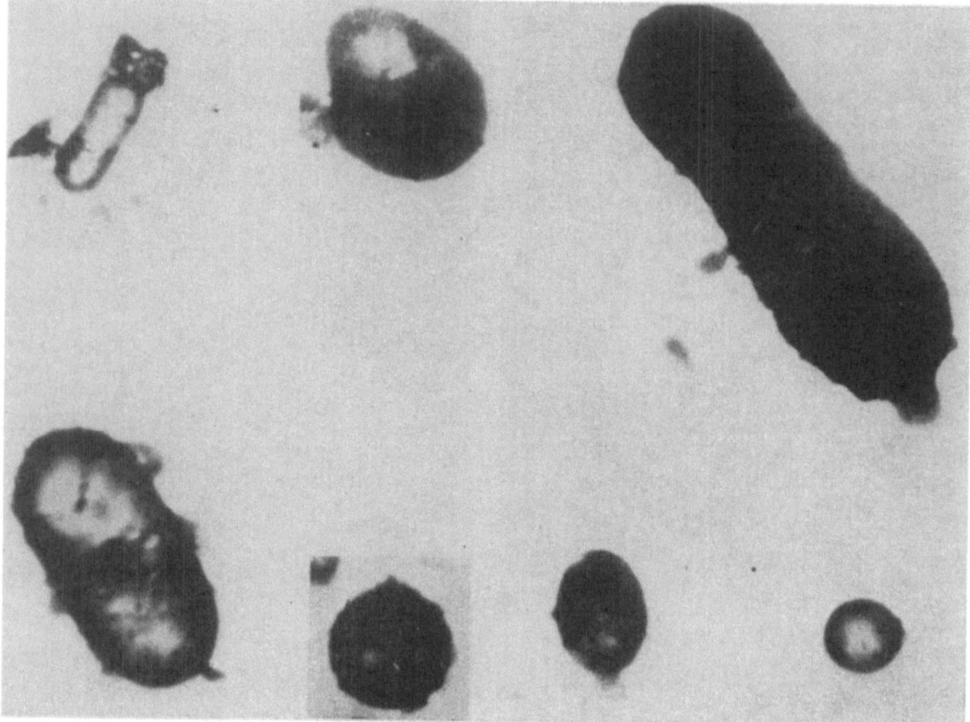

3e.

Fig. 3e.

into droplets. Of course liquid glass near to solidification has high viscosity and clearly this object froze-in before splitting up. It will be shown later that amongst the micro objects there is a complete sequence of shapes from waisted cylinder, through dumb-bell to parted droplets. Figure 2f is a case wherein a thin long cylinder has tried to split into three droplets, but froze-in.

B. GLASS CHIPS

The Lunar fines contain large number of fragmental pieces of glass, mostly brown in colour, giving the appearance of having been broken down by violent shock. Interferograms ($\times 500$) for two typical such fragments are shown in Figures 3a, 3b. One can very closely simulate such discontinuous fringe patterns by crushing ordinary glass rod with hammer blows. For example numerous fragments of such crushed shocked glass give interferograms such as Figure 3c and 3d. The typical discontinuous fracture structure in these so resembles that in the Lunar glass fragments, that it is clear that some violent mechanical shock has operated equally on this Lunar material.

One notable Lunar chip, 1 mm in length, shown in Figure 4a gave the striking monochromatic interference fronge pattern of Figure 4b. By alternatively using white-light it is established that the surface is spherically concave. Here is a remarkably smooth

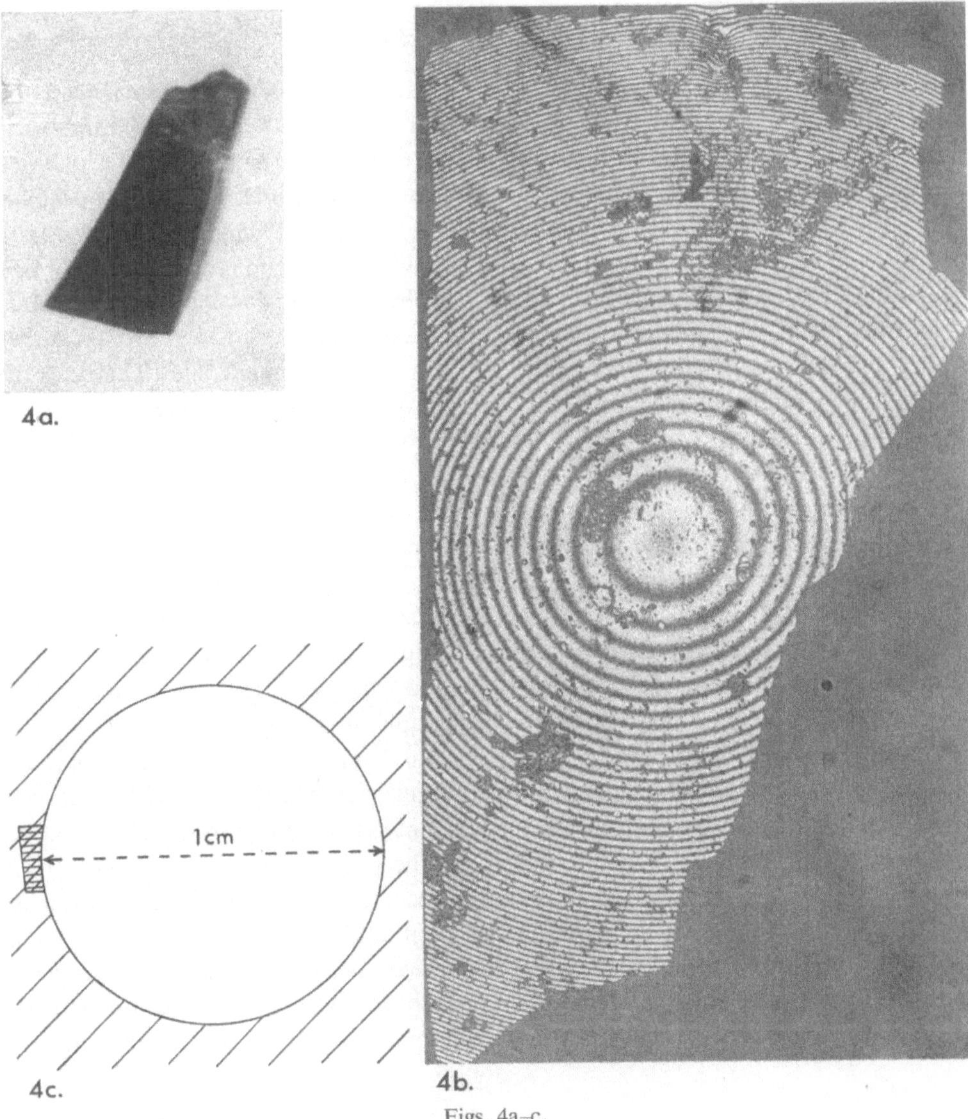

Figs. 4a–c.

almost perfectly spherical hollow, of radius of curvature, obtained from the fringes as almost exactly 0.5 cm. A reasonable conjectured explanation of this is shown by the diagram Figure 4c. It is suggested that a larger piece of glass formed containing within itself a near spherical vacuole of 1 cm diam. This broke up and the chip, under discussion, as shown, shows a section of the highly specular 1 cm diam vacuole.

As an alternative, it is to be noted that some of the fragments of laboratory crushed glass do show roughly spherical small hillocks on occasion, but they are irregular and local small pips. Only one small depression was found in a considerable number of crushed particles.

C. MICRO OBJECTS

It was possible to hand pick objects down to sizes of some 0.1 mm diam. Below this the glassy objects were only photographed. A surprisingly large number of quite small specular objects exist in the dust. Although a careful statistic has not been carried out, the impression is gained that there is a discontinuity in distribution. There appears, at first sight, to be a paucity of objects between the 100 μ and 50μ range, and the large numbers appear between some 50 μ and 10 μ, with proportionately less below this. Statistical sampling from a mere 5 gm total must be subject to considerable uncertainty and little importance should be placed on these estimates. As a rough average, down to the resolution of a 4 mm 0.9 NA lens, there have been found some 100 objects per approximate milligram smear of Lunar dust. The objects vary in character from spheres through to waisted cylinders and are best illustrated photographically as in Figure 3e. The size range covered is 0 to 20 μ in this illustration. Though most are amber in colour there is a colour spread from deep blue to black, transmission from almost transparency to opacity and a range from smooth uniformity to those containing clearly defined inclusions or vacuoles. These are variously spheres, egg-shapes, pear-shapes, dumb-bells and cylinders. One can track through the complete sequence expected from a fluid cylinder trying to break up into droplets and samples exist of a complete range of the fronzen-in states expected from such transitions. Some of the larger of these micro-objects have given reasonable interferograms.

When reviewing various mechanisms to account for the origin of these obviously pre-molten materials the possibility was considered that the objects could be in a sense artefacts. The question asked was whether the hot exhaust gases from the descending space-craft could have possibly melted some of the numerous small glass fragments in the dust and thus artificially created the spherules and other molten objects. A request to NASA was therefore put in asking for Apollo 12 material picked up as far as possible from the landing craft. This was granted and examination of this material completely excludes the idea that the rockets have anything at all to do with the objects. These objects are undoubtedly indigenous to the Moon prior to landing.

3. Apollo 12

Three 0.5 gm Apollo 12 samples were received, namely (a) contingency sample 12070.34 picked up some 10 m from the spacecraft landing site, (b) documented sample 12057.70 obtained from the bottom of the collecting box whose total material was collected over the range 80 and 330 m distant from the landing site, and (c) selected samples fines 12001.70 all picked up 120 m from the site.

The extractable objects from all three locations are very closely similar. Figure 5a and Figure 5b are representative typical groups. There is close identity between Apollo 11 and Apollo 12 objects. Even blue as well as amber objects were extracted. For briefness only the above two representations are included here. Not only is there close identity in shapes, the interferograms are also closely related.

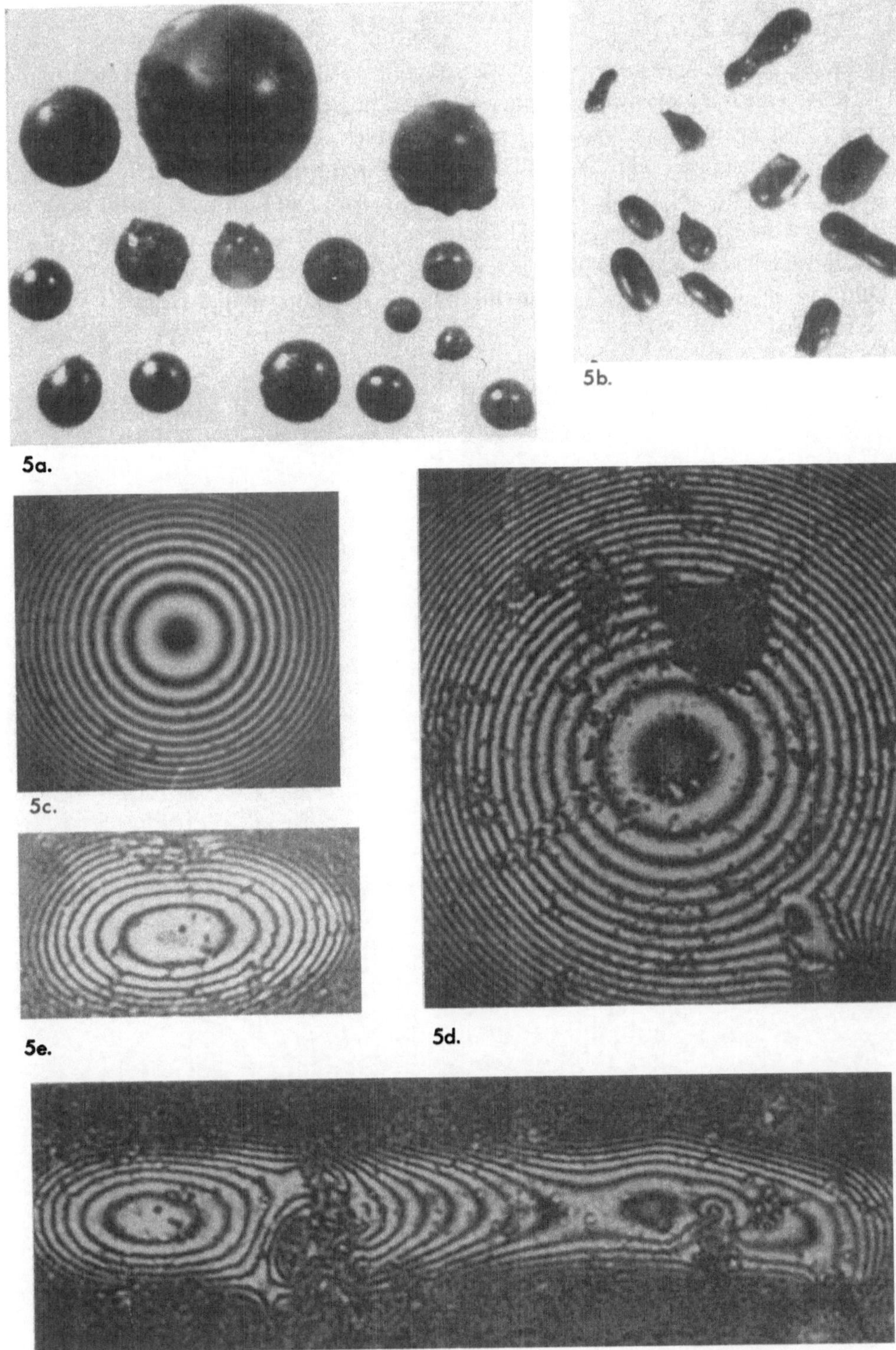

Figs. 5a–f.

4. Interferogram. Apollo 12

(1) Figure 5c is a finge pattern from the spherule of diameter 0.24 mm from sample 12070.34. This is a typical closely spherical object possessing a highly specular smooth surface. Yet again, as in Apollo 11 some spherules show surface cracks and other features. For instance Figure 5d is a 0.5 mm diam object (slightly elliptical, for principal radii of curvature obtainable from the fringes are 0.26 and 0.24 mm) and it shows a score mark and an impact crater. The score mark exhibits in the fringes the typical 'pile up' which glass reveals when it is scratched with a sharp hard edge. Such a furrow could have been formed on a fast moving spherule if landing on hard rocky or crystalline material.

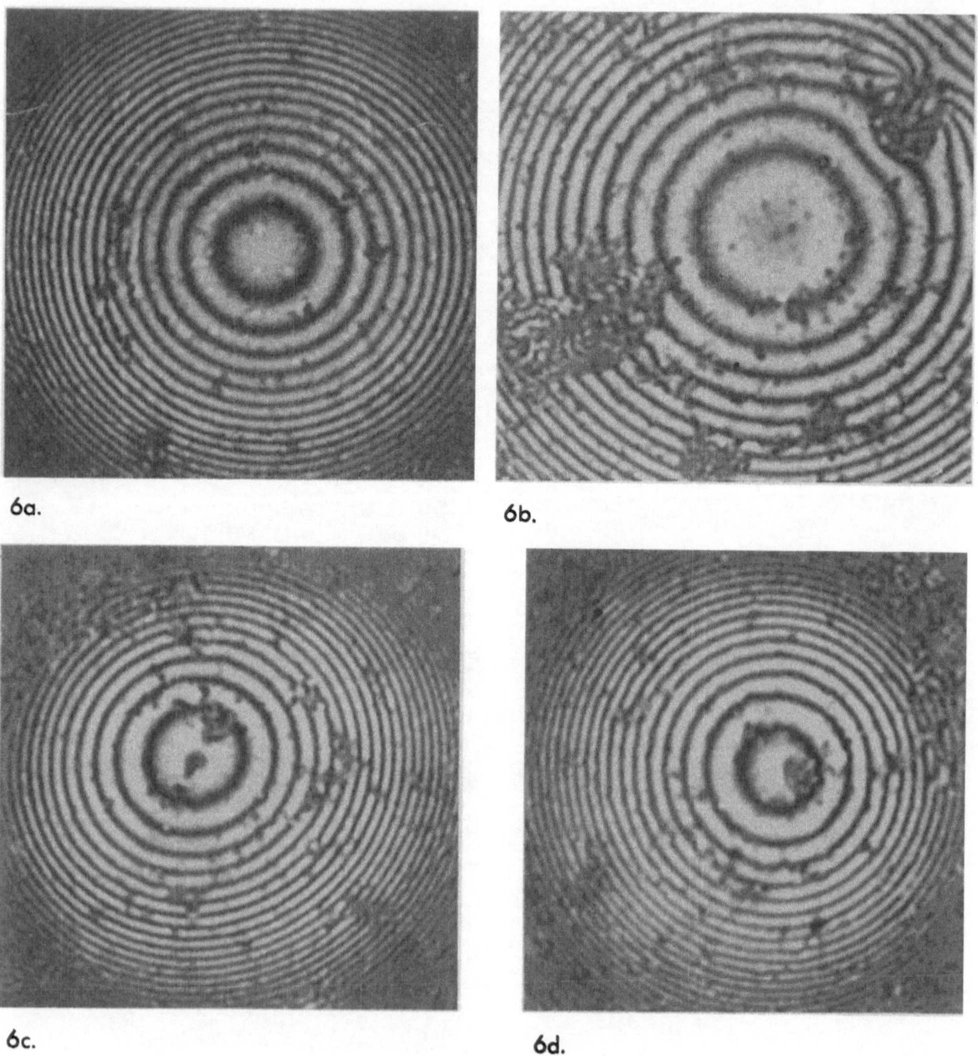

6a. 6b.

6c. 6d.

Figs. 6a–d.

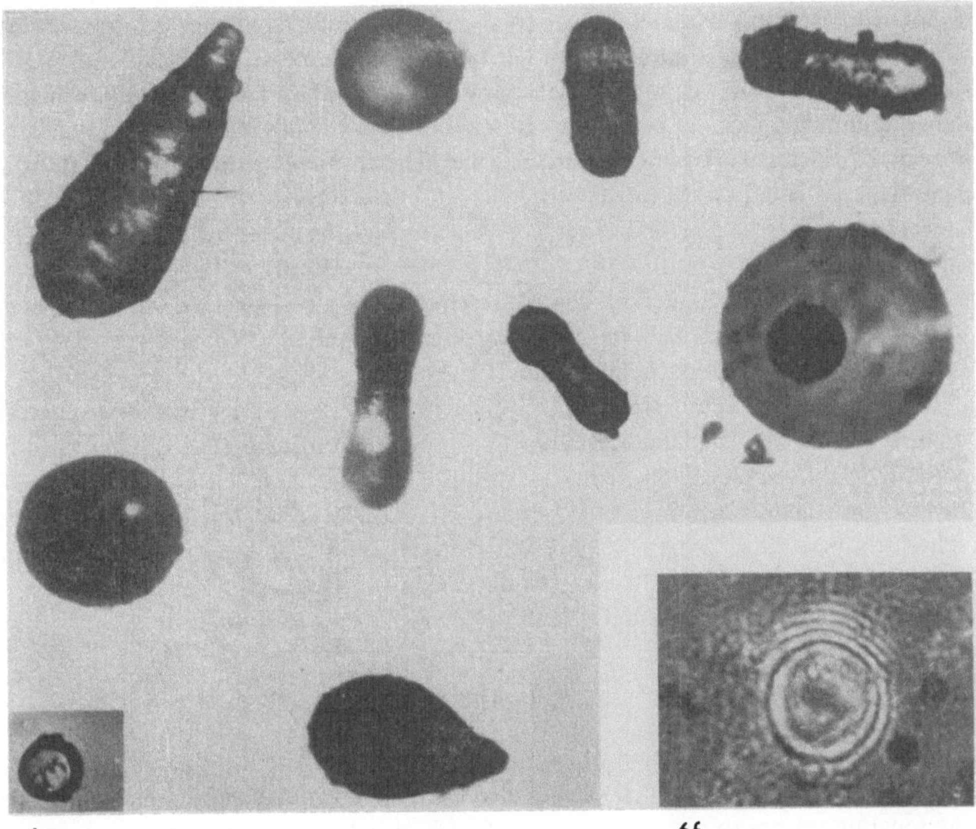

6e. 6f.

Figs. 6e–f.

Figure 5e is the interferogram from a characteristically egg-shaped object in which the principal radii are 0.18 and 0.05 mm respectively. Many other such object have been examined in detail. Figure 5f is the rough fringe pattern given by an elongated cylinder trying to pull itself into droplets but freezing in during the procedure, a feature often seen in Apollo 11 objects.

(2) Interferograms for objects from the second source (documented sample 12057.70 follow. Figure 6a is from a familiar high quality type of spherule. It has a diameter of 0.3 mm. Figure 6b is a striking pattern from a 0.6 mm spherule. So good is the fringe definition that there is no difficulty in easily seeing fringes well beyond the 50th order, indeed almost up to twice this. This is proof of a very smooth surface specularity. Some patches of adhering dust (it is very tenacious) still remain. A well defined microcrater of some two light waves in depth can be identified.

(3) The selected sample No. 12001.70 is a virtual repetition of what has already before. Thus Figure 6c is an example which illustrates how easily interferometry can show up even slight deviation from sphericity here a mere 4%, for the principal radii

of curvature are 0.153 and 0.147 mm respectively. Much smaller deviations can be evaluated. As a final example Figure 6d shows an object of diameter 0.24 mm which reveals what is quite a common feature. Although many of the objects are highly smooth and specular, yet a majority show some degree of surface wrinkling, which is however only an irregularity on a micro scale. This particular fringe pattern typically reveals such irregularities, which are hill and dale features of only a twentieth of a light-wave or less, in height or depth.

There are no real essential differences between all the Apollo 11 and Apollo 12 specimens.

(4) *Micro objects.* Figure 6e shows a selection of non-extracted micro objects (×1000) from a typical Apollo 12 dust smears. Sizes here vary from 80 to 15 μ. All have their close counterparts in Apollo 11 dust.

It is difficult to secure reasonably satisfactory interferograms from tiny curved objects. It is easy with plane objects such as are found on some glass fragments. But with highly curved objects the fringe localisation is not in coincidence with the curved surface and this creates difficulties. Figure 6f (×3000) is an interferogram from a spherical object 25 μ in diam. The object is slightly elliptical and only a few fringes are secured. Yet enough data have been obtained to show that most micro objects are reasonably smooth and specular.

5. Distribution

Comparison of the distribution of extracted objects from Apollo 11 and Apollo 12 is made below. There is too much statistical variability in the counts of micro objects in the separate smears to make differences of such objects very significant.

In Apollo 11 the objects total 207 distributed thus (sample 10084.15) (see Table I).

From this some 86% are largely spherical and 14% cylinders. The three Apollo 12 samples are so small (0.5 gm each) that it is wiser to group objects together as indicated in Table II.

Pro rata the Apollo 12 samples, together, are about twice as rich in objects as the Apollo 11 sample.

TABLE I

Object	Number
Very clear spherules	6
Red-brown spherules	32
Small spherules	56
Rough spherules	41
Broken spherules	6
Spherules with tails	12
Grey spherules	25
Cylinders	29

TABLE II

Sample	Object	Number	Total	% cylinders
Contingency	Spherules	39		
12070.34	Egg shapes	3	50	16%
	cylinders	8		
Documented	Spherules	25		
12057.70	Complete cylinders	14	45	45%
	Broken cylinders	6		
Selected	Spherules	24		
12001.70	Complete cylinders	7	35	32%
	Broken) cylinders	4		

There is a wide scatter in content in the smears. An attempt was made with reasonable success to produce on each slide studied a smear roughly in weight about 1 mg. On the average, each smear from Apollo 11 yielded something like 100 objects whilst similar smears from Apollo 12 were about twice as rich. The detailed content of the richest single smear found, which was from Apollo 12, in the selected sample No. 12001.70 is as follows. There are 230 small spheres of very good shape, 55 which can be described as egg-shaped and 30 which were cylinders, some with and some without obvious waists. Thus of this total of 315 objects only some $9\frac{1}{2}\%$ can be described as cylindrical. Clearly with very small objects, surface tension forces tend to favour sphericity. It is notable that this represents 3.15×10^9 objects per kg within range of optical resolution. An electron microscope study might yield numbers of still smaller objects.

6. Discussion

There seems to be a real difference between the extractable ('macro') objects and the much more abundant 'micro' objects. It is proposed that several distinct mechanisms might be involved. It can in the first instance be considered that the 'macro' objects are created through a meteoric impact which raises the temperature both of itself and of the rock struck to that of melting point. Consequently there would be violent splashing in possibly three distinct ways. Firstly droplets would be flung out violently and the smaller would freeze rapidly into hard glassy spherules. These would impact on landing producing the features already described. Fine threads or jets of glass could also be thrown out and these would break up into the observed cylinders. The larger blobs of glass would not have time to freeze and onlanding would both form flat 'pancakes' (as reported) and also possibly these would produce secondary splashing with more spherules etc. created.

It might be expected that this same impact which leads to drop formation could also pulverise surrounding areas and thus hurl up a cloud of small solid particles. It would be the passage of the frozen spherules through such a cloud which would explain the micro-craters seen on so many spherules.

This proposed mechanism can account both for the widespread distribution of glass lumps and glass spherules. It certainly does not explain the very high content of tiny fragmented pieces of glass in the Lunar soil and neither does it explain the enormous numbers of micro objects, nor the fact that many of them are broken. So a subordinate hypothesis is proposed. It is suggested that repeated multiple impacts (and clearly enormous numbers of impacts have happened) lead each to the passage of a violent shock wave across the surface of the Moon, sufficiently violent to break down existing glass to the vast number of small fragments observed. This explains too the shocked nature of so many glass fragments. Then, on occasion, it is conjectured that a more than usually violent impact happens and as a consequence the resulting shock front which is propagated has now a high temperature, sufficient to melt locally some of the small glass fragments. This then creates the micro objects which are predominantly of the very same colour and character as the neighboring shattered small glass fragments so abundant in the soil.

It is not surprising then if later impacts producing later shock waves can break up some micro objects, as is found. It is very difficult to account otherwise for broken micro spherules and broken micro cylinders. For even if these had been projectiles their momenta could hardly lead to shattering on impacting into the soft dust.

The smooth interferometric specularity observed often must be associated with the formation of the objects in the good vacuum conditions on the Moon. For an interferometric examination carried out on the tiny glass 'ballotini' made industrially for filters etc. by hot flames blasting onto glass fibres, in the Earth's atmosphere, shows fringe patterns which give evidence of crude rough surface wrinkling which bear no resemblance to Lunar spherule interferograms at all. Nor do Earth tektites show surface features resembling Lunar glassy objects.

A possible alternative mechanism for production of the glassy Lunar objects can be conjectured as follows. Let it be proposed that a really large violent meteoric impact melts a big region of rock, creating in effect a veritable extensive lake of very hot molten glass. Now let this be followed by a violent eruption of gas from below. This eruption will be explosively violent, into the effective high vacuum of the Moon's 'atmosphere'. Such a gas blow might be prolonged and could produce a veritable fountain of an enormous number of tiny glass droplets, hurled far and wide. Even the micro objects might also be envisaged as a consequence of sufficiently violent gas eruptions through a molten lake of glass.

However one still has to explain the existence of the fracture-reduced fragments of glass in the soil. Maybe the two mechanisms proposed could both operate.

If this hot lake mechanism is valid, one might expect that if future astronauts land inside a crater then they could find that its inner floor consisted of solidified glass, of the same constitution as the numerous typical spherules.

Finally it is notable that although two sites of the Apollo missions are some 800 miles apart, there is such close identity in the nature of the spherules. In addition to this is the fact that the retrieved 100 gm sample brought back by the Russian automatic retrieval rocket, although picked up from a site 1500 miles away from one of the Apollo sites, also shows at least one glass spherule.

There seems every indication that the whole Moon is probably covered with these objects and the numbers of them must be very vast.

THE VALENCE STATES OF 3*d*: TRANSITION ELEMENTS IN APOLLO 11 AND 12 ROCKS

ALVIN J. COHEN

University of Pittsburgh, Pennsylvania, U.S.A.

Abstract. The absorption spectra of Apollo 11 fine-grained rocks, 10017 and 10022 are due entirely to pyroxene minerals. Spectral bands due to Fe^{3+}, Fe^{2+}, Cr^{3+} and Ti^{4+} and Ti^{3+} are detected. Single crystals of olivine in rocks 12021 and 12018 show bands due to Fe^{3+}, Fe^{2+}, Cr^{3+}, Ti^{3+}, Mn^{3+}, and Mn^{2+}. Pyroxenes in the same rocks exhibit band maxima of the same cationic species as in the olivines. Spectral shifts are noted due to anisotrophy of the crystal structures.

Heating sections 10017, 10022, and 12018 from the rock interiors at 200–225°C for 2 h caused large decreases in the spectral intensity of Fe^{3+}, Cr^{3+} and Ti^{3+}, indicating the following reaction:

$$Fe^{3+} + Cr^{3+} + Ti^{3+} \rightarrow Fe^{2+} + Cr^{2+} + Ti^{4+}$$

This suggests that Fe^{3+}, Cr^{3+} and (a portion of) Ti^{3+} are not in equilibrium. It is most probable that they were produced subsequent to the formation of the rocks by a combination of secondary ionization processes following cosmic ray bombardment and by trace radioactivity present in the rocks.

An orange glass, 150 μ in diam and 50 μ thick contained in brecciated rock, 10048.44, exhibited 15 identifiable absorption bands related to Fe^{2+}, Cr^{3+}, Ti^{3+}, Mn^{3+} or Mn^{2+} ions.

Plagioclase in 12021.65 has perfect transmission over the region studied. The limit of Fe^{3+} is in the order of < 1 ppm and Fe^{2+}, 1000 ppm or less in this plagioclase single crystal of dimensions 0.6 mm $\times$ 0.2 mm $\times$ 30 μ.

1. Introduction

Although Fe^{3+} and Ti^{3+} have been detected in Apollo 11 rocks by spectrophotometric (Hapke *et al.*, 1960) and by EPR techniques (Weeks *et al.*, 1970), no mechanism for their origin in lunar minerals has hitherto been suggested. The absorption spectra of lunar pyroxene and olivine crystals are reported here in detail for the first time. Due to the random orientation of these crystals, in polished sections 30 μ thick, the anisotropy of the spectra was not investigated.

A. EXPERIMENTAL

The absorption spectral data were taken using a Cary Model 14 Spectrophotometer with IR-1 Modification and a halogen lamp of variable intensity as a light source in the visible and infrared regions investigated. The spectra of the small single crystals and glass were measured using matched circular slits of 2 mm diam in sample and reference beams of the instrument. All samples investigated were approximately 30 μ thick except for the orange glass from rock 10048.44 which was approximately 50 μ thick. All single crystals measured were 2 mm or greater in diameter except the pyroxene, 12018.50 which was a grouping of several smaller crystals. The orange glass was 150 μ in diam. surrounded by an opaque matrix of the breccia in which it was embedded.

The thinness of the specimens studied precluded detection of weak *d-d* transitions of the Fe^{3+} ion. The 30 micron thickness was necessary in order to study the intense Fe^{3+} charge-transfer bands in the ultraviolet region. The Fe^{3+} band peak could not

Urey and Runcorn (eds.), The Moon, 264–278.

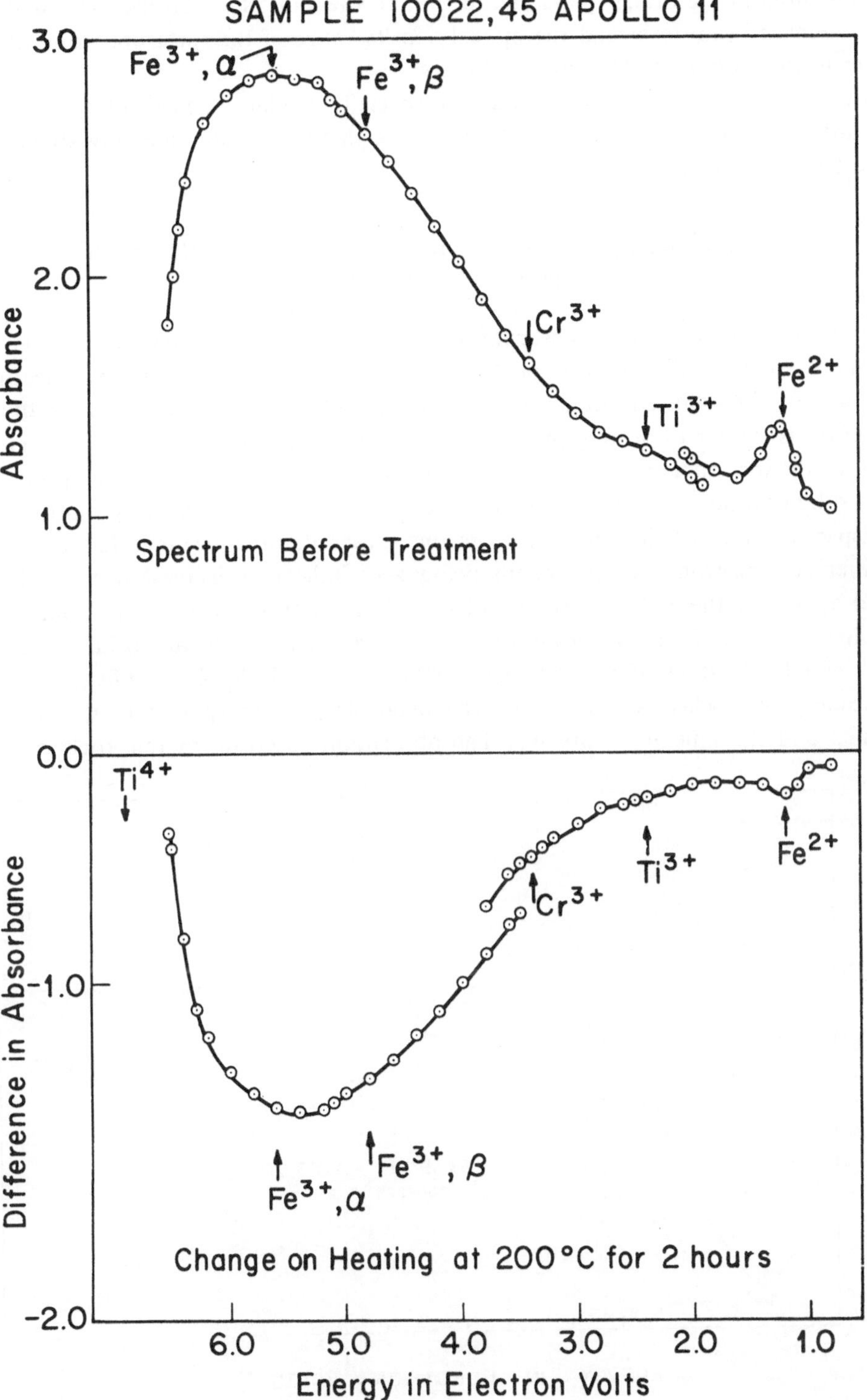
SAMPLE 10022,45 APOLLO 11
3.0
2.0
1.0
0.0
-1.0
-2.0
Absorbance
Difference in Absorbance
Fe3+, α
Fe3+, β
Cr3+
Ti 3+
Fe2+
Spectrum Before Treatment
Ti4+
Change on Heating at 200°C for 2 hours
6.0
5.0
4.0
3.0
2.0
1.0
Energy in Electron Volts

Fig. 1.

be measured in the orange glass due to the thickness of 50 μ. Screens were used in the reference beam to extend the optical density range of the spectrophotometer in order to measure the Fe^{3+} peak in all cases.

The polished sections were mounted on specially machined brass holders so that the spectrophotometer beam passed only through the sample under investigation, no other extraneous matter being present.

2. Absorption Spectra of whole Rock Before and after Heating at 200–225 °C for two Hours

The effect of heating on the absorption spectrum of a polished interior section of Apollo 11 rock, 10017 has been published earlier (Hapke *et al.*, 1970). The heating at 200 °C for 2 h followed irradiation with ultraviolet light and X-rays. After standing, the effect is similar to heating pristine 10022 or 12018 pristine interior rock. Figure 1 shows the spectrum of rock 10022.45 before heating and the change in the spectrum caused by heating at 200 °C for 2 h. The second spectrum is obtained by subtracting the spectrum taken before heating from one measured after heating. The result is a difference spectrum. Spectral bands below zero indicate a decrease in each given band appearing there. This is the result of a decrease in the specific absorbing ion causing the band. Thus one sees in Figure 1 a decrease in bands due to Fe^{3+}, Cr^{3+}, Ti^{3+} and Fe^{2+}. Rock 10022 (similarly 10017) contains little olivine but is rich in pyroxenes, plagioclase and ilmenite. The plagioclase is transparent in the region studied and the ilmenite is opaque. The absorption spectrum of this rock is due

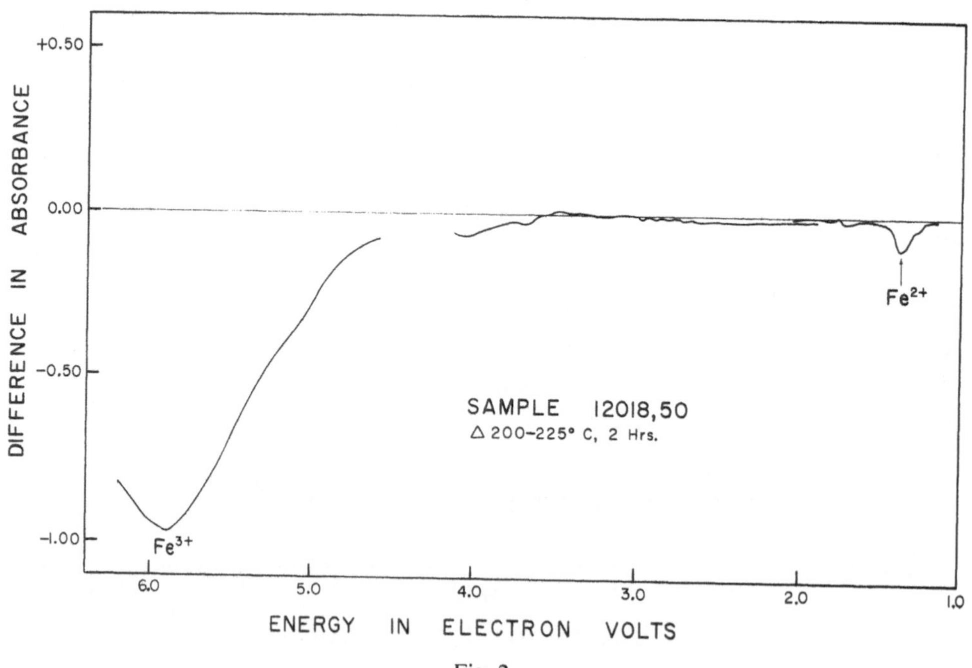

Fig. 2.

entirely to the pyroxene minerals present and changes upon heating are therefore changes in the pyroxene. The apparent decrease of the Fe^{2+} on heating is because of the interaction of neighboring Fe^{3+} and Fe^{2+}, probably across t2g orbitals. Thus decrease in the concentration of the Fe^{3+} causes an apparent decrease in the Fe^{2+} although the amount of Fe^{2+} is actually increasing slightly.

Figure 2 illustrates the difference spectrum of Apollo 12 rock 12018.50, an interior portion, upon heating at 200–225 °C for two hours. The major effects noted are decreases in Fe^{3+} and Fe^{2+} bands. This rock contains both pyroxenes and olivine and thus the difference spectrum is more complex than in the other two rocks studied. Table I summarizes the bands found to decrease upon heating at 200–225 °C in all

TABLE I

Negative difference spectra, after heating at 200–225 °C for 2 h, all specimens approximately 30 μ thick

Absorbing ion	Peak maximum in eV		
	Apollo 11 10017	Apollo 11 10022	Apollo 12 12018
Fe^{2+}	1.1	1.3	1.28, 1.37
Ti^{3+}	–	2.4	1.72, 2.46
Cr^{3+}	3.2, 3.53	3.5	1.82, 1.94, 3.20, 3.70, 4.06
Fe^{3+}	4.5	–	–
Cr^{3+}	4.8	4.8	5.28
Fe^{3+}	5.5	5.5	5.90
Ti^{4+}	tail > 6	tail > 6	tail > 6

three rocks. The transitions to which the bands are related will be discussed when the spectra of the individual minerals are presented. Two positive bands are observed in Figure 2 at 3.35 and 3.42 eV. A band due to Mn^{2+} appears in spodumene at 3.35 eV (unpublished work) and this may be a similar transition in lunar augite. The band at 3.42 may also be due to Mn^{2+}.

The decrease of Fe^{3+}, Cr^{3+} and Ti^{3+} upon heating at relatively low temperature has been interpreted as reduction of the Fe^{3+} and Cr^{3+} accompanied by oxidation of the Ti^3 according to the following equation:

$$Fe^{3+} + Cr^{3+} + 2Ti^{3+} \rightarrow Fe^{2+} + Cr^{2+} + 2Ti^{4+}$$

This reaction takes place in both pyroxenes and olivine as will be shown later.

An examination of the three rocks studied indicates that they have all cooled slowly. Under these conditions one expects to find the cations in the individual minerals in equilibrium. If this is true, one would not expect oxidation-reduction reactions to take place among the cations at a temperature as low as 200–225 °C. It is known that natural α- and β-radiation cause oxidation of Fe^{2+} in terrestrial minerals. β-radiation is more efficient than α-radiation. Cosmic rays are a rich source of secondary β-rays. Therefore it is suggested that both Fe^{2+} and Cr^{2+} are oxidized by secondary radiation from cosmic rays as well as by natural radioactivity present in the lunar rocks. A Ti^{4+} ion is reduced to Ti^{3+} for each Fe^{2+} and Cr^{2+} originally present in the rock that is oxidized to Fe^{3+} and Cr^{3+}. It is suggested that a major portion of these three ionic species are produced in the pyroxene and olivine of lunar rocks after the rocks have cooled and been subjected to subsequent radiation.

The conclusion that ferric iron was produced subsequent to the cooling of the rock is strengthened by the low oxygen fugacity found in lunar rocks and by the absence of magnetite.

The probable location of the ions, being oxidized or reduced, in the mineral structure will be suggested when the individual mineral is discussed.

3. Spectra of Pyroxene in Rocks 12021 and 12018

Rock 12021 has crystals of large size compared to Apollo 11 rocks. Some pyroxene grains have zones of differing color. These colored zones consist of a core of greenish yellow pigeonite rimmed with a brown zone of augite pyroxene. (Warner, 1970). The absorption spectrum of the pyroxene single-crystal measured in rock 12021.65 was a zoned crystal as described above. The spectrum is shown in Figures 3 and 4 as a dashed curve. The absorption band maxima attributed to Fe^{2+} are listed in Table II. The Fe^{2+} spectrum consists of two *d-d* transitions of Fe^{2+} in M_1 sites in the structure and two *d-d* transitions due to Fe^{2+} in M_2 sites. The configuration of the M_1 site is that of a regular octahedron and that of the M_2 site is that of a distorted

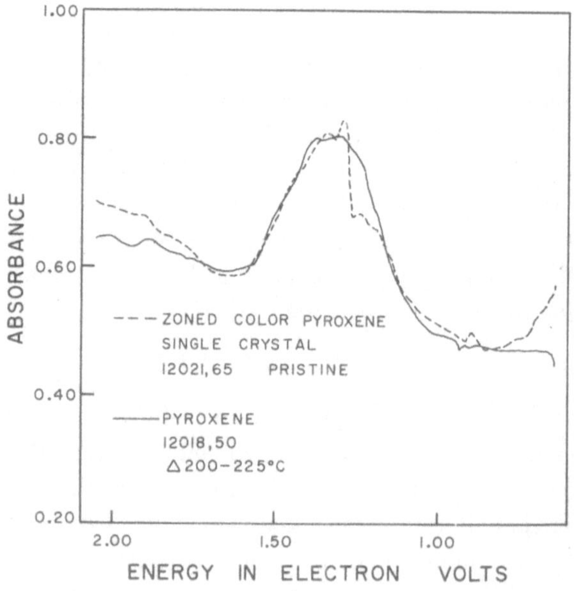

Fig. 3.

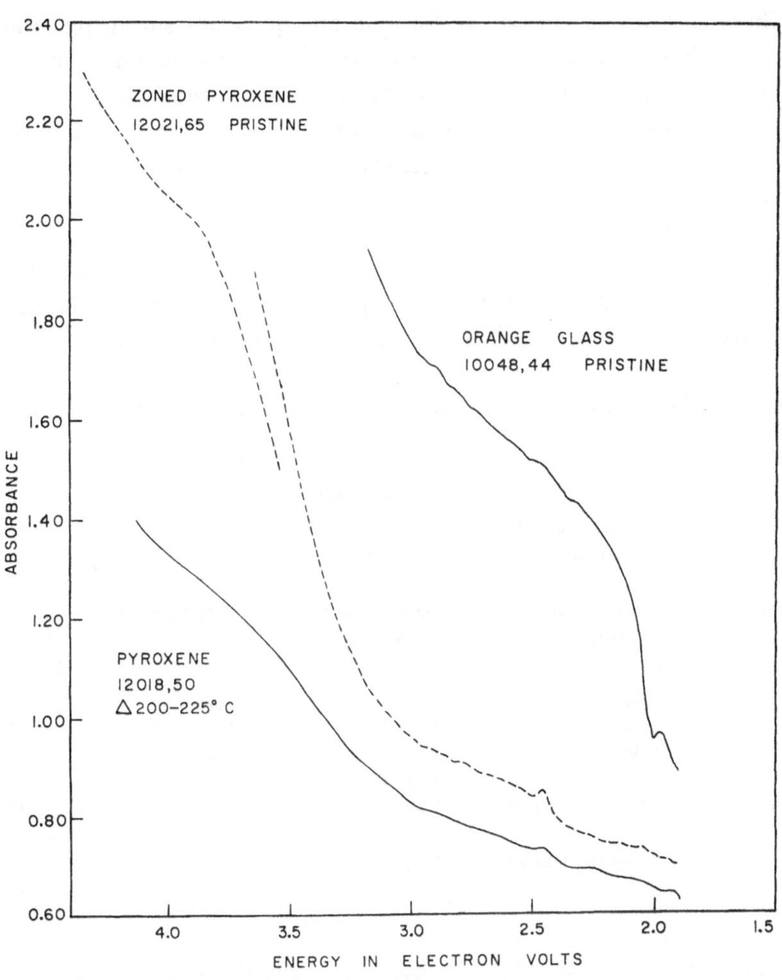

Fig. 4.

TABLE II

Ferrous absorption bands in zoned pyroxene crystal from rock 12021.65

Structural position	Band maximum in eV	Absorbance	Δ cm^{-1}
M_2–pigeonite	0.675	0.58	5 440
M_2–augite	0.75	0.49	6050
M_1–pigeonite	1.19	0.66	9600
M_1–augite	1.24	0.68	10000
M_1–pigeonite	1.29	0.85	10405
M_2–augite	1.34	0.81	10810
M_1	1.41	0.76	11370
charge transfer } Pigeonite	1.97	0.71	15890
$Fe^{2+} \rightarrow Fe^{3+}$ } Augite	2.01	0.72	16210

polyhedron. Since one would expect the Fe^{2+} in an equivalent site in pigeonite to absorb energy at a lower value (longer wavelength) than in titanaugite, the M_1 and M_2 absorption bands are doubled (except for the M_1 transition at 1.41 eV) for transitions of Fe^{2+} in each of the two regions of the structurally-zoned pyroxene crystal. The charge-transfer band is also doubled, because of an energy-shift in the two zones of the crystal.

In the pyroxene structure one would ordinarily expect Fe^{3+} to favor M_1 positions and Fe^{2+} to favor M_2 positions. However, if the Fe^{3+} is caused by radiation damage it is not yet possible to predict whether a Fe^{2+} in an M_2 or M_1 site will be favored or whether the oxidation of Fe^{2+} to Fe^{3+} will be a random process favoring neither structural position.

Table III lists the Fe^{2+} bands in a wine-brown pyroxene in Apollo 12 rock 12018.50

TABLE III

Ferrous and ferric absorption bands in a poly-crystalline aggregate of pyroxene from rock 12018.50 after heating at 200–225°C for 2 h

Ion	Structural position	Band maximum in eV	Absorbance	Δ cm^{-1}
Fe^{2+}	M_2	0.66	0.48	5325
Fe^{2+}	M_1	1.265	0.78	10205
Fe^{2+}	M_2	1.31	0.80	10565
Fe^{2+}	M_1	1.38	0.80	11130
Fe^{2+}	Charge transfer $Fe^{2+} \rightarrow Fe^{3+}$	1.94	0.67	15650
Fe^{3+}	Charge transfer	5.635	2.54	45450

after heating at 200–225°C for 2 h. The pyroxene crystals in this rock are also relatively large. In the polished section studied, the best region of 2 mm diam was a clump of pyroxene crystals of the same color. The absorption spectrum of this clump of pyroxene crystals is shown in Figures 3 and 4. These crystals are of the same thickness as the zoned-crystal in rock 12021. The intensity of the absorption is almost identical in the infra-red region as shown in Figure 3. However, the Fe^{2+} spectrum is not doubled. In comparing the absorption peak positions to those listed in Table II, one cannot choose whether the polycrystalline aggregate of crystals is pigeonite or augite.

The Fe^{3+} charge transfer band is so intense in the zoned pyroxene, it could not be measured. In the pyroxene in rock 12018 after heating, the intensity of the Fe^{3+} band has been reduced enough to be measured as shown in Table III. It is not illustrated but is similar to the α-Fe^{3+} band in Figure 1 before any heat treatment although of somewhat lower absorbance.

Tables IV and V tabulate the absorption maxima of bands related to manganese that are too weak to be seen readily in the difference spectra of the whole rock;

TABLE IV

Chromic, titanous, manganic and manganus bands in zoned pyroxene from rock 12021.65

Ion	Probable structural position	Transition	Band maximum in eV	Absorbance	Δ cm^{-1}
Cr^{3+}	M_1	*d-d*	1.85	0.65	14920
Cr^{3+}	M_1	*d-d*	1.90	0.68	15325
Cr^{3+}	M_1	Charge transfer	3.95	2.02	31860
Ti^{3+}	M_1	Charge transfer	2.45	0.85	19760
Mn^{3+}	M_1	*d-d*	2.28	0.76	18390
Mn^{2+}	M_1	*d-d*	2.78	0.91	22425
Mn^{2+}	M_1	*d-d*	2.89	0.92	23310
Mn^{2+}	M_1	*d-d*	2.98	0.96	24035

TABLE V

Chromic, titanous, manganic and manganous bands in pyroxene from rock 12018.50 after heating at 200–225°C for 2 h

Ion	Probable structural position	Transition	Band maximum in eV	Absorbance	Δ cm^{-1}
Cr^{3+}	M_1	*d-d*	1.80	0.62	14520
	M_1	*d-d*	1.89	0.64	15245
	M_1	*d-d*	2.01	0.66	16210
	M_1	Charge transfer	2.95	0.82	23795
	M_1	Charge transfer	3.9	1.29	31455
Ti^{3+}	M_1	Charge transfer	1.75	0.615	14115
	M_1	Charge transfer	2.45	0.735	19760
Mn^{3+}	M_1	*d-d*	2.28	0.68	18390
Mn^{2+}	M_1	*d-d*	2.95	0.81	23795

they are detectable in the spectra of the crystals of pyroxene and olivine. If the major portion of Fe^{3+} and Cr^{3+} are produced by oxidation of Fe^{2+} and Cr^{2+} respectively then the higher valence states which normally would favor M_1 positions may be in M_2 positions. The Fe^{3+} ($3d^5$) with zero crystal field stabilization energy would favor the M_1 site because of its smaller ionic size compared to Fe^{2+}. The Cr^{3+} ($3d^5$) ion would favor the M_1 sites both for crystal field and ionic size reasons. However Fe^{2+} ($3d^6$) and Cr^{2+} ($3d^4$) favor M_2 positions based on site distortion factors. Since the oxidized states are produced after the crystal has formed by radiation damage, one has Fe^{3+} in a Fe^{2+} site and Cr^{3+} in a Cr^{2+} site. These sites must undergo some local steric readjustment due to the change in cationic charge. The Ti^{4+} reduced to Ti^{3+} upon oxidation of the Fe^{2+} and Cr^{3+} may be in either silicon sites or M_1 sites or both, the M_1 site possibly being favored.

4. Spectra of Olivines in Rocks 12021 and 12018

The olivines in these two Apollo 12 rocks are the largest single crystals present. They are light yellow-green in color in 30 μ thickness. Figure 5 illustrates the Fe^{2+} spectrum of an olivine crystal in rock 12021.65 as found in the untreated rock and the spectrum of a single olivine crystal in rock 12018.50 after heating at 200–225 °C for two hours. Tables VI and VII tabulate the Fe^{2+} band data for these two crystals. The transitions are all *d-d* except the charge transfer bands near 2.0 eV. There is a Fe^{3+} charge transfer band in the ultraviolet region in the olivine crystal in rock 12021.65 at

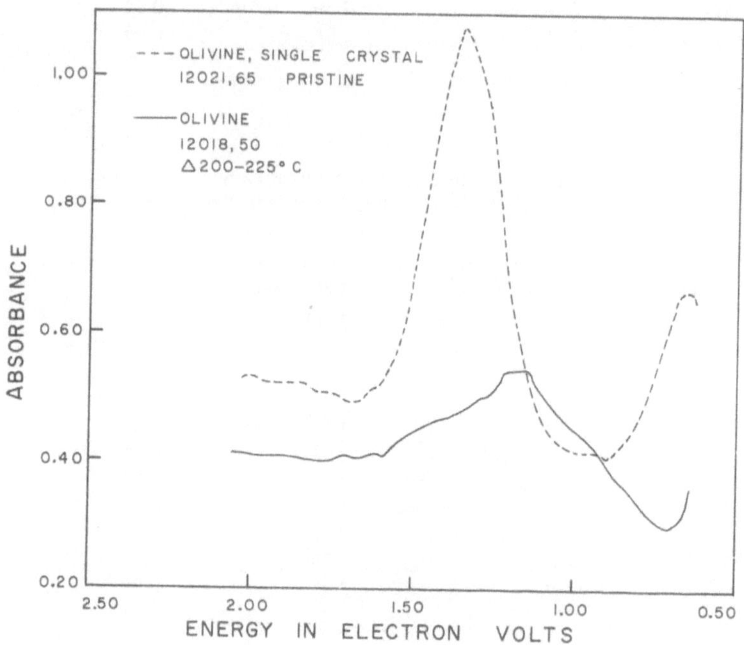

Fig. 5.

TABLE VI

Ferrous absorption band in olivine in rock 12021.65

Structural position	Band maximum in eV	Absorbance	Δ cm^{-1}
	0.66	0.68	5325
M_2	1.18	0.52	9520
M_1	1.34	1.08	10810
M_1	1.62	0.51	13065
Charge transfer $Fe^{2+} \rightarrow Fe^{3+}$	2.00	0.54	16130

TABLE VII

Ferrous absorption bands in olivine in rock 12018.50 after heating at 200–225 °C for 2 h.

Structural position	Band maximum in eV	Absorbance	Δ cm^{-1}
M_2	1.18	0.54	9520
M_1	1.28	0.50	10325
M_1	1.62	0.41	13070
Charge transfer $Fe^{2+} \rightarrow Fe^{3+}$	2.05	0.38	16535

TABLE VIII

Chromic, titanous, manganic and manganous bands in olivine crystal in rock 12021.65

Ion	Probable structural position	Transition	Band maximum in eV	Absorbance	Δ cm^{-1}
Cr^{3+}	M_1	*d-d*	1.77	0.51	14275
Cr^{3+}	M_1	*d-d*	1.86	0.52	15000
Cr^{3+}	M_1	Charge transfer	4.00	1.67	32265
Cr^{3+}		Charge transfer	4.59	2.18	37020
Ti^{3+}	M_1	Charge transfer $Ti^{3+} \rightarrow Ti^{4+}$	2.44	0.645	19680
Mn^{3+}	M_1	*d-d*	2.30	0.58	18550
Mn^{2+}	M_1	*d-d*	2.67	0.67	21535

TABLE IX

Chromic, Titanous, Manganic and Manganous bands in olive crystal in rock 12018.50 after heating at 200–225 °C for 2 h

Ion	Probable structural position	Transition	Band maximum in eV	Absorbance	Δ cm^{-1}
Cr^{3+}	M_1	*d-d*	1.92	0.41	15485
Cr^{3+}		Charge transfer	3.24	0.55	26135
Cr^{3+}		Charge transfer	4.59	3.06	37020
Ti^{3+}	M_1	Charge transfer	1.71	0.41	13790
Ti^{3+}	M_1	Charge transfer $Ti^{3+} \rightarrow Ti^{4+}$	2.52	0.44	20325
Mn^{3+}	M_1	*d-d*	2.38	0.40	19195
Mn^{2+}	M_1	*d-d*	2.72	0.44	21940
Mn^{2+}	M_1	*d-d*	3.06	0.50	24680

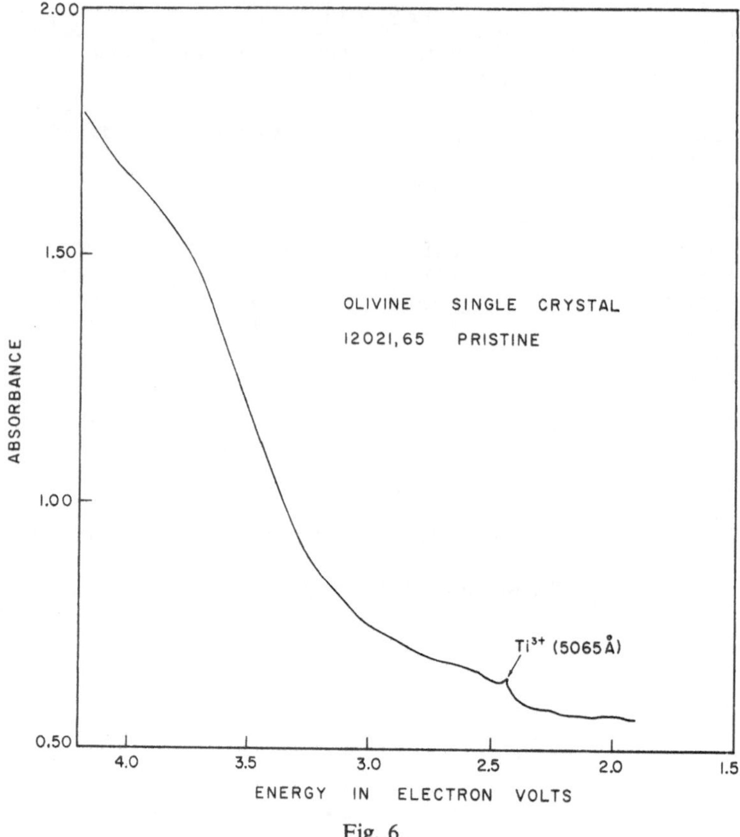

Fig. 6.

5.64 eV with an absorbance of 2.86. The Fe^{3+} band was too intense to measure in the heated olivine from rock 12018.50.

The M_1 structural position in olivine is that of a tetragonally distorted octahedron while the M_2 site has the configuration of a trigonally distorted octahedron. The Fe^{2+} ion tends to favor the M_2 position in olivine.

Tables VIII and IX list the peaks of absorption bands attributed to Cr^{3+}, Ti^{3+}, Mn^{3+} and Mn^{2+} in the two olivines measured.

Figure 6 illustrates the high intensity of the Cr^{3+} charge-transfer band at 4.00 eV (3100A) compared to the Ti^{3+} charge transfer band in olivine at 2.44 eV (5065A). The Cr^{3+} band not shown at 4.59 eV is still more intense. This figure indicates that Cr^{3+} is important in affecting the color on the ultraviolet side of the spectrum as well as Fe^{3+} in both olivine and pyroxene.

5. Comparison of Difference Spectrum of Rock 12018.50 to Spectra of Pyroxene and Olivine

In Table X, the negative peaks in the difference spectrum of rock 12018.50 after heating at 200–225 °C for 2 h is compared to the spectra of the individual pyroxene

TABLE X

Mineral source of difference spectral bands in rock 12018.50

Ion	Rock peaks in eV (negative)	Pyroxene peaks in eV	Olivine peaks in eV
Fe^{2+}	1.28	1.31	1.28
Fe^{2+}	1.37	1.38	1.41
Ti^{3+}	1.72	1.75	1.71
Cr^{3+}	1.82	1.80	–
Cr^{3+}	1.94	1.94	1.92
Ti^{3+}	2.46	2.45	2.52
Mn^{2+}(?)	2.99	2.95	3.06
Cr^{3+}	3.20	–	–
Cr^{3+}	3.70	–	–
Cr^{3+}	4.06	3.9	–
Cr^{3+}	5.28	–	4.59
Fe^{3+}	5.90 (positive)	5.64	–
Mn^{2+}(?)	3.35	–	–
Mn^{2+}(?)	3.42	–	–

and olivine crystals in the same rock after heating. All peaks in the whole rock match peaks of either mineral or both minerals except the charge transfer peaks of Cr^{3+} (5.28) and Fe^{3+} (5.90) are shifted to higher energy. Also there is no Cr^{3+} peak at 3.7 in the minerals. The shift to higher energy in the deep ultraviolet may be apparent rather than real due to the increase of Ti^{4+} in the vacuum ultraviolet, the tail of which overlaps the Fe^{3+} and Cr^{3+} bands in the region 5–6 eV, causing them to appear to be at higher energy. This is caused by the increase in Ti^{4+} band due to oxidation of the Ti^{3+} on heating. The entire reaction on heating at 200–225 °C appears to be:

$$Fe^{3+} + Cr^{3+} + 3Ti^{3+} (+Mn^{3+}) \rightarrow Fe^{2+} + Cr^{2+} + 3Ti^{4+} (+Mn^{2+}).$$

The data on Mn^{2+} is uncertain and the positive bands in the difference spectrum of the whole rock at 3.35 and 3.42 eV have not been positively identified.

6. Spectrum of Plagioclase in Rock 12021.65

The absorption spectrum of a single crystal of plagioclase in rock 12021.65 with dimensions 0.6 mm × 0.2 mm (× 30 μ thick) was measured. In the wavelength region 2000–17500 Å, there is no light absorption. The absence of the intense Fe^{3+} charge transfer band in the ultraviolet indicates that the Fe^{3+} ion in plagioclase is less than one ppm. The weakly absorbing Fe^{2+} could be present in the order of magnitude of 1000 ppm and not be detected. It is hoped to measure a thicker crystal of lunar plagioclase to see if any spectral bands can be observed.

7. The Absorption Spectrum of a Fragment of Oranges Glass in Brecciated Rock 10048.44

This 150 μ diam particle of orange glass in a polished section of breccia, 50 μ thick had the largest cross-section of any transparent material present. The glass fragment was roughly circular with a bright orange-red color in transmitted light. Figure 4 includes the spectrum of this glass in the visible region, the most prominent band is the $Fe^{2+} \rightarrow Fe^{3+}$ charge-transfer band with a maximum at 1.98 eV. Figure 7 shows

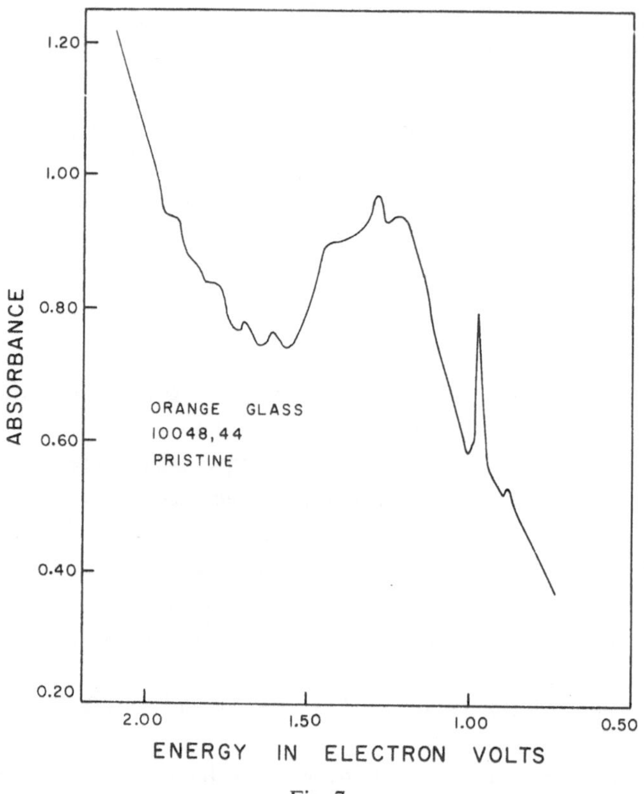

Fig. 7.

the absorption spectrum of the glass in the infra-red region. A prominent vibrational band is present at 0.98 eV. The source of this band has not been identified. Table XI lists the spectral bands present in this glass and the ions responsible. Due to the small size of the fragment, the spectrum could not be observed below 3.18 eV (3900A) because of its high intensity of absorption and the small amount of light transmitted by the small size of the fragment. Thus the ferric peak could not be observed although the 1.98 eV $Fe^{2+} \rightarrow Fe^{3+}$ charge-transfer band is present indicating that Fe^{3+} is present in the glass in sufficient quantity to interact with the Fe^{2+} ion.

TABLE XI

Absorption bands in orange glass fragment in brecciated rock 10048.44

Ion	Probable configuration	Transition	Band maximum in eV	Absorbance	Δ cm^{-1}
Fe^{2+}	octahedral	*d-d*	1.22	0.94	9840
Fe^{2+}	octahedral	*d-d*	1.29	0.97	10405
Fe^{2+}	octahedral	*d-d*	1.44	0.90	11615
Fe^{2+}	octahedral	*d-d*	1.61	0.77	12985
Fe^{2+}		Charge transfer	1.98	0.97	15970
Cr^{3+}	octahedral	*d-d*	1.80	0.84	14520
Cr^{3+}	octahedral	*d-d*	1.86	0.87	15000
Cr^{3+}	octahedral	*d-d*	1.93	0.94	15565
Ti^{3+}	octahedral	Charge transfer	1.69	0.78	13630
Ti^{3+}	octahedral	Charge transfer	2.48	1.52	20005
Mn^{3+}	octahedral	*d-d*	2.33	1.44	18795
Mn^{2+}	octahedral	*d-d*	2.18	1.33	10725
Mn^{2+}	octahedral	*d-d*	2.75	1.63	22180
Mn^{2+}	octahedral	*d-d*	2.85	1.67	22985
Mn^{2+}	octahedral	*d-d*	2.91	1.71	23390

7. Conclusions

Heating lunar basaltic rocks at low temperature in air causes reduction of Fe^{3+} and Cr^{3+} and oxidation of the Ti^{3+} in the pyroxenes and olivine present. It is probable that Mn^{3+} is present and reduced to Mn^{2+} by the mild heating at 200 to 225°C. It is suggested that Fe^{3+}, Cr^{3+}, and Mn^{3+} are reduced by such a mild heat treatment because they are not in equilibrium in the mineral structure. They are all in sites of the reduced species, viz. a Fe^{2+} site is the location of Fe^{3+} in the structure. The structural site must readjust to a smaller cation with different bonding. Electroneutrality is preserved in the crystal by a Ti^{3+} being oxidized for each cation reduced by radiation. It is suggested that the most likely sources for the oxidation-reduction reactions are β-rays produced as secondaries following cosmic ray bombardment. Another source of β's is natural radioactivity present in the minerals. Since the rocks studied are from the interior of the rock, the effects of solar X-rays, ultraviolet light, and solar wind protons and electrons are negligible.

Absorption bands of Cr^{2+} were not detected in any of the specimens studied. Indications of Ti^{4+} were indicated by increase in the difference spectra of rocks in the deep ultraviolet upon oxidation of the Ti^{3+} present.

Deep-seated lunar rocks should be lower in Fe^{3+}, Cr^{3+}, and Ti^{3+} than surface rocks if the main source of the oxidation-reduction reactions is cosmic ray bombardment. The concentrations of these species should also increase with time of radiation by cosmic rays.

Acknowledgments

This work was supported by grant NAS 9-9942 of the National Aeronautics and

Space Administration. Mr Grover Moreland of the National Museum of Natural History, Washington, D.C., is thanked for preparation of the polished sections of lunar rock used in this work. Mr. Ed Skopinski prepared the final drawings of the figures.

Professor S. Keith Runcorn is thanked for the invitation to present this paper at the Moon Symposium at the University of Newcastle upon Tyne.

Note added in Proof: Olivine, single crystal in rocks 12021.65 should read: pigeonite, single crystal in rocks 12021.65, throughout the text, figures and tables.

References

Hapke, B. W., Cohen, A. J., Cassidy, W. A., and Wells, E. N.: 1970, 'Solar Radiation Effects on the Optical Properties of Apollo 11 Samples', *Proceedings of the Apollo 11 Lunar Science Conference*, **3**, 2199, Pergamon Press.

Warner, J.: 1970, Apollo 12 Lunar-Sample Information NASA Technical Report R-353, p. 94–96.

Weeks, R. A., Kolopus, J. L., Kline D., and Chatelain, A.: 1970, 'Nuclear Resonance of ^{27}Al and Electron resonance of Fe and Mn', *Proceedings of the Apollo 11 Lunar Science Conference*, **3**, 2467, Pergamon Press.

LUMINESCENCE EXCITATION BY PROTONS AND ELECTRONS, APPLIED TO APOLLO LUNAR SAMPLES

J. E. GEAKE and G. WALKER
UMIST, Manchester, England

and

A. A. MILLS
University of Leicester, England

Abstract. A proton accelerator, modified for work on lunar samples, is described; 60 keV protons at about 1 μA cm^{-2} are used. Luminescence emission spectra in the visible and near IR regions are shown for lunar samples from Apollo 11 and 12, and preliminary results are given for Apollo 14 samples. Lunar samples are compared with terrestrial and meteoritic materials. Plagioclase is found to be the most efficient luminescent material present in the lunar samples, and the activator for its dominant green peak is found to be Mn^{2+}.

Apparatus is described for taking colour photographs of the luminescence emission from rock chips under 6 keV electron excitation. Most of the lunar samples investigated show only plagioclase emission, but one breccia shows a wide variety of colours from different luminescent constituents.

1. Introduction

All of our lunar fines and rock samples showed some luminescence when excited by either protons or electrons. Visually, the fines showed faint bluish-white emission, whereas the rocks and breccias showed brighter emission, but in most cases only from their plagioclase parts which appeared bluish-white. Only one breccia sample showed a wider variety of luminescence colours. The luminescence emission from the samples was recorded photoelectrically as spectral scans, and photographically as colour photographs of the emission. All the samples were also inspected visually under UV lamps giving mainly 2537 Å and 3650 Å emission, but negligible luminescence was observed.

Our interest in the luminescence of the lunar samples is twofold: firstly, to use the characteristic emission spectra of the minerals concerned (mainly plagioclase) to elucidate the mechanism involved in terms of the crystal field situation, and to identify any doping agents involved; secondly, to use luminescence photography to explore the distribution of the luminescent components, and in particular any variations within individual crystals in the hope of gaining some information about the process, rate and direction of crystallisation. The equipment used for these two types of investigation will now be described, and some of the preliminary results will be summarised.

2. Proton-Excitation Equipment

Equipment already existing at Manchester [1] was up-graded for the lunar sample investigation, mainly by cleaning up the vacuum system. The accelerator is shown

Urey and Runcorn (eds.), The Moon, 279–297.

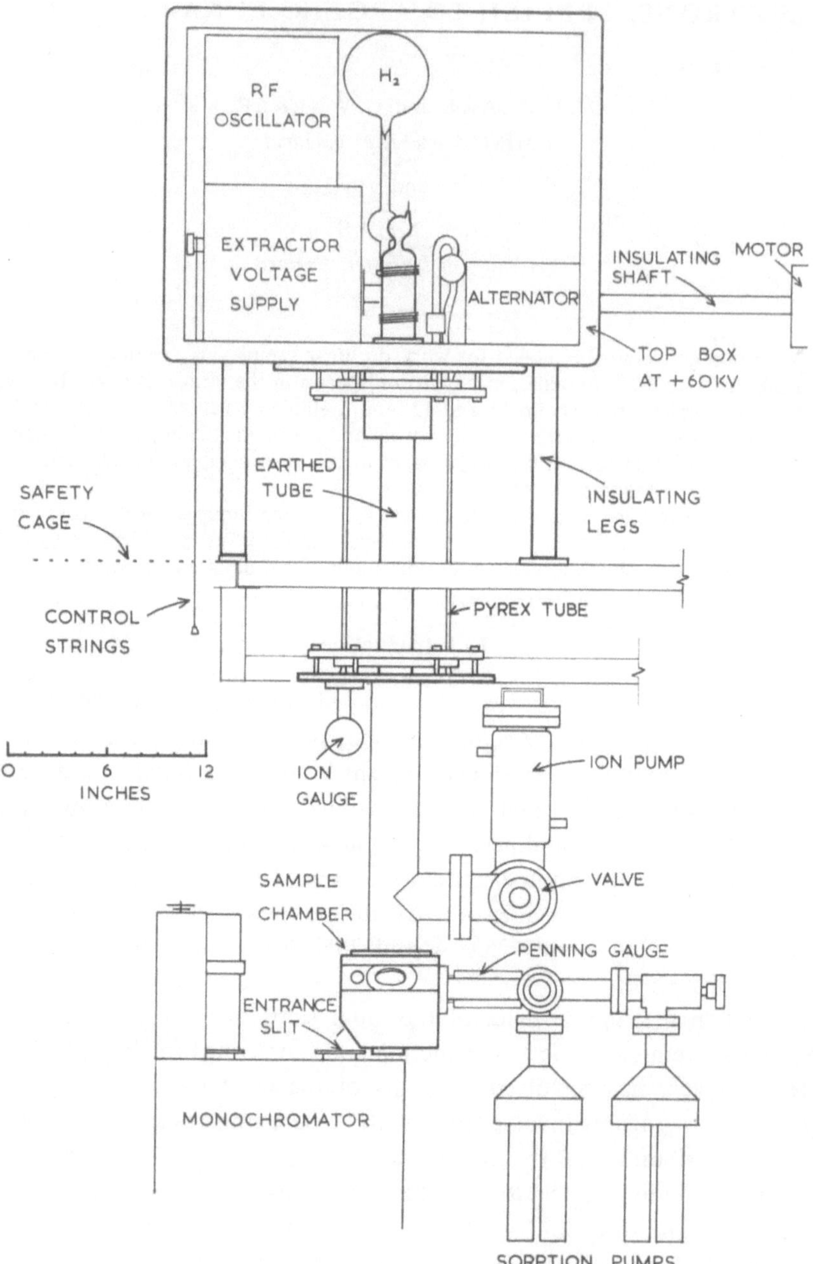

Fig. 1. The proton source and accelerator. An anti-corona collar round the top of the accelerator column and the second photomultiplier to monitor total light are both omitted for clarity.

Fig. 2. The sample chamber, pumps and monochromator.

diagrammatically in Figure 1, and a view of the sample chamber, pumps and monochromator is shown in Figure 2. Protons are produced by ionising spectroscopically pure hydrogen gas at low pressure by means of an RF field at about 20 Mhz, and they leave the ion bottle through a $\frac{1}{16}$ in. diam canal. The whole of the top box containing the ion source is at a high positive potential, usually 60 keV, and the protons are accelerated by the potential difference between the exit canal of the ion source and the upper end of an axial stainless steel tube at Earth potential; they then coast down inside this tube and hit the target in the sample chamber at the bottom. The samples are placed, four at a time, on the turret shown in Figure 3

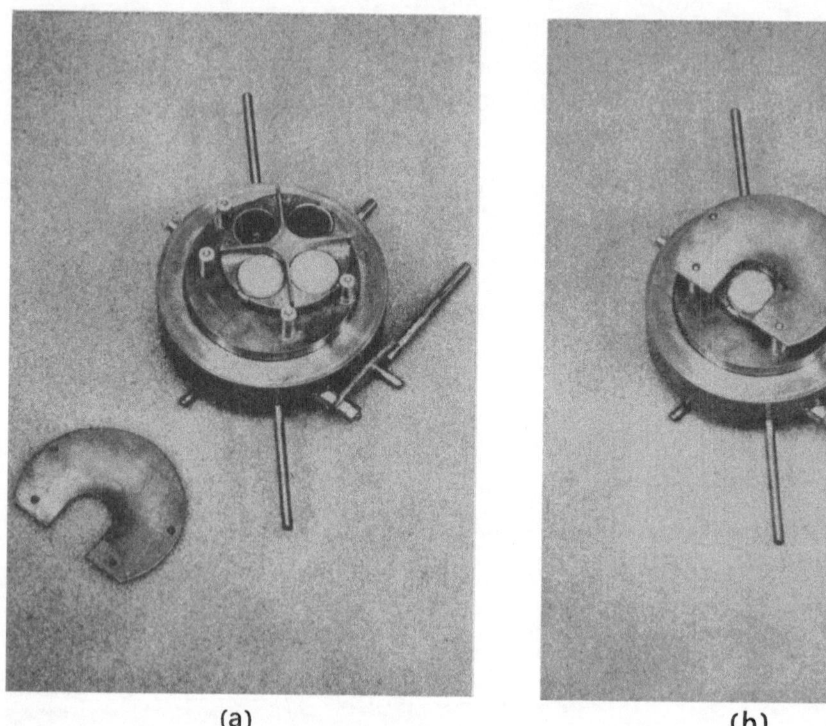

(a) (b)

Fig. 3. The sample turret (a) with the lid off, for loading, (b) with the lid on, ready for use.

which is inserted in the base of the sample chamber; this turret facilitates the intercomparison of samples under the proton beam. A flap valve seals the top of the sample chamber and enables the turret to be removed while keeping the rest of the system under vacuum; it has an insulated shaft and also serves as a collector for beam-current measurement. The field optics of the system results in a beam about 1 in. in diameter at the target, with a current density of about 1 μA cm^{-2}.

In order to reduce the possibility of contaminating lunar samples all the Neoprene O-rings in the system have been replaced by Viton ones of lower vapour pressure, and the oil diffusion and backing pumps have been replaced by an electrostatic getter-ion pump and sorption roughing pumps. The former is a $2\frac{1}{2}$ in. diam triode pump based on the design of Herb *et al.* [2], but with the addition of a grid of a type suggested by Cross [3].

The pump consists essentially of a stainless steel tube surrounded by a water-cooling jacket, an axial anode and a source of electrons. It operates by the injection of electrons from a tungsten filament into the electrostatic field between the earthed tube, which is the cathode, and the axial anode at +5 keV consisting of $\frac{1}{16}$ in. diam tungsten rod carrying a $\frac{1}{4}$ in. diam titanium slug. The electrons follow spiral orbits about the central anode, causing ionisation of the residual gas; eventually they hit the titanium slug, causing it to become white hot and to sublime onto the walls of the pump, thereby 'gettering' the active gases. Inert gases are pumped by ionisation followed by physical burial of the positive ions in the pump wall by titanium. The pumping speed for nitrogen is about 150 litre s^{-1}. The action of the grid wires placed between the anode and the cathode is to improve the sticking probability of the inert gas ions which hit the pump wall, by ensuring that they arrive with an energy of at least 1 keV. In order to increase the active life of the pump between overhauls the filament is duplicated. It is found to be advantageous to keep the pump itself under as good a vacuum as possible between runs; it is therefore connected to the system through a large all-metal isolating valve.

Two four-tube sorption pumps cooled in liquid nitrogen are used to pump the system down initially, to about 10^{-3} torr, at which point the ion pump valve is opened and the sorption pumps are isolated and re-activated. Although the ion pump itself is capable of a very low ultimate pressure, in the present rather long and complicated vacuum system the lowest pressure usually attained, before introducing hydrogen into the ion source, is about 10^{-6} torr; after admitting hydrogen, through a needle valve, an operating pressure of about 10^{-5} torr is reached. This is adequate to permit the proton beam to pass. The vacuum gauges used consist of a Penning gauge on the sample chamber and an ion gauge on the accelerator column; the former is used mainly to monitor the roughing pump pressure and the latter as an aid to adjusting the hydrogen leak on the ion source. Ultimately, the hydrogen leak and the RF oscillator frequency are both adjusted for a maximum beam current, as intercepted by the flap valve.

3. Luminescence Emission Spectra

Light emitted by the sample under proton excitation is reflected by two mirrors, one inside the sample chamber and one outside its window, onto the entrance slit in the horizontal top plate of an Ebert monochromator [4]. This uses a 5×4 in. Bausch and Lomb grating with 1200 lines mm^{-1}, and has a dispersion of 8 Å mm^{-1} in the first order. The spectrum is scanned by tilting the grating, by means of a synchronous motor and a gear-driven micrometer screw. Light leaving the exit slit is recorded, using an EMI 9558B trialkali photomultiplier for the visible region and a Mullard 150 CVP Ag–AgO–Cs photomultiplier, with its photocathode cooled to liquid–

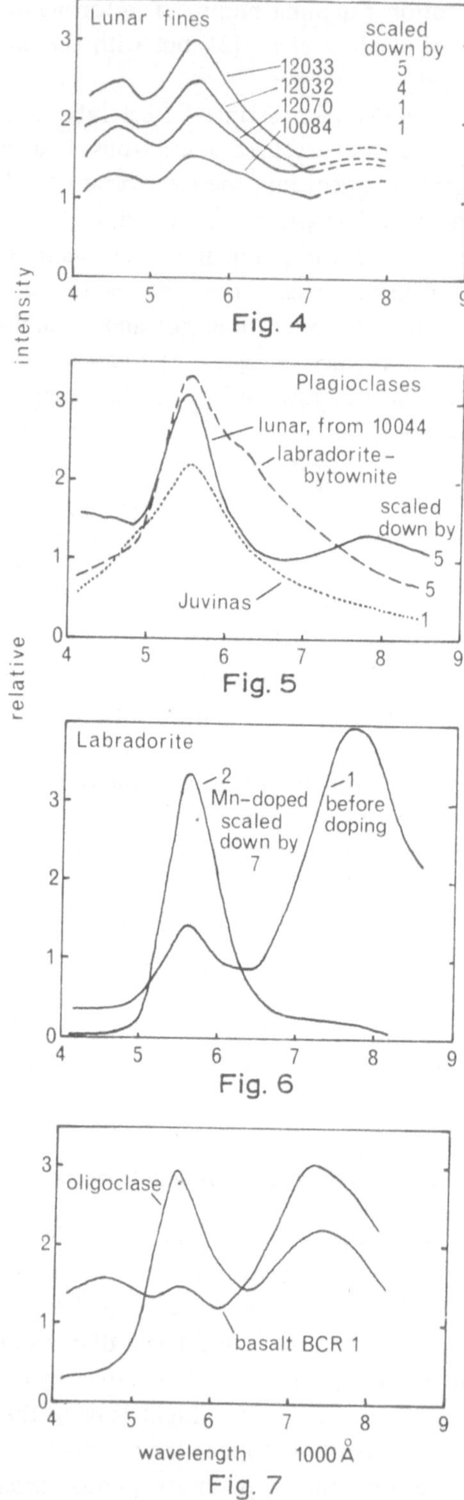

Figures 4–7. Luminescence emission spectra with 60 keV proton excitation, using an instrumental profile width of about 200 Å.

nitrogen temperature, for the near IR region. A further photomultiplier monitors the total light and corrects the output for time fluctuations and for efficiency fall-off caused by proton damage. A piece of Polaroid is placed permanently over the entrance slit to minimise the effect of grating anomalies. A Wratten 2B filter is used for visible region scans and a Wratten 25 filter for IR scans, to remove errors due to second-order light. Spectral response calibration is achieved by comparison with a tungsten lamp of known colour temperature.

Figure 4 shows corrected emission spectra for all three of our Apollo 12 fines samples (12032.39, 12033.60 and 12070.113) [5]; our previously published spectrum for Apollo 11 fines sample 10084.6 is also shown for comparison. We have two types of Apollo 12 fines – 12070.113 is very dark in appearance and resembles the Apollo 11 fines 10084.6, whereas 12032.39 and 12033.60 are much lighter because they contain more plagioclase [(Na, Ca) $(Al, Si)_4O_8$] and less ilmenite [$FeTiO_3$] ,and they are probably ray material [7]. However, all four samples show very similar emission spectra, with a weak 4500 Å peak, a stronger peak at 5600 Å, and some emission in the near IR, each of which we ascribe to plagioclase. It is only the efficiencies that are different for the four samples, and these seem to be related to the plagioclase content. The efficiency increases faster than the plagioclase content, probably because there are two effects: as plagioclase is the most efficient luminescing component, more plagioclase means more emission, but it also means a lower content of absorbing material (eg. ilmenite), as indicated by the albedo, so that the emitted light can get out more easily and can better survive scattering by surface roughnesses, giving an even greater apparent efficiency. The true efficiency of the actual luminescing material is probably about the same for all four samples. Preliminary results for two samples of Apollo 14 fines (14157 and 14263) show that their emission spectra are very similar to those for Apollo 11 and 12 samples, as shown in Figure 4, and that their efficiencies are between those for 12070 and 12032.

The emission spectra of some terrestrial plagioclases are shown in Figures 5, 6 and 7. Apart from the relative intensity of the IR emission, as discussed later, the main difference is that their overall efficiencies are 10–100 times that of the lunar fines. However, the plagioclase found in lunar *rocks* does have an efficiency comparable with that of terrestrial plagioclase; thus the separated plagioclase from lunar rock sample 10044.43 has about the same efficiency as labradorite-bytownite, as shown in Figure 5. The low efficiency of the lunar fines is probably due to the admixture of the plagioclase with opaque material, and also possibly to the coating of individual grains with a thin opaque layer, as suggested by Hapke *et al.* [8, 9] and others.

4. The Cause of Plagioclase Emission

Reasoning from crystal field theory suggested that the prominent green peak shown by both lunar and terrestrial plagioclase might be caused by the presence of Mn^{2+}. To test this, the proton-excited emission spectrum of a typical terrestrial plagioclase–labradorite–was scanned, as shown in Figure 6, curve 1. It was then doped with about 0.1% of Mn by heating it to 1050 °C for 30 min in argon with $MnSO_4$. The result was a 16-fold increase in the height of the green peak, as shown by curve 2, accompanied by a halving of the blue and IR emission. In order to check that simply heating the plagioclase was not producing the change (perhaps by converting the low-temperature form of anorthite–Ca-rich plagioclase–to the high-temperature form) a further sample was given the same heat treatment without any $MnSO_4$: its emission spectrum was found to be unchanged.

Plagioclases usually contain more than 100 ppm of Mn, as determined by X-ray fluorescence analysis. The plagioclase separated from lunar rock 10044 was found to contain about 200 ppm of Mn (about the same as an average terrestrial sample); its emission spectrum is shown in Figure 5–it has a strong green peak like that for the Mn-doped labradorite in Figure 6, but rather more blue and IR emission as well. Sippel and Spencer [10] have pointed out that albite (Na-rich plagioclase) does not show a green peak, and they conclude that when it is present this peak is caused by a divalent activator substituting for Ca^{2+} in the lattice. Our results seem consistent with Mn^{2+} being the activator responsible. The situation may be compared with that of the Mn^{2+} emission peak which is found for α-$CaSiO_3$: Mn (α-wollastonite) and which also occurs at about 5600 Å [11]. We have discussed the formation and location of this green peak in detail elsewhere [5] in terms of the crystal field situation, and have shown that on theoretical grounds Mn^{2+} is a likely activator.

The other main emission bands of plagioclase are one in the blue and one in the near IR. All plagioclases show some blue emission, as do iron-free enstatites, forsterites and many other silicates [12]. The emission band is broad, with a somewhat variable peak wavelength. It is likely that this blue emission band has a similar origin in all these relatively iron-free silicates, and that it is due to a particular type of lattice defect rather than to any impurity activator. Sippel and Spencer [10] have found that some heavily shocked plagioclases and silica mineral phases show an enhanced blue emission, which also implies that lattice defects are responsible.

Most terrestrial plagioclases show a prominent emission band in the near IR, as shown for labradorite in Figure 6, curve 1, and for basalt BCR-1 in Figure 7. However, sometimes this band is only of comparable intensity to the green band, as for oligoclase (Figure 7), and occasionally it is virtually absent as for labradorite-bytownite (Figure 5). We differ here from Sippel and Spencer [10] who suggest that all terrestrial plagioclases show a strong IR emission band. The lunar materials we have examined tend to show a weak emission band in the IR, as shown by the four samples of lunar fines in Figure 4 and by the separated lunar plagioclase in Figure 5. We also find that meteorites in which plagioclase is the main luminescent phase, e.g. hypersthene and bronzite chondrites and pyroxene-plagioclase achondrites, do not show an IR peak–see, for example, Juvinas (a pyroxene-plagioclase achondrite) shown in Figure 5.

The wavelength of the IR emission peak is variable, ranging from about 7300 to 7700 Å, and the band is usually broader than the green band. It is often the dominant emission band for terrestrial plagioclase, as for the labradorite examined, and it does not appear to be associated with manganese since the addition of manganese was found to reduce its intensity, as shown in Figure 6. If this emission is due to an impurity activator than it should be possible by selective doping with suspected activators to identify the element responsible, which is probably present naturally in amounts of not less than 100 ppm. As most plagioclases contain up to 1% of iron, and as Fe^{2+} often gives rise to absorption bands in the near IR in silicates, this might be a possible candidate, although Fe^{2+} also acts as a luminescence 'killer' in silicates. Preliminary tests in which iron was added to labradorite (by heating it to 1050 °C in argon with 0.1% hydrated ferrous sulphate) did not show any increase in the (already high) IR emission – in fact there appeared to be a slight reduction. A program of selective doping and of comparative analysis of different plagioclases is now under way, in the hope of identifying the cause of the IR emission.

5. Cathodoluminescence

The distribution of the luminescent constituents of our lunar rock and breccia chips has been investigated by taking colour photographs of the emission. As electrons cause much less damage than protons, giving brighter and more constant emission for the same energy, electron excitation is to be preferred when using visual or photographic methods which require long exposures.

Several workers have in the past investigated electron-excited luminescence ('cathodoluminescence'); the brilliant luminescence exhibited by diamonds, rubies, and other gems when exposed to cathode rays was discovered in the 1870's by Crookes [13], and investigated with the assistance of Maskelyne. Their technique was simply to seal the stones into a glass discharge tube opposite a cathode which could be connected to an induction coil. Besides the bright and sometimes varying luminescence, Crookes also reported the generation of polarised luminescent light, the loss of efficiency on continuous bombardment, and the brown stain which readily forms on any surface subjected to such bombardment.

The identical technique was applied to meteorites by Buddhue [14] in 1941. He found enstatite to be the most active mineral, usually giving a blue luminescence but with occasional red grains. He examined the spectrum of the glow (apparently with a direct-vision spectroscope) and states that phosphorescent glows were also observed for periods up to a minute. The luminescence was found to be brighter in cathode rays than in beams of ions.

The combination of discharge tube excitation with microscopic observation seems to have been first accomplished by Gallup [15]. He describes a simple device for the direct investigation of manganese-activated calcium silicate phosphors.

It was, however, the development and general use of the analytical electron microprobe which really kindled interest in cathodoluminescence. Rocks and minerals could be seen glowing in the electron beam, and otherwise invisible growth structures were sometimes delineated [16]. It was quickly realised that for visual investigation of morphology the complication and expense of a microprobe was unnecessary, and Sippel [17] described apparatus whereby specimens could be moved and viewed beneath a microscope while being irradiated with an electron beam generated by a glow discharge. Simple electron-irradiation devices were also made by other workers, and successfully applied to problems in terrestrial mineralogy [18, 19].

Rising interest in transient lunar phenomena and the nature of the lunar surface led to Derham and Geake's [20] demonstration of the luminescence of meteorites when bombarded with protons. Enstatite was identified as the constituent generally responsible, luminescing red or blue according to the amount of activating manganese present [24]. Cathodoluminescence and microprobe studies by other workers have confirmed the attribution to manganese.

Sippel and Spencer [10] have applied their electron excitation technique to lunar crystalline rocks and breccias, while other workers have investigated the proton excitation of these materials [5, 6, 21, 22].

6. Electron–Excitation Apparatus for Luminescence Photography

Apparatus based on Sippel's design was constructed by Mills in 1966 in order to examine the cathodoluminescence of meteorites and tektites. This apparatus has now been used at Leicester for the present work on lunar samples, and one arrangement of it is shown in Figure 8. Important features of the design are:

(a) A vacuum chamber large enough to accommodate either centimetre-size rock fragments or thin sections, and reversible so that samples can be irradiated from either above or below.

(b) Provision for examination of the sample by either transmitted or surface luminescence or incident light, with or without crossed polarizers.

(c) Provision for x and y positioning of the sample by means of push-rods working through 0-ring seals.

(d) Provision for using either air or hydrogen (at a pressure of about 2×10^{-2} torr) as the discharge gas; the use of hydrogen minimises background illumination from the glow discharge.

(d) A cylindrical depression, 3 mm in diam and 1 mm deep, in the centre of the aluminium cathode; this helps to form and stabilise the electron beam.

A 30 keV power supply, with a 33 MΩ series stabilising resistance chain, results in a potential difference of about 6 keV across the discharge tube in normal operation. The current density at the sample is about 8 μA cm^{-2}. The anode and the sample chamber are at earth potential. The anode has a central hole surrounded by a crown of needle points which help to start the discharge; the electron beam passes through this hole and coasts through field-free space to the sample. A sheet-lead X-ray shield surrounds the discharge tube during operation, and the viewing windows are of lead glass. A rotary oil pump is sufficient to produce the vacuum required; it is fitted with a back-streaming trap containing activated alumina and Linde 5A molecular sieve material. The use of a 'soft' vacuum simplifies construction and operation, and obviates the charging problems encountered when non-conducting materials are irradiated in a hard vacuum.

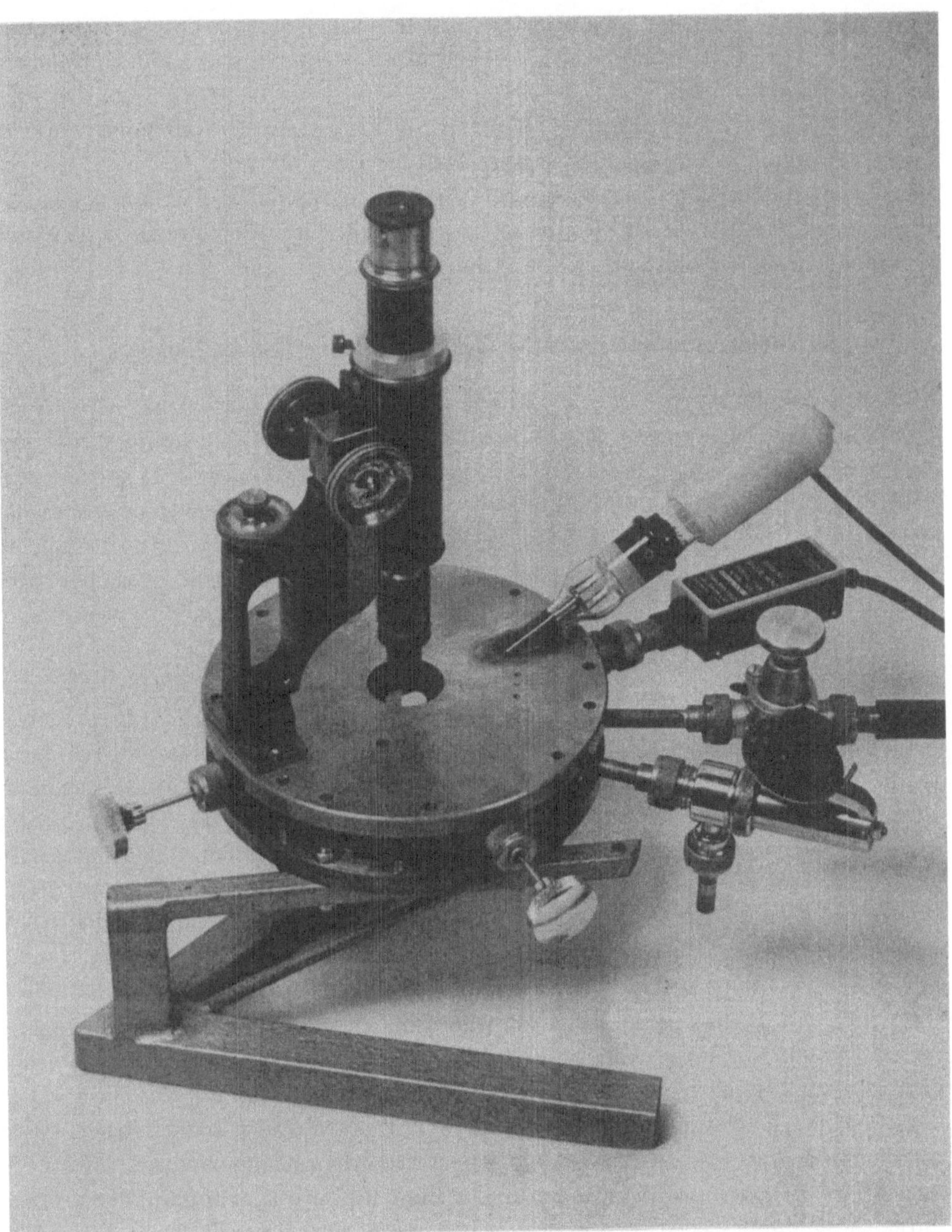

Fig. 8. Apparatus for observing luminescence under electron excitation, shown with the microscope attached, but without the camera. The discharge tube irradiates the sample from above in this arrangement; clockwise from the discharge tube are the Penning gauge head, the vacuum line, the gas inlet needle valve and the two sample-positioning rods.

An alternative discharge tube is under construction, of the type developed by Dugdale [23]; it employs a hollow-anode electron-beam source, and should produce a higher beam intensity.

Figure 8 shows the arrangement of the apparatus as used for our lunar work, in which the sample is both irradiated and observed from above. It is shown with the microscope attached, for visual inspection; for photographic use a 35 mm single-lens reflex camera is either attached to the microscope or else it is used directly, with extension tubes, to take macroscopic photographs.

7. Luminescence Distribution Within Rock and Breccia Samples

All of our Apollo 11 and 12 rock and breccia chips have been inspected with the cathodoluminescence apparatus, and in all cases where significant luminescence was observed colour photographs of the emission were taken. A typical exposure was 30 s at f/2 with ×1.85 magnification, on Kodak High-Speed Ektachrome (Daylight) reversal film; some microphotographs were taken with exposures of up to half an hour. The flatter parts of the rough chips were chosen, as the depth of focus obtainable was a major limitation. The results cannot be reproduced here in colour, but some of them are shown in black and white in Figures 9 to 12.

All our rock chips showed bluish-white luminescence from their plagioclase parts, and little else. For example, rock chip 10058.38, which is shown in Figure 9, was observed visually and so arranged that one could 'blink' from luminescence emission to direct illumination. It was evident that of its three main constituents, the white or clear plagioclase parts were luminescing bluish-white or blue, whereas the brown pyroxene and black ilmenite were completely non-luminescent. One small lath luminesced red.

Other rocks showed coarse or fine patterns of plagioclase emission, consistent with their stated grain sizes. Figure 10 shows the luminescence emission from two Apollo 12 rocks: (a) is a medium-grained rock, 12051, showing fan-shaped groups of plagioclase laths, and (b) is of rock 12002 showing smaller and more irregular plagioclase crystals. Colour variations from bluish-white to blue across some plagioclase crystals, and from one crystal to another, probably indicate variations in the concentration of the manganese activator, an effect that we have also noticed with manganese-activated enstatite in meteorites [24]. We have observed that when plagioclase samples are inspected visually while being scanned in our luminescence spectrophotometer, then those whose luminescence appears nearly white give spectral profiles having a moderate green peak; when this peak is weak the emission appears blue, and when it is very strong, as for the Mn-doped labradorite shown in Figure 6 and for the plagioclase-rich meteorite Juvinas shown in Figure 5, the appearance is yellow-green. Several of the lunar rock chips also showed occasional small pink-luminescent grains. Some chips showed very bright blue-luminescent specks, but these we take to be contamination, probably by saw material.

Two breccias were available, both from Apollo 11, and their luminescence appear-

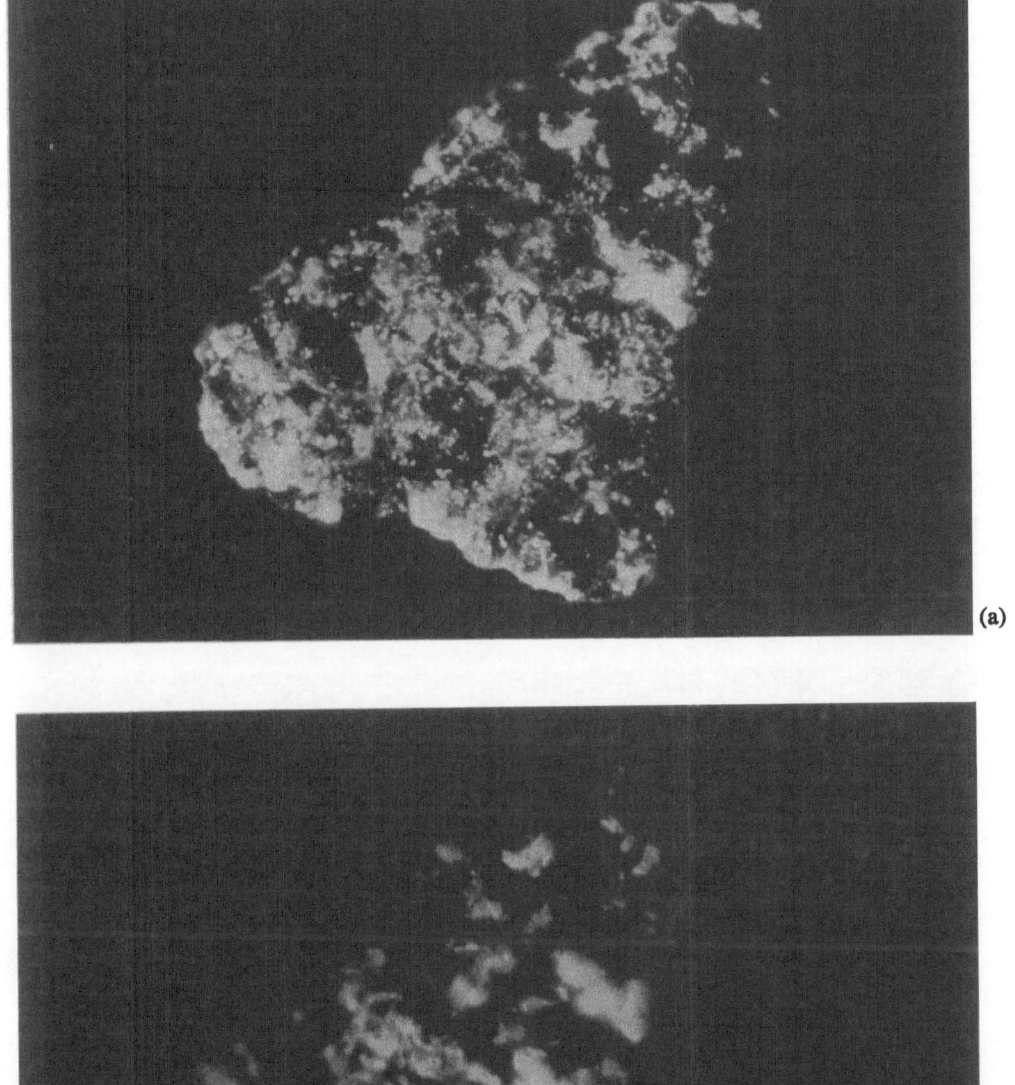

Fig. 9. Rock sample 10058.38 (a) in white light, (b) luminescence emission. The height of each print represents about 1 cm.

(a)

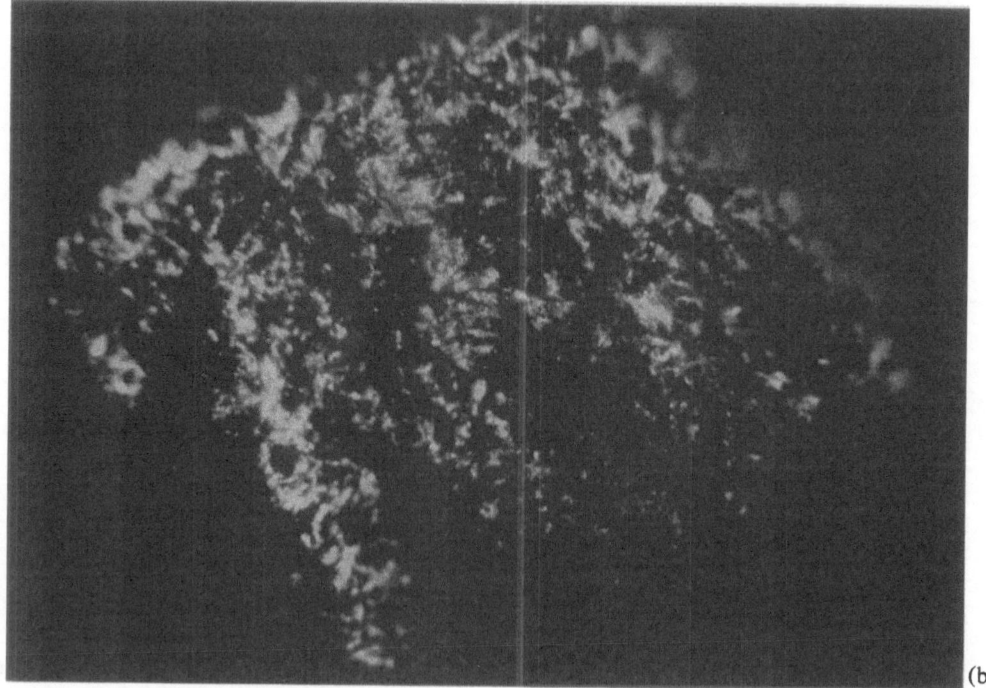

(b)

Fig. 10. Luminescence emission of two Apollo 12 rock samples: (a) 12051.16, (b) 12002.102. The height of each print represents about 1 cm.

Fig. 11. Breccia sample 10023.8, (a) in white light, (b) luminescence emission. The height of each print represents about 1 cm

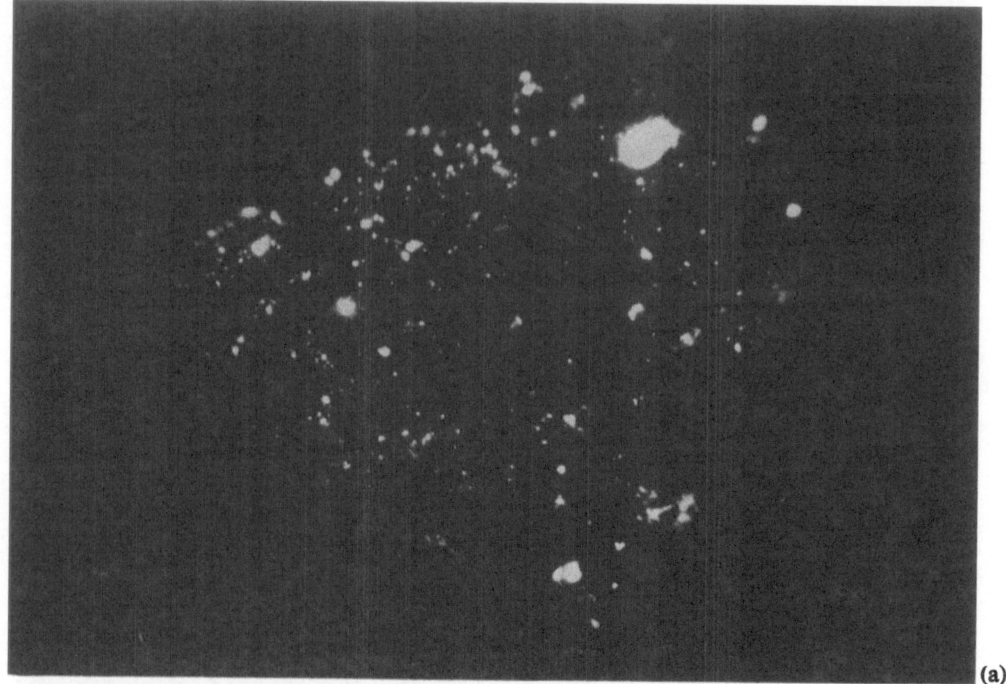

(a)

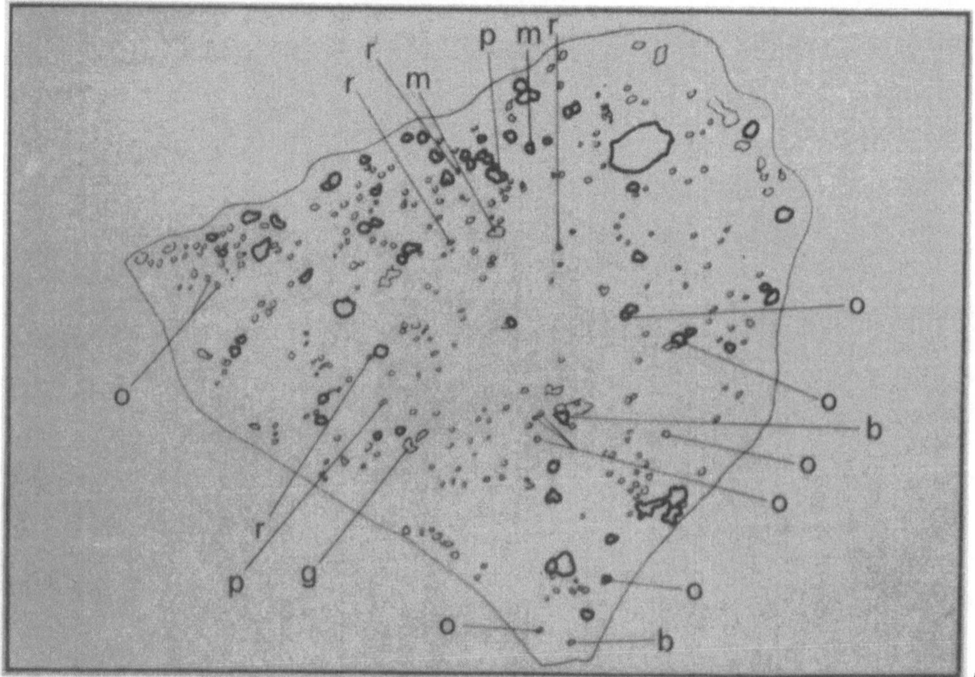

(b)

Fig. 12. (a) Luminescence emission of breccia sample 10059.36, (b) a map indicating of some of the luminescence colours seen: r = red, p = pink, o = orange, g = green, b = blue, m = mauve. Most of the other emitting regions appear white or bluish white. The height of the print represents about 1 cm.

ance was quite different from that of the rocks, showing many luminescent grains embedded in a non-luminescent matrix. Sample 10023.8 (Figure 11) showed white-luminescent grains (probably plagioclase), and little else, but sample 10059.36 (Figure 12) showed a surprising variety of emission colours including red, orange, yellow, green, blue and a range of mauves and pinks which usually vary in hue across the grains concerned. Luminescent grains of all sizes from 1 mm or so down to a few μm are seen. This one breccia sample appears to contain so much luminescence information that it would be useful to identify the materials responsible, but this will only be practicable with a thin section of the same sample on which electron microprobe analysis can be carried out, using the luminescence photograph as a map. Luminescence may then provide a quick and almost non-destructive method of exploring the surface of a rock, and of identifying at least some of its constituents, without making sections. We have noticed that most of the rock chips show a few grains of some luminescence colour other than that of plagioclase.

It is already possible to say that the red luminescence may be due to apatite. We have shown by electron microprobe analysis that some luminescent regions in two hypersthene chondrite meteorites (Appley Bridge and Mangwendi) contained Ca and P, but no Fe, indicating apatite. The emission varied from red to orange across the grains, probably due to a variation in the Cl/F ratio.

Two samples of lunar fines (10084.6 and 12032.39) have also been observed and photographed under electron excitation. They were much fainter than the luminescent parts of the rocks: they both appeared generally mauve, with brighter white-luminescent specks and larger grains, which were probably plagioclase. No other colours were seen. This low efficiency in comparison with similar constituents in rock samples is probably due to the surface damage and grain-coating effects discussed earlier.

8. Conclusions

We conclude that for the Apollo lunar material we have examined the main luminescent constituent is plagioclase, that the main emission of lunar plagioclase is a green peak at about 5600 Å, and that the cause of this peak is about a hundred parts per million of manganese in the form of Mn^{2+} substituting for Ca^{2+} in the plagioclase lattice. The less efficient blue emission at about 4500 Å is probably caused by a lattice defect and is common to most silicates; the weak IR peak at about 7700 Å is probably caused by an impurity activator yet to be identified. Most terrestrial plagioclases show much stronger IR emission than the lunar material.

Electrons and protons in the keV region are both effective in exciting luminescence in lunar material; UV is ineffective. The overall efficiencies of lunar rock and breccia samples depend largely on their content of plagioclase, which itself has an efficiency of about 10^{-3}. Samples of lunar fines also have efficiencies depending on their plagioclase content, but they are at least an order of magnitude less efficient than the rock samples possibly because the grains have a thin surface layer which is either opaque or damaged. Two light-coloured Apollo 12 fines samples, thought to be ray material, are an order of magnitude more efficient than the dark fines material from Apollo 11 and 12, which has an efficiency of about 10^{-5}.

Lunar rock and breccia samples generally show blue-white luminescence distributed texturally as their plagioclase component, with a few grains of other luminescent materials one of which may be apatite. One breccia (10059) shows a wide variety of luminescence colours from components yet to be identified.

Acknowledgements

We are grateful to M. L. Gould for technical assistance; to F. Kirkman of UMIST, Leicester University photographic section, Hull University photographic section, N. W. Scott, D. Rendell and Colour 061 Ltd. for photographic assistance; to G. F. J. Garlick of Hull University Physics Dept. for the loan of the terrestrial plagioclase samples and for carrying out the Mn doping; to J. Zussman, A. C. Dunham and J. Esson of Manchester University Geology Dept. for advice and for Mn determination by X-ray fluorescence analysis, also for the loan of the sample of separated lunar plagioclase; to P. Suddaby of the Geology Dept. Imperial College, University of London, for help with the electron micro-probe investigation of meteorites; to C. J. E. Kempster for advice on the structure of plagioclase; to The British Museum (Natural History) for the loan of meteorite samples; to The Science Research Council for financial support, and finally to NASA, and especially to all the Apollo astronauts, for providing the lunar samples.

References

[1] Derham, C. J. and Geake, J. E.: 1963, Technical Note No. 2, USAF Contract AF61(052)-379.
[2] Maliakal, J. C., Limon, P. J., Arden, E. E., and Herb, R. G.: 1964, *J. Vac. Sci. Tech.* **1**, 54.
[3] Cross, J.: private communication.
[4] Geake, J. E. and Lumb, M. D.: 1963, Technical Note No. 1, USAF Contract AF61(052)-379.
[5] Geake, J. E., Walker, G., Mills, A. A., and Garlick, G. F. J.: 1971, *Proc. Second Lunar Science Conference, Geochim. Cosmochim. Acta Suppl. 2* **3**, 2265.
[6] Geake, J. E., Dollfus, A., Garlick, G. F. J., Lamb, W., Walker, G., Steigmann, G. A., and Titulaer, C.: 1970, *Proc. Apollo 11 Lunar Science Conference, Geochim. Cosmochim. Acta Suppl. 1* **3**, 2127.
[7] Dollfus, A., Geake, J. E., and Titulaer, C.: 1971, *Proc. Second Lunar Science Conference, Geochim. Cosmochim. Acta Suppl. 2* **3**, 2285.
[8] Hapke, B. W., Cohen, A. J., Cassidy, W. A., and Wells, E. N.: 1970, *Proc. Apollo 11 Lunar Science Conference, Geochim. Cosmochim. Acta Suppl. 1* **3**, 2199.
[9] Hapke, B. W., Cassidy, W. A., and Wells, E. N.: 1971, *Second Lunar Science Conference, Houston* (unpublished).
[10] Sippel, R. F. and Spencer, A. B.: 1970, *Proc. Apollo 11 Lunar Science Conference, Geochim. Cosmochim. Acta Suppl. 1* **3**, 2413.
[11] Lang, H. and Kressin, G.: 1955, *Z. Phys.* **142**, 380.
[12] Geake, J. E. and Walker, G.: 1966, *Geochim. Cosmochim. Acta* **30**, 927.
[13] Crookes, W.: 1879, *Phil. Trans. Roy. Soc.* **170**, 641.
[14] Buddhue, J. D.: 1941, *Am. J. Sci.* **239**, 839.
[15] Gallup, J.: 1936, *J. Opt. Soc. Am.* **26**, 213.
[16] Smith, J. V. and Stenstrom, R. C.: 1965, *J. Geol.* **73**, 627.
[17] Sippel, R. F.: 1965, *Rev. Sci. Inst.* **36**, 1556.

[18] Long, J. V. P. and Agrell, S. O.: 1965, *Min. Mag.* **34**, 318.
[19] Greer, R. T., Staley, W. G., and Vand, V.: 1967, *Amer. Ceramic Soc. Bull.* **46**, 829.
[20] Derham, C. J. and Geake, J. E.: 1964, *Nature* **201**, 62.
[21] Blair, I. M. and Edgington, J. A.: 1970, *Proc. Apollo 11 Lunar Science Conference, Geochim. Cosmochim. Acta Suppl. 1* **3**, 2001.
[22] Nash, D. B. and Greer, R. T.: 1970, *Proc. Apollo 11 Lunar Science Conference, Geochim. Cosmochim. Acta Suppl. 1* **3**, 2341.
[23] Dugdale, R. A.: 1966 *J. Materials Sci.* **1**, 160.
[24] Geake, J. E. and Walker, G.: 1967, *Proc. Roy. Soc.* **296**, 337.

THE SOLAR IRRADIATION RECORD IN LUNAR DUST GRAINS

J. BORG and B. VASSENT

Centre de Spectrométrie de Masse du C.N.R.S.-91 ORSAY-France

Abstract. Comparative studies of the distribution of latent and etched tracks in lunar grains from five different size fractions of three lunar fine samples and of six lunar dust samples taken at different depths in core tube 12028 have been performed by using transmission and scanning electron microscopies. Two very different sets of etching conditions were used: a slight etching was applied for transmission microscopy but a much stronger etching was used for scanning microscopy. We observed: (1) a definite stratigraphy in the core tube, both in the latent and etched track distributions; (2) striking differences between the densities of the latent and slightly etched tracks ($\geqslant 10^{10}$ tracks cm^{-2}) and those of the strongly etched tracks, ranging from $\simeq 10^8$ to 5.10^9 tracks cm^{-2}; (3) a lack of correlation between the grain size and the density of strongly etched tracks observed on the external surface of grains from sample 12032; (4) no variation of the density of the tracks with the depth inside a grain. Some implications of the present results concerning the ancient low energy solar cosmic rays and the fabric of the lunar soil will be briefly discussed.

1. Introduction

This paper describes our work on fossil nuclear particle tracks in lunar dust grains which is a part of the Orsay group investigations concerning the galactic and solar irradiation records in extraterrestrial matter.

Before our transmission electron microscope studies of meteoritic and lunar samples [1, 2, 3, 4, 5], only etched tracks were studied [6] by scanning electron microscopy and optical microscopy in view of understanding the lunar regolith formation and determining some characteristics of the ancient solar flare cosmic rays.

In the present work, both latent and etched tracks have been studied in silicate grains extracted from five size fractions – 100, 200, 325, 400 and 400 Mesh residue – of the Apollo 11 and 12 lunar dust and core tube samples. Latent and very slightly etched tracks were observed directly by 100 keV and 1 MeV transmission electron microscopy in the finest feldspar and pyroxene grains. In the coarser grains the density of etched tracks was determined by high resolution scanning electron microscopy first on the external surface of the grains and then on internal surfaces obtained by polishing.

The main purposes of these studies were: (1) to discuss dynamic processes acting in the lunar regolith and more particularly to look for eventual differences in the irradiation history of the dust grains as a function either of their size or of their depth in core tube 12028; (2) to detect different types of solar radiations and more particularly to try to find evidence for the existence of low energy heavy ions, related to the suprathermal solar protons recently discovered by Franck [7] in the interplanetary space; 3. to correlate the irradiation history of the grain to some of their bulk characteristics such as their carbon and methane contents.

Urey and Runcorn (eds.), The Moon, 298–308.

2. The Solar Flare Record, Observed at Depth Greater than One Micron

Observations were made with a scanning electron microscope on etched grain surfaces, after what we define as a '*strong*' etching [8], enlarging the track diameter up to about 2000 Å. The track densities were measured in grains from the 100 to 400 Mesh fractions extracted either from different samples of fines (10084, 12032, 12070) or at different depths in core tube 12028. The track densities measured at the center of the grains ranged from 10^7 to 10^9 tracks cm^{-2} (Table I). No significant increase in these densities was observed when going from the 100 Mesh to the 400 Mesh grains, in the various samples. For a given grain, after the *strong* etching, the track densities measured on the external surface and at the center of a grain were about the same; this confirms that a high proportion ($\gtrsim 80\%$) of the grains do not show the homogeneous edge zoned track distribution – or track gradient – which can be expected for grains individually exposed in space to the very heavy (VH) nuclei of the solar flare cosmic rays and characterized by a drop in the track density by a factor of 2 to 3 on a depth of about 20 μ. Even the most favorable track gradient we found was not homogeneous

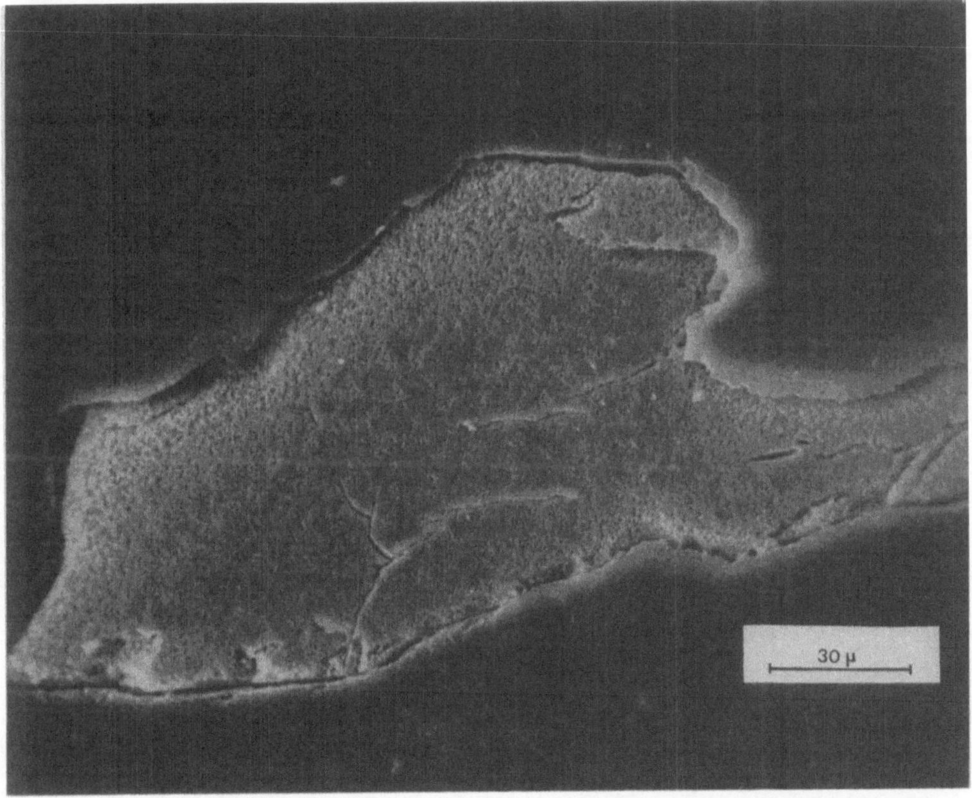

Fig. 1. 200 Mesh lunar pyroxene grain (sample 12028,55) showing a typical edge zoned track distribution. Less than 20% of the grains showed such a gradient generally considered as resulting from the 'individual' irradiation of the grains in solar flare cosmic rays.

TABLE I

Comparative studies of different size fractions of lunar dust samples

	400 Mesh residue		400 Mesh		325 Mesh		200 Mesh			100 Mesh		
	% coated	max $\times 10^8$	ext $\times 10^8$	int $\times 10^8$	ext $\times 10^8$	int $\times 10^8$	ext $\times 10^8$	int $\times 10^8$	% grad	ext $\times 10^8$	int $\times 10^8$	% grad
10084	67	$\geqslant 10^3$		20–30			0.6–50	0.2–12	$\leqslant 15$	1–6	0.3–12	$\leqslant 30$
							$\geqslant 50$[a]					
12032	15	$\geqslant 10^3$			0.4	1.4	0.1–0.3	0.1–4		0.6–5	0.5–6	
12070	41	$\geqslant 10^3$		9–20			1–7	1–16	$\leqslant 7$	1–13	0.5–3	$\leqslant 20$
55	39	$\geqslant 10^3$		1.3–45	4–40	2–30		0.8–7	$\leqslant 20$	0.12–14	0.4–7	$\leqslant 25$
62	7	$\simeq 10^3$		0.01–0.5		0.3–0.5	0.02–2	0.02–0.5	$\leqslant 20$	0.01–1	0.2–0.4	$\leqslant 20$
75	–	–					3–17	0.2–16		0.4–100	0.2–2.5	
98	5	$\leqslant 10^2$		2–11		3–4		2–25			0.1–20	
155	–	–		0.18–17				0.03–3		0.2–4	0.02–3	
203	35	$\geqslant 10^3$	30–100	5–14		2–45	3–20	3–15	$\leqslant 10$	1–80	0.7–23	

[a] Values obtained after the slight etching of the grains.

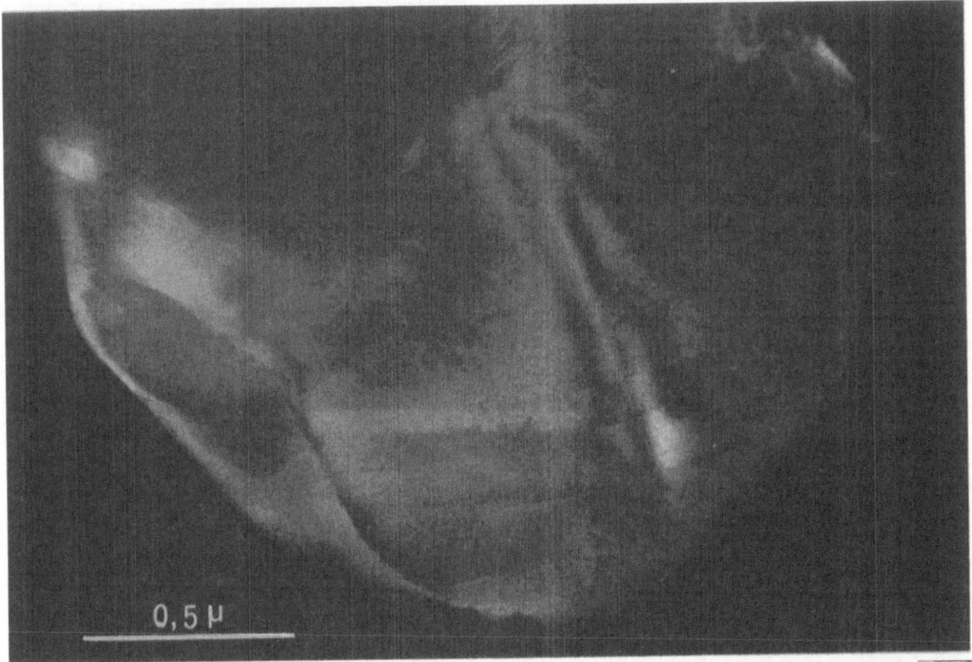

Fig. 2. 1 MeV dark field micrograph of an Apollo 12 uncrushed dust grain containing a high density ($\sim 10^{11}$ tracks cm^{-2}) of latent nuclear particle tracks appearing as the lines of dark contrast.

Fig. 3. 100 keV micrograph of an Apollo 12 uncrushed dust grain, obtained thanks to the courtesy of the JEOL Company. This grain was 'slightly' etched and the tracks appear as the white shallow canals visible only on the edge of the grain.

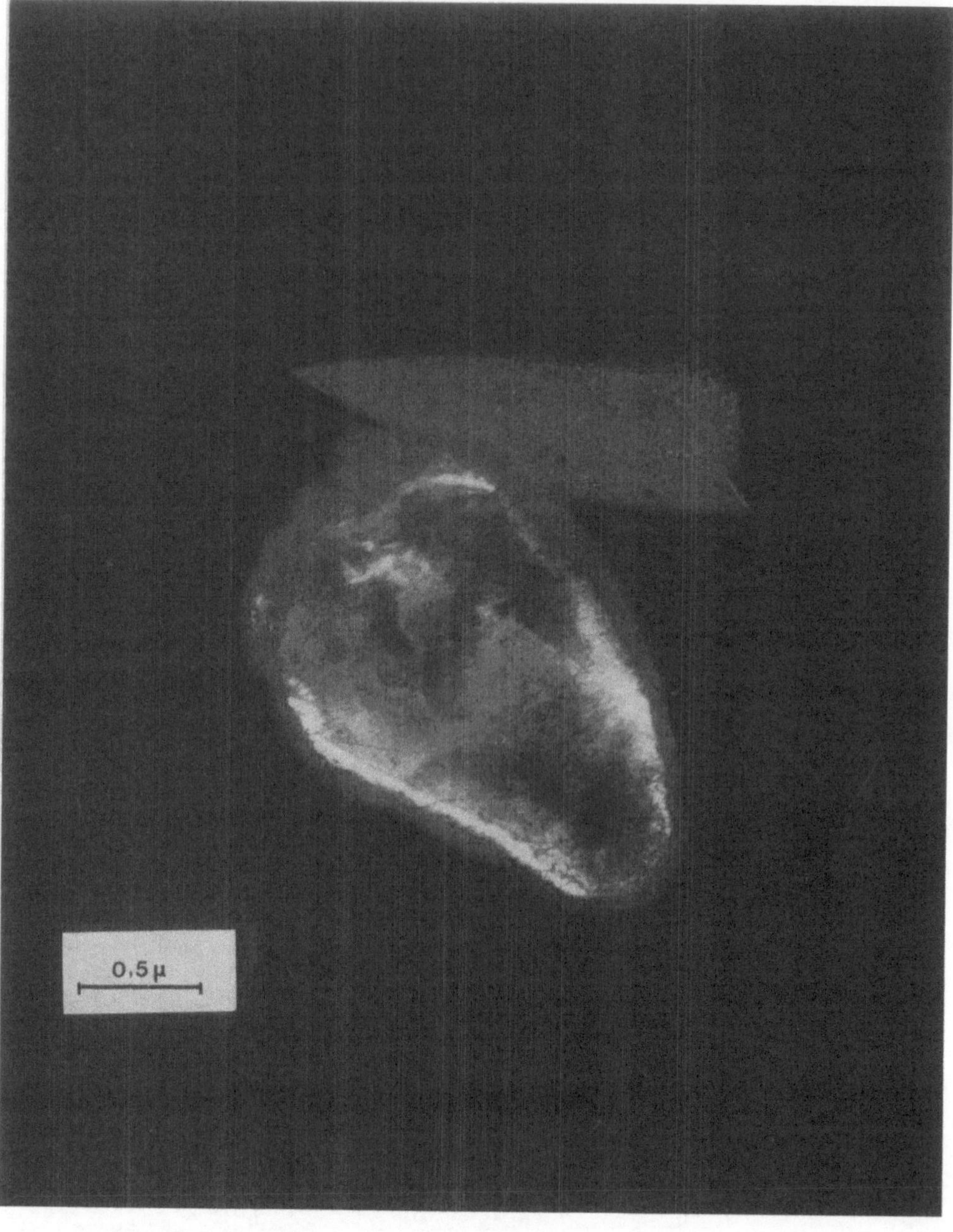

Fig. 4. 1 MeV dark field micrograph of an Apollo 12 uncrushed dust grain, showing the superficial amorphous coating observed in crystals containing the highest track densities.

(Figure 1) both along the edge of the grain and in its depth extension, and this does not facilitate the study of ancient solar flare VH nuclei, as discussed elsewhere [9].

No correlation of the track density with the depth was observed in core tube 12028 but a marked stratification was observed particularly distinct in sample 12028, 62.

Comstock *et al.* [10] and Crozaz *et al.* [11], who studied the track density distribution in the same core tube found only 20% of the grains showing a track gradient, as we did. On the contrary Arrhenius *et al.* [12] found measurable track gradients in a high proportion of the 100 Mesh grains showing the highest track densities ($\gtrsim 5.10^8$ tracks cm^{-2}). These authors find also an evident stratification in the core tube, and no correlation of the track density with the depth.

3. The Ultramicroscopic Record in Uncrushed Micron Sized Dust Grains

Before any etching the finest crystalline grains in the various dust samples generally contain high nuclear particle track densities which can exceed 10^{11} tracks cm^{-2} [1, 2, 3, 4, 5, 13) and which are only observable by high voltage (1 MeV) trransmission electron microscopy: they appear as lines of dark contrast in Figure 2. Furthermore, the grains are frequently rounded and covered with a superficial coating of amorphous material of about 500 Å in thickness [1, 2, 3] appearing as the dark lining surrounding the grain in Figure 3.

After a '*slight*' chemical etching first developed by Barber *et al.* [13] the latent tracks are transformed in shallow etched canals with diameters of about 200 Å (Figure 4) that C. Jouret has been able to observe currently with a 100 keV electron microscope.

4. The Ultramicroscopic Record in the Micron Sized Superficial Layer of the Coarser Grains

We intended to verify if the high density of tracks observed in the finest grains were also registered in a micron-sized superficial layer in the coarser grains. Indeed during the 'strong' etching used for ordinary scanning electron microscope studies this layer could have been rubbed away thus preventing the observation of high track densities.

For this purpose, two types of experiments were performed: (1) five different size fractions of sample 12070 were crushed into micron sized fragments subsequently observed by transmission electron microscopies. We noted a marked increase in the probability of observing high latent and slightly etched track densities when going from the coarsest to the finest size fractions and this 'size effect' supports the hypothesis that the high track densities are registered in the most superficial layer in the grains; (2) external surfaces of 200 Mesh feldspar and pyroxene grains from lunar sample 10084 were etched with the slight etching conditions used for transmission electron microscopy. Then we observed the external surface of the grains with a high

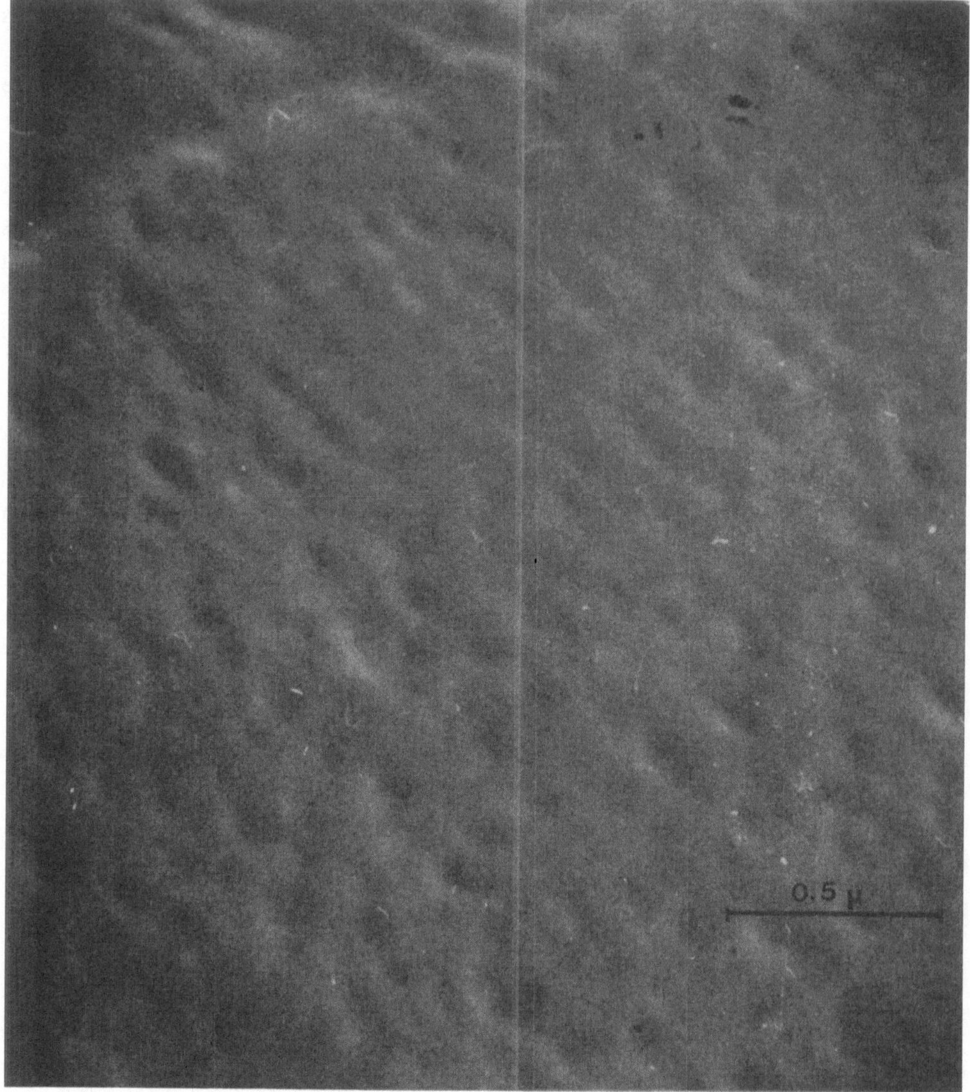

Fig. 5. High resolution scanning picture of the surface of a 200 Mesh grain (sample 10084) which has been slightly etched. The track density is $\sim 2.10^{10}$ tracks cm^{-2}.

resolution scanning electron microscope (JEOL) and detected high densities of tracks as those reported in Figure 5 and ranging from 5.10^9 to 2.10^{10} tracks cm^{-2}. Finally, some of these grains, either polished down or not, were then more strongly etched for 'ordinary' scanning electron microscopy and they definitively showed the lower track densities already noted by several groups for such etching conditions. In Figure 6, we reported our first preliminary attempts in plotting the variations of the track density with the depth in the most superficial micron-sized layer of the grains.

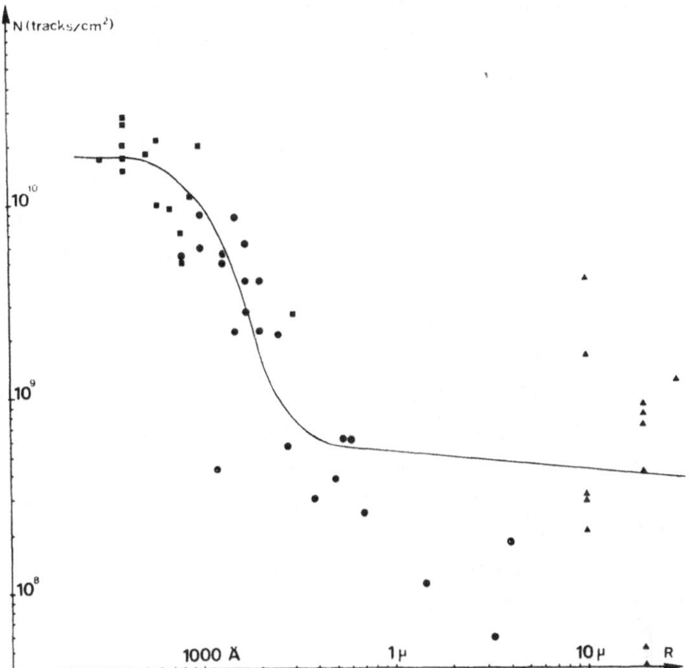

Fig. 6. Track density as a function of the depth R in 200 Mesh grains of sample 10084. The squares and the dots represent the values measured on the external surfaces of the grains after a 'slight' and a 'strong' etching respectively. For these points we plotted the track density as a function of the track diameter which gives a rough estimate for R. The triangles figure the values obtained for another set of grains on internal polished surfaces, at depth R directly measured on the SEM micrographs.

5. Dynamic Processes at the Surface of the Moon

Arrhenius *et al.* [12], Comstock *et al.* [10] and Crozaz *et al.* [11] have suggested that the solar flare irradiation of the dust grains most likely started when they were released in the lunar soil. In a given layer the variations observed between the track densities are then explained by differences in residence times at the top of the regolith, with the samples showing a gradient having been the closest to the surface.

At the present time we slightly favor another possible irradiation history for the grains suggested to us by M. Maurette and occurring in two successive steps:

(1) first, the grains were mostly loaded with tracks at depths further than 1 μ when they were part of the top surface of a rock. Then, the variations in track densities in the lunar dust grains could be explained by the variations in the depths at which they got 'excavated' from the rocks most likely by micrometeorite impacts. This conclusion is supported by the greater proportion of glassy spherules showing a track gradient [14] and by roughly the similar track density distribution (ranging from 10^7 to 10^9 tracks cm^{-2}) we obtained by crushing a mm sized fragment artificially 'excavated' from the top surface of lunar rock 10047. This interpretation also explains why the 400 Mesh grains are not more loaded with 'volume' tracks than the 100 Mesh grains, but it

cannot account easily for the marked decrease in the track densities in sample12028,62.

(2) then, the grains spend a too short time – 10^3 to 10^4 yr only – in the most superficial layers of the regolith to get a solar flare track contribution exceeding that acquired before their release in the lunar soil, thus explaining the 'lack' of track gradient. But during this short time, they were bombarded superficially with solar wind nuclei responsible for the amorphous coating on the grain [1, 2, 3, 4, 5] and with heavy solar ions with energies intermediate between those of the solar wind nuclei and those of the more energetic solar flare cosmic rays producing high track densities in the superficial micron sized layer in the grains.

The simultaneous implantation of solar wind and low energy solar ions is evidenced by the correlation between the high track densities and the existence of an amorphous coating in the smallest grains (Table I) and argues against a cosmic dust origin for the finest fraction as suggested by Barber *et al.* [13] because it is very difficult to explain that such a regular coating could have survived an impact with the Moon. To check further the validity of this irradiation scheme for the dust grains, it will be interesting to apply the methods described in the next chapter to verify if there is a correlation between the high track density measured in the most superficial layer of the 200 Mesh grains and those measured at depths greater than one micron in the same grains.

6. Evidences for Suprathermal Ions in the Interplanetary Space

Barber *et al.* [13] favor the hypothesis that the micron sized dust grains are cosmic dust particles accreted by the Moon and loaded in the interplanetary space by solar flare cosmic rays tracks. They regard it as "highly unlikely that suprathermal heavy ions were responsible of these high track densities" – as we much earlier suggested [4, 5] – because: (1) the energy of suprathermal ions is below the sensibility threshold for track registration; (2) the high track densities are registered in the volume of the grains and therefore no steep surface track gradient can be observed.

We want to make the following comments concerning these criticisms:

(1) The curve reported in Figure 6 shows that a steep track gradient is indeed observed in the most superficial micron-sized layer of the grains. By supposing that the tracks are registered on the whole range of the ions, the curve in Figure 6 represents a rough integrated range spectrum for the particles : $N(R>R_0)=f(R_0)$, where N represents the tracks with range R greater than R_0 and from which we can deduce a differential energy spectrum $dN/dE=f(E)$ by computing the quantity $(dN/dR)\times$ $\times(dR/dE)$. This differential spectrum is shaped like a pulse. Therefore it looks similar to that measured for suprathermal protons by Franck [6] but is very different from the power law spectrum which can be extrapolated from the most recent data of Armstrong *et al.* [15] and Stone [16], for solar flare VH nuclei with energy $\gtrsim 1$ MeV/amu.

(2) In a paper which will be published elsewhere [17] we will show that nickel ions with energies down to 0.02 MeV/nucleon can produce tracks in silicate grains and therefore the most severe objection to the existence of heavy suprathermal ions is no longer valid.

7. Correlation with Other Results

The Orsay group has already discussed some correlations between the microscopic irradiation features in the grains and the bulk mechanical and optical properties of the regolith [4, 5]. We will mention here a new development in these correlation studies.

Cadogan *et al.* [18] have shown that part of the indigenous methane and ethane contents in the lunar fines seems to be correlated to the solar flare irradiation of the grains exposed unshielded on the lunar surface as described in the work of Arrhenius *et al.* [12] and to the fraction of grains with an amorphous coating as suggested by us. These results seem to indicate that part of the indigeneous methane in the lunar dust could have been synthesized by an ancient solar wind implantation in the grains. The Orsay group is currently verifying these important conclusions in analyzing the irradiation record in dust samples simultaneously studied by the Bristol group and in conducting artificial solar wind type implantations in micron-sized fragments obtained by crushing internal chunk of various types of lunar rocks.

Acknowledgements

All the transmission microscope observations were made by L. Durrieu and C. Jouret at the Institut d'Optique Electronique du CNRS, Toulouse, France. We acknowledge the generous help of the JEOL Company for the scanning electron microscope observations. We are indebted to Dr R. Bernas for his very active support and interest. We wish to express our gratitude to Dr M. Maurette for many stimulating discussions.

References

[1] Borg, J., Dran, J. C., Durrieu, L., Jouret, C., and Maurette, M.: 1970, *Earth Planetary Sci. Letters* **8**, 379.

[2] Dran, J. C., Durrieu, L., Jouret, C., and Maurette, M.: 1970, *Earth Planetary Sci. Letters* **9**, 391.

[3] Borg, J., Dran, J. C., Durrieu, L., Jouret, C., and Maurette, M.: 1970, 'High Voltage Electron Microscope Studies of Extraterrestrial Matter', 7th Int. Colloquium Corpuscular Photography and Visual Solid Detectors, Barcelona, Paper No. 41.

[4] Borg, J., Durrieu, L., Jouret, C., and Maurette, M.: 1971, 'The Ultramicroscopic Irradiation Record of Micron-Sized Lunar Dust Grains', Second Lunar Science Conference (unpublished proceedings).

[5] Borg, J., Durrieu, L., Jouret, C., and Maurette, M.: 1971, *Geochim. Cosmochim. Acta* (in press).

[6] We give only the references of the most recent contributions actually in press in *Geochim. Cosmochim. Acta* (October 1971): Arrhenius, G., Liang, S., MacDougall, D., Wilkening, L., Bhandari, N., Bhatt, S., Lal, D., Rajagopalan, G., Tamhane, A. S., and Venkatavaradan, V. S.: 'The Exposure History of the Apollo 12 Regolith'.
Barber, D. G., Cowsik, R., Hutcheon, I. D., Price, P. B., and Rajan, R. S.: 'Solar Flares, the Lunar Surface and Gas-Rich Meteorites'.
Borg, J., Durrieu, L., Jouret, C., and Maurette, M.: 'Ultramicroscopic Features in Micron-Sized Lunar Dust Grains and Cosmophysics'.
Crozaz, G., Walker, R. M., and Woolum, D.: 'Nuclear Track Studies of Dynamic Surface Processes on the Moon and the Constancy of Solar Activity'.
Fleischer, R. L., Hart, H. R., and Comstock, G. M.: 'Very Heavy Solar Cosmic Rays: Energy Spectrum and Implication for Lunar Erosion'.

[7] Frank, L. A.: 1970, *J. Geophys. Res.* **75**, 707.
[8] Bastin, G., Borg, J., Dran, J. C., Maurette, M., and Vassent, B.: 1970, 'Etude de l'enregistrement de traces d'interactions nucléaires dans des pyroxènes et des feldspaths d'origine lunaire, météoritique et terrestre', Proceedings of the VIIth International Colloquium on Corpuscular Photography and Visual Solid Detectors, Barcelona, July 1970.
[9] Dran, J. C., Duraud, J. P., and Maurette, M.: 1972, this volume, p. 309.
[10] Comstock, G. M., Evwaraye, A. O., Fleischer, R. L., and Hart, H. R. Jr.: 1971, 'The Particle Track Record of Lunar Soil', General Electric Company, Report No. 71-C-073.
[11] See for example the work of Crozaz *et al.* in [6].
[12] See for example the work of Arrhenius *et al.* in [6].
[13] Barber, D. J., Hutcheon, I., and Price, P. B.: 1971, *Science* **171**, 372.
[14] Crozaz, G., Haack, U., Hair, M., Maurette, M., Walker, R. M., and Woolum, D.: 1970, *Geochim. Cosmochim. Acta Suppl. I* **3**, 2051.
[15] Armstrong, T. P. and Krimigis, S. M.: 1970, 'A Statistical Study of Solar Protons, Alphas and $Z > 3$ Nuclei in 1967–68', Johns Hopkins Univ. Preprint.
[16] Stone, E.: 1971, 'Low Energy Solar Cosmic Rays', Caltech Preprint.
[17] Borg, J., Maurette, M., and Vassent, B.: 1971, 'Low Energy Solar Nuclear Particles: New Detection Methods and New Results', Paper to be presented at the 12th Int. Conf. on Cosmic Rays. Hobart, Australia (August 1971).
[18] Cadogan, P. H., Eglinton, G., Maxwell, J. R., and Pillinger, C. T.: 1971, *Nature* **231**, 29.

LOW ENERGY SOLAR NUCLEAR PARTICLE IRRADIATION OF LUNAR AND METEORITIC BRECCIAS

J. C. DRAN, J. P. DURAUD, and M. MAURETTE
Centre de Spectrométrie de Masse du CNRS–91 ORSAY–France

Abstract. We have studied by combined high voltage and scanning electron microscopies both the latent and etched track distributions in lunar fragmental rocks as well as in solar type gas-rich meteorites. We have used a new experimental approach for studying gas-rich meteorites to avoid problems with the interpretation of the results. Some implications of the present results concerning the ancient solar flare cosmic rays and the origin of the track distributions in gas-rich meteorite crystals will be discussed.

1. Introduction

In 1960, R. M. Walker opened up a completely new field of research on the premise that lunar material should contain nuclear particle tracks produced by energetic heavy nuclei both of solar and galactic origin [1]. In 1963, Suess *et al.* [2] suggested that the large amount of rare gas found in the *dark parts* of 'solar' type gas-rich meteorites could be due to an ancient implantation of very low energy solar wind nuclei in their constituent grains, before the compaction of the grains into the meteorite dark parts. Since these 2 original suggestions concerning the possibility of finding natural mineral grains irradiated in *solar* nuclear particles of various types, a lot of work has been done mainly on lunar material [3], but also on gas-rich meteorites [4, 5, 6] and more recently on Antarctica dust [7], in view of finding such grains and exploiting their solar irradiation records for different purposes, in particular to determine the characteristics of the ancient solar flare cosmic rays as well as to decipher the origin of gas-rich meteorites.

In this paper we will present a new experimental approach for the problem of finding mineral grains 'individually' irradiated in space before their 'compaction' into lunar and meteoritic rocks considered as breccias. In particular we have first defined the microscopic irradiation record in lunar dust grains – which certainly get individually irradiated in solar nuclear particles – by using not only scanning electron microscopy and optical microscopy to study etched tracks but also transmission electron microscopy to look for that part of the high resolution irradiation record not appearing with the previously used other techniques. Then we observed possible changes in this record as the grains get artificially heated or naturally 'sintered' into lunar breccias 10046 and 10059. Finally with the same methods we searched for grains showing a 'lunar type' solar irradiation record in the dark parts of several gas-rich meteorites (Weston, Pantar I, Breitschied, Kapoyeta).

In the first part of this paper we will point out some factors which could obscure the solar record in gas-rich meteorites. Then we will describe how we minimized these factors not taken into serious consideration before our work by presenting our main

Urey and Runcorn (eds.), The Moon, 309–323. *All Rights Reserved.*

observations of lunar and meteoritic grains by scanning and transmission electron microscopies. Finally we will discuss the present possibilities in finding extraterrestrial mineral grains with a good solar irradiation record allowing the determination of some characteristics of ancient solar nuclear particle fluxes and we will question the possibility of clearly deciphering the origin of gas-rich meteorites by analyzing their irradiation record.

2. Experimental Approach

Before the study of lunar samples and our transmission electron microscope observations of extraterrestrial matter [8, 9] it was generally considered that the following criteria [1, 4, 5] should be satisfied in identifying grains individually irradiated in space before their 'compaction' into the meteorites:

(1) a clear enrichment of the most superficial layer of the grains in solar type rare-gas should be observed;

(2) the grains should contain a high density, ϱ, of *long* nuclear particle tracks well above the maximum contribution ($\varrho \lesssim 10^7$ tracks·cm^{-2}) expected from the exposure of the meteorite in the VH nuclei of the *galactic* cosmic rays;

(3) the track length distribution in the high ϱ grains should correspond to that expected from solar flare VH nuclei in being peaked at about 5 to 10 μ;

(4) a 'homogeneous' edge zoned track distribution, characterized by a sudden drop of the track density on a depth of about 10 to 20 μ should be observed in the high ϱ grains; this track gradient would simply reflect the 'attenuation' with the depth into the grains of a composite beam of VH nuclei containing more low energy nuclei than more energetic ones.

There are several limitations in applying these track criteria:

(1) the high track density criterion alone is only valid if the tracks are long and if an important proportion of the grains (at least a few percent) contain high ϱ values. Otherwise it can also be expected that the tracks have been produced by spallation recoils or that the grains are uranium bearing phases because it is known that in meteoritic [10] and lunar [11] materials spontaneously fissionable elements are indeed veryinhomogeneously distributed in minor mineral phases which could then get loaded with high fission track densities in a relatively short period of time;

(2) track length greater than about 1 μ can only be directly measured with an optical microscope and this is not easy when the track density is $>10^8$ tracks·cm^{-2} or when they are found in an edge zoning in the grains; therefore it is difficult to verify if the tracks in the high ϱ grains are those expected from solar flare VH nuclei;

(3) before using the track zoning criterion which is apparently the easiest to apply one must be sure that the actual external surfaces of the grains are the true 'premeteoritic' surfaces exposed in space and that no effect else than a solar VH beam attenuation with the depth can produce a track gradient.

These last limitations concerning the most frequently used [4, 5, 6] 'fossil track zoning' seem to be the most severe for the following reasons:

(1) by using 1 MeV and 100 keV electron microscopies we observed very frequently

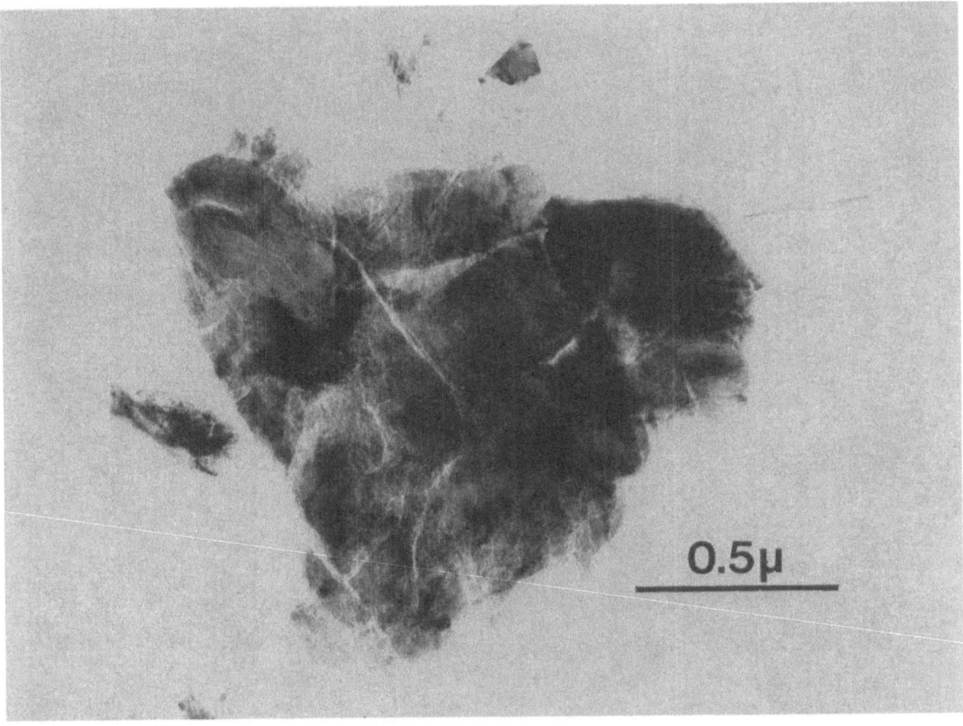

Fig. 1. 100 keV dark field micrograph of a slightly etched pyroxene grain, extracted from the dark part of the Kapoyeta meteorite and showing the extensive microfracturing in the grain.

an extensive microfracturing (Figure 1) in the grains extracted from the dark parts of several gas-rich meteorites. Therefore for track zoning studies we decided to drop the 'hand picking' techniques previously applied [4, 5, 6] because it is likely that during such an extraction the highly fractured grains would lost parts from their external surfaces. Instead we applied to lunar and meteoritic breccias a technique already developped for lunar igneous rocks [11] and where a polished section of a chunk of meteorite was first prepared by using epoxy resin; then the section was etched to reveal the fossil tracks and observed with a scanning electron microscope (SEM) with results similar to those reported in Figure 2 for a dark part fragment* of Kapoyeta; this etched section shows an extensive grain fracturing which again raises serious doubts about the validity of any 'hand picking' technique either to preserve the external surface of the grains or to search for grains with 'rounded' habits indicative of their 'premeteoritic' erosion. Furthermore, even by using this technique several *small* grains were still removed from the surface after the etching. Therefore we decided to etch a crushed residue from a dark part and to search for latent and slightly etched

* Another interesting advantage of this technique consists in the possibility of analysing in great detail the boundary between a dark and a light part, by polishing a fragment containing these 2 different parts.

Fig. 2. Scanning electron micrograph of a polished section from a dark part of the Kapoyeta meteorite. This section was irradiated with fission fragments and etched 20 min in a boiling solution of sodium hydroxyde. Fission tracks and a high density of fossil tracks can be observed in the grains identified by single and double arrows respectively.

tracks in the micron-sized fragments by using transmission electron microscopy in order to minimize the possibility of missing the irradiation record in the smallest grains (we further applied this technique to examine grains from peculiar small parts in the dark parts including 'very dark' and 'white cloudy' parts, etc. ...);

(2) track zonings are *frequently* not related to a beam attenuation in the pyroxenes. Indeed we have observed fossil zonings in grains extracted from the deep interior of several *igneous* lunar rocks (and therefore well shielded from any VH solar flare irradiation) and also artificial zonings when studying the sensitivity for track registration in pyroxene and feldspar crystals. In this experiment [12] the fossil tracks were first erased by heating. Then the grains were polished, covered with a thin iron foil and exposed to artificial fluxes of 156 MeV protons and α-particles. The registration of the tracks of the recoil nuclei induced during the nuclear interactions of the incident particles with the iron nuclei was subsequently studied in the grains. The feldspars did not register these low energy iron recoil nuclei but out of about 20 pyroxenes grains we observed one grain with a high track density homogeneously distributed in the volume of the crystal (Figure 3a) and 2 grains with an edge zoned track distribution, inhomogeneous both along the edge of the grains and in its depth extension

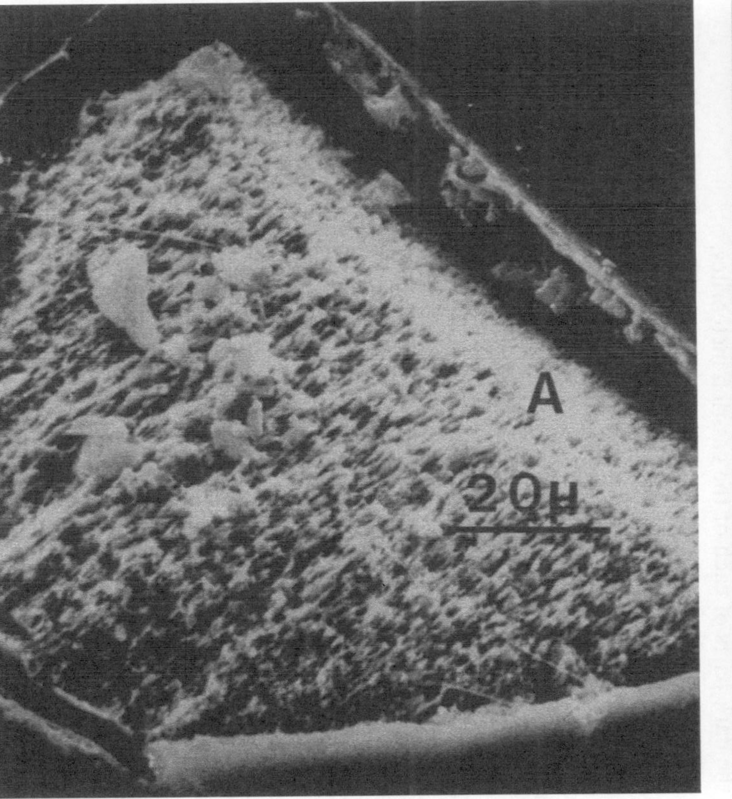

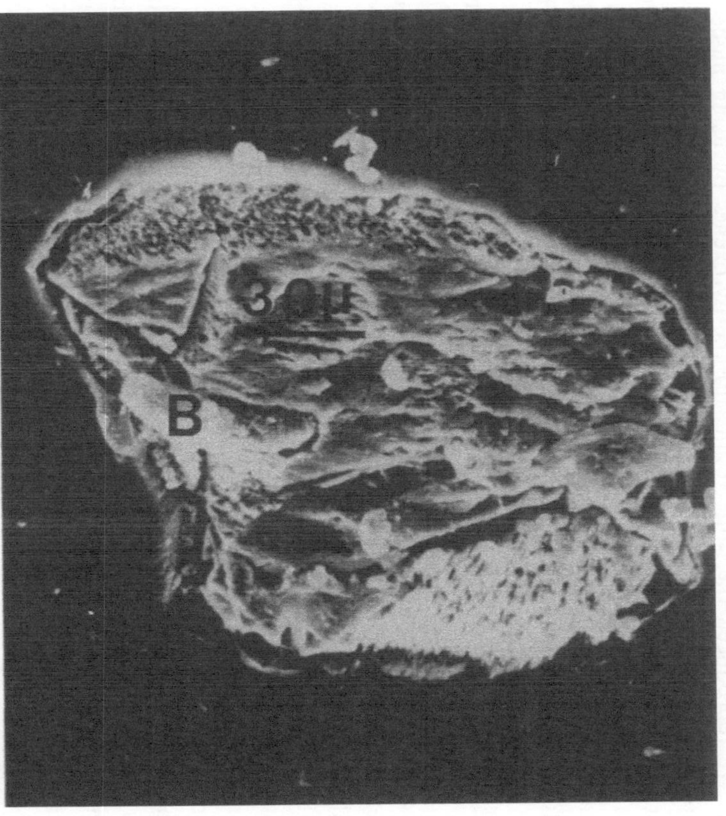

Fig. 3. Scanning electron micrographs of iron interaction tracks in lunar dust pyroxene grains.

(Figure 3b). We reproduced the same results for the pyroxenes by using higher Z material and 3 GeV protons. For the feldspars irradiated in the same higher Z conditions a very homogeneous track distribution was generally observed. We interpreted these striking and disturbing results as being due to a chemical zoning in the pyroxene grains modifying either their sensitivity for track registration or their etching rate. Therefore it is quite evident that feldspar crystals have to be selected in priority for identifying an 'individual' irradiation of meteoritic grains in space on a track gradient basis. If they are too rare or too small sized for an easy study then the pyroxenes could be used as a second choice material, but with great care *especially if they contain iron as a major constituent*. If the pyroxene zonings are unfrequent and inhomogeneous then an 'iron converter foil' experiment should be conducted to verify if the artificial zoning is *not* superimposed on the fossil zoning, otherwise the fossil tracks could be due to fossil recoils produced during the nuclear interactions of the galactic cosmic rays with medium heavy elements in the grains, after their 'compaction' into the meteorites.

We also developed [13] an additional criterion based on the transformation of the fossil irradiation record made of tracks and rare gas atoms when the lunar dust grains get either *artificially* heated or *naturally* sintered into lunar breccias. In these grains the tracks are replaced by inclusions which grow and do not disappear when the annealing temperature is increased and which are observable by transmission electron microscopy (Figure 5). Therefore we searched also for inclusions in the 'sintered' grains from the dark parts of each of the gas-rich meteorites.

In the next chapter we will report our first and incomplete search for gas-rich meteorite crystals 'individually' irradiated in space and we will describe in more detail how we minimized various experimental uncertainties.

3. The Ultramicroscopic Irradiation Record from Lunar Dust Particles to Solar Type Gas-Rich Meteorites

A. LUNAR DUST GRAINS

Our transmission and scanning electron microscope observations are given in more detail and exploited for other purposes elsewhere [13, 14]. The main features of the irradiation record are:

(1) in the *finest* and *uncrushed* crystalline dust grains a high proportion ($\sim$80%) of the grains contain a very high density of nuclear particle tracks exceeding frequently 10^{10} tracks$\cdot$cm^{-2};

(2) in the coarser dust grains the high track density seems to be registered in a superficial micron-sized layer; at depth$\gg 1\ \mu$ the track density drops to much lower values in the range 10^8–10^9 tracks$\cdot$cm^{-2} [13, 14];

(3) with the SEM a surprisingly *low* proportion of the coarser grains show edge zoned track distributions which even in the most favorable case (Figure 4) are not completely homogeneous;

(4) when the track bearing grains are heated in a furnace they get loaded with

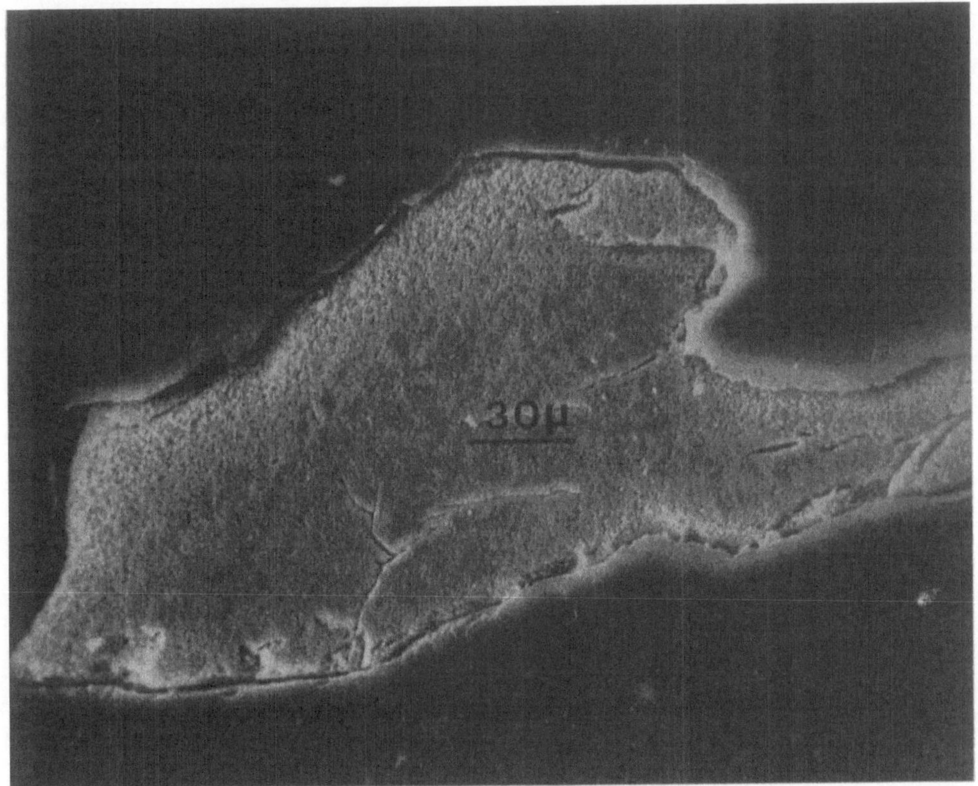

Fig. 4. Scanning electron micrograph of the best fossil zoning so far observed, in a lunar dust pyroxene grain.

high density of inclusions observable by high voltage electron microscopy (Figure 5a) and looking like the radiation induced phase transformations observed by Price and Walker [15] in their much earlier observation of artificially anealed fission fragment tracks in terrestrial mica;

(5) after a slight etching first developped by Barber *et al.* [16] short etched tracks are observed in the grains by 100 keV electron microscopy.

B. LUNAR BRECCIAS 10046 AND 10059

With the 100 keV and 1000 keV electron microscopes we already reported [7, 13] the following observations for the crystalline grains:

(1) a small proportion of the grains (~1%) contain high densities of latent tracks before etching;

(2) the grains are frequently loaded with fossil inclusions (Figure 5b);

(3) a small proportion of the grains (~10%) show high densities of etched nuclear particle tracks with the 100 keV electron microscope.

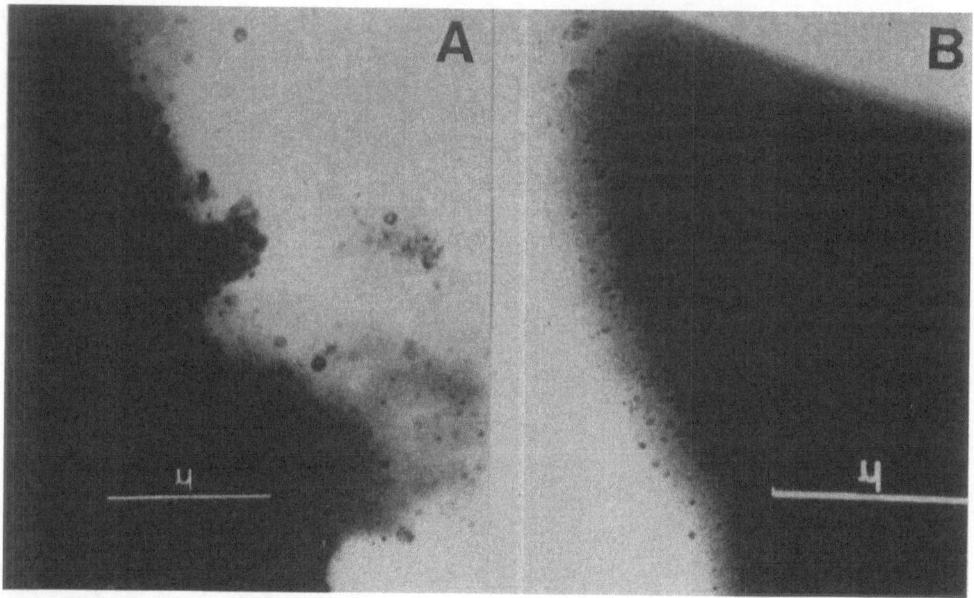

Fig. 5. 1 MeV dark field micrographs of lunar dust grains. In B the grain was extracted from lunar breccia 10046 and has been 'naturally' metamorphized; in A the crystal was artificially heated at 800°C, during 1 hr, under vacuum.

The SEM observation of etched polished sections reveals 3 different groups of grains:

(1) in about 1% of the grains with size smaller than about 30 μ a reaction layer, probably indicative of high track densities registered in the whole volume of the grains, is observed (Figure 6a);

(2) all the largest glassy objects representing about 20% of the coarser grains contain very well formed etch pits attributable to nuclear particle tracks with densities up to 10^8 tracks.cm^{-2} (Figure 6b);

(3) about 10% of the crystalline grains show shallow etch pits (Figure 6c) with densities up to $\sim 10^9$ tracks·cm^{-2} and looking strikingly like those resulting from partially annealed VH tracks or from short spallation recoil tracks; by looking at more than 100 such grains in breccia 10046 we have not observed a track gradient similar to that reported in Figure 4.

C. DARK PARTS OF SOLAR* TYPE GAS-RICH METEORITES

For each meteorite (Weston, Pantar I, Breitschied, Kapoyeta), we first observed by high voltage electron microscopy, about 100 grains obtained by crushing small fragments hand picked in the dark parts of the meteorites. Then we etched $\sim$500 grains in each crushed residue to observe slightly etched tracks by 100 keV electron

* We also studied 'planetary' type gas-rich meteorites (Orgueil [8, 9], Mighei, Murray, Renazzo, Mokora) but this work will be reported later on [17].

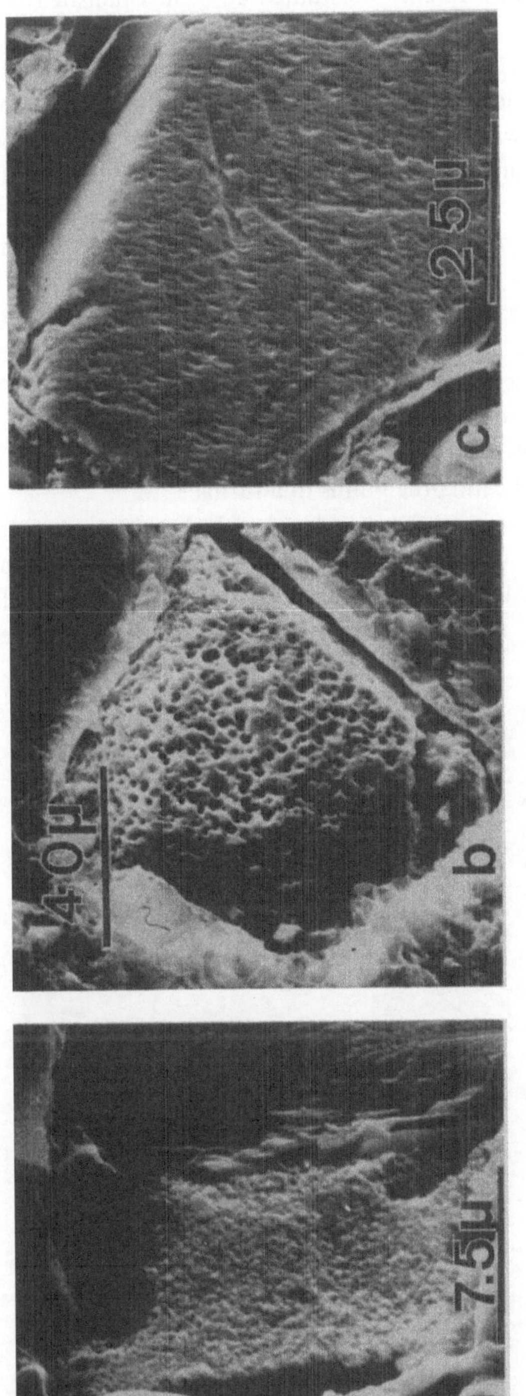

Fig. 6. Scanning electron micrographs obtained during the study of an etched, polished section from lunar breccia 10046 and showing 3 different kinds of grains.

microscopy in the same meteorites. Finally we studied an etched polished section of Kapoyeta with the SEM.

With the 1 MeV electron microscope we did not observe either grains with high ϱ values or inclusions similar to those reported in Figure 5a. Therefore the proportion of grains with high track density was certainly smaller than 1%. Furthermore the grains showed an extensive microfracturing which was less frequent in the light parts of the meteorites.

With the 100 keV electron microscope we discovered etched tracks only in Pantar I, where 2 grains showed etched track densities of about 2×10^9 and 6×10^9 tracks $\cdot$ $\cdot \text{cm}^{-2}$; but these tracks were certainly much longer than those constituting the high track density pattern in lunar dust grains and are likely of a different origin. Furthermore the striking microfracturing was beautifully revealed by such an etching (Figure 1)–on some area of this figure the microcracks could be easily confused with nuclear particle tracks as they are similarly etched but we have checked that contrarily to the tracks they are not annealed by heating the grains in a furnace.

With the SEM we observed Kapoyeta, because hand picked grains were extensively studied with this instrument by other groups [4, 5, 6]. We first irradiated the polished sections with a dose of fission fragment tracks of about 10^7 tracks $\cdot \text{cm}^{-2}$, well below the level expected from a solar flare irradiation. Then we etched for the feldspars

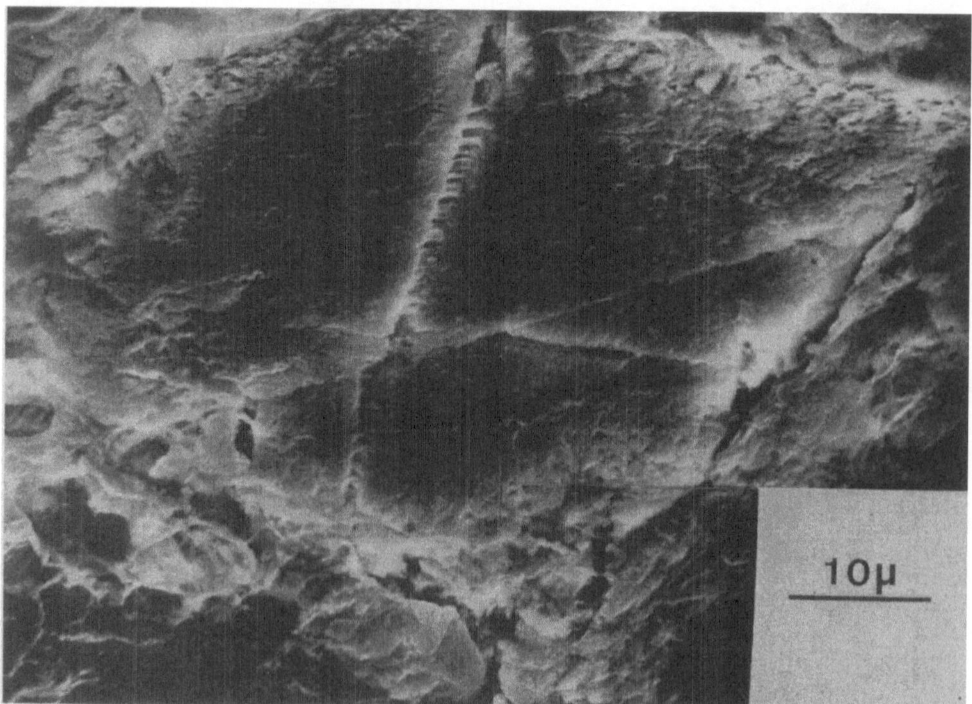

Fig. 7. Scanning electron micrograph of an *inhomogeneous* fossil track edge zoning in a pyroxene crystal extracted from a dark part of Kapoyeta.

and determined the ratio between the number of grains showing track densities in excess of 10^8 tracks·cm^{-2} and that of grains with fission tracks. In Figure 2 this proportion is about $\frac{1}{7}$ but indeed by scanning a larger area we determined that it was smaller than $\frac{1}{30}$. Then we studied the possible track gradients and microprobed the grains with high ϱ values to check further their nature, looking in particular for P and Zr which are generally considered as tracer for high uranium bearing phases [11]. Finally we repeated these observations–but we also microprobed for iron–for the pyroxenes which were much more abundant and much larger than the feldspars (in Figure 2 the pyroxenes are essentially all the grains without fission tracks). Our very priliminary results are: (1) out of 30 feldspar grains only one had a ϱ value $\gtrsim 10^8$ tracks·cm^{-2} but *no* gradient was observed; (2) out of 63 pyroxene grains 14 had ϱ values in excess of 10^8 tracks·cm^{-2} but only one grain contained a fossil zoning (Figure 7) which was very inhomogeneous; (3) the tracks in the pyroxenes were generally much shorter than the fission fragments tracks simultaneously etched in the grains.

Very recently we also conducted a much faster survey of a Kapoyeta polished section etched for pyroxene grains and already studied with the SEM, by using Nomarski phase contrast in reflected light with an optical microscope. This survey confirmed very clearly the SEM observations thus showing that reflected light optical microscopy can be first used to unravel the irradiation record of gas-rich meteorite in identifying the high ϱ grains to be further studied with the SEM.

4. Discussion

A. Comparison of the Present Results to Those of Other Workers

Our results with gas-rich meteorites disagree either with previous works concerning the study of etched tracks with the SEM in hand-picked grains [4, 5, 6] or with the transmission electron microscope observations of Barber *et al.* [3] reported later than our already unsuccessful preliminary attempt [8] to find very high ϱ values in gas-rich meteorites with a high voltage electron microscope: contrarily to the other groups, we have no clear evidence from track studies alone, for an 'individual irradiation' of the constituent grains of the dark parts of solar type gas-rich meteorites. Therefore we want strongly to point out that the following discussion has to be considered as a very preliminary one as long as the discrepancies between the different groups will not be understood.

These main discrepancies are:

(1) contrarily to the groups of Lal and Pellas which report by oral communications the existence of feldspar grains with homogeneous track gradients in Kapoyeta, we still have not discovered one feldspar grain with a gradient in this meteorite by using a more carefully experimental approach which can handle the very small sized feldspar grains appearing in Figure 2 without the uncertainties attached to any hand picking technique;

(2) Pellas reported a high proportion ($\sim 30\%$) of 'irradiated' pyroxene grains in

Kapoyeta. This proportion only agrees with that of about 20% determined in the present work for pyroxene grains showing high ϱ values ($\gtrsim 10^8$ tracks$\cdot$cm^{-2}). But unfortunately the proportion of these high ϱ grains showing a track gradient was much smaller ($\sim 2\%$) and all the gradients we so far studied were quite inhomogeneous both along the edge of the grains and in their depth extension;

(3) in the high ϱ pyroxenes we observed generally short and not long tracks;

(4) from their transmission electron microscope observations, Barber *et al.* [3] seem to have foud a lot of very high ϱ grains in Fayetteville and Kapoyeta as they said "Although the meteorite studies are still at a preliminary stage, our observation of track densities comparable to those in lunar grains and at least 20 times greater than had been originally reported in track-rich meteorite grains [4, 5] are highly significant because they remove the previous major distinction [8] between lunar grains and meteoritic grains and suggest the possibility of a similar origin". However our earlier [8] and present observations do not verify such a statement because we find that the proportion of grains with *latent* track densities in excess of 10^{10} tracks$\cdot$cm^{-2} is certainly much smaller ($< 1\%$) than in the lunar dust particles and that the proportion of grains with high densities of *slightly etched* tracks is $\sim 1\%$ in the most favorable case of Pantar I where the maximum ϱ value is still much smaller than those in the lunar dust grains.

Several tentative explanations for the discrepancies can be proposed:

(1) it is quite certain that a major difficulty in determining a proportion of grains with a gradient of etched tracks in the highly fractured gas-rich meteorites is to take really into account all the single grains in which the etching could have revealed the tracks and not for example count 5 grains for *one* highly fractured grain. We minimized this difficulty by preirradiating the polished sections with fission fragments and by performing microprobe and back scattered electron works with the same apparatus on the same area. The statistics of the Lal and Pellas groups are certainly different from ours because by hand picking they selected the largest crystals and we presume that they did not take into consideration in their statistics the more abundant smaller grains in their 'gently' crushed residue;

(2) our reported lack of grains with an homogeneous edge zoned track distribution is not surprising because such grains are very rare even in those extracted from the lunar dust and breccias. Therefore contrarily to the other investigators we did not expect to find a better zoning in gas-rich meteorites except if the constituent grains of these meteorites get processed by irradiation, erosion or 'sintering' mechanisms totally different from those acting in the lunar regolith;

(3) the discrepancies between our results and those of Barber *et al.* [3] are much more difficult to understand. Indeed by comparing our high voltage dark field micrographs of lunar dust grains loaded with latent tracks it was quite evident that these tracks were much easily observed with our 1 MeV microscope than with their 650 keV microscope generally operated at 500 keV. One of the explanation for the better contrast and resolution was the smaller chromatic aberration at 1 MeV. As a consequence our estimate of the proportions of lunar dust crystalline grains loaded with

latent tracks was generally greater than theirs. Therefore it seems unlikely that 'suddenly' we would have missed the high latent track densities in the gas-rich meteorite grains. Another possibility is that the Berkeley group has foud a still unreported criterion to detect irradiated grains by a quick survey of polished sections?

B. FABRIC OF THE SOLAR TYPE GAS-RICH METEORITES AND LUNAR BRECCIAS

The track irradiation record in gas-rich meteorites is probably different from that in other types of meteorites [1]. However we think that at the present time this record is not sufficiently well understood and is too far from the ideal record expected from grains individually exposed to energetic solar VH nuclei to conclude from track evidence alone that the dark parts of the solar type gas-rich meteorites have been formed by the 'sintering' of grains individually irradiated in space. Feldspar grains which are quite rare in the gas-rich meteorites have to be systematically studied with the technique we described and the beam attenuation origin for any possible zoning has to be checked each time with a converter foil experiment. Our results also cast serious doubts upon 2 earlier conclusions by Pellas *et al.* [5]: (1) the grains have certainly not 'been irradiated all around' and therefore it is no longer necessary to argue that the irradiation took place when the grains were freely floating in space at an early stage in the solar system history; (2) the distorsion appearing on X-rays diffraction patterns is not due to radiation damage *unobserved by transmission electron microscopy* but most likely to the extensive microfracturing in the grains.

The main evidence still supporting this hypothesis of a 'premeteoritic' irradiation is based on the great superficial enrichment of the dark part grains in rare-gas with abundance patterns and isotopic ratios similar to those produced by a solar wind implantation in aluminum foils deployed on the Moon surface during the Apollo 11 and 12 missions and subsequently brought back on the Earth [18]. However it should be definitively checked if the high densities of etchable 'track like' microfractures in the grains could not interfere with the conclusion–based on etched rare-gas release experiments [19]–that the rare-gas atoms are implanted in the most superficial layer of the grains.

Mason and Melson [20] have recently suggested that some solar type gas-rich meteorites are breccias which have been *less* metamorphized than lunar breccias. It will certainly be fruitful to check further the validity of this hypothesis by correlating the most striking texture features we observed in the dark parts (the extensive microfracturing and the absence of amorphous or highly disordered grains) to the following observations: (1) the microfracturing is less marked but still observable in the light parts of the same meteorites; (2) the lunar dust material most likely processed by micrometeorite and meteorite impacts appears as a two component matrix made of a totally amorphous minor fraction mixed to a major fraction of well ordered crystals in which microfracturing is very unfrequent and these characteristics are also observed in lunar breccias; (3) in the lunar breccia grains a major proportion of the tracks have been erased and replaced by inclusions; (4) Shergotty which is considered as a typical example of non gas-rich shocked meteorite, contains $\sim$20% of highly disordered grains but the microfracturing is less extensive than that appearing in Figure 1.

The track irradiation record in the lunar breccias so far studied has probably been 'metamorphized' by a heat sintering of the grains. This conclusion is supported by the observation of 'thermal like' inclusions and by the marked decrease in the proportion of grains with etched and latent tracks. Therefore we hope that the present methods when applied to the Apollo 14 rocks which are mostly complex 'breccias within breccias' will greatly help in understanding their origin.

C. CHARACTERISTICS OF ANCIENT SOLAR NUCLEAR PARTICLE FLUXES

It seems to us that the optimistic view of considering that the irradiation record in the gas-rich meteorites will allow the study of ancient *solar flare* cosmic rays, very early in the solar system history [4, 5, 6], is far from reach at the present time. We think that it is much easier to find and exploit a solar record in lunar material.

But we also pointed out [21] that *good* track gradients to conduct such studies are very rare in the grains extracted from the lunar dust and lunar breccias. Therefore it appears that the best method to determine the characteristics of 'ancient' solar flare cosmic rays is to use polished sections containing the 'top' surfaces of lunar *igneous* rocks and to study the depth variation of the track density in the feldspar grains observed in these sections, by using methods described in the work of Crozaz *et al.* [3].

However, if our preliminary transmission [13] and high resolution scanning electron microscope [14] observations are confirmed by further works, then a high proportion of the lunar dust grains also contain 2 other types of *homogeneous* solar irradiation records which can be potentially exploited by methods described elsewhere [13, 21]. These records are registered in the most superficial layers of the grains and would be due: (1) to a solar wind implantation producing a superficial metamictized layer looking like a coating surrounding the grains, with a thickness of about 500 Å; (2) to solar 'suprathermal' ions building up high track densities in excess of 10^{10} tracks $\cdot$ $\cdot cm^{-2}$ in a micron-sized superficial layer in the grains.

Acknowledgements

All the transmission microscope observations were made by L. Durrieu and C. Jouret at the Institut d'Optique Electronique du CNRS, Toulouse, France. We acknowledge the generous help of the JEOL Company for the SEM observations. One of us (M. Maurette) is deeply indebted to Dr R. Bernas for his very active and enthusiastic support and interest. We also thank Drs R. Hutchinson (British Museum, London), B. Mason (Smithsonian Institution, Washington D.C.) and A. Widatalla (Geological Survey Dept., Sudan) for the loan of meteoritic samples. We wish to express our gratitude to B. Vassent for her efficient and clever assistance with the experimental work and to J. Caro for her help in the manuscript preparation.

References

[1] Walker, R. M.: 1972, 'Fossil Track Studies in Extraterrestrial Materials', *Rad. Effects* (in press).

[2] Suess, H. E., Wanke, H., and Wlotzka, F.: 1964, *Geochim. Cosmochim. Acta* **28**, 595.

[3] We only give the references of the most recent contributions in press in *Geochim. Cosmochim. Acta* (1971):
Arrhenius, G., Liang, S., Mac Dougall, D., Wilkening, L., Bhandari, N., Bhatt, S., Lal, D., Rajagopalan, G., Tamhane, A. S., and Venkatavaradan, V. S.: 'The Exposure History of the Apollo 12 Regolith'.
Barber, D. G., Cowsik, R., Hutcheon, I. D., Price, P. B., and Rajan, R. S.: 'Solar Flares, the Lunar Surface and Gas-Rich Meteorites'.
Borg, J., Durrieu, L., Jouret, C., and Maurette, M.: 'Ultramicroscopic Features in Micron-Sized Lunar Dust Grains and Cosmophysics'.
Crozaz, G., Walker, R. M., and Woolum, D.: 'Nuclear Track Studies of Dynamic Surface Processes on the Moon and the Constancy of Solar Activity'.
Fleischer, R. L., Hart, H. R., and Comstock, G. M.: 'Very Heavy Solar Cosmic Rays: Energy Spectrum and Implication for Lunar Erosion'.

[4] Lal, D. and Rajan, R. S.: 1969, *Nature* **223**, 269.

[5] Pellas, P., Poupeau, G., Lorin, J. C., Reeves, H., and Audouze, J.: 1969, *Nature* **223**.

[6] Lorin, J. C.: 1969, *J. Physique* **30**, C3-102.

[7] Durrieu, L., Jouret, C., Leroulley, J., and Maurette, M.: 1971, *Jernkontorets Ann.*, in press.

[8] Borg, J., Dran, J. C., Durrieu, L., Jouret, C., and Maurette, M.: 1970, *Earth Planetary Sci. Letters* **8**, 379.

[9] Dran, J. C., Durrieu, L., and Jouret, C.: 1970, *Earth Planetary Sci. Letters* **9**, 391.

[10] Fleischer, R. L.: 1968, *Geochim. Cosmochim. Acta* **32**, 983.

[11] Burnett, D., Monnin, M., Seitz, M., Walker, R. M., and Yuhas, D.: 1971, *Geochim. Cosmochim. Acta*, in press.

[12] Bastin, G., Borg, J., Dran, J. C., Maurette, M., and Vassent, B.: 1970, *Proceeding of the VIIth International Colloquium on Corpuscular Photography and visual solid detectors, Barcelona, July 1970*

[13] See for example the work of Borg *et al.* in reference [8].

[14] Borg, J., and Vassent, B.: 1972, this volume, p. 298.

[15] Price, P. B. and Walker, R. M.: 1962, *J. Appl. Phys.* **33**, 3410.

[16] Barber, D. J., Hutcheon, I., and Price, P. B.: 1971, *Science* **171**, 372.

[17] Dran, J. C., Duraud, J. P., Durrieu, L., Jouret, C., Legressus, R., and Maurette, M.: 1971, to be submitted to *Geochim. Cosmochim. Acta.*

[18] Geiss, J., Eberhardt, P., Bühler, F., Meister, J., and Signer, P.: 1970, *J. Geophys. Res.* **75**, 5972.

[19] Eberhardt, P., Geiss, J., and Groglern, N.: 1965, *J. Geophys. Res.* **70**, 4375.

[20] Mason, B. and Melson, W.: 1970, *The Lunar Rocks*, Wiley-Interscience.

[21] Borg, J., Durrieu, L., Jouret, C., and Maurette, M.: Second Lunar Science Conference, unpublished proceedings.

THE THERMOLUMINESCENCE OF LUNAR SAMPLES

G. F. J. GARLICK and IRENE ROBINSON
Department of Physics, University of Hull, England

Abstract. Studies of thermoluminescence emission from lunar samples show that at lunar day surface temperatures and even at 18°C the trapped electron responsible are decaying to their ground states by a non-thermal process. Data from various workers show similar characteristics with respect to this process. They show that thermoluminescence cannot be used as a technique for dating lunar samples (or terrestrial plagioclases) over any significant geological period. A quantitative model for the non-thermal process is proposed.

1. Introduction

Minerals in which there are luminescent components lend themselves to the technique of dating by means of thermoluminescence experiments. The electrons originally excited into metastable states or traps in the material by radioactive emissions or by other source such as cosmic rays may be released and return to normal ground states with emission of visible or near visible radiation if a sample is heated. If heating is at a uniform rate (or at some other controlled rate) then the so called 'thermoluminescence' curves can be quantitatively interpreted.

Since electron traps can themselves be due to defects in solids induced by high energy particles (radiation damage) the distribution of thermal activation energies (or trap depths) and the thermal stability of the traps themselves can be explored as a tool for analysis of the past radiation history of the sample. These aspects of thermoluminescence in minerals were thoroughly discussed at the Spoleto conference in 1968 [1]. In this paper we are concerned with the application of thermoluminescence studies to lunar material recovered in the Apollo 11 and 12 missions.

It was found by several different groups [2, 3, 4, 5, 6] that the light sum of thermoluminescence obtained by warming Apollo 11 samples in the dark, and with no laboratory excitation, was many orders of magnitude below that expected from the accumulated dose of excitation from radioactive and cosmic ray sources on the lunar surface. Luminescence in lunar materials is of low efficiency ($\approx 10^{-5}$ W output per watt input) but even this cannot account for the low thermoluminescence yields. It therefore became evident that trapped electrons in lunar samples were continuously 'leaking away' from the trapping states. Some authors attributed this to the thermal activation processes normally responsible for escape of trapped electrons which follow a Boltzmann type probability relation:

$$p = s \exp(-E/kT) \tag{1}$$

s being the probability of escape per second, a constant ($\approx 10^8$–10^9 for many silicates) E the thermal activation energy (often termed the trap depth), k being Boltzmann's constant and T the absolute temperature. Thus electrons in deep traps (E large) will

Urey and Runcorn (eds.), The Moon, 324–329.

be much more stable than those in shallow traps (E small). While the high day temperature of the lunar surface ($\approx 120\,^{\circ}\mathrm{C}$) could be invoked as a strong factor in the leakage process we were impressed by the fact that the rate of leakage was similar at terrestrial ambient temperatures around 20 °C even in the case of deep traps. We therefore attempted to analyse the various data from our own measurements and those of other workers. This work and a theoretical analysis of the data form the substance of the following sections.

2. Experimental Investigations

A. EXPERIMENTAL PROCEDURES

In our thermoluminescence experiments samples of lunar material and of terrestrial plagioclases were used since the thermoluminescence characteristics and trapped electron leakage effects were found to be similar for the latter to those of lunar samples. Each selected specimen was mounted as a thin layer of dust in a light tight vacuum system on a 'finger' the temperature of which could be raised from room temperature

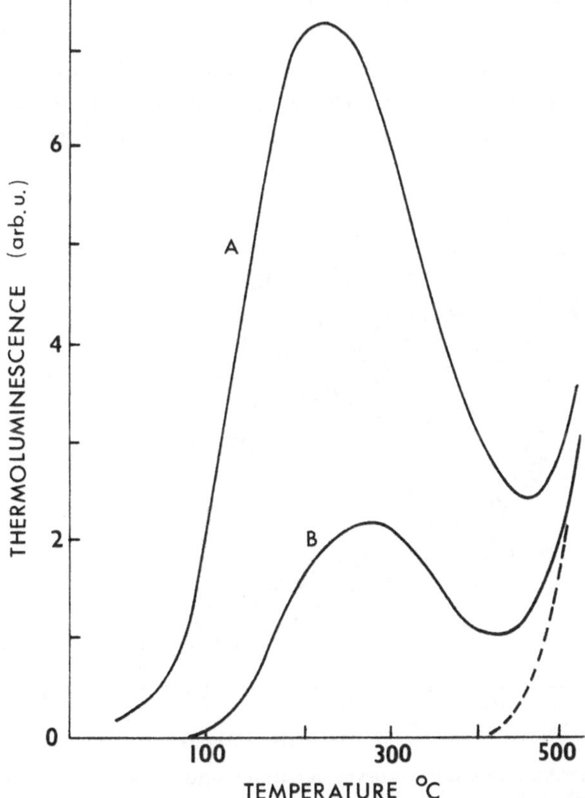

Fig. 1. Thermoluminescence curves of the plagioclase labradorite showing leakage of trapped electrons. A–Curve obtained after 16min in dark at 18°C after 10^7 rad of 30kV X-irradiation. B–Curve obtained after 45 h in dark at 18°C after 10^7 rad of 30kV X-irradiation.

at a uniform rate of 1 K per s. The samples were viewed by a selected photomultiplier with visible or near infrared response. As an excitation source a 30 keV X-ray beam was used and all samples received the same dose of 10^7 rad. After excitation samples were held in the dark for a specified time and then warmed to obtain from the thermoluminescence curves the number of trapped electrons still remaining in the traps of different depths. These are the kinds of experiment arrangement and procedure adopted by the other workers whose data we incorporated in our analysis.

B. EXPERIMENTAL RESULTS

Since there are plenty of examples of the kinds of thermoluminescence curves obtained for Apollo 11 and 12 samples after laboratory excitation by X-rays or β-rays [2, 3, 4] reference may be made to the respective publications. We give, however, an example of decay of trapped electrons after excitation of a terrestrial plagioclase, labradorite, by presenting in Figure 1 the thermoluminescence curves obtained by waiting for 16 min and for 45 h in the dark at 18 °C respectively after the same initial excitation of 10^7 rad. of 30 keV X-rays. The loss of trapped electrons is very evident. Even the higher temperature regions of the curve (i.e. deep traps) are reduced.

In order to present in a concise way the various data from our experiments and those of other workers we have measured the height of a thermoluminescence curve at selected positions on the temperature abscissa as a function of the time the sample was held in the dark (at a selected lower temperature) before warming it to produce the curve. Figure 2 gives the curve height vs. time graphs so obtained. It is very notice-

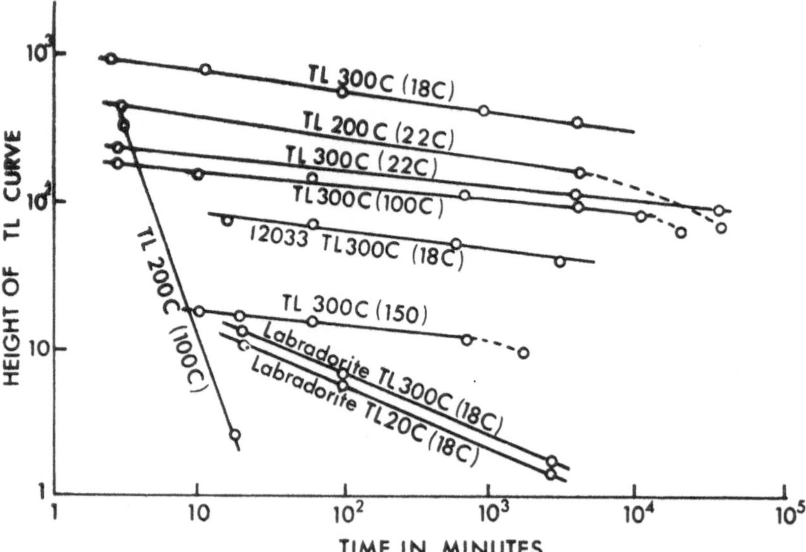

Fig. 2. Decay of thermoluminescence curves due to retention of samples at selected fixed temperatures for different times before warming. Point on thermoluminescence curve shown by first temperature on each curve: retention temperature shown in brackets. Curves for decay at 22° and 100°C are from Dalrymple and Doell[3]: curve for decay at 150°C is from Hoyt *et al.*[2]: both data are for Apollo 11 samples.

able that in all cases but one the slopes are very similar for the lunar samples and independent of the thermoluminescence temperature chosen. In the one exceptional case of the decay for the 200 °C curve region when the sample is held at 100 °C the slope is large and decay rapid. We conclude that this is the consequence of a predominant thermal activation process.

The decay curves of Figure 2 for the terrestrial labradorite show similar behaviour but the larger slopes indicate a more rapid leakage of trapped electrons.

3. Discussion of Results

From the data of Figure 2 we conclude that there are two processes responsible for the loss of trapped electrons in lunar samples and in terrestrial plagioclases, one the thermal activation process (represented by Equation (1)) and the other a non-thermal leakage which affects electrons in traps over a wide range of depths or activation energies. The occurrence of the latter process means that we cannot use the thermoluminescence experiments as a dating technique in any straightforward way since only very small residual amounts of trapped electrons are found by the time the sample is available after retrieval from the lunar surface ($\approx$9 weeks).

In order to provide theoretical data on the residues of trapped electrons to be expected when only thermal activation processes are present and when distributions of trap depths are very large we have made a theoretical analysis for a uniform distribution of trap depths. If we use Equation (1) and assume first order kinetics the fraction n/n_0 of trapped electrons remaining at a given time after excitation ceases, and at a given sample temperature is given by

$$n/n_0 = \exp\{-s\exp(-E/kT)\cdot t\}. \tag{2}$$

Families of curves for n_0 constant with E are given in Figure 3 assuming the constant $s = 10^9\ \mathrm{s}^{-1}$. It is a simple matter to shift such curves to accommodate other s values.

The striking feature of such a relation is that for a given temperature and decay time all electrons in traps shallower than a given depth have escaped the other trapping states of greater depth remaining saturated. This means that for thermal activation alone the thermoluminescence curve decay shown in Figure 1 should be markedly different at different points along the temperature abscissa whereas Figure 2 shows for various samples little variation in the rates of decay with the temperature unless the thermoluminescence temperature is near to the temperature at which decay takes place. This supports our conclusion that a non thermal leakage process is present. The next task is to explain why a linear decay (i.e. power law) is displayed by he straight line graphs of Figure 2.

Non-thermal leakage of trapped electrons is most easily explained if a model as shown in Figure 4 is adopted. Any overlap of wave functions of trapped electrons and those applicable to the ground state energy level will give a finite transition probability. If each trap is simply associated with a neighbouring ground state then the

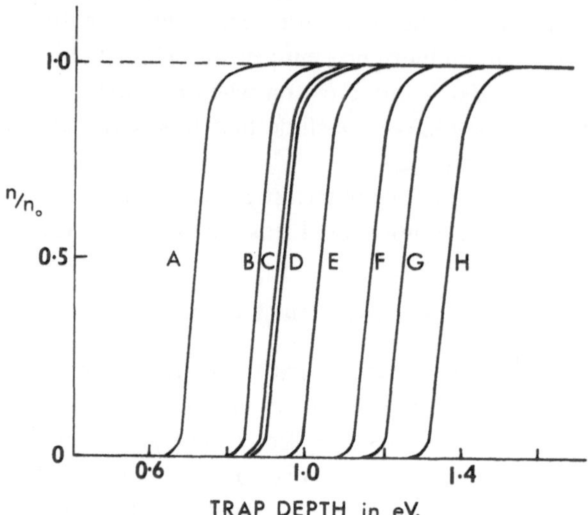

Fig. 3. Change in distribution of trapped electrons with time for different sample temperatures according to Equation (1) with $s = 10^9\ s^{-1}$ A–After 10 min at 300 K (same curve is 10 weeks at 300 K for $s = 10^5\ s^{-1}$); B–After 1 week at 300 K; C–After 10 weeks at 300 K; D–After 10 min at 400 K; E–After 10 yr at 300 K; F–After 1 week at 400 K; G–After 10 weeks at 400 K; H–After 10 yr at 400 K.

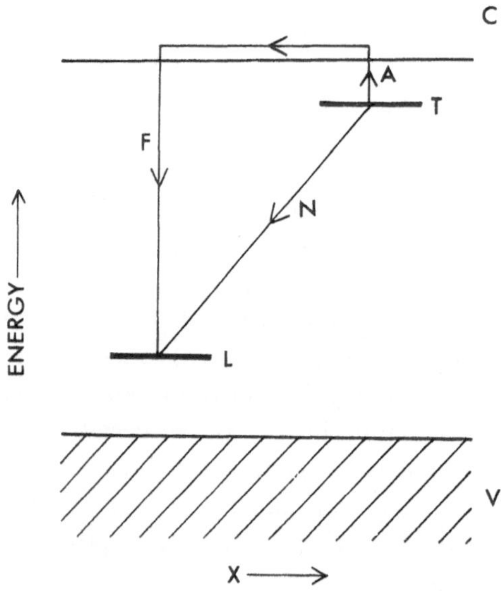

Fig. 4. Energy band model for non-thermal loss of trapped electrons in plagioclase. A–Thermal activation process; C–Conduction band of host crystal lattice; F–Fluorescence transition in luminescence centre; L–Luminescence centre; N–Non-thermal recombination of trapped electron with luminescence centre; T–Electron trap; V–Valence band of host crystal lattice; X–Spatial coordinate in host crystal lattice.

decay of trapped electrons will be given by first order kinetics

$$dn/dt = -\alpha n \tag{3}$$

where α is the decay constant determined by the matrix element for the transition. However, if there is in the sample a random spatial distribution of trapping states and ground states the decay will be a complicated sum of exponentials. A similar situation has already been quantitatively treated by Thomas *et al.* [7] in explaining the luminescence emission spectra of GaP and other solids, known as the 'distant pair' theory. We shall not reiterate that analysis here but simply state that the resulting decay will follow a simple power law relation.

Turning to another facet of thermoluminescence studies on lunar samples it has been found both for Apollo 11 and Apollo 12 core samples that the residual thermoluminescence increases with the depth of the core sample below the lunar surface [2, 8]. This might imply that thermal activation is playing a dominant part in the decay of trapped electrons in surface layers. However, the albedo of such core samples (measured in the laboratory) also shows an increase with depth of core sample [9]. This would give an apparent increased thermoluminescence since optical absorption is decreasing with the depth from which the sample was taken. Since thermoluminescence also depends on the intrinsic luminescence efficiency of a sample the latter may also be higher for depths of more than a cm or so due to the smaller damage in the sample from cosmic radiation. It is clear that interpretation of thermoluminescence for core samples must necessitate corrections for differences in sample opacity and luminescence efficiency.

References

[1] McDougall, D. J. (ed.): 1968, *Thermoluminescence of Geological Materials*, Academic Press.

[2] Hoyt, H. P., Kardos, J. L., Miyajima, M., Seitz, M. G., Sun, S. S., Walker, R. M., and Wittels, M. C.: 1970, *Proc. Apollo 11 Lunar Sci. Conf. Geochim. Cosmochim. Acta Suppl. 1* **3**, 2269.

[3] Dalrymple, G. B. and Doell, R. R.: 1970, *Proc. Apollo 11 Lunar Sci. Conf. Geochim. Cosmochim. Acta Suppl. 1* **3**, 2081.

[4] Geake, J. E., Dollfus, A., Garlick, G. F. J., Lamb, W., Walker, G., Steigmann, G. A., and Titulaer, C.: 1970, *Proc. Apollo 11 Lunar Sci. Conf. Geochim. Cosmochim. Acta Suppl. 1* **3**, 2127.

[5] Nash, D. B. and Greer, R. T.: 1970, *Proc. Apollo 11 Lunar Sci. Conf. Geochim. Cosmochim. Acta. Suppl. 1* **3**, 2341.

[6] Blair, J. M. and Edgington, J. A.: 1970, *Proc. Apollo 11 Lunar Sci. Conf. Geochim. Cosmochim. Acta. Suppl. 1* **3**, 2001.

[7] Thomas, D. G., Hopfield, J. J., and Colbow K.: 1964, *Radiative Recombination in Semiconductors* 67.

[8] Hoyt, H., Kardos, J., Miyajima, M., Walker, R., and Zimmerman, D.: *Proc. Apollo 12 Lunar Sci. Conf.*, MIT Press (in press).

[9] Nash, D. B. (unpublished work).

THE PARTICLE TRACK RECORD OF THE LUNAR SURFACE

G. M. COMSTOCK*

General Electric Research and Development Center, Schenectady, New York, U.S.A.

Abstract. Information about lunar surface history revealed by fossil particle tracks is summarized. Such tracks are the result of damage left in dielectric materials by highly ionizing charged particles including heavy solar and galactic cosmic ray nuclei, heavy nuclei recoiling from cosmic ray induced spallation reactions and induced- and spontaneous-fission fragments. From the distribution of cosmic ray and spallation tracks in the lunar rock, surface residence times of 1 to 30 million yr and rock erosion rates of 1 to 10 Å/yr have been determined. Particle tracks also record surface orientation and depth history of the rocks and contain information about ancient solar activity. The distribution of particle tracks in lunar soil is found to be consistent with a model which includes repeated excavation, layering and burial. With this model one core 12025 + 28 soil layer can be identified as unmixed and weakly irradiated; the others contain soil which has been better and better mixed and more and more irradiated.

1. Introduction

The study of fossil particle tracks in lunar material has yielded much information concerning the history of lunar rocks and soil as well as ancient levels of energetic particle fluxes (Barber *et al.*, 1971a, b; Bhandari *et al.*, 1971; Borg *et al.*, 1971a, b; Comstock *et al.*, 1971; Crozaz *et al.*, 1970, 1971; Fleischer *et al.*, 1970b, c, 1971a, b; Lal *et al.*, 1970; Price *et al.*, 1970b, 1971; Arrhenius *et al.*, 1971).

In this paper we summarize some of the results of particle track studies in lunar materials and develop further the conclusions that can be drawn from track densities in soil grains. We first summarize briefly the possible sources of lunar particle tracks. The production rates of cosmic ray and spallation tracks are then discussed and interpretation in terms of rock exposure ages and erosion rates are summarized. The last part of the paper is devoted to a discussion of the layering, mixing and irradiation history of the soil, based on the track distributions observed in the Apollo 12 double core.

2. Sources of Lunar Particle Tracks

Fossil particle tracks mark the damaged region left in dielectric materials by the passage of highly ionizing charged particles. If the particles are sufficiently ionizing they cause permanent disruptions of the lattice structure (Fleischer *et al.*, 1967). Laboratory evidence (Fleischer *et al.*, 1970d) indicates that in lunar crystals the annealing or erasure of tracks due to temperatures normally experienced on the Moon can be considered negligible. When introduced to the proper chemical etchant the damaged tracks can etch faster than the general surface resulting in a cone-shaped cavity. For a given etching time the shape of the cone depends on the charge and energy of the incident particle; in particular the cone length depends strongly on the rate of primary ionization (Price *et al.*, 1967).

In a given material there is an ionization rate, or registration threshold below which

* Now at Centre de Spectrométrie de Masse du C.N.R.S., 91 – Orsay, France.

Urey and Runcorn (eds.), The Moon, 330–352.

there will not be sufficient damage to result in an etchable track. In lunar materials this threshold allows the registration of nuclei heavier than $Z \sim 20$ (calcium) (Plieninger and Krätschmer, 1971). These nuclei are therefore observable over that portion of their range where the rate of primary ionization exceeds the registration threshold.

In Table I we have listed the possible sources of particles capable of leaving etchable tracks in lunar material. This list also applies to meteorites for which many of the

TABLE I

Possible sources of tracks on the Moon

(1) Cosmic ray tracks – Solar and galactic nuclei with charge $Z > 20$–23.
(2) Spallation tracks – Heavy nuclei recoiling form spallation reactions induced by primary and secondary cosmic ray protons, neutrons and alpha particles.
(3) Fission tracks – Fragments form spontaneous fission of U^{238}, Pu^{244} and possibly super-heavy elements.
(4) Induced fission tracks – Fission induced by cosmic ray protons on Pb, Th or U or by secondary neutrons.

Also

(5) Meson jets induced by very high energy cosmic rays (rare).
(6) Dirac monopoles (hypothetical).

techniques used to study particle tracks were developed (Fleischer *et al.*, 1967; Price *et al.*, 1968a; Pellas *et al.*, 1969; Walker, 1970). Monopoles, if they exist (Fleischer *et al.*, 1970a), would have to leave tracks $\gtrsim \frac{1}{2}$ cm long in order not to be confused with the heaviest cosmic ray tracks (Fleischer *et al.*, 1967; Barber *et al.*, 1971a).

Spontaneous and cosmic-ray-induced fission tracks are important constituents of particle tracks on the Moon, although the only reported cosmic-ray-induced fission events are those found in a sample of lead-bearing filter glass that was part of the Surveyor 3 spacecraft (Fleischer *et al.*, 1971b). The study of spontaneous fission tracks in terrestrial materials has led to the very fruitful field of fission-track dating (reviewed by Fleischer and Hart, 1970d).

On the Moon, however, fission track densities are generally much less than cosmic ray and spallation track densities except in certain uranium-rich minerals (Burnett *et al.*, 1971) and perhaps at great depths (which have been shielded from cosmic rays). Lunar fission tracks can be studied by mapping uranium-rich inclusions (Crozaz *et al.*, 1970; Fleischer *et al.*, 1970c) or by investigating excess track densities along crystal cleavages which often represent grain boundaries where heavy elements concentrate (Bhandari *et al.*, 1971). Bhandari *et al.* report that these excess tracks tend to be longer (13–25 μm) than the abundant iron tracks (10–13 μm) and interpret their results as evidence for the primordial existence of Pu^{244} and possibly of super-heavy elements.

By far the most abundant particle tracks on the Moon are the cosmic ray and spallation recoil tracks (as used here the term *cosmic ray* includes solar as well as extra-solar

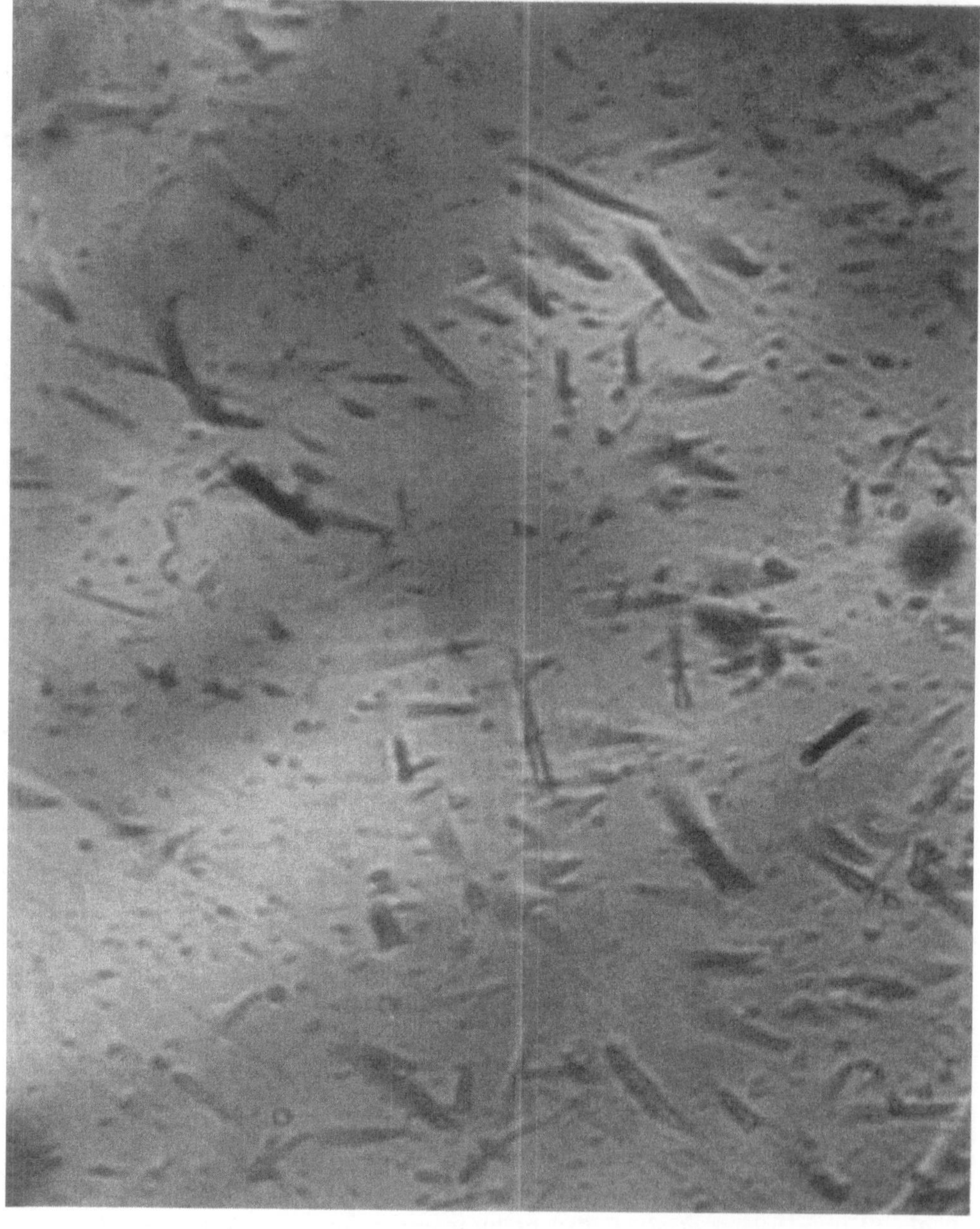

Fig. 1. Optical photograph of cosmic ray tracks (comet-shaped) and spallation tracks (dots) in a lunar crystal.

or galactic particles.) An optical photograph (transmitted light) of a sample of these tracks is shown in Figure 1. The long comet-shaped objects – *cosmic ray tracks*, $\lesssim 11\ \mu m$ – were produced primarily by cosmic ray iron nuclei, the most abundant species which will register tracks. Their density here is only a few million per cm^2 which is relatively low for lunar material. The dots – *spallation tracks* – are very short tracks left by heavy nuclei recoiling from cosmic-ray-induced spallation reactions.

In addition to the dominant iron tracks there occasionally occur longer tracks resulting from heavier cosmic rays. Price *et al.* (1971) have reported such tracks, including two tracks most likely from lead- or uranium-like nuclei. Such long tracks may allow us to study the ancient relative abundances of extremely heavy cosmic ray nuclei, provided we can determine the depth history of the rock (the heavier nuclei are more strongly attenuated with depth).

3. Cosmic Ray and Spallation Tracks

In order to understand what the cosmic ray and spallation track record can tell us about lunar surface history, we first must know the depth dependence of the production rate of these tracks. For the cosmic ray tracks we need the average energy spectrum of the heavy cosmic ray nuclei, predominately the iron group, shown in Figure 2. The spectrum is composed of two parts: a lower-energy solar contribution resulting from solar flare activity and a galactic contribution at higher energies.

The relative nuclear abundances and shape of the galactic spectrum have been measured in satellite and balloon experiments (Comstock *et al.*, 1969; Freier and Waddington, 1968; Price *et al.*, 1970a; Garcia-Munoz and Simpson, 1970) during times of minimum solar activity. We have corrected this shape for the average effect of solar modulation over the 11-yr solar cycle using a model by Wang (1970). The contribution of each species to the production of etchable tracks is proportional, among other things, to the etchable track length $\Delta R(Z)$ and the relative abundance $A(Z)$. The abundance $A(Z)\approx 0$ for species with $Z>26$ and $\Delta R(Z)=0$ for $Z<20$–23 for lunar material (Bhandari *et al.*, 1971b; Plieninger and Krätschmer, 1971). Measured $A(Z)$ and plausible values of $\Delta R(Z)$ are listed in Table II.

The other factors which contribute to the track production rate (see Equation 2) have much weaker charge dependences over the dominant charge interval. Hence we may define an equivalent iron-like flux given by

$$\left(\frac{dN}{dT}\right)_E = \left(\frac{dN}{dT}\right)_{Fe} \sum_{\text{all } Z} \left(\frac{\Delta R(Z)}{\Delta R(\text{Fe})}\right)\left(\frac{A(Z)}{A(\text{Fe})}\right) \tag{1a}$$

where (dN/dT) is the differential particle flux. The values listed in Table II yield

$$\left(\frac{dN}{dT}\right)_E = 1.63\left(\frac{dN}{dT}\right)_{Fe} = 0.52\left(\frac{dN}{dT}\right)_{VH} \tag{1b}$$

Equation (1b) gives the equivalent flux, corrected for average solar modulation,

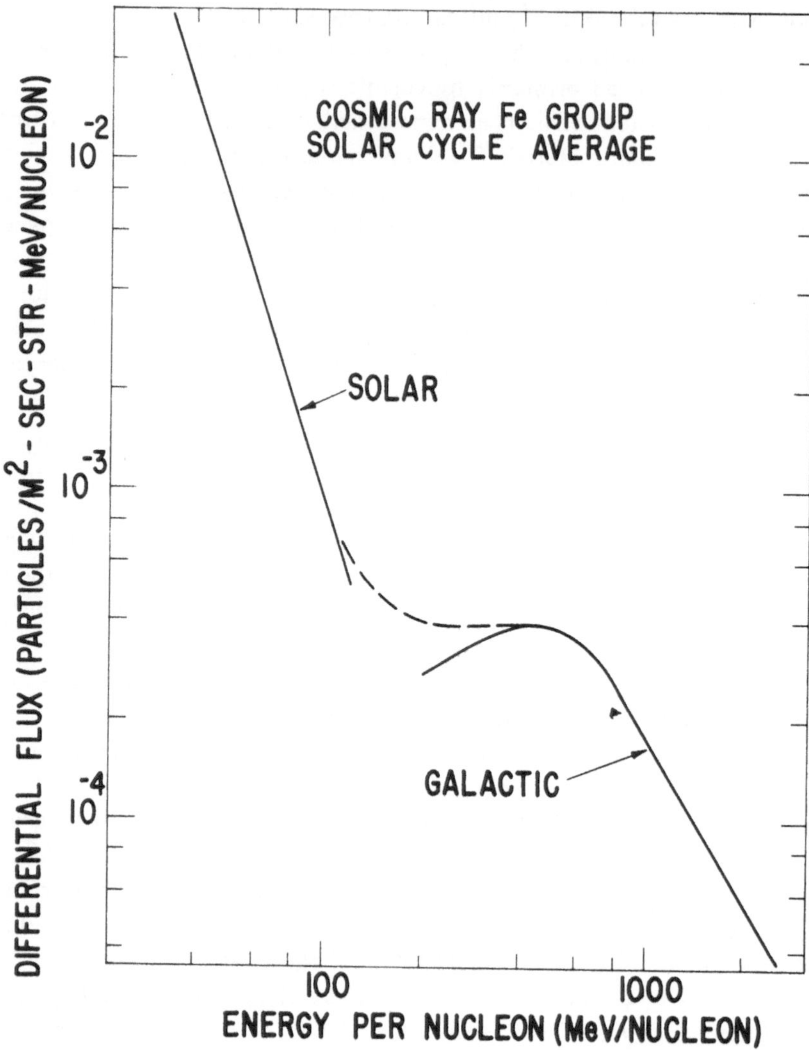

Fig. 2. Average differential energy spectrum of equivalent iron-like flux capable of leaving etchable tracks in lunar material (see text and Equation (1b)).

which is plotted for the galactic contribution in Figure 2. This flux is about a factor of two less than that given by Fleischer *et al.* (1967) and Crozaz *et al.* (1970), which is the total flux of the larger VH charge group ($Z \geqslant 20$) treated as entirely iron, so that

$$\left(\frac{\mathrm{d}N}{\mathrm{d}T}\right)_E = \left(\frac{\mathrm{d}N}{\mathrm{d}T}\right)_{\mathrm{VH}}. \tag{1c}$$

Note that the lower flux of Equation (1b) will imply correspondingly greater galactic cosmic ray exposure times than values based on the higher flux of Equation (1c). Plieninger and Krätschmer (1971) report that the registration threshold may in fact

TABLE II

Species	Z	$A(Z)$[a]	$\Delta R(Z)$[b]
–	< 23	–	0
vanadium	23	0.20	2.5 μm
chromium	24	0.55	5.5
manganese	25	0.44	8.5
iron	26	1.00	11.5
–	> 26	≈ 0	–

[a] Average of Price *et al.* (1968b, 1970a) and Garcia-Munoz and Simpson (1970).
[b] Bhandari *et al.* (1971b) (but see Plieninger and Kratschmer, 1971, and Rajan and Price, 1971, who report larger values of $\Delta R(Z)$).

be as low as $Z=20$ (Ca) in which case the equivalent flux $(\mathrm{d}N/\mathrm{d}T)_E$ will be greater than that given by Equation (1b). However, the contribution from the species Ca-V will be small because of their lower values of $A(Z)$ and $\Delta R(Z)$ (see note added in proof).

For the solar contribution we rely on the energy spectrum derived from cosmic ray tracks observed in a piece of Surveyor 3 filter glass (Garber *et al.*, 1971a; Crozaz *et al.*, 1971; Fleischer *et al.*, 1971b). The registration thershold for this glass is somewhat lower than for lunar minerals, however, the relative abundances of solar flare species P-Mn is so low that the charge interval effectively recorded by the Surveyor 3 glass is essentially the same as that recorded by lunar material (Fleischer *et al.*, 1971b). The major uncertainty in applying the Surveyor 3 flux to lunar material is that the Surveyor 3 spacecraft was on the Moon for only a fraction of the present solar cycle, about 2.6 yr, during the period of maximum solar activity. It is not certain how the Surveyor 3 flux relates to the average solar contribution over many solar cycles. If the present cycle is typical then the average solar flux should be about half of the observed Surveyor 3 flux, as plotted in Figure 2. Crozaz *et al.* (1971) use a higher flux.

From this composite energy spectrum we calculate the production rate of cosmic ray tracks in lunar material, given by

$$\dot{\varrho}(\mathbf{r}) = \alpha \int\limits_{\Omega} \left(\frac{\mathrm{d}N}{\mathrm{d}T}\right)_E \left(\frac{\mathrm{d}T}{\mathrm{d}R}\right)_{\mathrm{Fe}} \Delta R(\mathrm{Fe})\, e^{-\chi(\theta,\Phi)/\lambda}\, p(\theta,\Phi)\, \mathrm{d}\Omega \tag{2}$$

where α is the etching efficiency, $(\mathrm{d}N/\mathrm{d}T)_E$ is the equivalent iron-like flux defined earlier, evaluated at T which is the incident kinetic energy per nucleon corresponding to range $\chi(\theta, \Phi)$ in lunar rock or soil, $(\mathrm{d}T/\mathrm{d}R)_{\mathrm{Fe}}$ is the rate of energy loss of iron in lunar minerals evaluated at T, λ is the mean particle loss path length due to nuclear interactions, $p(\theta, \Phi)$ is a projection function which takes into account the orientation of the etched surface with respect to rock or soil surface (Fleischer *et al.*, 1967), and $\chi(\theta, \Phi)$ is the particle path length from the sample point to the surface in the direction of the solid angle increment $d\Omega$. Since the original orientation of the etched surface is often unknown, we compute the average production rate for a set of randomly oriented crystals for which case $p(\theta, \Phi) = \frac{1}{2}$.

The cosmic ray track production in lunar soil derived in this way is shown in Figure 3. It is clear from this curve that we expect a steep gradient in track density within a millimeter of the exposed surface. This steep production rate extends down to within ~1 μm of the surface as discussed at this conference by Borg *et al.* (1971b).

The second curve shown in Figure 3 is the spallation track production rate determined from the high energy cosmic ray proton flux and from laboratory production experiments (Kohman *et al.*, 1967), including calibration of the response of individual lunar grains (Fleischer *et al.*, 1971a – experimentally, the magnitude of the produc-

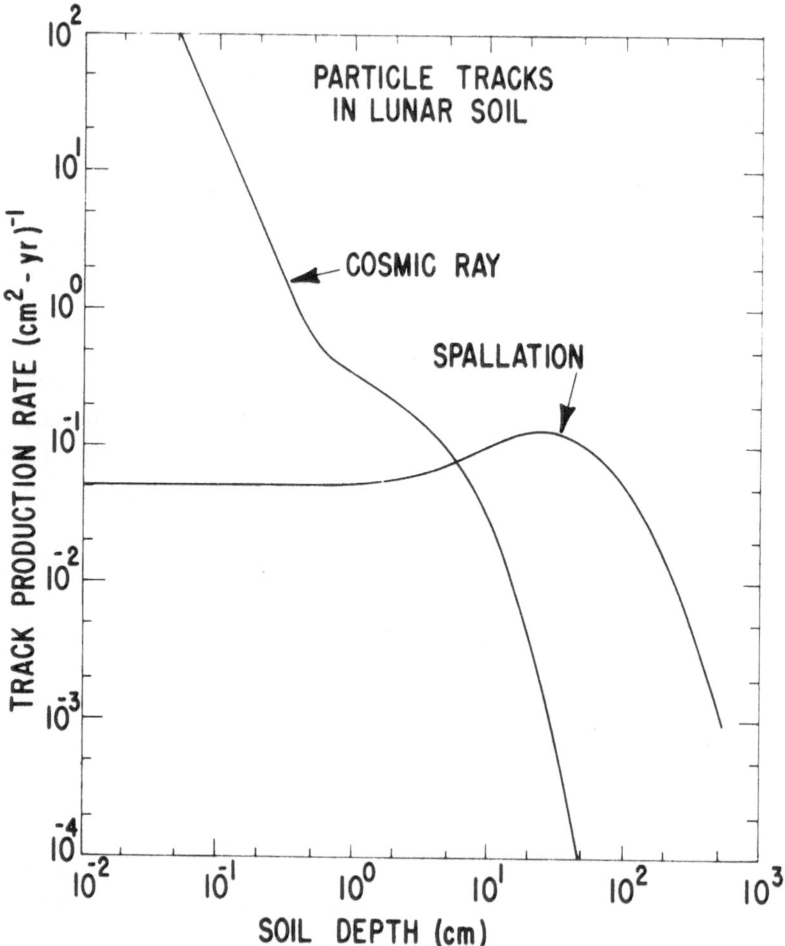

Fig. 3. Calculated rate of production of cosmic ray tracks (from Equation (2) with $\Delta R(\mathrm{Fe}) = 11.5$ μm) and empirical production rate of spallation tracks, shown for semi-infinite lunar soil (density 1.6 gm cm^{-3}). These curves apply also to semi-infinite rock if the depth scale is decreased by a factor of about 2.

tion rate in individual grains was found to vary by ±70% from the average value indicated in Figure 3). Because of the production of various secondary particles capable of inducing spallation reactions, the production of spallation recoil nuclei capable of leaving etchable tracks rises to a maximum at about 25 cm of soil.

We emphasize that these two production curves have very different shapes so that each yields different information about the history of a sample. Spallation tracks record the time spent in about the upper two meters of soil. Most of the cosmic ray tracks, on the other hand, were formed while the rocks or soil grains were situated on or within a few centimeters of the surface.

4. Lunar Rocks and Erosion Rates

We define for the rocks two spallation track residence times: the 'surface age' determined by assuming that irradiation occurred entirely on the surface, and a 'minimum age' which assumes that the irradiation occurred at the maximum rate. Typical track ages are given in Table III along with radiometric spallation ages determined by various workers.

TABLE III

Fossil track ages (10^6 yr) of lunar rocks

Sample number	Cosmic ray surface residence times top/bottom		Spallation recoil tracks [c]		Radiometric spallation ages [d]
	(High age) [a]	(Low age) [b]	Surface age	Minimum age	
10017	11.2 total	5.7 total	420	170	200–640
10044	8.4 total	4.3 total	270	110	56–100
10049	57 total	29 total	21	8.5	22.5–25
12002	47/0	24/0	55	20	50–145
12017	1.4/2.0	0.7/1.0	105	40	–
12021	25/25	13/13	740	300	300
12065	27/0	14/0	170	70	160–200

[a] From Equation (1b) and $\Delta R(\mathrm{Fe}) = 11.5\ \mu\mathrm{m}$.
[b] Fleischer *et al.* (1970c, 1971a) using Equation (1c). Note that the Low age is also consistent with Equation (1a) and $\Delta R(\mathrm{Fe}) \approx 20\ \mu\mathrm{m}$ (Rajan and Price), so that the High and Low ages represent the limits of uncertainty in $\Delta R(\mathrm{Fe})$. (See note added in proof.)
[c] Fleischer *et al.* (1971a).
[d] Spread of values from various authors, see Fleischer *et al.* (1971a) for references.

A typical rock studied has a radius of a few centimeters, so that the cosmic ray track density at its center is due primarily to galactic cosmic rays. If the rock erosion rate is not too great these central tracks may be used to find the 'cosmic ray surface residence time' T_s by assuming

$$T_s = \varrho(R)/\dot{\varrho}(R) \tag{3}$$

where $\varrho(R)$ is the observed track density at the center of the rock and $\dot{\varrho}(R)$ is calculated taking into account the shape and orientation of the rock. In Table III we have listed some values if T_s based on both Equation (1b) (High age) and on Equation (1c) (Low age) (see note added in proof).

The spallation and radiometric ages in Table III show the same trend from rock to rock. Most of these ages are greater than the cosmic ray surface residence times, indicating that the rocks have resided for some time below the surface but within the top 2 m. For rocks 10049 and 12002 the spallation surface ages and radiometric ages are comparable to the cosmic ray surface residence times indicating a simpler history. Rocks 12002 and 12065 have been irradiated on one side only.

Close to the surface of the rock, for a side which was facing up at some time, one finds a cosmic ray track gradient due to the solar flare particles (Figure 4). A similar gradient on the opposite side indicates that the rock has had more than one orientation on the lunar surface. The observed gradient is generally less than the gradient predicted

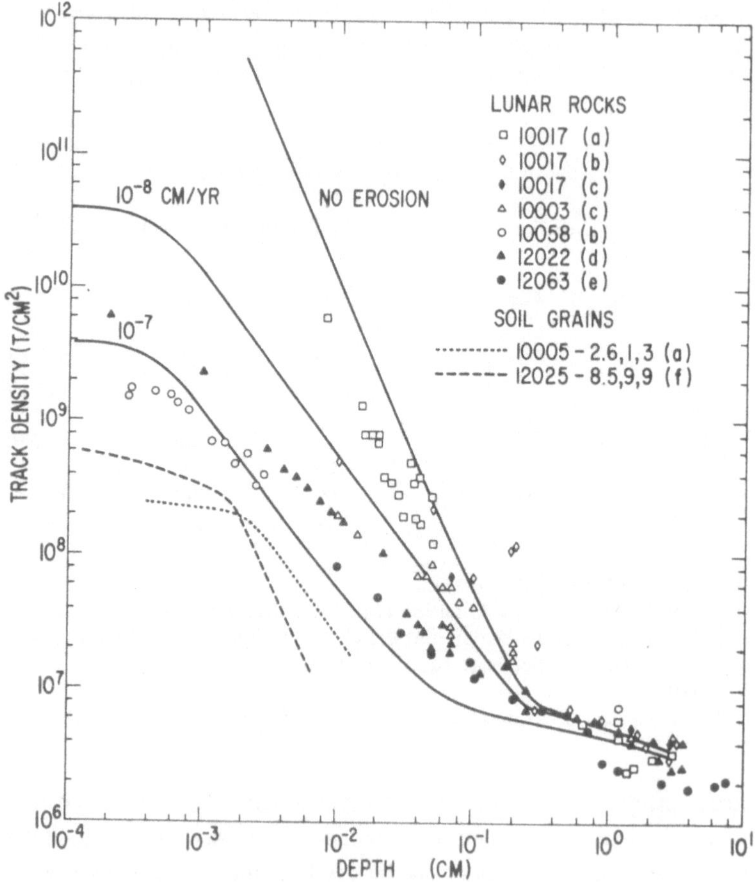

Fig. 4. Cosmic ray track gradients observed in several lunar rocks and two soil grains. The solid curves are calculated for a typical rock undergoing no erosion, and for two finite erosion rates. (a) Fleischer *et al.* (1970c), (b) Crozaz *et al.* (1970), (c) Price and O'Sullivan (1970b), (d) Barber *et al.* (1971a), (e) Crozaz *et al.* (1971), (f) Fleischer *et al.* (1971a) and Figure 5.

by Equation (2) which has an exponent of about -2.4. This is generally regarded as an effect of rock erosion (Crozaz *et al.*, 1970, 1971; Fleischer *et al.*, 1971b; Barber *et al.*, 1970b, 1971a, b.)

In Figure 4 the solid curves refer to a vertical cross section through a spherical rock of final radius 3 cm which has resided in one orientation on the surface for $T_s = 15$ m.y. The no-erosion curve is derived from Equation (1b), (2) and

$$\varrho(d) = \dot{\varrho}(d)\, T_s\,. \tag{4}$$

The curves for finite erosion rates $e = 10^{-8}$ cm yr^{-1} and 10^{-7} cm yr^{-1} are derived from Equation (1b), (2) with a time-dependent radius $r = R + e(T_s - t)$ and

$$\varrho(d, e) = \int_0^{T_s} \dot{\varrho}(d, et)\, \mathrm{d}t = \frac{1}{e} \int_R^{R+eT_s} \dot{\varrho}(d, r)\, \mathrm{d}r\,, \tag{5}$$

where d is the final depth below the rock surface, e is the erosion rate and R is the final radius.

These curves are not meant to fit any particular rock. For ϱ near the center of a given rock we must take into account its shape and size, the time it has spent in each orientation on the lunar surface and the possible contribution from irradiation before direct surface exposure. For example, rock 12063 has spent less time on the surface than the others. Below about 1 mm, however, the track density is essentially independent of the size and shape of the rock. Still closer to the surface the track density in an eroding rock becomes independent of the surface residence time as well, as the production rate comes into equilibrium with the eroding surface. More precisely, as the surface residence time T_s increases the track density at each depth approaches a maximum or equilibrium value given by

$$\varrho_E(d, e) = \frac{1}{e} \int_R^{r} \dot{\varrho}(d, r)\, \mathrm{d}r\,. \tag{6}$$

For surface residence times on the order of 10 m.y. the track densities derived from Equation (5) coincide with the equilibrium values at depths less than a few hundred microns for erosion rates $e \leqslant 10^{-7}$ cm yr^{-1}. Greater depths (to $d \sim e\, T_s$) would also be in equilibrium if it were not for the effects of the galactic contribution and the finite size of the rock. Below a few hundred microns, therefore, the magnitude of the track density depends only on the erosion rate and the average flux of solar flare particles (Equation (6)). The slope of the track gradient in this near-surface region depends on the exponent of the solar flare energy spectrum but may be modified by irregular erosion or chipping, unevenness of the surface near the sample point and possible loss of surface material during transport from the Moon. For example if rock 10058 has lost $\sim 10\,\mu$ (one layer of rather small grains) during transport then it should be plotted coincident with rock 12022 in Figure 4.

The data shown in Figure 4 (see references in figure caption) indicate erosion rates varying from $e < 10^{-8}$ cm yr^{-1} for 10017 to 3×10^{-8} cm yr^{-1} for 12022 and $\sim 5 \times$ $\times 10^{-8}$ cm yr^{-1} for 12063. Crozaz *et al.* (1971) argue that the average solar particle flux may be a factor of 2 greater than we have used in deriving the curves in Figure 4 and hence obtain proportionally higher erosion rates. The low value implied for 10017 may indicate that there have been wide fluctuations in the micrometeoroid flux at the Moon's surface. 10017 may have spent most of its surface exposure time during a period of meager micrometeoroid activity, then was buried and brought to the surface again less than 0.3 m.y. ago (corresponding to $< 100\ \mu$ of erosion at the rate of 3×10^{-8} cm yr^{-1}).

Unique opportunities to study solar flare activity in the past without the uncertainties of rock erosion are provided by rocks which have been recently covered with a glass coating; the best example of this is rock 12017 (Fleischer *et al.*, 1971a). Although cosmic ray tracks in the center of the rock indicate that 12017 has been on the surface for at least 1.7 m.y., cosmic ray tracks found in grains imbedded within the glaze itself indicate that the glaze has been exposed on the rock for only 9000 yr. These tracks show the decrease with depth arising from solar flare particles. The glass coating is much less retentive of tracks than the mineral crystals; tracks in the glaze should fade away after about 500 yr on the lunar surface due to thermal annealing (Fleischer *et al.*, 1971a). The track density in the glaze also decreases with depth but has a lower magnitude than the track density in imbedded crystals, consistent with the ~ 500 yr retention time and the present flux of solar flare particles.

5. Lunar Soil

The record of particle tracks in the individual soil grains indicates a history even more complex than that of the rocks. (Hereinafter, the phrase 'track density' will refer to cosmic ray tracks, not spallation tracks, unless otherwise noted.) For the purposes of discussion we may divide the soil grain samples into three groups according to track density: (1) $\varrho \gtrsim 5 \times 10^8$ cm^{-2}, (2) ϱ with a strong gradient, and (3) $\varrho \lesssim 5 \times 10^8$ cm^{-2}.

(1) Grains with high track densities include very many micron-sized grains and micron-deep coatings on larger grains (Borg *et al.*, 1971a, b) containing 10^{10}–10^{12} cm^{-2}. Very high densities have also been reported throughout some larger grains ($\sim 100\ \mu$ radius) (Crozaz *et al.*, 1971; Barber *et al.*, 1971a, b). These track densities must be the result of direct exposure to space for $T_s \sim 10^4$–10^5 yr for the micron-sized grains and considerably longer for the larger grains.

(2) The second class of samples includes grains of ~ 100 micron-radius which contain a steep track density gradient extending inward from one or two edges. These have track densities typically approaching $\varrho \sim 10^9$ cm^{-2} in the outer 10 μ dropping to $\varrho \sim 2 \times 10^7$ cm^{-2} in the center or low-density edge. A contour map of one such gradient is shown in Figure 5. Contours such as this may be strongly affected by the shape of the grain but in general they are what one expects if the grain had been exposed on the lunar surface in a particular orientation and covered by no more than several microns of dust.

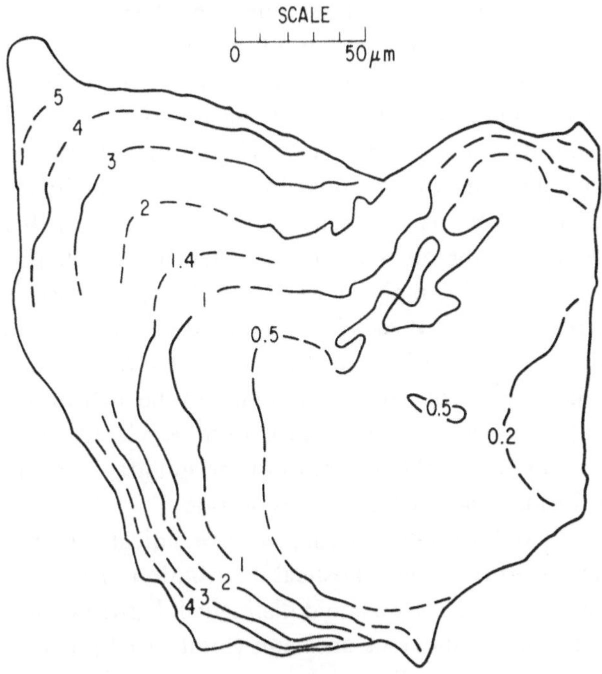

Fig. 5. Contour map of cosmic ray track density in a pyroxene soil grain with strong track gradient. The grain was found 8.5 cm below the lunar surface.

Track density profiles measured at 7500× along the steepest track gradients of two such grains are shown in Figure 4. There are two possible conditions under which tracks may have been accumulated to produce such gradients: (1) the grains were tossed randomly to the very surface as part of an ejecta blanket and later buried, or (b) the tracks accumulated while the grains were still part of an eroding rock, eventually being exposed, chipped out and buried in the soil.

Consider first case (a). The dominant erosion processes are discrete events, well separated in time, which will tend to destroy or bury a soil grain. Hence while the grain is on the surface (case a) the acquired track gradient should follow a no-erosion curve given by Equation (4) for some surface residence time T_s. In this case we interpret the sharp breaks in these curves at $\approx 20\ \mu$ as being due to an (uncorrected-for) shielding of several microns of dust, which is reasonable considering the cohesiveness of the soil. For the soil grains plotted in Figure 4 we infer $T_s \sim 10^4$ yr., which is consistent with the rate of disturbance of a surface grain by micrometeoroid bombardment (Shoemaker *et al.*, 1970).

If case (a) is the predominant condition for soil-grain gradient production, then we should expect a wide range of track densities from grain to grain corresponding to an exponential distribution in surface residence times. This in turn would yield information on the rate of disturbance of the surface.

In case (b) the track gradient in the grain represents the gradient which existed in the upper $\sim 100\ \mu$ of the parent rock at the time when the grain was chipped out. (The grain could not have come from the interior of the rock since in that case the observed gradient would be too steep.) Hence the track gradients for the soil grains in Figure 4 should be fitted by equilibrium curves satisfying Equation (6) for whatever erosion rates were experienced by the parent rocks. The implied erosion rate is $\approx 3 \times 10^{-7}$ cm yr^{-1} for both grains shown, or about an order of magnitude greater than the average rate observed in the rocks presently on the surface. This higher rate presumably would be related to the greater rate of meteoroid bombardment early in the Moon's history which would imply an early epoch for the formation of these soil grains from their parent rocks. If case (b) is the predominant condition for soil-grain gradient production, then we should expect a relatively narrow range of track densities from grain to grain ($\sim$ factor of 10) corresponding to the distribution of erosion rates experienced during the period of soil formation. Rock 10017 indicates that the erosion rate may vary considerably but most of the soil gradients originating in rocks will be produced during times of higher erosion rates.

Both conditions (a) and (b) should occur; we have indicated how the dominate condition can be determined from the statistical distribution of track densities in grains with track gradients. In principle we could distinguish between conditions (a) and (b) by appealing to the slopes of the track gradients (Figure 4). In practice these slopes are strongly effected by the shape of the grain, by uneven shielding during irradiation and by loss of material by soil abrasion. Careful mapping of many grains, however, may yield a meaningful average slope. In the above discussion we have assumed that the average level of the solar flare particle flux is the same now as it was when the track gradients were recorded. There is no reason to suppose that this is not true. However long-range variation in flare activity remains a possible, though less likely alternative to variation in erosion rate.

(3) The third broad class of soil grains consists of those which have a uniform track density $\varrho < 5 \times 10^8$ cm^{-2}. The lack of a track gradient or very high track density indicates that these grains have never been within a few hundred microns of the surface. Their cosmic ray track record is the result of exposure within the top ~ 10 cm of soil (Figure 3) and hence is determined by the history of soil activity.

In the discussion of soil activity which follows we combine the third class of grains with the track densities measured in the *center* of soil grains having track gradients (the latter being about 20% of the combined population). These two groups have similar frequency distributions for both *cosmic ray* and *spallation* track densities (Comstock *et al.*, 1971) and hence have a similar sub-surface history.

In Figure 6 we have indicated the distribution of cosmic ray track densities in individual soil grains with $\varrho < 5 \times 10^8$ cm^{-2} for several depths in the Apollo 12 double core (Comstock *et al.*, 1971). Arrhenius *et al.* (1971) have measured similar distributions

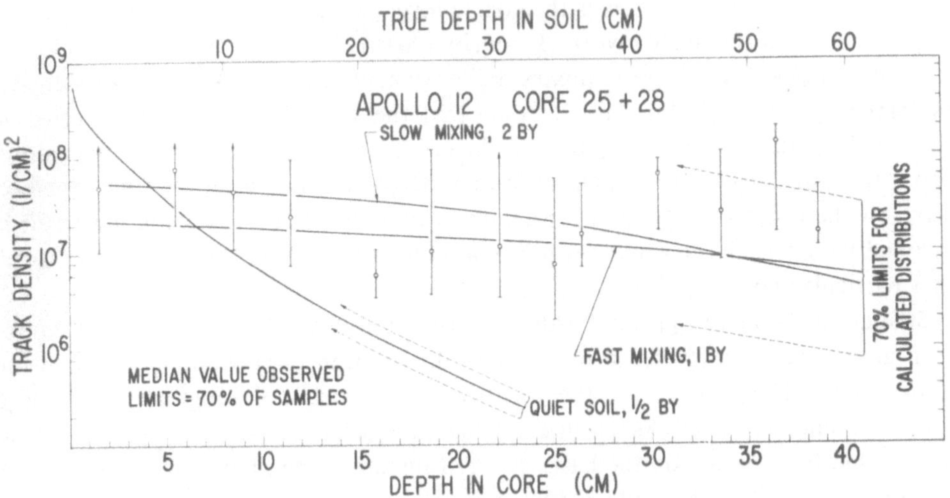

Fig. 6. Cosmic ray track distribution in double core 12025 + 28. The data points indicate the observed spread in track density from grain to grain. The curves have been calculated for some models of soil history discussed in the text (see Comstock *et al.*, 1971, for a complete discussion of the data and calculations.) The true soil depths are given by Carrier *et al.* (1971).

at additional depths. The data points give the median track density observed at each depth and the limit bars show the spread of the 70% of the samples closest to the median. The statistical error in the measurement of these track densities is typically $\lesssim 10\%$, so the wide spread in track densities is due to a complex irradiation history for each depth. The 'Quiet soil' curve (Figure 6) is calculated from Equations (2) and (4) for an irradiation time of 500 m.y. Clearly this curve is not consistent with the observations – considerable soil movement has taken place.

As one approximation to this movement we might assume that the soil has suffered continual depth-dependent mixing by meteoroid impact (Shoemaker *et al.*, 1970). We have worked out computer models based on this assumption (Comstock *et al.*, 1971), and the results of two such models are shown in Figure 6 by the curves marked 'slow mixing' and 'fast mixing'. Fast mixing refers to soil mixed to depths d_c in time periods τ (m.y.) $= d_c$ (cm); slow mixing is ten times slower. The mixing time derived by Shoemaker *et al.* (1970) for soil is $\tau \approx 1.6\, d_c$ for $d_c \sim 10$–100 cm. The solid curves in Figure 6 give the calculated median values with the 70% spread of hypothetical samples and irradiation times as indicated. It is seen that the magnitude and overall distribution of observed track densities is well reproduced by these models.

On the other hand the track density distributions do appear to contain some structure, which is even more evident with the greater statistics given by Arrhenius *et al.* (1971). In addition the Apollo 12 double core has some well defined visual layers, perhaps as many as 13, through the 60 cm depth of the sampled soil (LSPET, 1970; Sellers *et al.*, 1971). At least one of these layers correlates strikingly with the structure in the track distributions. This is the coarse layer at a true soil depth of ≈16–20 cm which contains a relatively low track density (Arrhenius *et al.*, 1971).

Another model which approximates the movement of the soil has been investigated by Arrhenius *et al.* (1971). They assume that the soil has been steadily buried by deposited layers, with no mixing between layers and no irradiation prior to deposition. From the track densities in each layer they derive surface residence times for each layer on the order of 0–60 m.y. Arrhenius *et al.* find that some mixing must still be invoked within each layer in order to explain the distribution of track densities within that particular layer.

Spallation tracks can give us further information on the history of the soil grains. In Figure 7 we have plotted a preliminary distribution of spallation track densities in the Apollo 12 double core. Here again we see a spread, although not as broad as for the cosmic ray tracks. As pointed out by Comstock *et al.* (1971) the pattern shown here would be expected on the basis of a soil mixing model, but a burial model with no previous irradiation would predict a spallation track density that continued to increase with depth by a factor of ~10 in the top 60 cm., due to the high spallation production rate (Figure 3.) The spallation tracks in Figure 7 force us to conclude that

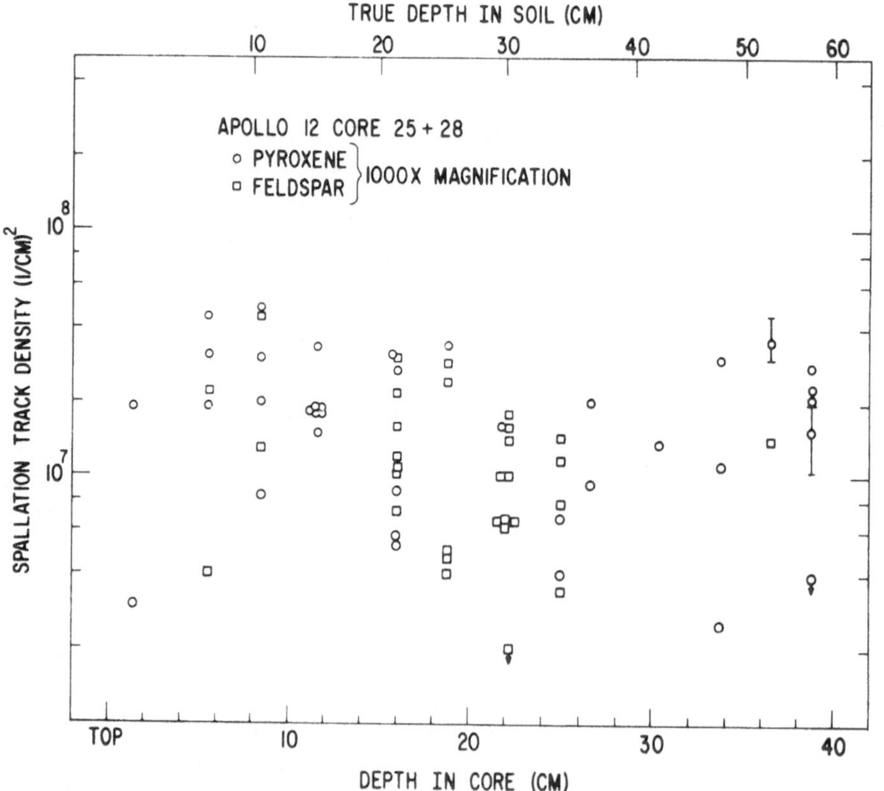

Fig. 7. Preliminary spallation track distribution in double core 12025 + 28 (see Comstock *et al.*, 1971, for acomplete discussion of the data.)

most of the grains have been pre irradiated in the top 2 m of soil, or 1 m of rock, for an integrated time of ~100 m.y. before excavation, deposition and burial at the core site. This implies a more complex history than simple burial after a single excavation from virgin (trackless) material, for most of the layers.

We know that the Moon is continually bombarded by meteoroids, so that some depth-dependent mixing must be taking place (Öpik, 1969). The nature of these events is such that material is removed from an excavation and deposited in a thin layer around the site. At a given point on the Moon this layering will continue to build up the soil level until an impact occurs near that point, re-excavating some of the deposited material. We know further that there is a correlation between deposition distance from the excavation site and original stratigraphy at the excavation site (Schmitt and Sutton, 1971). The layer deposited at a given point comes from a rather restricted depth interval and thus represents a mixing of only a few of the original layers. Hence each layer may have a distinctly different history. At a given site some layers will have been excavated from deep, previously unirradiated material, while others will be composed of a more mature, better mixed population of soil.

The question which we can ask is: what can we say about the mixing history of each observed layer? Our approach to this question appeals again to the mixing model calculations. These calculations (Comstock *et al.*, 1971) describe material which has been exposed for some time at various depths in the upper 25 cm of soil, mixed occasionally with nearly trackless material exposed only at greater depths. As a first approximation this should be similar to the mixing of layers with different exposure histories. Comstock *et al.* (1971) assumed that the entire double core sample represents one homogeneously mixed population. We now assume that the previous history of

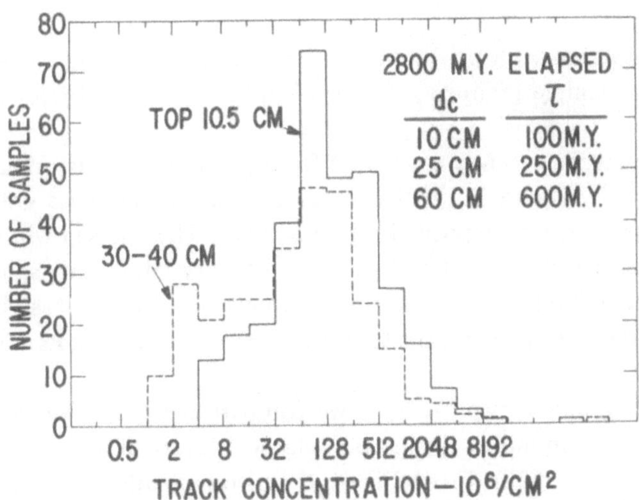

Fig. 8. Calculated distributions of cosmic ray track density after 2.8 b.y. of soil mixing and irradiation. Hypothetical samples were periodically mixed down to three characteristic depths *dc* with time periods τ. These distributions are characteristic of well-mixed soil.

a given layer (i.e., before deposition and burial at its present site) is reproduced by a mixing model for some mixing age appropriate to that particular layer.

A layer which is composed of what we shall refer to as 'well mixed' material should have a track density distribution similar to that shown in Figure 8. This distribution is the result of Monte Carlo calculations for hypothetical soil grains subjected to depth-dependent mixing and irradiation for 2800 m.y. It is seen that most of the samples are included in an order-of-magnitude spread around $\varrho \approx 10^8$ cm^{-2} and many samples occur with $\varrho \gg 10^8$ cm^{-2}. A track density distribution calculated for a soil population which has been mixed for only 1200 m.y. is shown in Figure 9. Here we have

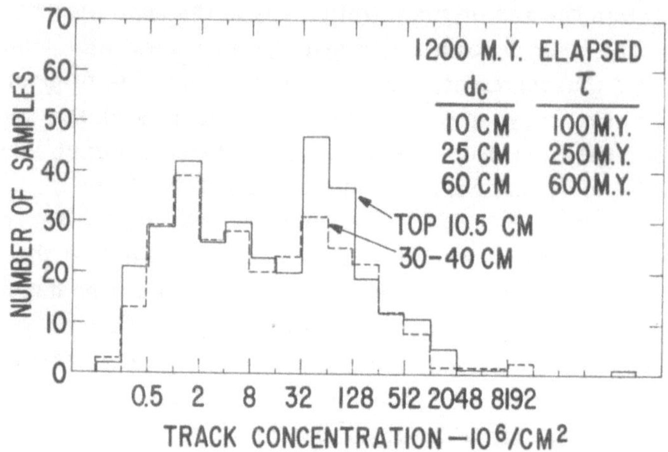

Fig. 9. Calculated distributions of cosmic ray track density after 1.2 b.y. of soil mixing and irradiation. These distributions are characteristic of partially-mixed soil.

a broader distribution with $\varrho \sim 10^6$–10^8 cm^{-2} but still including some samples with $\varrho \gg 10^8$ cm^{-2}. A sample grain population of this kind we shall refer to as 'partially mixed'.

Soil which has been recently excavated from previously undisturbed material will have $\varrho \gg 10^6$ cm^{-2} (Figure 3). Whatever cosmic ray tracks these grains have were formed after the layer was deposited on the surface. These layers will have predominantly low track densities $\varrho \sim 10^6$–10^7 cm^{-2} with very few, if any samples with $\varrho \gtrsim 10^8$ cm^{-2}. We will refer to such layers as 'unmixed' although it is understood that some very shallow mixing within the layer must take place during its surface residence time.

Using Figures 8 and 9 as prototypes we compare these definitions with the track density distributions found in the various physical layers observed in the Apollo 12 double core (LSPET, 1970). From Figure 6 and the detailed distributions given by Arrhenius *et al.* (1971) we identify a mixing history with each physical layer in Table IV. The low track density reported for layer VI identifies it as material from an unmixed population (the samples at ≈ 20 cm in Figure 6 may be associated with this).

TABLE IV

Layer designation [a]	Maximum soil depth [b]	Mixing history
X	1.6 cm	well mixed
IX	3.1	partially mixed
VIII	11.8	well mixed
VII	16.1	well mixed
VI	19.6	unmixed
V	24.0	well mixed
IV	29.7	partially mixed
III-4	35.2	well mixed
III-3	38.8	well mixed
III-2	43.2	partially mixed
III-1	52.2	well mixed
II	58.8	well mixed
I	$\geqslant 60.5$	partially mixed

[a] LSPET (1970); Bhandari *et al.* (1971)a.
[b] Carrier *et al.* (1971); LSPET (1970).

For layers II, III-1, VII, VIII and X the track density distributions given by Arrhenius *et al* (1971) show sharp peaks in the interval $\varrho \sim 3 \times 10^7$ to 10^8 cm^{-2} corresponding to the derived distribution for well-mixed soil (Figure 8). The track distributions for layers I and IV are much broader, similar to the distribution in Figure 9 calculated for partially-mixed soil. The other layers, III-2, III-3, III-4, V and IX, appear to be intermediate between Figures 8 and 9.

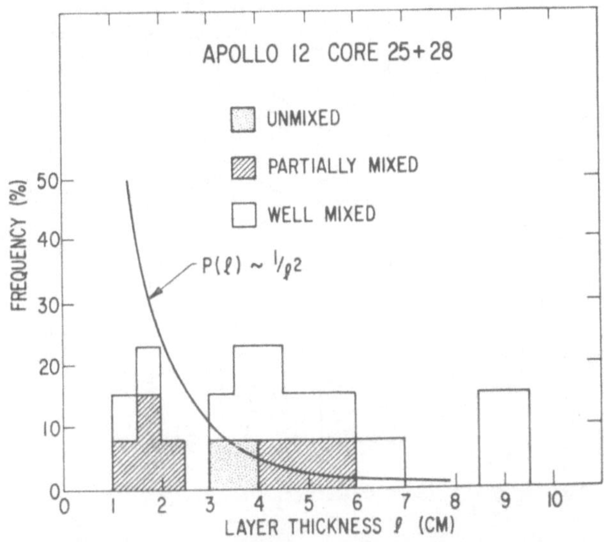

Fig. 10. Frequency distribution of true soil thicknesses of the visual layers in double core 12025 + 28. Assigned mixing histories are indicated (Table IV). The curve $P(l)$ represents an expected frequency distribution.

When compared with physical features (Sellers *et al.*, 1971) we find that those layers identified as well mixed tend to be medium grey, with no internal structure, generally finer-grained with unclear boundaries. This is consistent with a longer history of physical activity (excavation, deposition, abrasion) and a greater homogeneity. The layers identified as partially mixed generally have discernible boundaries and some differences in physical properties. The unmixed layer VI is easily distinguishable from the others and is coarser-grained, consistent with recent excavation from 'virgin' material.

In general, the better-mixed soil has been subjected to more physical activity and more admixture from different localities and depths, tending to be more homogeneous with more of the grains well-irradiated by cosmic rays.

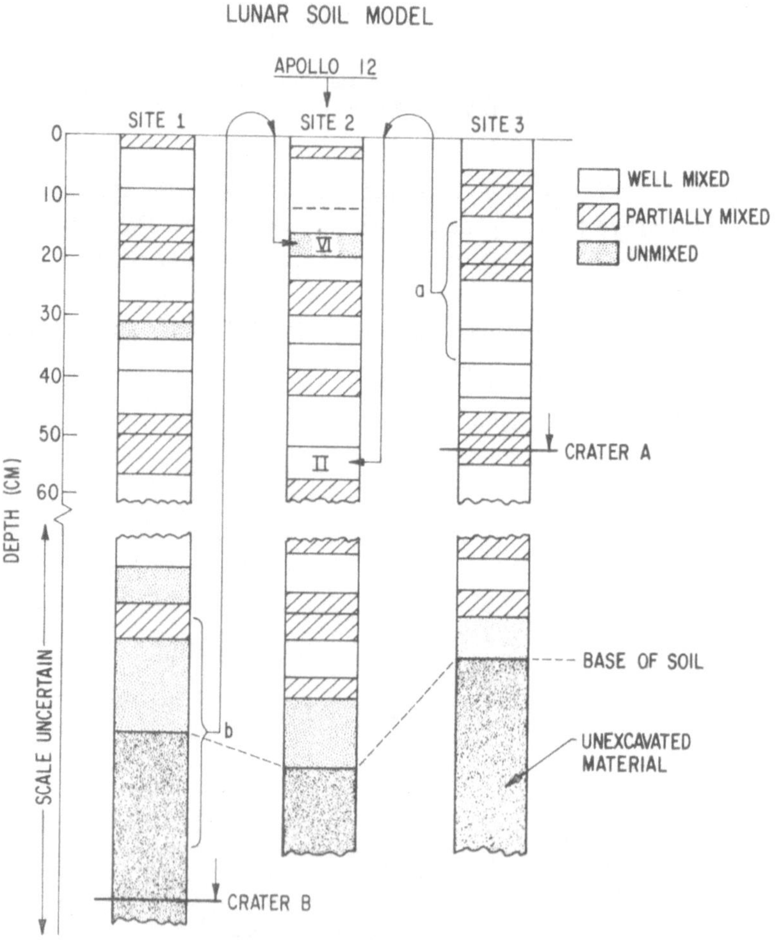

Fig. 11. Hypothetical lunar soil illustrating the formation of Apollo 12 double core layers II and VI according to a model described in the text. Arrangement of the hypothetical parent layers at sites 1 and 3 is not unique.

The identification of mixing histories allows us to make another inference about the physical layers. In Figure 10 we have plotted the distribution of approximate thicknesses of the visual layers, corrected for compaction (Carrier *et al.*, 1971). In addition we have drawn a curve $P(l) \sim 1/l^2$ representing the expected distribution of thicknesses. This curve was derived by assuming that excavated material is deposited outside the crater in a smoothly thinning blanket (Öpik, 1969) within 1–2 diam of a crater site and taking into account the size frequency distribution of cratering events (Shoemaker *et al.*, 1970). Figure 10 suggests an overabundance of thick layers. However, the thickest layers have been identified as well mixed; hence they may have been built up from several smaller layers of well-mixed material which are not distinguishable because they are too homogeneous.

Our picture of lunar soil, based on the distribution of cosmic ray particle tracks, may be summarized by reference to Figure 11 which shows the hypothetical formation of two of the Apollo 12 double core layers. Figure 11 shows three hypothetical sites, the upper part of site 2 is drawn to represent the Apollo 12 double core 25+28. The three sites are shown at three different times, site 3 at an early time T_1, site 1 at a later time T_2 and site 2 at the present. Each site has been built up by repeated layering (presumably of ejecta blankets). Specifically, we suppose that at time T_1 an impact occurred at site 3 excavating material down to some depth such as that marked 'Crater A'. Excavated material from some depth interval (marked a in Figure 11) was deposited on the surface at site 2 and become what we know as layer II. Since this material was already fairly well mixed at site 3 it forms a well-mixed layer at site 2. If the original layers in depth interval *a* at site 3 had been predominately partially mixed, or included an unmixed layer, then the material deposited at site 2 would have been only partially mixed. Between times T_1 and T_2 other layers were deposited at site 2 on top of layer II. Note that the final layering sequence was not necessarily deposited contiguously, that is, several layers may have been deposited on top of layer II, then some of these re-excavated by shallow cratering events and replaced by other layers.

At the later time T_2 we suppose that an impact event occurred at site 1, excavating material to a great depth (Crater B in Figure 11). Unirradiated soil or subsoil material (depth interval *b* in Figure 11) was deposited on the surface at site 2, forming the coarse layer VI. Further layering and possibly shallow excavations occurred at site 2 between time T_2 and the present when the observed layer sequence was sampled.

We have shown that most of the observed layers may come from partially- or well-mixed material, so that most soil grains have experienced several excavations and surface exposures. Each particular surface residence time may therefore be much less than the integrated surface residence time. Since most of the cosmic ray tracks are acquired near the surface, the surface residence times derived by the method of Arrhenius *et al.* (1971) for the individual layers will be similar to average integrated surface residence times for the soil grains in those layers.

The existence of layers is not inconsistent with mixing by excavation because the stratigraphy below a given depth is not destroyed by excavation of material above that depth. Frequent excavation above a depth d means that the net *burial* rate at depth d in general will be less than the average *layering* rate on the surface. It is expected that spallation tracks in the soil, with their high rate of production in the upper 2 meters, can provide information about this net burial rate, especially for layers with a simple cosmic ray irradiation history (e.g. Apollo 12 double core layer VI).

A relevant physical process which has not been discussed above is the possibility of surface transport (e.g., Gold, 1971). Surface transport over great distances will require that soil grains remain on the surface for relatively long times and can be expected to acquire quite high cosmic ray track densities (the center of a 100-μm-diam grain on the soil surface will acquire cosmic ray tracks at a rate of $\dot{\varrho} \sim 10^3$ cm^{-2}-yr). Such transport probably can be ruled out for the grains in layer VI (Table IV) which have $\varrho \sim 10^7$ cm^{-2} (Arrhenius *et al.*, 1971). On the other hand large grains which have very high track densities, $\varrho > 10^9$ cm^{-2} (Barber *et al.*, 1971a, b; Crozaz *et al.*, 1971), represent prolonged exposure and the possibility of surface transport for these grains cannot be eliminated.

6. Conclusions

The particle track record of lunar surface history for both rocks and soil can be summarized in three general depth domains (Figure 3).

(1) Direct surface irradiation by solar flare heavy nuclei of the first 0 to ~0.2 cm results in steep track density gradients which have recorded rock erosion rates of about 10^{-8} to 10^{-7} cm yr^{-1} (Figure 4). Track gradients in soil grains may indicate a parent rock erosion rate of $\sim 3 \times 10^{-7}$ cm yr^{-1} early in the Moon's history. When the exposure time and erosion rate are known, such as for the glass coating on 12017, then the level of solar particle flux in the past can be determined.

(2) The second depth domain, ~0.2 cm to ~10 cm, involves irradiation primarily by galactic cosmic ray heavy nuclei. These particle tracks have provided surface residence times for the rocks of ~1 to 50 m.y. (Table III). Cosmic ray tracks (Figure 6) acquired predominately at these depths record the mixing and layering history of lunar soil. Track distributions in the soil are consistent with a soil history which includes repeated excavation, layering and burial, such that one core 12025+28 soil layer is unmixed and weakly irradiated whereas the others contain soil which has been better and better mixed while being more and more irradiated (Table IV).

(3) The third depth domain, 0 to ~200 cm, is recorded by the tracks of spallation-recoil nuclei. These tracks indicate that both the rocks and soil grains have a wide range of residence times in the top 2 m of the lunar surface (Table III, Figure 7).

Acknowledgements

The author is indebted to R. L. Fleischer and H. R. Hart, Jr. for many helpful comments. This work was supported in part by NASA under contract NAS 9-7898.

Note added in proof: Recent data on cosmic ray abundances (Cartwright *et al.*, 1971; Binns *et al.*, 1971; Webber *et al.*, 1971) show that Fe is significantly more abundant relative to the VH group than had been indicated by the earlier data used in Table II, especially at relativistic energies. The improved abundances $A(Z)$ change Equation 1b to

$$\left(\frac{\mathrm{d}N}{\mathrm{d}T}\right)_E \approx 0.6 \left(\frac{\mathrm{d}N}{\mathrm{d}T}\right)_{\mathrm{VH}} \tag{1b'}$$

at low energies. Using better (but still uncertain) values of $\Delta R(Z)/\Delta R(\mathrm{Fe})$ for $Z \geqslant 20$ (Plieninger and Kratschmer, 1971) will further change this to

$$\left(\frac{\mathrm{d}N}{\mathrm{d}T}\right)_E \approx 0.75 \left(\frac{\mathrm{d}N}{\mathrm{d}T}\right)_{\mathrm{VH}} \tag{1b''}$$

which yields values for cosmic ray surface residence times intermediate between the 'high' and 'low' ages given in Table III, for $\Delta R(\mathrm{Fe}) \approx 12\ \mu m$. However there are still uncertainties in $\Delta R(\mathrm{Fe})$ and in the time average of $(\mathrm{d}N/\mathrm{d}T)_{\mathrm{VH}}$ at low energies.

References

Arrhenius, G., Liang, S., Macdougall, D., Wilkening, L., Bhandari, N., Bhat, S., Lal, D., Rajagopalan, G., Tamhane, A. S., and Venkatavaradan, V. S.: 1971, *Proc. Second Lunar Science Conf.* **3**, 2583 (MIT Press).

Barber, D. J., Cowsik, R., Hutcheon, I. D., Price, P. B., and Rajan, R. S.: 1971a, *Proc. Second Lunar Science Conf.* **3**, 2705 (MIT Press).

Barber, D. J., Hutcheon, I. D., and Price, P. B.: 1971b, *Science* **171**, 372.

Bhandari, N., Bhat, S., Lal, D., Rajagopalan, G., Tamhane, A. S., and Venkatavaradan, V. S.: 1971, *Proc. Second Lunar Science Conf.* **3**, 2599 (MIT Press).

Binns, W. R., Fernandez, J. I., Israel, M. H., Klarmann, J., and Mewaldt, R. A.: 1971, *12th Int. Conf. on Cosmic Rays*, Hobart, OG–73.

Borg, J., Maurette, M., Durrieu, L., and Jouret, C.: 1971a, *Proc. Second Lunar Science Conf.* **3**, 2027 (MIT Press).

Borg, J. and Vassent, B.; Durrieu, L., Dran, J. C., and Maurette, M.: 1971, this volume, pp. 298; 309.

Burnett, D., Monnin, M., Seitz, M., Walker, R., and Yuhas, D.: 1971, *Proc. Second Lunar Science Conf.* **2**, 1503 (MIT Press).

Carrier, W. D., III, Johnson, S. W., Werner, R. A., and Schmidt, R.: 1971, *Proc. Second Lunar Science Conf.* **3**, 1959 (MIT Press).

Cartwright, B. G., Garcia-Munoz, M., and Simpson, J. A.: 1971, *12th Int. Conf. on Cosmic Rays*, Hobart, OG–63.

Comstock, G. M., Fan, C. Y., and Simpson, J. A.: 1969, *Astrophys. J.* **155**, 609.

Comstock, G. M., Evwaraye, A. O., Fleischer, R. L., and Hart, H. R., Jr.: 1971, *Proc. Second Lunar Science Conf.* **3**, 2569 (MIT Press).

Crozaz, G., Haack, U., Hair, M., Maurette, M., Walker, R., and Woolum, D.: 1970, *Geochim. Cosmochim. Acta, Suppl. I* **3**, 2051.

Crozaz, G., Walker, R., and Woolum, D.: 1971, *Proc. Second Lunar Science Conf.* **3**, 2543 (MIT Press).

Fleischer, R. L., Price, P. B., Walker, R. M., Maurette, M., and Morgan, G.: 1967, *J. Geophys. Res.* **72**, 355.

Fleischer, R. L., Hart, H. R., Jr., Jacobs, I. S., Price, P. B., Schwarz, W. M., and Woods, R. T.: 1970a, *J. Appl. Phys.* **41**, 958.

Fleischer, R. L., Haines, E. L., Hanneman, R. E., Hart, H. R., Jr., Kasper, J. S., Lifshin, E., Woods, R. T., and Price, P. B.: 1970b, *Science* **167**, 568.

Fleischer, R. L., Haines, E. L., Hart, H. R., Jr., Woods, R. T., and Comstock, G. M.: 1970c, *Geochim. Cosmochim. Acta, Suppl. 1* **3**, 2103.
Fleischer, R. L. and Hart, H. R., Jr.: 1970d, *Proc. Burg Wartenstein Conf. on Calibration of Hominoid Evolution*, July 1971, and General Electric Research and Development Center Report, No. 70-C-328.
Fleischer, R. L., Hart, H. R., Jr., Comstock, G. M., and Evwaraye, A. O.: 1971a, *Proc. Second Lunar Science Conf.* **3**, 2559 (MIT Press).
Fleischer, R. L., Hart, H. R., Jr., and Comstock, G. M.: 1971b, *Science* **171**, 1240.
Freier, P. S. and Waddington, C. J.: 1968, *Phys. Rev.* **175**, 1641.
Garcia-Munoz, M. G. and Simpson, J. A.: 1970, '11th Internat. Conf. on Cosmic Rays', *Acta Phys. Acad. Sci. Hung.* **29**, Suppl. 1, 317.
Gold, T.: 1971, this volume, p. 55.
Kohman, T. D. and Bender, M. L.: 1967, in B. S. P. Shen (ed.), *High Energy Nuclear Reactions in Astrophysics*, Benjamin, Inc., pp. 169–245.
Lal, D., Macdougall, D., Wilkening, L., and Arrhenius, G.: 1970, *Geochim. Cosmochim. Acta, Suppl. 1*, **3**, 2103.
LSPET (Lunar Sample Preliminary Examination Team) 1970, *Science* **167**, 1325.
Öpik, Ernst J.: 1969, in *Ann. Rev. Astron. Astrophys.* **7**, 473 (Annual Reviews, Inc., Palo Alto, Calif.).
Pellas, P., Poupeau, G., Lorin, J. C., Reeves, H., and Audouze, J.: 1969, *Nature* **223**, 272.
Plieninger, T. and Krätschmer, W.: 1971, (Abstract) *Trans. Amer. Geophys. Union* **52**, 268.
Price, P. B., Fleischer, R. L., Peterson, D. D., O'Ceallaigh, C., O'Sullivan, D., and Thompson, A.: 1967, *Phys. Rev.* **164**, 1618.
Price, P. B., Fleischer, R. L., and Moak, C. D.: 1968a, *Phys. Rev.* **167**, 277.
Price, P. B., Peterson, D. D., Fleischer, R. L., O'Ceallaigh, C., O'Sullivan, D., and Thompson, A.: 1968b, *Phys. Rev. Letters* **21**, 630.
Price, P. B., Peterson, D. D., Fleischer, R. L., O'Ceallaigh, C., O'Sullivan, D., and Thompson, A.: 1970a, *Hung. Phys. Acta* **29**, *Suppl. 1*, 417.
Price, P. B. and O'Sullivan, D.: 1970b, *Geochim. Cosmochim. Acta, Suppl. 1* **3**, 2351.
Price, P. B., Rajan, R. S., and Shirk, E. K.: 1971, *Proc. Second Lunar Science Conf.* **3**, 2621 (MIT Press).
Rajan, R. S. and Price, P. B.: 1971, to be published.
Schmitt, H. H. and Sutton, R. L.: 1971, *Presented at Second Lunar Science Conf.*, Houston, Texas (unpublished).
Sellers, G. A., Woo, C. C., Bird, M. L., and Duke, M. B.: 1971, *Proc. Second Lunar Science Conf.* **1**, 665 (MIT Press).
Shoemaker, E. M., Hait, M. H., Swann, G. A., Schleicher, D. L., Schaber, G. G., Sutton, R. L., Dahlem, D. H., Goddard, E. N., and Waters, A. C.: 1970, *Geochim. Cosmochim. Acta, Suppl. 1* **3**, 2399.
Walker, R. M.: 1970, *Radiat. Eff.* **4**, 239.
Wang, J. R.: 1970, *Astrophys. J.* **160**, 261.
Webber, W. R., Damle, S. V., and Kish, J.: 1971, University of New Hampshire, preprint UNH-71-21, and *12th Int. Conf. on Cosmic Rays*, Hobart, OG-66.

G. THE LUNAR INTERIOR

LUNAR MAGNETIC FIELD MEASUREMENTS, ELECTRICAL CONDUCTIVITY CALCULATIONS AND THERMAL PROFILE INFERENCES

D. S. COLBURN

Ames Research Center, NASA, Moffett Field, California, U.S.A.

Abstract. Steady magnetic field measurements of magnitude 30 to 100 γ on the lunar surface impose problems of interpretation when coupled with the non-detectability of a lunar field at 0.4 lunar radius altitude and the limb induced perturbations of the solar wind reported by Mihalov *et al.* at the Explorer orbit. The lunar time varying magnetic field clearly indicates the presence of eddy currents in the lunar interior and allows calculation of an electrical conductivity profile. The problem is complicated by the day-night asymmetry of the Moon's electromagnetic environment, the possible presence of the TM mode and the variable wave directions of the driving function. The electrical conductivity is calculated to be low near the surface, rising to a peak of $6 \times 10^{-3}\ \Omega^{-1}\ \mathrm{m}^{-1}$ at 250 km, dropping steeply inwards to a value of about $10^{-5}\ \Omega^{-1}\ \mathrm{m}^{-1}$, and then rising toward the interior. A transition at 250 km depth from a high conductivity to a low conductivity material is inferred, suggesting an olivine-like core at approximately 800°C, although other models are possible.

1. Introduction

The measurement of lunar magnetic fields has long been of interest because of the promise that these measurements would provide information about the lunar interior: a significant dipolar field would imply a dynamo action in the lunar interior similar to the Earth's dynamo, while the presence of induced fields would imply a value for the electrical conductivity in the lunar interior.

The first magnetic experiments were made by the U.S.S.R. on the Luna spacecraft. These set an upper limit on any lunar magnetic field and also indicated a magnetospheric region surrounding the Moon. More definitive measurements of the lunar environment came with the launching of the Explorer 35 lunar orbiting satellite in 1967. Magnetometers on this satellite detected no lunar bow shock and no evidence at periselene of a lunar field (Colburn *et al.*, 1967; Ness *et al.*, 1967). The upper limit for a lunar centered dipole moment was set at 10^{20} G cm^3 corresponding to a maximum dipolar field component of 4 γ (1 $\gamma = 10^{-5}$ G) on the lunar surface (Behannon, 1968). Larger surface fields were not ruled out if they were of quadrupole or higher order. Definitive measurements of lunar fields commenced with the deployment of the Lunar Surface Magnetometer on Apollo 12 and the Lunar Portable Magnetometer on Apollo 14.

In this paper we discuss first the measurement of a steady magnetic field at three points on the lunar surface. We consider next the predictions of other lunar magnetic regions inferred from satellite observations. Finally we consider the calculations of the electrical conductivity profile of the lunar interior determined by the Moon's electromagnetic response, and its implications for a thermal model.

Urey and Runcorn (eds.), The Moon, 355–371.

2. The Instruments

Explorer 35 was placed into lunar orbit on July 19, 1967 and is still in operation. The periselene is at an altitude of 0.4 R_M ($R_M = 1740$ km) with the aposelene altitude $4.4 R_M$ The orbital plane is tilted 11° out of the ecliptic plane. Among the experiments are a plasma probe and two magnetometers. The Ames magnetometer that supplied the data reported here, is a three-axis vector fluxgate with 0.4 γ ($1\ \gamma = 10^{-5}$ G) resolution employing spin demodulation and filtering to avoid aliased data (Colburn *et al.*, 1967; Mihalov *et al.*, 1968). In this system the high bandwidth signals from the two sensors in the spin plane of the spinning spacecraft are multiplied by sines and cosines of the spin frequency and mixed appropriately to furnish a high bandwidth set of the three components of the vector field in an inertial (nonspinning) frame (Sonett, 1965). These are filtered to preserve the Nyquist criterion for alias free data before being sampled at the sampling rate (1 vector per 6.14 s) and telemetered to Earth (Sonett, 1968). The maintenance of spectral purity is vital to the use of power spectral density estimates in determining an electrical conductivity profile.

The Apollo 12 Lunar Surface Magnetometer (LSM) is a triaxial fluxgate emplaced on the lunar surface at the Apollo 12 site at the eastern edge of Oceanus Procellarum, coordinates 3.0° S, 23.4° W. The instrument telemeters to Earth a vector measurement of the field at a rate of 3.3 samples per s. The site survey mode that was commanded three days after emplacement rotated the sensors to each of the three coordinate directions in turn. The three sensors are mounted at the ends of three mutually orthogonal 100 cm booms, permitting a measurement of the local field gradient in the horizontal plane (Dyal *et al.*, 1970b). Since no measurable gradient was found, a lower limit was placed on the distance to the field source, the distance also depending on the source configuration. A similar Lunar Surface Magnetometer was emplaced at the Apollo 15 site, with a third one scheduled for Apollo 16.

At the Apollo 14 site the field was measured by the Lunar Portable Magnetometer (LPM). This instrument was set up at two locations by the astronauts who then sent back the readings by voice channel. A 20 s time constant filter averaged out the high frequency fluctuations to measure the steady state field component at the site. A similar Lunar Portable Magnetometer is scheduled to be operated on Apollo 16.

Another magnetic measurement of the Moon has become possible by the magnetometer aboard the subsatellite launched into lunar orbit by the Apollo 15 command module (P. J. Coleman, Jr., principal investigator). It is expected this instrument and the similar one on Apollo 16 will be able to map locally magnetized areas on the lunar surface when on the lunar dark side or in the Earth's magnetic tail. The low orbital altitude, typically 100 km or less, will allow detection of surface feature at detail unobtainable at the orbit of Explorer 35. The Explorer measurement is very important in providing a measurement of the background field surrounding the Moon and is essentially unperturbed by the Moon's presence except when Explorer 35 is almost directly in the lunar shadow.

3. The Surface Field

Data from the Apollo 12 LSM showed a steady field component at the site of 38 γ and it was known at the outset that this was not due to a dipole centered in the Moon, for a centered dipole of such strength would have been observed at the orbit of Explorer 35. It likewise could not have been due to a dipole source closer than 200 m to the instrument because of the null result of the gradient measurement. The possibility of a localized surface field highly variable with location suggested magnetization of the lunar crustal regions and prompted the design and approval of the LPM that made its measurements only slightly more than one year after inception.

Table I shows the vector field measurements reported for the two Apollo 14 sites

TABLE I

Magnetic field measurements at lunar sites (Dyal *et al.*, 1971)

Site		Apollo 12	Apollo 14-A	14-C′
Coordinates				
S		3.0°	3.7°	3.7°
W		23.4°	17.5°	17.5°
Location		LSM site	170 m from LM	Cone crater rim
Field Magnitude		$38 \pm 3\ \gamma$	$103 \pm 5\ \gamma$	$43 \pm 6\ \gamma$
Components	Up	$-24.4 \pm 2.0\ \gamma$	$-93 \pm 4\ \gamma$	$-15 \pm 4\ \gamma$
	East	$+13.0 \pm 1.8\ \gamma$	$+38 \pm 5\ \gamma$	$-36 \pm 5\ \gamma$
	North	$-25.6 \pm 0.8\ \gamma$	$-24 \pm 5\ \gamma$	$-19 \pm 8\ \gamma$
Gradient		$< 133\ \gamma$/km	$> 54 \pm 7\ \gamma$/km	

and the Apollo 12 site (Dyal *et al.*, 1971; Dyal *et al.*, 1970a). The vectors are also displayed in Figure 1. The Apollo 12 and 14-C′ (Cone Crater) magnitudes are approximately 40 γ while the other Apollo 14 site has a magnitude factor of 2.5 greater. The vectors point down and southerly: the maximum angle between any two is 84°. The lower limit on the field gradient at Apollo 14 is defined as the average gradient between the two sites or 54 $\pm 7\ \gamma$/km, while the upper limit for Apollo 12, 133 γ/km is determined by the site survey measurement, over a distance span on the order of 1 m.

Models for the steady fields have been constructed but none have been completely satisfactory. Since it is known from Explorer 35 measurements that the surface field caused by any centered dipole must be less than 4 γ, more localized field sources are hypothesized, which would presumably rule out the mechanism of a planetary dynamo. Alternatively, the measurements could represent the bulk magnetization of surface rocks. The remanent field in a rock slab which has been uniformly magnetized has a highly nonuniform field, the magnitude being greatest near the edges and direction altering radically with position near the edges. The magnitude becomes relatively

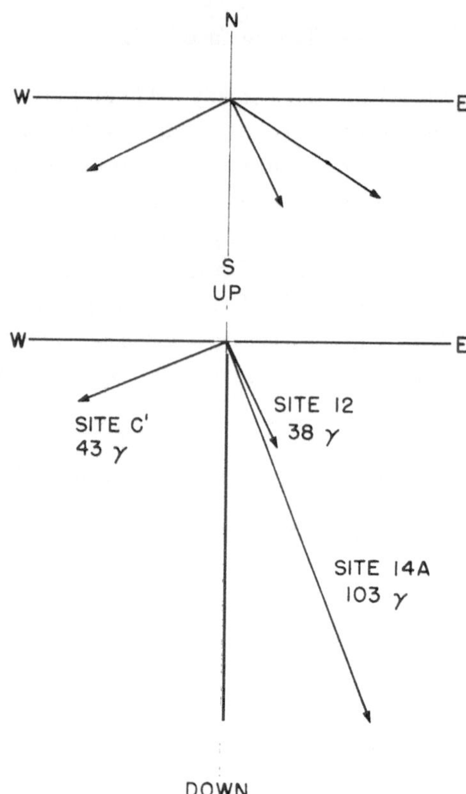

Fig. 1. Steady magnetic field vector at three locations on the Moon. Site 12 is the Apollo ALSEP location; sites 14*A* and *C′* are the two measuring sites on the Apollo 14 moonwalk. The maximum angle between any two vectors is 84°. (From Dyal *et al.*, 1971).

smaller as one approaches the broad flat region of the magnetized slab. The edge delineated by Cone Crater, as pointed out by Dyal, or perhaps of a larger Mare region hidden by surface features (Runcorn and Quaide, private communication) would explain the large difference in the two Apollo 14 measurements. In view of these differences and the proposed models, the significance of all measurements lying in a single quadrant is not clear. More data would be required to establish the direction of the original magnetizing field.

Remanent magnetism in rock areas consistent with magnetic measurements on Apollo rock can explain the steady field measurements (Runcorn *et al.*, 1970). This implies, however, that large rock volumes cooled through the Curie point under the influence of a field of more than 1000 γ. This creates difficulty in lunar thermal history models, as will be discussed later.

4. Magnetic Map Inferred from Explorer 35

Mihalov *et al.* (1971) have found perturbations in Explorer 35 magnetometer measurements that appear to map the lunar surface. To describe the method, it is necessary to review the magnetic environment at the Explorer 35 orbit.

The interplanetary magnetic field has a magnitude on the order of 5 γ but can become several times greater under disturbed conditions. The preferred direction is outward or inward along an Archimedes spiral determined by the solar wind and the Sun's rotation (Parker, 1963) although the field can assume any direction from time to time. The distinction between outward and inward can be made statistically, and for the period under question generally encompasses one outward and one inward sector (occasionally more) every solar rotation period of 27 days (Wilcox and Colburn, 1970). The solar wind is not deflected significantly by the Moon, as evidenced by the lack of a measurable bow wave, and hence the particles are believed to strike the lunar surface and be neutralized there, while directly behind the Moon is a cavity in which solar wind is essentially absent and the magnetic field is slightly enhanced because of the diamagnetism of the solar wind (Colburn *et al.*, 1971). At the border of the cavity a characteristic dip in magnetic field magnitude is observed. Beyond the Mach angle defining the rarefaction wave, however, a peak in the magnetic field is occasionally observed. These are believed to be due to a lunar solar wind interaction occurring at the limb and propagating outward along the Mach angle. The mechanism is postulated to be the deflection of solar wind by local magnetic field. No magnetic effect of the Moon is seen over the large majority of the Explorer 35 orbit lying forward of this Mach angle; the magnetic field is as if measured from a spacecraft very far from the Moon. Because the solar wind travels at a speed faster than known magnetosonic and Alfvenic propagation speeds, the information signaling the presence of the Moon cannot travel upstream to these regions. From time to time the magnetic field peak exterior to the rarefaction wave becomes unusually large. There were 100 large peaks representing 7.4% of the possible times of occurrence. Mihalov has catalogued these occurrences and related them to regions on the Moon at the limb by associating each occurrence with the point on the lunar surface closest to the solar wind velocity vector

TABLE II

Regions on the lunar surface where the concentration of source locations was tested (Mihalov *et al.*, 1971). *F* is the fraction of observations showing the anomaly and *P* is calculated probability that there is no concentration at the location. The observations are from the orbiting Explorer 35 satellite. Mihalov *et al.* conclude that local magnetization is the favored explanation for the phenomena.

Seleno-centric latitude	Seleno-centric longitude	Area (X 10^5 km^2)	*F*	*P*	Nearby future
5°–20°S	135°E–165°W	10	0.5	$\ll 10^{-5}$	Mare X̄ (Gagarin)
6°N	88°W	2	0.4	$< 10^{-5}$	'Montes d'Alembert'
0°–20°N	60°–115°E	10	0.3	3×10^{-3}	Mare Marginis
5°S	138°W	0.7	0.3	6×10^{-3}	Crater 244 (Vavilov)
2°N	35°W	1	0.2	$< 10^{-5}$	Eneke-Kepler
5°N	0°	0.9	0.1	5×10^{-4}	Pallas
0°	25°E	0.8	0.1	4×10^{-5}	Delambre

drawn through the spacecraft location at the time of the observation (Mihalov *et al.*, 1971; Sonett and Mihalov, 1972). The mapping has been shown to be statistically significant. In Table II are shown the areas outlined by Mihalov as being connected with the exterior peaks observed by Explorer 35.

The table shows that the highlands are favored, and also that the far side is favored. While mechanisms other than local magnetic fields could be postulated as causes of the field increases seen by Explorer 35, Mihalov *et al.* find cogent reasons for believing that the local magnetic field is the most likely cause. On the theoretical side, Barnes *et al.* (1971) have investigated the possible frequency of magnetic structure on the surface. They find that regions of field such as indicated by the Explorer 35 perturbations must have a length scale of at least 10 km and a compressed field strength of more than 10 γ, and that thousands of these could be present on the lunar surface.

5. Electrical Conductivity Profile Calculations

The lunar surface magnetic field measurement was early thought of as a means of probing the interior of the Moon (Sonett, 1966). If we consider the Moon as a sphere whose electrical conductivity is a function of radius only, we find that there are two modes in which it can respond to the time varying electrical and magnetic field associated with the solar wind: the transverse magnetic (TM) and the transverse electric (TE) (Schubert and Schwartz, 1969). For the TM response the driving function is an electrical field due to the solar wind. Because of the high electrical conductivity of the plasma constituting the solar wind, the electric field must be considered to be zero in a frame at rest in the solar wind plasma. A simple transformation to a frame at rest with respect to the moon shows an electric field of $E = v \times B$ where B is the interplanetary magnetic field and v is the velocity of the Moon with respect to the solar wind. This response should extend all the way down to zero frequency. The current system of the response must be such that the current flows through the Moon and its crust and closes through the highly conducting solar wind. If this response were to be significant, it should cause a bow shock; a phenomenon not observed by Explorer 35. It is generally concluded that the relatively high resistance of the Moon's crustal layers effectively cuts off the TM mode (Sonett and Colburn, 1967). The high resistance is deduced from the probable electrical conductivity of lunar material at the mean surface temperature, which must be assumed by material more than a few meters deep considering the low thermal conductivity of rock materials. An exhaustive statistical analysis of magnetometer measurements may yet show a modicum of TM response, but it must remain a secondary effect.

We turn now to the TE mode, which is excited by $\dot{B}$, the rate of change of the interplanetary magnetic field. The response of a planetary body to this mode is the establishment of eddy currents within the body that do not need to flow through the relatively nonconducting crust. This mode does not occur in the steady state but is observable at frequencies commensurate with lunar magnetic field measurements. Preliminary findings on the TE response of the Moon were reported by Sonett *et al.* (1971b, c).

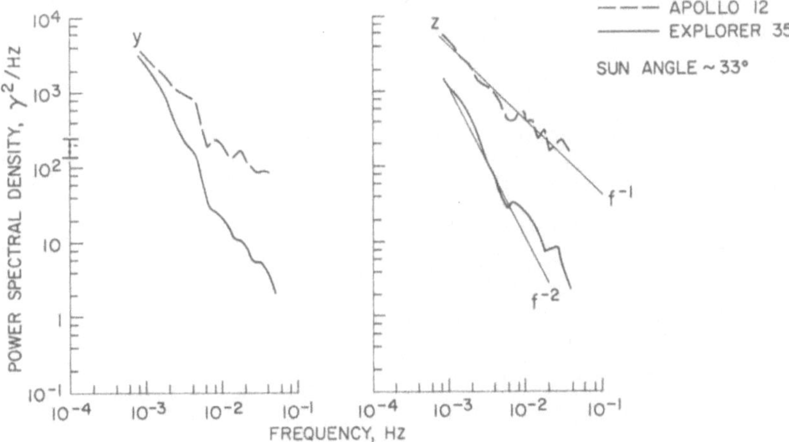

Fig. 2. Power spectral density of the interplanetary field near the Moon (Explorer 35) and the lunar surface field (Apollo 12) for the same 2-h time period. The two horizontal components are shown: Z is north and Y is east at the Apollo 12 site. Other 2-h segments show variations from this example but all show an amplification in the horizontal components increasing with frequency.

Figure 2 shows a typical plot of power spectral density versus frequency for the horizontal components of the lunar surface field (Apollo 12) and for the same time period the corresponding components of the free stream solar wind field near the Moon. The time period was 2 hr and the Apollo 12 magnetometer location 33° from the Moon-Sun line. The Moon was outside the Earth's bow shock in the solar wind. The power in the interplanetary field drops off with frequency approximately as f^{-2}, a typical frequency dependence for interplanetary spectra. The power in the surface components is larger, particularly at the higher frequencies.

Autocorrelation techniques instead of cross correlation techniques are used because of the complication of the variable Doppler shift between the locations of the two measurements. The analysis is based on the assumption that the free stream field is the sole driving function for the lunar response. This is made credible by the examination of the amplification of particular sine wave cycles or transients such as were published in Sonett *et al.* (1971c).

Data from the first lunation were examined to find periods in which both magnetometers were reading simultaneously with no data gaps. Seven 2-h and seven 1-hr segments were used, with the Explorer 35 data transformed into LSM coordinates. The Apollo 12 data were filtered and decimated, to approximate the same filter and bit rate as the Explorer 35 data. Autocorrelation and power spectral density estimates were made for several frequencies. For the 2-hr swaths 50 frequencies were used, linearly spaced from 0.83 mHz to 41.5 mHz. For the 1-h swaths the number of frequencies was 25 (1.66 to 41.5 mHz) – The amplification A was defined as the square root of the ratio of power density at the lunar surface to power density in the free stream. The subscript x, y, or z is added to signify respectively the vertical, east, and north components of the vectors as viewed from Apollo 12.

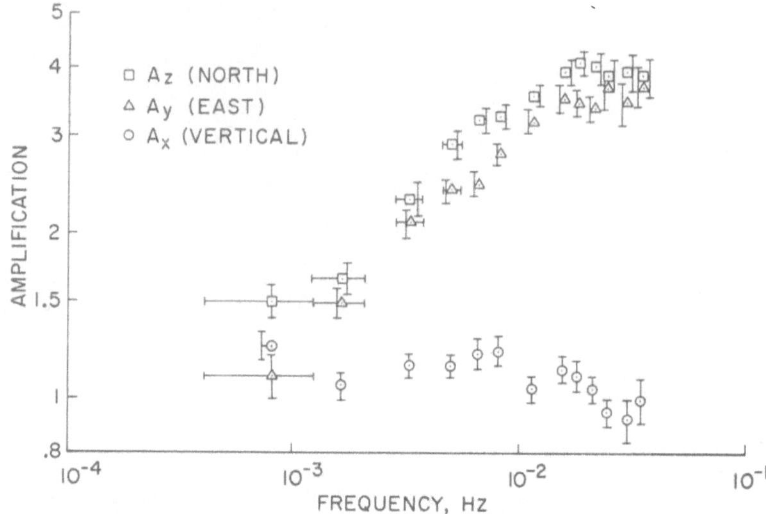

Fig. 3. Amplification, the ratio of amplitudes of lunar surface and free stream magnetic fields, shown as a function of frequency. The amplification is defined as the square root of the ratio of power spectral densities. The error bars for frequency are the windows defined by the lags in the autocorrelation calculation. The error bars in amplification are the one standard deviation limits determined from the means of 14 data spectra. Amplification occurs for the north and east components but remains near unity for the vertical component.

Values of A_x, A_y, and A_z are shown in Figure 3 at representative frequencies common to the 14 cases. The averages of 14 values are plotted with error bars designating the one standard deviation error in the mean. (The point at 8.3 mHz is an average of only seven cases.) The amplification for the horizontal components starts near unity at the lowest frequency and rises significantly with frequency to values approaching 4 at the highest frequencies. For the vertical component the amplification remains near unity.

The results in Figure 3 were anticipated by theoretical analyses. Blank and Sill (1969) derived a model for the TE mode and Schubert and Schwartz (1969) derived a model dealing consistently with the combined TE and TM modes in a wave field. In both treatments a thin current layer was assumed to surround the Moon such that outside of the current layer the field had the free stream value, the normal component of which (since $\nabla \cdot B=0$) was conserved across the layer. The approximation is appropriate on the sunlit side where the solar wind plasma is moving into the Moon at a speed both supersonic and super Alfvenic, so that in the magnetohydrodynamic approximation no information regarding the Moon's presence may be conveyed upstream. The confinement permits very large amplifications of horizontal components. The current layer is not appropriate on the dark side; however, model calculations show that measurements on the sunlit side are not altered more than a few percent by the dark side portion of the assumed current system, and the symmetry is necessary for tractibility.

Attenuation of the vertical component at a given frequency would be substantially complete at a depth where the electrical skin depth for that frequency is small. If the Moon were immersed in a vacuum, the attenuation would drop off as r^3 and be observed at the surface. The near unity values for the vertical amplification imply that a current layer is indeed present above the lunar surface and much closer than the underlying volume of high electrical conductivity.

The scatter in the A_x data has not yet been satisfactorily explained but may be due to plasma noise effects and departure from spherical symmetry.

The magnetic field amplification at the lunar surface, observed experimentally, is believed to give information about the electrical conductivity in the lunar interior. The Moon is considered here to have an electrical conductivity $\sigma(r)$ varying only with radial distance. The magnetic field oscillation in the solar wind is defined as

$$H = \hat{\eta} H_0 \exp\left\{2\pi i\left(\frac{\zeta}{\lambda} - ft\right)\right\}$$

The cartesian coordinate system (ξ, η, ζ) with unit vectors $\hat{\xi}$, $\hat{\eta}$, $\hat{\zeta}$ is fixed relative to the Moon, which moves through the solar wind in the negative ζ direction with speed v relative to the wave front. (For transverse waves, neglecting propagation speeds, v is the aberrated solar wind speed.) The amplitude is H_0, the wavelength λ, and the frequency of the oscillation is f. In the lunar interior the solution of Maxwell's equations for the TE mode is represented by a potential Ω satisfying

$$\nabla^2 \Omega + k^2 \Omega = 0$$

where

$$k^2 = 4\pi^2 f^2 \mu\varepsilon + i2\pi f \mu\sigma$$

and μ and ε are taken as free space values. The potential is found as a sum of product solutions of the form

$$\Omega = \sin\phi \sum_{l=1}^{\infty} F_l(r)\, \varrho_l^1(\cos\Theta)$$

where r, Θ, and ϕ are spherical polar coordinates with ζ the polar axis and ϱ_l^1 are associated Legendre polynomials. The boundary condition for the TE mode is the matching of the lunar surface normal component of the magnetic field to the free stream value, as discussed earlier, confining the lunar disturbance to the immediate proximity of the Moon.

Equations for the calculation of the fields are given elsewhere (Sonett *et al.*, 1971d); alternatively they may be written for the TE case by taking the solution to the Earth induction problem (Lahiri and Price, 1939, or see Chapman and Bartels, 1962, Section 22.13, Equations 23–31) and making proper modifications for the changed boundary conditions. For the work reported here, the transfer function is defined as the ratio of the magnitudes of transverse components for the first mode of the modal expansion.

Using the values of A_y and A_z averaged from 14 swaths (of 1- and 2-h duration) Sonett *et al.* have calculated a curve of amplification versus frequency. Because of the significant differences in A_y and A_z, not yet explained, fits were made separately for A_y^2, A_z^2, and $\bar{A}=[0.5(A_y^2+A_z^2)]^{1/2}$.

The inversion method consists of starting with a conductivity profile, calculating an A versus f curve and minimizing by the least squares method the deviation of the calculated A with $\bar{A}$, A_y, or A_z. In order to do this the conductivity profile was characterized by eight parameters, namely the conductivity at $r=800$, 1200, 1400, 1450, 1490, 1510, 1550, and 1740 km. The conductivity for $0 \leqslant r \leqslant 800$ km was considered constant, the method not being very sensitive to conductivities at these depths. At all other values of r, $\log r$ is defined as a linear interpolation from neighboring $\log r$ values. In each iteration the amplification for the model is calculated at eight frequencies, namely 0.83, 1.7, 5, 12, 17, 22, 25, and 35 mHz, and the least squares fit at those frequencies between the model and the data is minimized by successive computer iterations using the Newton Raphson method.

The method showed significant convergence in five iterations. Figure 4 shows the resulting fit of a model amplification curve in which the rms value $\bar{A}$ was fitted. Figure 5 shows the conductivity profile found by the routine that produces this result and profiles found by matching A_y or A_z independently. In each case the profile is found to have a maximum near $r=1500$ km. The conductivity starts at a low value at the surface as is expected because of the mean temperature of $-30\,°\mathrm{C}$ and the strong temperature dependence of candidate lunar materials. Because the current system is effectively cut off, the method is imprecise at low conductivities. The conductivity then rises with depth to a maximum of about $6\times10^{-3}\ \Omega^{-1}\ \mathrm{m}^{-1}$ at $r=1500$ km and then decreases by some 2 orders of magnitude in 100 km before resuming a rise with depth.

The temperature profile can be inferred from the conductivity profile only indirectly,

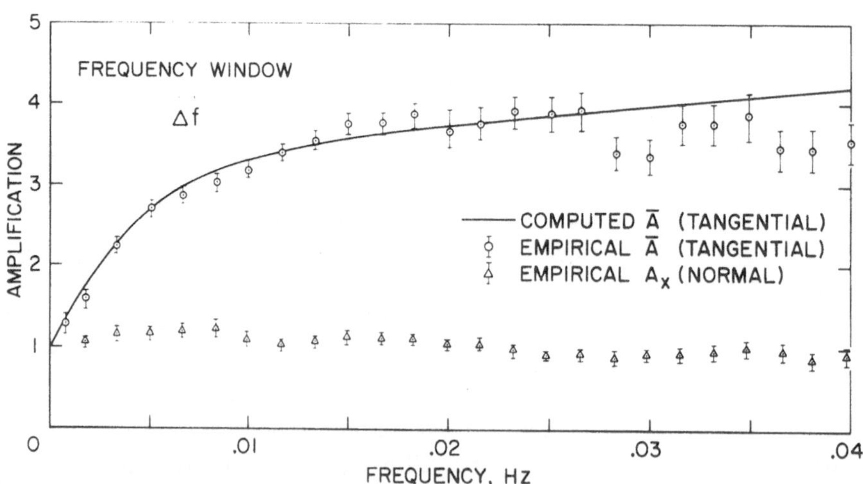

Fig. 4. Amplification as a function of frequency. The data are as in Figure 3, with the tangential amplifications combined by the relation $\bar{A}=[0.5(A_y{}^2+A_z{}^2)]^{1/2}$. The solid line is the amplification calculated from a model described in the text. The amplification of the model is fitted to $\bar{A}$ values at frequencies of 0.83, 1.7, 5, 12, 17, 22, 25, and 35 mHz.

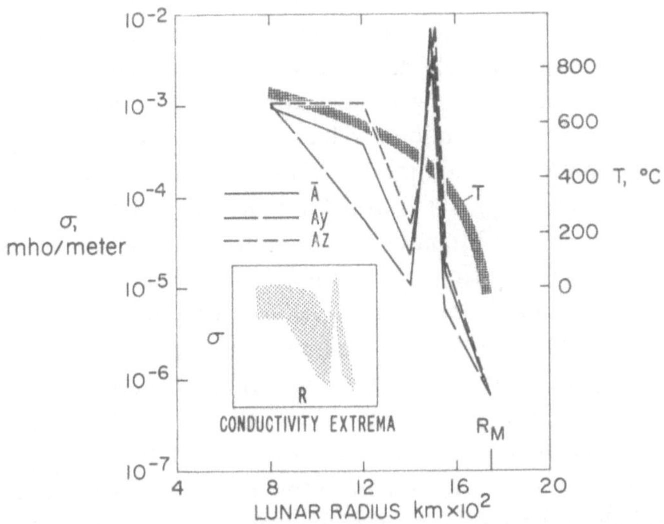

Fig. 5. Lunar electrical conductivity as a function of radial distance, according to calculations based on the amplifications A_y, A_z, and $\bar{A}$. As depth increases the conductivity rises to a peak at $r = 1500$ km, decreases to a minimum at $r = 1400$ km and then continues to rise. Six other cases were run, using the one standard deviation limits of the As for calculating the conductivities; these all contain the peak and lie in the shaded area shown. Plotted also is a tentative version of a lunar thermal profile. The assumption of a relatively smooth thermal profile and known conductivity functions requires a core conductivity similar to that of olivine with a transition to a more highly conducting material at $r = 1500$ km.

since it also depends on composition and conductivity-temperature functions. If candidate materials are ordered according to conductivity, one of the more conducting is a basalt such as the Lunar Sample 10024.22 analyzed by Nagata *et al.* (1970; see also Schwerer *et al.*, 1971). One of the less conducting is an olivine (see England *et al.*, 1968). The conductivity gradient from the surface to $r = 1500$ km is approximated by a 2°C/km temperature gradient for the basalt function and a 4°C/km gradient for the olivine, with most other candidate materials lying somewhere in between. It is not likely that the temperature can decrease with depth between $r = 1500$ km and $r = 1400$ km because any such gradient occurring during the Moon's formation would have smoothed out during the lunar history, and localized present-day heat sources at that depth are unlikely. If this reasoning is correct, the transition from 1500 to 1400 km must be accompanied by a modest increase in temperature as well as a large decrease in conductivity. The transition is then plausible if the conductivity function at 1500 km is relatively high, like that of a Nagata basalt, and the function at $r = 1400$ km is low, like that of an olivine. This reasoning leads to a moderate temperature profile such as sketched in Figure 5, with a temperature on the order of 700°C at $r = 800$ km and approximately 800°C at the center. Since the sensitivity of the model is low near the center, the central temperature is inferred from the fit of lunar thermal history models to the profile outside of the boundary $r = 800$ km

(Fricker *et al.*, 1967). The thermal profile curve is consistent with a lunar thermal history model with 25% chondritic radioactive concentration; i.e., one must assume that the Moon had been relatively deficient in radioactive concentration compared to the chondritic meteorites during its formation to avoid higher temperatures in the present epoch, assuming traditional values for thermal conductivity. Acceptance of a solid state thermal conduction mechanism such as proposed by Runcorn (1962; see also Turcotte and Oxburgh, 1969) would not alter the present-day thermal profile estimate but would allow the Moon to have been hotter at an earlier time.

In a parallel investigation using a different set of magnetic data Dyal and Parkin (1971) derived a monotonic lunar profile. The data were restricted to periods when the LSM was in the lunar night and consequently separated from the confining pressure of the incoming solar wind. Any lunar induced magnetic perturbation on the dark side is free to propagate into the essentially plasma-free cavity in the solar wind shadow. The geometry is complicated by the fact that the field direction is generally oblique to the axis of the shadow region. No unified model of the Moon's reaction that includes both the sunward confinement and the cavity region has been developed. For tractibility, just as Sonett *et al.* (1971a, d) assumed a spherical current layer for the dayside analysis, Dyal and Parkin assumed the Moon to be immersed in a vacuum, since the distance from the lunar measurement site to the confining currents is large.

Dyal and Parkin analyzed the vertical component of many step function transients observed in the dark side. They fitted the data to a two-layer model moon of radius R_M, with $r=R_1$ denoting the boundary between the inner layer of conductivity σ_1 and the outer nonconducting layer. Theory predicts that if the vacuum field surrounding the model moon is homogeneous and undergoes a unit step increase, the vertical component of the surface field rises as (Dyal and Parkin, 1971)

$$B = 1 - \frac{6}{\pi^2}\left(\frac{R_1}{R_m}\right)^3 \sum_{s=1}^{\infty} \frac{1}{S^2} \exp\left(\frac{-S^2\pi^2 t}{\mu_0\sigma_1 R_1^2}\right)$$

The initial value at $t<0$ is $B=0$, and for $t\gtrsim 0$, $B=1-(R_1/R_M)^3$. The final value is unity. The dominant time constant is $\tau=\mu_0\,\sigma_1 R_1^2/\pi^2$, controlling $\sim$60% of the series at the onset and dominating more and more as the higher order terms die out.

In Figure 6 is shown the data for one of the many steps analyzed by Dyal and Parkin; in this case the step was negative-going. The dashed line shows the fit of their model, with $0.95\ R_M < R_1$, $\sigma = 1.5\times10^{-4}\ \Omega^{-1}\ \mathrm{m}^{-1}$ and the time constant $\tau\sim 55$ s. The curve marked 'step response' is the response under the same conditions of a model similar to that of Figure 5, a single current layer profile in which the conductivity is considered negligible except at $r=1505$ km where a layer of thickness Δr has a conductivity σ characterized by $\sigma\Delta r = 70\ \Omega^{-1}$. The initial fractional drop is $1-(R_1/R_M)^3$, for this model as well as for the two-layer model, and for $R_1=1505$ km the initial drop does not conform to the data, as is shown in the figure. Schubert and Colburn (1971) suggest, however, that the input function is imperfectly known but

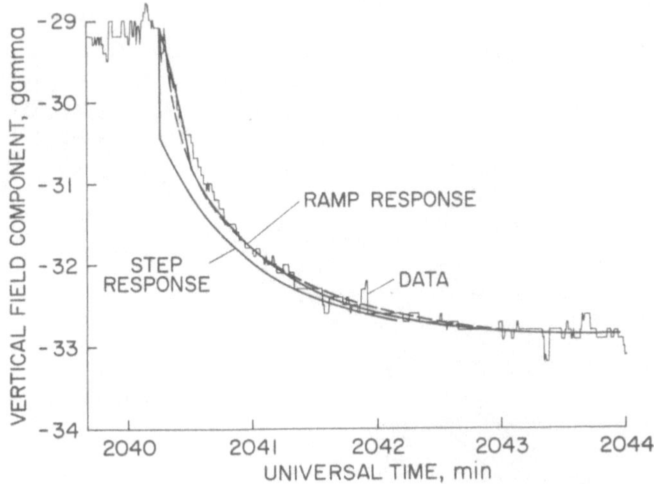

Fig. 6. Transient observed in the vertical component of lunar surface magnetometer data while the magnetometer site was in lunar night. Eddy currents in the interior inhibit rapid changes in this component. The dashed line is the response of the Dyal and Parkin model obtained from many of these transients (Dyal and Parkin, 1971). Another model, described in the text, has the responses shown depending on whether the input function is best described as a step or a ramp (Schubert and Colburn, 1971).

that since most transients in the solar wind are frozen into the plasma, the entire Moon does not see the field change at the same time. A more appropriate concept is that of a wave front sweeping over the Moon in a finite length of time. Schubert and Colburn approximate the effect of the moving wave front by a ramp function by which the surrounding field, assumed homogeneous, rises to its final value in 15 s. For this driving function the current layer model responds according to the curve labeled 'ramp response', which is satisfactorily close to the data.

The data of Figure 6 represent but one case of the 10 used by Dyal and Parkin to obtain a size and conductivity for the core: $\sigma = 1.7 \pm 0.4 \times 10^{-4}\ \Omega^{-1}\ \mathrm{m}^{-1}$, $R_1 > 0.95\ R_\mathrm{M}$. Consequently, while a fit to the data of Figure 6 is not definitive Schubert and Colburn make two points: (1) it is important to consider the correct driving function for the Moon, and (2) if the ramp approximation is valid, the profile of Figure 5 and also two-layer profiles with $R_1 < 0.95\ R_\mathrm{M}$ are candidates for fitting the transient data.

Dyal and Parkin (1971) report a long tail on their transient response curves representing time decays longer than 4 min. The fit to these data implies a region of still higher conductivity deeper in the Moon, represented in the model by a third layer of conductivity $\sigma_2 > 10^{-2}\ \Omega^{-1}\ \mathrm{m}^{-1}$ and $R_2 > 0.6\ R_\mathrm{M} = 1040$ km. The implications for the thermal profile are a higher inner temperature, as high as $1240\,\mathrm{K} = 967\,°\mathrm{C}$.

It has been questioned whether the conductivity maximum in the conductivity profile is necessary to fit the data because a monotonic profile is simpler and would appear to follow the experimental data fairly closely. A comparison of the fits of

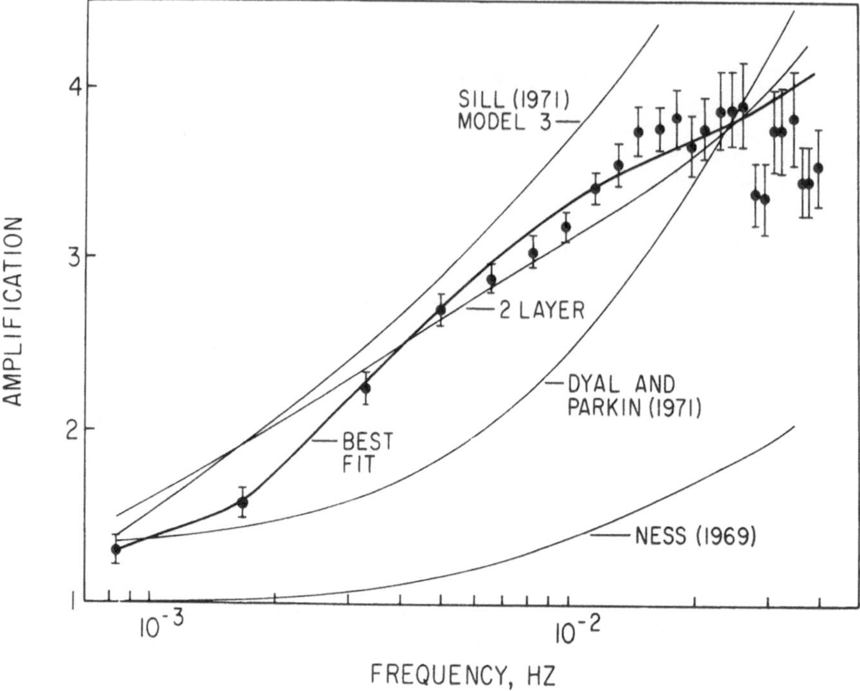

Fig. 7. Amplification as a function of frequency for several lunar electrical conductivity models. The data and the curve labeled 'best fit' are as shown in Figure 4. The curve labeled 'two-layer' is a best fit under the constraints of a two-layer model when $R_1 = 1560$ km. The conductivity peak in the model is necessary in order to come close enough to the three lowest frequency data points.

various models is shown in Figure 7. A two-layer model is shown, with a core of radius $R_1 = 1560$ km and a conductivity $\sigma = 7.6 \times 10^{-4}\ \Omega^{-1}\ \mathrm{m}^{-1}$. This model was attained as a best fit to the A data at the given radius and is close to the model of Kuckes (1971) for which $R_1 = 1580$ km and $\sigma = 6 \times 10^{-4}\ \Omega^{-1}\ \mathrm{m}^{-1}$. The difference between the fits of the two-layer and the conductivity peak model is principally at the three lowest frequencies: a two-layer model fitting the middle range cannot dip sharply enough to approach the data at the lower frequencies. Other curves shown for comparison are those of model 3 of Sill (1971) and Ness (1969). Other models investigated by Sill do not provide a better fit than his model 3. The model of Ness had $R_1 = 1426$ km and $\sigma = 8 \times 10^{-5}\ \Omega^{-1}\ \mathrm{m}^{-1}$. This model was based on an interpretation of a transient event observed by Explorer 35. Sonett *et al.* (1971e) have since proposed an alternate explanation for the data that would remove the basis for that model. It was concluded that Explorer 35 was too far from the Moon to measure a lunar response.

A difficulty with model fits for frequencies greater than 10^{-2} Hz lies in the wavelength of the driving function (Schubert and Schwartz, 1972). Convection in the solar wind dominates; wave propagation velocities in the frame of reference of the solar wind plasma are generally much less than the bulk velocity of the solar wind so that the wavelength can be approximated by

$$\lambda = (n \cdot v)/f$$

where n is the unit normal to the wave front and v the solar wind bulk velocity. For frequencies below 10^{-2} Hz, $\lambda/R_M \gg 1$ for nearly all wave directions. Model fits by Sill and Kuckes are based on infinite wavelength. For the work reported here, $n \cdot v$ is taken to be 400 km s^{-1}, introducing differences in the high frequency response and also in the definition of the driving function using a modal expansion.

It is possible to extend the model to include the higher order spatial harmonics. The extended model, more accurate at high frequencies, has been presented by Schubert and Schwartz (1971). The use of the model at frequencies above 10^{-2} Hz will depend on the feasibility of sorting out correct wave vector directions in the experimental data.

6. Conclusion

Measurement of a steady field at three places on the lunar surface implies that the surface layer was magnetized during cooling by a field estimated to have been at least of 1000 γ amplitude (Helsley, 1970). The source of this magnetizing field is not clear (cf. Sonett and Runcorn, 1971). Had a dynamo once been active in the lunar interior, using normal values of thermal conductivity, the Moon would now possess a molten core. This appears unlikely from the electrical conductivity calculations, and also from rigidity arguments. If the field were an extension of the photospheric field, its value at the photosphere would be very large; under conditions where the field pressure dominated the solar wind, the field would decrease as r^{-3}, requiring 10^5 G at the photosphere. If the solar wind had sufficient momentum density to shape the magnetic field, the radial field would follow an r^{-2} law (Parker, 1963). The field requirement would then be more modest, 600 G, but the solar wind density would be orders of magnitude greater than the present day value, a situation not expected to have occurred after a possible T-Tauri stage of solar evolution (Sonett *et al.*, 1968). Polarity reversals due to the Sun's rotation would also make it unlikely that the Sun's field could magnetize a rock layer slowly cooling through the Curie point.

Immersion of the Moon in the Earth's field to attain a 1000 γ magnetizing field implies a synchronously rotating Moon at an orbital distance dangerously close to the Roche limit, unless the earth's field had been far greater than present day values (Dyal *et al.*, 1970a). A satisfactory explanation for the required magnetizing field has not yet been found.

The conductivity profiles obtained by the dayside data imply a discontinuity at $r = 1500$ km. The conductivity maximum implies a change at that depth from a more to a less electrically conducting material. Alternatively the presence at that depth of a relatively thin highly conducting layer as suggested by Urey (see Urey *et al.*, 1971) would also fit the electrical conductivity data. In this case the $\sigma \Delta r$ product would be $\sim 10^2\ \Omega^{-1}$ and the lunar interior temperature might be lower than suggested if a rock other than olivine is postulated for the deep interior.

The barrier at a depth of some 250 km appears even for models with monotonic conductivity profiles, although there is some question from the darkside transient data as to whether a shallower depth is more appropriate. A barrier implies a change of composition or state at that depth, possibly the lower limit of a surface melting process. The process may be tied to the surface melting required to explain the ages of lunar samples. Reynolds *et al.* (1971) have considered models for the surface heating process by a combination of accretional and radioactive means. They find certain combinations that fit the timetable of lunar rock ages. Wright (1971) suggests that the conductivity profile can be explained by oxygen depletion.

Except for the Explorer 35 survey the positive magnetic measurements of the moon have been confined to three locations for the steady field and one location for the time varying field. The successful deployment of the Apollo 15 LSM in August 1971 promises additional detail, and together with the Apollo 16 LSM will allow consideration of possible lunar asymmetry (Schubert and Schwartz, 1971a). Other refinements to be investigated are the separation of the TM and TE modes and the determination of the lunar response for various k vectors.

References

Barnes, A., Cassen, P., Mihalov, J. D., and Eviatar, A.: 1971, *Science* **172**, 716.
Behannon, K. W.: 1968, *J. Geophys. Res.* **73**, 7257.
Blank, J. L. and Sill, W. R.: 1969, *J. Geophys. Res.* **74**, 736.
Chapman, S. and Bartels, J.: 1962, *Geomagnetism*, Oxford University Press, London.
Colburn, D. S., Currie, R. G., Mihalov, J. D., and Sonett, C. P.: 1967, *Science* **158**, 1040.
Colburn, D. S., Mihalov, J. D., and Sonett, C. P.: 1971, *J. Geophys. Res.* **76**, 2940.
Dyal, P. and Parkin, C. W.: 1971, *J. Geophys. Res.* **76**, 5947.
Dyal, P., Parkin, C. W., and Sonett, C. P.: 1970a, *Science* **169**, 762.
Dyal, P., Parkin, C. W., and Sonett, C. P.: 1970b, *IEEE Transactions on Geoscience Electronics* GE-8, 203.
Dyal, P., Parkin, C. W., Sonett, C. P., DuBois, R. L., and Simmons, G.: 1971, in *Apollo 14 Preliminary Science Report*, NASA SP-272, U.S. Government, Washington, D.C.
England, A. W., Simmons, G., and Strangway, D.: 1968, *J. Geophys. Res.* **73**, 3219.
Fricker, P. E., Reynolds, R. T., and Summers, A. L.: 1967, *J. Geophys. Res.* **72**, 2649.
Helsley, C. E.: 1970, *Science* **167**, 693.
Kuckes, A. F.: 1971, *Nature* **232**, 249.
Lahiri, B. N. and Price, A. T.: 1939, *Phil. Trans. Roy. Soc.* **A237**, 509.
Mihalov, J. D., Colburn, D. S., Currie, R. G., and Sonett, C. P.: 1968, *J. Geophys. Res.* **73**, 943.
Mihalov, J. D., Sonett, C. P., Binsack, J. H., and Moutsoulas, M. D.: 1971, *Science* **171**, 892.
Nagata, T., Rikitake, T., and Kono, M.: 1970, in *Space Research X*, North Holland Publishing Co., Amsterdam.
Ness, N. F.: 1969, NASA Goddard Space Flight Center Report X-616-69-191.
Ness, N. F., Behannon, K. W., Scearce, C. S., and Cantarano, S. C.: 1967, *J. Geophys. Res.* **72**, 5769.
Parker, E. N.: 1963, *Interplanetary Dynamical Processes*, Interscience, New York.
Reynolds, R. T., Fricker, P. E., and Summers, A. L.: 1971, *AIAA Progress Series Volume, Lunar Thermal Characteristics*, American Institute of Aeronautics and Astronautics, New York (in press).
Runcorn, S. K.: 1962, *Nature* **195**, 1150.
Runcorn, S. K., Collinson, D. W., O'Reilly, W., Stephenson, A., Greenwood, N. N., and Battey, M. H.: 1970, *Science* **167**, 697.
Schubert, G. and Colburn, D. S.: 1971, *J. Geophys. Res.* **76**, 8174.
Schubert, G. and Schwartz, K.: 1969, *The Moon* **1**, 106.
Schubert, G. and Schwartz, K.: 1971, *Cosmic Electrodyn.* **2**, 244.

Schubert, G. and Schwartz, K.: 1972, *J. Geophys. Res.* **77** (in press).
Schwerer, F. C., Nagata, T., and Fisher, R. M.: 1971, *The Moon* **2**, 408.
Sill, W. R.: 1971, *J. Geophys. Res.* **76**, 251.
Sonett, C. P.: 1965, in *Space Research VI*, Spartan Books, Washington, 280.
Sonett, C. P.: 1966, *Minutes of Apollo Lunar Surface Science Symposium* **1**, E-1, unpublished.
Sonett, C. P.: 1968, *IEEE Transactions on Geoscience Electronics* GE-6, 126.
Sonett, C. P. and Colburn, D. S.: 1967, *Nature* **216**, 340.
Sonett, C. P. and Mihalov, J. D.: 1972, *J. Geophys. Res.* **77** (in press).
Sonett, C. P. and Runcorn, S. K.: 1971, *Comm. Astrophys. Space Phys.* **3**, 149.
Sonett, C. P., Colburn, D. S., and Schwartz, K.: 1968, *Nature* **219**, 924.
Sonett, C. P., Colburn, D. S., Dyal, P., Parkin, C. W., Smith, B. F., Schubert, G., and Schwartz, K.: 1971a, *Nature* **230**, 359.
Sonett, C. P., Dyal, P., Colburn, D. S., Smith, B. F., Schubert, G., Schwartz, K., Mihalov, J. D., and Parkin, C. W.: 1971b, in C. de Jager (ed.), *Highlights of Astronomy 1970*, D. Reidel Publ. Co., Dordrecht, p. 173.
Sonett, C. P., Dyal, P., Parkin, C. W., Colburn, D. S., Mihalov, J. D., and Smith, B. F.: 1971c, *Science* **172**, 256.
Sonett, C. P., Schubert, G., Smith, B. F., Schwartz, K., and Colburn, D. S.: 1971d, in *Proceedings of the Second Lunar Science Conference* **3**, MIT Press, Cambridge, Massachusetts, 2415.
Sonett, C. P., Mihalov, J. D., and Ness, N. F.: 1971e, *J. Geophys. Res.* **76**, 5172.
Turcotte, D. L. and Oxburgh, E. R.: 1969, *Nature* **223**, 250.
Urey, H. C., Marti, K., Hawkins, J. W., and Liu, M. K.: 1971, in *Proceedings of the Second Lunar Science Conference* **2**, MIT Press, Cambridge, Massachusetts, 987.
Wright, D. A.: 1971, *Nature Phys. Sci.* **231**, 169.
Wilcox, J. M. and Colburn, D. S.: 1970, *J. Geophys. Res.* **75**, 6366.

THERMAL GRADIENTS IN THE OUTER LUNAR LAYERS

J. A. BASTIN, S. J. PANDYA, and D. A. UPSON

Queen Mary College, University of London, England

Abstract. During the next Apollo mission Apollo 15, it is planned to fix thermocouples at various depths up to 3 m below the lunar surface. It seems likely that the resulting temperature measurements will show a positive temperature differential with depth resulting from a net outward heat flux. It is the purpose of this paper to examine experiments already carried out which indicate a temperature gradient. Since the thermal flux is of direct importance in fundamental problems of lunar origin and the nature of the lunar interior, the relation between this quantity and the temperature gradient will be examined.

1. Thermal Balance in the Outer Surface Layers

The thermal transfer relaxation time for a system of linear dimension L is given by:

$$\tau \sim L^2 \left(\frac{\varrho c}{k}\right) \tag{1}$$

in which ϱ is the density, c the specific heat and k a thermal transfer coefficient which we can formally introduce to take into account the effects of convection, radiative transfer and true lattice conduction. It is well known that, if we introduce only thermal lattice conductivity into Equation (1) taking the values of diffusivity $k/\varrho c$ from the measurements of lunar basalts measured at 230 K (Horai *et al.*, 1970) then Equation (1) shows that the thermal relaxation time τ for the Moon as a whole is greater than t_m, the age of the Moon, by a factor of about 10^2. However, over smaller distances of the order of 50 km and less the reverse is the case and Equation (1) indicates that the heat flux over consecutive layers is virtually constant for all layers within this distance scale. In this case, if r_1 and r_2 are the radii, measured from the lunar centre, of two relatively close spheres, then

$$r_1^2 k_1 \left(\frac{\partial T}{\partial r}\right)_{r=r_1} = r_2^2 k_2 \left(\frac{\partial T}{\partial r}\right)_{r=r_2}$$
$$r_1 - r_2 \ll \left[\left(\frac{k}{\varrho c}\right) t_n\right]^{1/2}. \tag{2}$$

We now apply Equation (2) to two layers, one in the regolith whose thermal gradient is to be measured by Apollo 15 and the second in the bedrock some kilometers below the lunar surface. The limit of sensitivity of a thermocouple measurement is of the order of 10^{-2} K and it is at once clear from Equation (2) that the Apollo 15 mission will only be able to measure the temperature gradient with worthwhile fractional accuracy if the regolith layer has a thermal conductivity which is very much less than that of the bedrock below. The conductivity of the fines (Cremers *et al.*, 1970; Gough *et al.*, 1971) measured in vacuo in the laboratory is nearly a factor of 10^3 less than

Urey and Runcorn (eds.), The Moon, 372–376.

that for type A basalt samples (Horai *et al.*, 1970). If we now, as a first approximation, identify type A basalt with the bedrock and the regolith with surface fines of constant density, we can determine the thermal gradient in the regolith providing we know the thermal gradient in the bedrock below. Estimates of this bedrock gradient may be obtained from evolutionary models (e.g., McDonald, (1959), 4×10^{-3} km^{-1}) but a more direct estimate comes from the magnetometer experiment (2×10^{-3} K m^{-1}; Sonett *et al.* (1971)). The two approaches agree within an order of magnitude, and together with laboratory rock measurements already quoted show that we should expect from Equation (2) a regolith gradient of 2 K per meter, a value which should be measureable with high accuracy by the Apollo 15 experiment.

Perhaps the most questionable assumption made in the above approximate treatment is the identification of the conductivity of the lunar surface layer with that of fines measured in the laboratory, and this will now be considered in detail.

2. Thermal Conductivity of the Lunar Regolith Layers

We first review briefly the direct thermal conductivity measurements made in the laboratory with lunar fines.

Perhaps the most reliable method is the line source method employed by Cremers *et al.* (1970 and 1971). The method uses a straight wire as a heat source within the powdered fines and the wire itself acts as its own resistance thermometer. The method shows the expected increase of effective conductivity with temperature resulting from the radiative contribution to the thermal transfer: although there is some experimental scatter, it seems unlikely that the measurements are more in fractional error than about 0.2. The method gives a value of 1.9×10^{-3} Wm^{-1} K^{-1} at the mean lunar temperature 230 K.

A more recent method reported by Gough *et al.* (1971) relies essentially on subjecting the base of a thin cylinder of fines to a step function change of temperature and measuring the radiation from the top surface as a function of time so as to determine the time variation of effective temperature of the top surface. The method is not so well developed as the line source method although it should be relatively accurate. At 320 K a value of 6.0×10^{-3} Wm^{-1} K^{-1} is obtained which, in view of the increase of conductivity with temperature, is in fair agreement with the line source method.

Whilst both the above methods probably give reasonably accurate values of the conductivity of the fines under the laboratory conditions, the measured values may for several reasons differ from those of the lunar surface layer. Firstly, in both cases the material was packed into a container so that the important intercrystalline contacts could well differ considerably from those of material on the lunar surface; secondly, although the measurements were made under high vacuum conditions, the samples had in both cases been exposed to air prior to the experiment. For these reasons it is of considerable interest that a laboratory measure of the far infrared absorption coefficient (Bastin *et al.*, 1970; Ade *et al.*, 1971) makes it possible to determine the conductivity of the top few centimeters of the lunar surface. Unlike the thermal conductivity the infrared absorption coefficient would be expected to be relatively insensitive to transportation and packaging. This measurement, together with microwave measurements at different frequencies of the lunar brightness temperature throughout a lunation, enable the quantity $k\varrho/c$ to be determined and this relates directly to the top few centimetres of the lunar surface since it is only this layer which contributes to the microwave measurements. Since ϱ and c are known with considerable accuracy, the method enables k to be determined directly and the value obtained is considerably lower than that found from the direct laboratory determination (10^{-4} $\mathrm{Wm^{-1}\,K^{-1}}$).

Although the top few centimeters of the lunar surface probably have very low conductivity, it is by no means sure that such conductivities extend into the regolith. Gough *et al.* (1971) have shown that compression of fines causes a considerable increase of conductivity and the various thermal and shock annealing processes known to exist in the regolith might well be expected to increase the conductivity still further – perhaps by an order of magnitude or more so that it approaches more the conductivity of the bedrock.

The experimental evidence for this comes from measurements of diffusivity of breccia (5.85×10^{-7} $\mathrm{m^2\,s^{-1}}$) by Horai *et al.* (1970). Taking a value for density as 2.21×10^3 $\mathrm{kg\,m^{-3}}$ and for specific heat as 5.40×10^2 $\mathrm{J\,kg^{-1}\,K^{-1}}$ the thermal conductivity at the mean lunar temperature of about 200 K turns out to be 0.7 $\mathrm{W\,m^{-1}\,K^{-1}}$. This value is not appreciably less than the conductivity of basalt (1.6 $\mathrm{W\,m^{-1}\,K^{-1}}$) Horai *et al.* (1970).

Considering breccia to be made from compressed fines, the result shows a marked increase in conductivity as the inter-grain contact is increased either by pressure or by melting. This shows that the thermal gradient would rapidly approach the value of the bedrock as one goes deeper in the regolith.

In conclusion, from direct measurements on lunar samples, it would thus seem that the regolith layer may have a thermal differential over its depth of several degrees K per meter, if not more. In view of the relatively hard and stable nature of this layer, such a result may seem intuitively unlikely and we therefore will look at other experiments which also indicate an appreciable gradient in the regolith.

3. Temperature Gradients in the Regolith

Although in the proposed Apollo 15 experiment there may be some problems with ensuring that the thermocouple junctions are strictly at the temperature of the nearby rock, there is no doubt that the experiment should be very reliable. It is, however, of interest at this time to consider the other methods by which this gradient has been inferred.

The first method utilises measurements of the lunar brightness temperature T_λ at wavelengths in the range 0.05–0.50 m. Throughout this range lunar rock has a relatively low electromagnetic absorption coefficient so that radiation received at the Earth comes not from the surface but is a sum of contributions from depths ranging up to many metres below the surface. If T is the thermodynamic temperature at depth z then it follows from the theory of radiative transfer that for the center of the Moon's disc

$$T_\lambda = (1 - R_0) \int_{z=0}^{\infty} k_\lambda \varrho T_z \, e^{-k_\lambda \varrho z} \, dz \tag{3}$$

where k_λ and ϱ are the electromagnetic mass attenuation coefficient and density respectively. R_0 is the reflection coefficient at normal incidence. If now we assume a uniform temperature gradient b with T_0 as the surface temperature

$$T_z = T_0 + bz \tag{4}$$

then Equation (3) gives the measured brightness temperature

$$T_\lambda = (1 - R_0)\left[T_0 + \frac{b}{k_\lambda \varrho}\right]. \tag{5}$$

Radar reflection measurements and measurements on lunar samples show that $(1 - R_0)$ lies between 0.90 and 0.97 and does not vary appreciably with wavelength. The differential of Equation (5) with λ would therefore be expected to give a direct measure of the thermal gradient which can be found absolutely if k_λ and ϱ are known. A discussion by Ade *et al.* (1971) indicates that

$$\frac{1}{k_\lambda \varrho} = 2 \cdot 6\lambda \tag{6}$$

so that (5) and (6) give

$$\frac{\partial T_\lambda}{\partial \lambda} = 2 \cdot 3b\,. \tag{7}$$

(It will be noted that the factor on the left hand side of (7) is dimensionless, i.e. it has the same numerical value whatever units the other quantities are expressed in. This would be expected since both b and $\partial T_\lambda / \partial \lambda$ have the dimensions of temperature divided by distance.)

The result given in Equation (7) strictly refers only to the center of the lunar disc. A more detailed treatment following a different approach but averaging over the whole disc has been given by Krotikov and Troitski (1964). The results of this analysis only modify Equation (7) by relatively small factors. The absorption coefficients for the lunar fines (Ade *et al.*, 1971) show the absorption coefficient of the lunar rock at 2mm wavelength to be greater than anticipated before the return of rock samples.

To some extent this discrepancy has been removed by the discovery of a scattering component in the fines (Clegg *et al.*, 1972).

If we take the values of the brightness temperature measured as a function of wavelength collected both by Troitskii (1967) and Linsky (1966) values of about 35 K m^{-1} are obtained for the thermal gradient.

The thermoluminescence data taken from various depths in the core samples (Hoyt *et al.*, 1971) also give evidence of a strong thermal gradient. The rising curve in the first few centimeters is without doubt the effect of high temperature on the top part of the surface due to the diurnal thermal input. From 8–40 cm, however, superimposed on random fluctuations, there is a marked slope which would be accounted for by a positive temperature gradient.

When allowance is made for the variation of ionisation with depth the method gives a value of 2 ± 2 K m^{-1} for the gradient.

All the work cited so far has shown a positive temperature differential with depth. However, a recent publication by Salisbury and Fernald (1971) using Arecibo Radio Telescope in the metre wavelength range indicates an inverse temperature gradient with depth. The reason given for this effect was the transport of heat by latent heat changes during the evaporation of mercury.

Our preliminary calculations based on the available data of the mercury content of the lunar soil gives a very negligible gradient due to this effect and we believe a probable error in telescope calibration could have caused this effect.

Since the writing of this paper, Apollo 15 experiment (Langseth, 1972) has shown a gradient of 1.75 K m^{-1} at depth below which thermal wave has negligible effect. The experiment also measured the conductivity at depths of the order of 1 metre giving a value of 1.7×10^{-2} W m^{-1} K^{-1}. These two results together with the work of Sonett *et al.* (1971) imply a surprisingly high conductivity for the deep bedrock.

References

Ade, P. A., Bastin, J. A., Marston, A. C., Pandya, S. J., and Puplett, E.: 1971, *Proc. Second Lunar Science Conference* **3**, 2203.

Bastin, J. A., Clegg, P. E., and Fielder, G.: 1970, *Proc. Apollo XI Lunar Science Conference* **3**, 1987.

Clegg, P. E., Pandya, S. J., Bastin, J. A., and Foster, S.: 1972, 'Far Infrared Properties of Lunar Rock', *Proc. Third Lunar Science Conference*, to be published.

Cremers, C. J., Birkebak, R. C., and Dawson, J. P.: 1970, *Proc. Apollo XI Lunar Science Conference* **3**, 2045.

Cremers, C. J. and Birkebak, R. C.: 1971, *Proc. Second Lunar Science Conference* **3**, 2311.

Gough, G. O., Pandya, S. J., and Bastin, J. A.: 1969, Private Communication.

Horai, K., Simmons, G., Kanamori, H., and Wones, D.: 1970, *Proc. Apollo XI Lunar Science Conference* **3**, 2243.

Hoyt, H., Jr., Miyajima, M., Walker, R., Zimmerman, D., Zimmerman, J., Britton, D., and Kardos, J. L.: 1971, 'Radiation Dose Rates and Thermal Gradients in the Lunar Regolith: Thermoluminescence and DTA of Apollo XII Samples' **3**, 2245.

Krotikov, V. D. and Troitskii, V. S.: 1963, *Soviet Phys. Uspekhi* **6**, 841.

Langseth, M. G. Jr.: 1972, *Third Lunar Science Conference* **3**, 475.

Linsky, J. L.: 1966, *Icarus* **5**, 606.

MacDonald, G. J. F.: 1959, *J. Geophys. Res.* **64**, 1967.

Sonett, C. P., Colburn, D. S., Dyal, P., Parkin, C. W., Smith, B. F., Schubert, G., and Schwartz, K.: 1971, *Nature* **230**, 359.

CONVECTION IN THE MOON

S. K. RUNCORN

Department of Geophysics and Planetary Physics, School of Physics University of Newcastle upon Tyne, England

Abstract. It is natural to inquire whether thermal convection is occurring in the Moon through solid state creep processes. The primary evidence is the departure of the Moon from the figure of hydrostatic equilibrium, but certain difficulties in the thermal history of the Moon are eased by assuming heat transfer by convection. If convection exists in the Moon it must have a second harmonic pattern, otherwise the lunar moments of inertia would not differ.

Two important predictions of the marginal theory of convection: the existence of a core of radius 0.06–0.3 of the lunar radius (for a second ergee harmonic) and the value of 0.4 for the ratio of the dynamical to surface ellipticities now have support, the latter from the data of the heights of the lunar surface. The former prediction is compatible with the value of the moment of inertia factor now found if the Moon's interior is 'hot'.

Further the existence of a fluid iron core 3400 m.y. ago seems required as a result of the remanent magnetization of the crystalline rocks of the maria basins inferred from the remanent magnetization of the returned Apollo samples and the fields measured by the Apollo 12 and Explorer magnetomers.

1. The Convection Hypothesis

Convection was first suggested in the lunar interior to explain its non-hydrostatic figure by Runcorn (1962, 1967). He pointed out that the discrepancy between the dynamical and the surface ellipticities of the Moon is significant and that it is naturally explained by convection because the density in the rising column is necessarily less than in the falling current. He further showed that the observed differences in the moments of inertia, apart from those corresponding to the hydrostatic model, require that the convection pattern has a second harmonic component. He also argued that if this was the main component the surface would be distorted into an ellipsoid, the long axis of which would lie in the centre of the rising or falling current.

Roberts (1965) has calculated, assuming the marginal convection theory, the ratio of the dynamical to the surface ellipticity. He used Baldwin's figure of 0.0012 for the latter and therefore he stated that this ratio (about 0.5) did not agree with theory and that in consequence convection in the Moon was unlikely. However as is shown in Table II the theoretical values Roberts finds depend on the boundary conditions and in his paper he favoured a free boundary for the outer surface of the convecting region. Study of continental drift, sea floor spreading and plate motions in the earth have given us a better insight into the mechanical behaviour of the solid interior of planets over long times (10^6–10^8 yr). The crust or lithosphere behaves like a rigid body with finite strength even over such periods whereas in the mantle below, flow over such times occurs under small stress differences. This sharply contrasting mechanical behaviour results from the occurrence of a Boltzmann factor in solid state processes, including solid state creep, the large temperature gradient near the surface determining the depth at which this rather sudden change in mechanical properties occurs.

Urey and Runcorn (eds.), The Moon, 377–383.

In the Earth convection currents move the continents and plates: the boundary condition is a complex one but may be more nearly a free one than a rigid one. Each plate is large so that it is probable that the boundary condition cannot simply be a zero slip or zero stress. On the other hand there have clearly been no relative movements of parts of the lunar crust: no 'continental drift' has occurred on the Moon since the formation of the crust. Thus the appropriate boundary condition for the outer surface is certainly a rigid one, i.e. zero radial velocity gradient. Table II shows that, using the newer data, the fit of observations to the prediction of convection theory is good.

To a good approximation the ellipticity e of the gravitational equipotential surface of the Moon is $(\frac{3}{5})(C\text{-}A)/C$. Thus it can be inferred that the process which has caused the present non-equilibrium figure of the Moon does not predate the filling of the maria basins with lava but occurred some time later. This conclusion follows from Table I in which the best fit to ellipsoids of the points on the uplands and points on the maria, taken together and separated into those with mascons and those without, all yield ellipsoids of the same shape. With a reasonable accuracy the differences between the semi-axes are about the same.

TABLE I

Axes of ellipsoids fitted to uplands, all maria, mascon maria and other maria (km)

	Semi-axis toward Earth a_x	Equatorial semi-axis in plane of sky a_y	Polar semi-axis a_z	$a_x-(a_y+a_z)/2$	a_y-a_z	Number of points
Uplands	1740.2	1737.3	1736.5	3.3	0.8	532
All maria	1739.1	1736.0	1734.7	3.8	1.3	385
Mascons	1738.7	1734.5	1734.0	4.5	0.5	95
Other maria	1739.3	1736.1	1735.2	3.7	0.9	290

TABLE II

Comparison of ratios of dynamical to surface ellipticities

Height of bulge	Baldwin-Roberts (1949), (1965)	Recent values [a]	Theory (Roberts, 1965)	
Dynamically	1 km	1.3 km	free	rigid
From geometrical libration	2 km	3.3 km	boundary	boundary
Ratio	0.5	0.39	0.2006	0.397

[a] Runcorn and Gray (1967); Runcorn and Shrubsall (1968); Runcorn and Hofmann, this volume, p. 22.

2. Lunar Palaeomagnetism

The magnetic studies on the Apollo 11 and 12 samples have shown that the crystalline rocks possess varying but appreciable magnetic stability, comparable with those terrestrial igneous and sedimentary rocks used in palaeomagnetic investigations. A natural remanent magnetization was found by Runcorn *et al.* 1970 to be possessed by the Apollo 11 crystalline rock of about 6.10^{-6} emu gm^{-1}, and it has been shown to be almost certainly acquired before it was returned to the Earth. Domains of iron were found to be responsible for the magnetic properties and these would have been likely to carry the remanent magnetization of the rock. As the blocking temperature of these particles is of the order of 600 °C, far higher than the maximum temperature reached by the rocks during the lunar day, it was concluded that the remanent magnetization was acquired when the rocks cooled from above the Curie point (780 °C) or soon afterwards as a result of chemical change. This clearly occurred when the rocks were part of the bedrock beneath the regolith, when they were part of the lava flows which filled Mare Tranquillitatis and Oceanus Procellarum. These lavas cooled from their melting point 3700 m.y. and 3400 m.y. ago respectively. Estimates of the field in which these lavas cooled is 1000 γ($1\ \gamma = 10^{-5}$ G). This reasoning leads to the theory that the Moon possessed a field of internal origin in the first 1000 m.y. or so of its existence of about 1000 γ at the surface which it has since lost, as the field now is less than 2 γ (Runcorn *et al.*, 1970, 1971).

This argument has been greatly strengthened by the discovery that a mean field of 36 γ (in addition to the fluctuating magnetic field of the solar wind) exists at the Apollo 12 landing site on Oceanus Procellarum. The magnetometer left there after the landing incorporated a gradiometer and as the field gradient was less than 10^{-8} G cm^{-1}, Sonett concluded that the mean field must arise from an extended source of at least 1 km in dimension. As the regolith is only a few metres thick and as Orbiter photographs suggest that the maria basins are filled with thin and extensive lava flows lying in series, it is reasonable to infer that the mean field arises from uniformly magnetized lava flows, in which case they have intensity of about 10^{-5} emu gm^{-1}. But this is true only if it is assumed that different flows are magnetized in similar directions, for if lava flows covering large formations of this maria basin are each magnetized in directions more or less randomly distributed, and if the depth of the basin is small compared with its diameter, then the net fields at the surface due to each will tend to cancel out. We conclude therefore that the magnetizing field remained constant not only over the time of cooling between Curie point and blocking temperature of one flow but for most of those in the basin. On this theory the intensity of magnetization found in the Apollo samples and the mean field measured by the Apollo 12 magnetometer are consistent. This argument strongly reinforces the explanation of the remanent magnetization by a field of internal origin and not one of external origin, for the monthly rotation of the Moon requires that only the component of an external field parallel to the axis of rotation of the Moon 'can be effective'. This rules out a solar wind field even if it were reasonable to suppose it to be much enhanced early in the history of the solar system. Further were the geomagnetic field to be the cause, the Moon would have had to be close to the Roche limit, within 2 or 3 Earth radii, if the geomagnetic field then was about the same strength as at present. While some discussions of the early history of the Earth-Moon system entertain the possibility of such a close approach, dynamical considerations only allow the Moon to be close to the Earth for a very short time. Thus the significant difference of age between Apollo 11 and 12 crystalline rocks and the recent indications that the highlands are also magnetized, requires there to have been a field present for a considerable period in the Moon's early existence. The hypothesis that the Earth's field is responsible for this rather widespread magnetization of the lunar crust is not very attractive.

In order to explain the existence of a lunar magnetic field of internal origin in the early history of the Moon which has since disappeared it is necessary first to postulate the existence of an iron core. As is shown in Figure 1, Runcorn (1967) postulated the existence of an iron core in order that the convection in the Moon

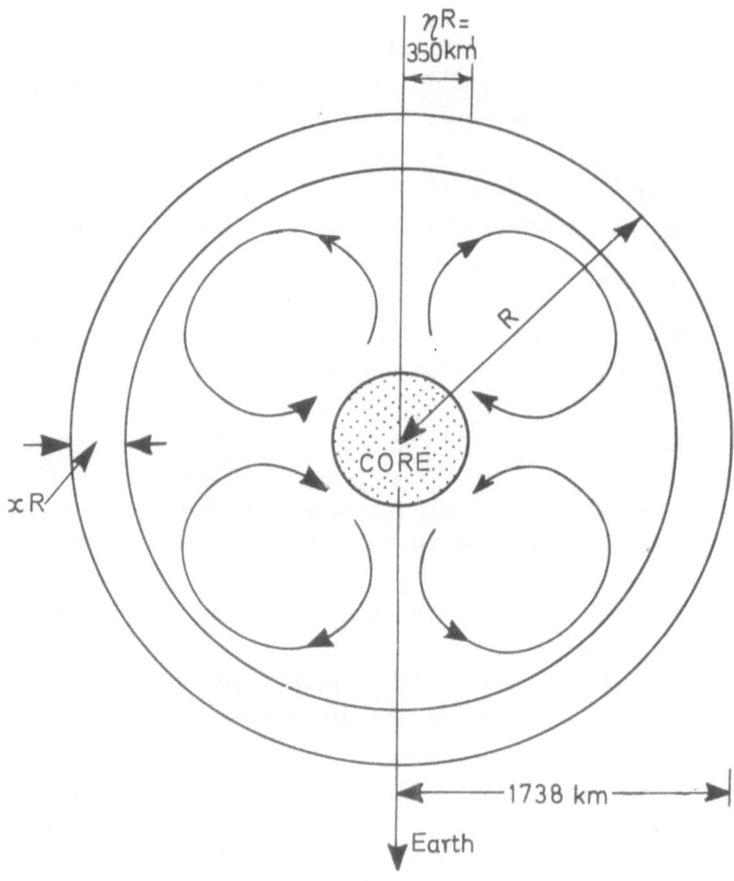

Fig. 1.

should be a second degree harmonic pattern. According to the marginal theory of convection in spherical shells the harmonic which establishes itself depends on the ratio of the radii of the inner and outer boundary surface and also on the boundary conditions. He therefore concluded that the radius of the core would have to lie between 0.06 and 0.3 of the radius of the Moon. Because of the very small pressure gradient in the Moon, the adiabatic gradient is also small so that it seems reasonable in discussing the problem of convection in the Moon to assume that the viscosity is uniform and the results of a rather simplified convection theory can be applied. In such a core, supposed molten, convection might once have been vigorous enough to cause the magnetic field to be generated according to dynamo theory. Its later disappearance might be due to the magnetic Reynolds number falling below the critical value for dynamo action (about 10) or due to cooling of the Moon resulting in the solidification of the iron core, or due to the decrease in the angular velocity of rotation of the Moon as it recedes from the Earth. In each case the field could die away quite quickly within a period of the order of the free decay time which is about a few thousand years for a core of this size. The first alternative is rather attractive because in a small core it is not easy to make the magnetic Reynolds number great enough for dynamo action – if one supposes the radius of the iron core to be 300 km then the magnetic Reynolds number is about $10^3 \times$ the velocity of the convective motions in cm s^{-1}. As the motions in the Earth's core are about 1 cm s^{-1}, motions not much less than this would generate dynamo action. Cooling of the Moon is also suggested by the evidence that in its early history a great amount of lava was produced, whereas in recent times there appears not very much evidence for very extensive lava flows.

3. The Existence of a Lunar Core

Runcorn (1967) concluded from the convection theory that a small iron core exists in the Moon of between 0.06 and 0.3 of the external radius: if no core exists theory suggest that a first rather than a second harmonic convection pattern would be expected and this would neither give a bulge nor give differences in the moments of inertia. The argument that the Moon once possessed a magnetic field requires such a core, and so differentiation must have taken place very early in lunar history and the central temperature 3000 m.y. ago must have been above M.P. of iron.

Runcorn showed that such a core was compatible with the Moon's mean density but now the question arises whether the recently determined moment of inertia factor $f = (C/Ma^2)$ of the Moon allows the existence of a core of these dimensions. It has always been clear that a core of much greater than the above could not be allowed on density considerations: but in any case a convection pattern of higher degree than the second fails to give differences in the moments of inertia.

The thickness of the lunar crust is probably greater than the Earth's lithosphere but if the temperature gradients are comparable then, due to the low pressures in the Moon, the density in the lunar crust decreases slightly with depth in the convecting region the density is nearly constant with radius because of the very small values of the adiabatic gradient. Thus it is possible to use a simple model of the distribution of density within the Moon to test whether a lunar core is allowable.

Let ϱ_m be the mean density of the Moon.
ϱ_c be the density of the core.
ϱ_1 be the density of the mantle apart from the crust.
η be the ratio of the core radius to Moon radius.
x be the radio of the crust thickness to Moon radius.
T be the temperature at the base of the crust.
α be the volume coefficient of thermal expansion of the mantle.

$$\varrho_m = \varrho_1 + \eta^3 (\varrho_c - \varrho_1) + 3x\alpha T\varrho_1/2 \tag{1}$$

$$\varrho_m (1 - \eta^3 \varrho_c/\varrho_m) = (1 - \eta^3)\,\varrho_1 + 3x\alpha T\varrho_1/2 \tag{2}$$

Thus

$$f = \tfrac{2}{5}[1 - (\eta^3 - \eta^5)(\varrho_c/\varrho_1 - 1) + x\alpha T] \tag{3}$$

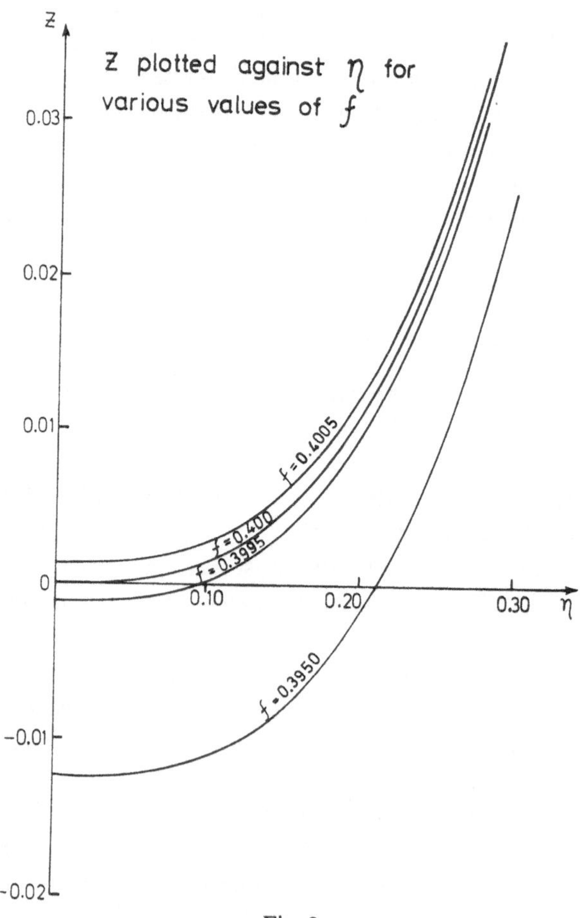

Fig. 2.

ϱ_1 can be eliminated by combining Equations (1) and (3). Expand neglecting small terms. Thus

$$x\alpha T = \frac{[(\varrho_c - \varrho_m)\eta^3(1-\eta^2)/\varrho_m + 5f/2 - 1]}{(1 + 3\varrho_c(5f/2 - 1)/2(\varrho_c - \varrho_m))}.$$

Figure 2 shows $z = x\alpha T$ plotted against η for various values of f putting $\varrho_m = 3.34$ and $\varrho_c = 7.9$.

It seems that for the most recent value of f given by Michael *et al.* (1970) 0.4005, a core with $\eta = 0.2$ is possible.

References

Baldwin, R. B.: 1949, *The Face of the Moon*, Chicago University Press.

Dyal, P., Parkin, C. W., and Sonett C. P.: 1970, *Science* **169**, 762.

Michael, W. H.: 1970, *The Moon* **1**, 484.

Roberts, P. H.: 1965, *Mathematics* **12**, 128.

Runcorn, S. K.: 1962, *Nature* **195**, 1150.

Runcorn, S. K.: 1967, *Proc. Roy. Soc.* **A296**, 240.

Runcorn, S. K. and Gray, B. M.: 1967, *The Mantles of the Earth and Terrestrial Planets*, John Wiley, New York.

Runcorn, S. K. and Shrubsall, M. H.: 1968, *Phys. Earth Planet. Interiors* **1**, 317.

Runcorn, S. K., Collinson, D. W., O'Reilly, W., Stephenson, A., Greenwood, N. N., and Battey, M. H.: 1970a, *Science* **167**, 697.

Runcorn, S. K., Collinson, D. W., O'Reilly, W., Battey, M. H., Stephenson, A., Jones, J. M., Manson, A. J., and Readman, P. W.: 1970b, *Magnetic Properties of Apollo 11 Lunar Samples, Geochim. Cosmochim. Acta, Supp. 1* **3**, 2369.

Runcorn, S. K., Collinson, D. W., O'Reilly, W., Stephenson, A., Battey, M. H., Manson, A. J., Readman, P. W.: 1971, *Proc. Roy. Soc. London A* **325**, 157.

POSSIBLE THERMAL HISTORY OF THE MOON

PETER E. FRICKER*, RAY T. REYNOLDS, and AUDREY L. SUMMERS
Ames Research Center, NASA, Moffett Field, California, U.S.A.

Abstract. The possible thermal history of the Moon is investigated by means of theoretical models. The calculations include the effects of melting and time-dependent redistribution of radioactive heat sources. The known constraints can best be satisfied by a model which is characterized by relatively high initial temperatures close to the melting range; melting and, consequently, fractionation and redistribution of radionuclides would occur during the first 1.5×10^9 yr and would then be followed by an effective cooling process. Heat flow measurements on the lunar surface should permit a distinction between such a completely fractionated model and a non-fractionated model or a model with restricted fractionation in the outer few hundred kilometers.

1. Introduction

In this contribution, a reconstruction of the thermal history of the Moon is attempted on the basis of theoretical models which are discussed in terms of present physical and geological evidence.

Calculations on the thermal history of the Moon have been performed for a wide range of assumed conditions by Urey (1952, 1962), MacDonald (1959, 1962), Kopal (1962), Levin (1962), Maeva (1964), Phinney and Anderson (1965), Fricker *et al.* (1967), and Reynolds *et al.* (in press). The results of these calculations indicate that the temperature distribution in the interior of the Moon is mainly dependent on the initial temperature, that is the temperature field upon completion of the formation of the Moon, and on the abundance and distribution of radioactive heat sources. Since these and other important parameters are uncertain, the range of possible theoretical models varies considerably. The calculations rely on the equation of heat conduction for a spherically symmetric solid body with internal heat sources

$$\varrho C_p \frac{\partial T}{\partial t} = \frac{1}{r^2} \frac{\partial}{\partial r} \left(r^2 K \frac{\partial T}{\partial r} \right) + A$$

where density ϱ, heat capacity C_p, temperature T, thermal conductivity K, and rate of heat production A are functions of radius r and time t.

In the event of melting in the lunar interior, the subsequent thermal history will be changed. Levin (1962) accounted for the latent heat of fusion and calculated the onset of melting for a number of models. Reynolds *et al.* (1966) showed the importance of melting phenomena and, for purposes of calculation, defined a stage of partial melting at the melting temperature. At the stage of partial melting, the heat of fusion would be absorbed before the onset of complete melting or released before resolidification. If complete melting occurs, increased heat transfer by fluid convection could take place on a large scale; Reynolds *et al.* (1966) have simulated the

* Current address: Swiss National Science Foundation, 3000 Bern, Switzerland.

Urey and Runcorn (eds.), The Moon, 384–391.

limiting case of highly efficient fluid convection by removing all the excess heat energy above the liquidus into the adjacent outer radius interval.

As a consequence of melting, differentiation would set in (McConnell *et al.*, 1967). In connection with the thermal evolution, redistribution of radioactive materials is of particular significance. Goldschmidt (1954) and others have pointed out that the principal long-lived radioactive heat sources U, Th and K, having large ionic radii, show a pronounced upward concentration in the Earth since they are rejected by the close-packed silicate lattices of the mantle. The fractionation trend of the radioactive elements in the upper layers of the Earth, in a physical environment comparable to that of the lunar interior, suggests that the interior of the Moon would also become depleted of radioactive materials if melting occurs. This suggestion is supported by the measured U, Th and K contents of lunar Apollo 11 and 12 samples (Compston *et al.*, 1970; LSPET, 1969; LSPET, 1970; Wakita *et al.*, 1970).

A numerical procedure for investigating the consequences of fractionation by time-dependent concentration of radionuclides toward the surface was developed by Fricker *et al.* (1967). With the onset of melting, the radioactive heat sources U, Th and K start to move into the adjacent outer radius interval after an initial stage of uniform radioactive distribution. The redistribution process is described by two variable parameters, the residual fraction *RF* and the moving factor *MF*. The residual fraction refers to the mass fraction of the radioactive isotopes retained in a given radius interval after the occurrence of partial or complete melting. The amount of radioactive isotopes in excess of the residual fraction is removed into the adjacent radius interval at a rate determined by the moving factor.

The effects of melting and fluid convection are included in the present calculations; the consequences of redistribution of radionuclides are also taken into consideration. The numerical procedure follows the outline given by Reynolds *et al.* (1966, in press) and Fricker *et al.* (1967).

The numerical values for the thermal parameters are listed in Tables I and II. The surface temperature boundary condition was held constant at 0°C. In line with age determination for lunar material from the Apollo 11 and 12 landing sites (Albee *et al.*, 1970, 1971; Tatsumoto, 1970; Turner, 1970, 1971) a lunar age of 4.6×10^9 yr was assumed. The liquidus of anhydrous basalt (Cohen *et al.*, 1967), which is very

TABLE I

Parameters for the thermal calculations

Radius, r	1.738×10^8 cm
Density, p	3.34 g/cm^{-3}
Heat capacity, C_p	1.2 J g^{-1} deg
Lattice conductivity, c	7.89×10^5 J cm^{-1} yr deg
Index of refraction, n	1.7
Opacity, ε	20 cm^{-1}
Surface temperature, T	273 K
Age	4.6×10^9 yr

TABLE II

Abundances, heat generation, and decay constants of radioactive heat sources

Radioactive nuclide	Abundance in 10^{-8} g/g		Radioactive heat generation B_j, J g^{-1} yr	Decay constant λj, $\times 10^{-10}$ yr^{-1}
	Meteoritic model [a]	Lunar model		
U^{233}	1.84	3.27	2.97	1.54
U^{235}	0.0133	0.024	18.0	9.71
Th^{232}	5.80	13.16	0.82	0.499
K^{40}	5.0	0.98	0.94	5.5

[a] Urey and MacDonald, 1970 (Table 11,4)

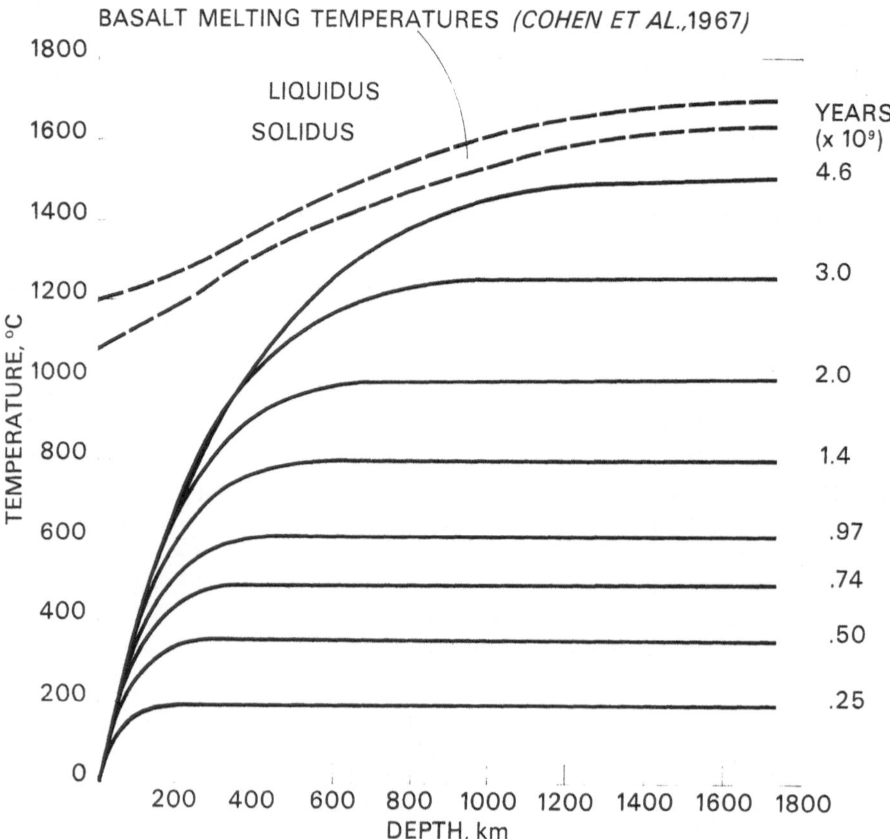

Fig. 1. Temperature distribution within initially cold solid Moon having meteoritic radioactivity (model 1).

close to the liquidus of the Apollo 11 basalts (Akimoto *et al.*, 1970; Ringwood and Essene, 1970) and to the solidus of peridotite (Kushiro *et al.*, 1968), was used as the effective melting temperature (Figure 1). Two models for radioactive abundances have been considered for the present calculations (Table II). Following the estimate for the average U content of the Earth's mantle by Wasserburg *et al.* (1964), an average initial U concentration of 3.29×10^{-8} g/g was adopted for the 'lunar' model. The measured abundances of radioactive heat sources in the Apollo 11 and 12 samples were taken into account by employing a K/U ratio of 2500 and a Th/U ratio of 4. The meteoritic model is based on the abundances of radionuclides in Type III carbonaceous chondrites as proposed by Urey and MacDonald (to be published). In the present models, the initial temperature distribution accounts for possible early heat contributions from accretional energy, decay of short-lived radionuclides, unipolar electric currents induced by interaction with the solar wind (Sonett and Colburn, 1968), and other short-term energy sources. The energy contribution by tidal interaction is very uncertain and has not been specifically considered. Large-scale solid state convection for the Moon has been proposed by Runcorn (1962). Fricker *et al.* (1967) suggest that the 'cool' outer layers and the small size of the Moon relative to the Earth would minimize the effect of this process.

2. Development of Thermal History Models

The temperature distribution in three different thermal history models of the Moon is presented in Figures 1 to 3. For a comparison of these models with actual evidence, an important constraint provided by crystallization age analyses has to be considered; work by Albee *et al.* (1970, 1971), Turner (1970, 1971) and others indicates that melting of basaltic maria material at the Apollo 11 and 12 landing sites occurred some 3.3 to 3.8×10^9 yr ago. It is likely that these basalts were formed by lava flows of internal origin (Ringwood and Essene, 1970).

Model 1 (Figure 1) shows the internal temperature distribution for a uniform Moon which has remained solid throughout its history because of the assumed low initial temperature of 0°C. After 4.6×10^9 yr, the internal temperatures at depths of around 900 km are approaching the solidus of anhydrous basalt. The present surface heat flow is 9.1×10^{-7} J cm^{-2} s. Although the melting temperatures of low-melting components may be reached at a late stage, the occurrence of lava flows of internal origin some 3.5×10^9 yr ago is not compatible with such a model. Even if an outer initially molten and fractionated layer of 200 km is considered for the 'cold' model, the internal temperatures after 10^9 yr are well below the solidus of basalt because of the efficient cooling process in the outer part of a lunar-sized body (Fricker *et al.*, 1970; Reynolds *et al.*, in press). One might add that the existence of a molten and fractionated outer layer at the initial stage has not been established; further age analyses of lunar samples should shed more light on this question.

In contrast to model 1, model 2 (Figure 2) is characterized by high initial temperature conditions. It was assumed that the Moon was melted throughout and that

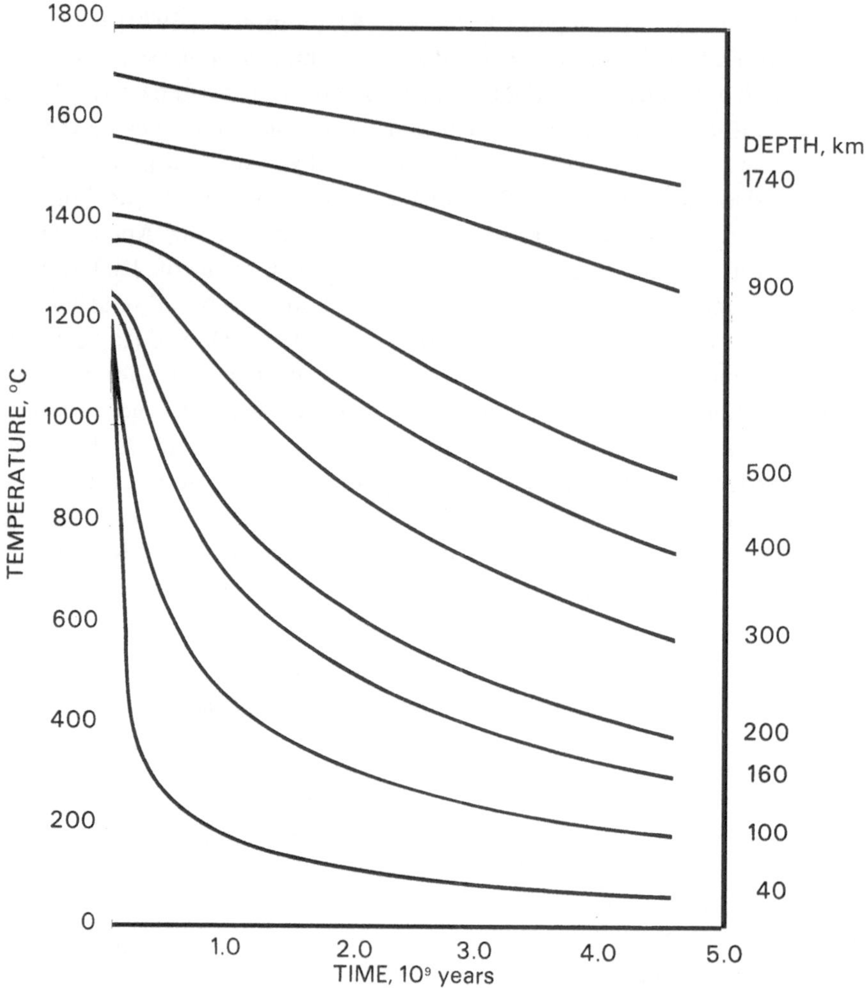

Fig. 2. Temperature distribution within initially molten and completely fractionated Moon having 'lunar' radio-activity (model 2).

all the radioactive heat sources were concentrated in the outermost radius interval during the period of formation. Subsequently, the outer layers are affected by a rapid cooling process. Thus it is difficult to account for the occurrence of volcanic activity 3.8 to 3.3×10^9 yr ago. The present surface heat flow is 18.2×10^{-7} J cm^{-2} s.

Model 3 (Figure 3) represents an intermediate case between the initially cold model and the initially molten model. The initial temperature has been defined so that the internal temperatures reach a maximum during the presumed period of maria formation about 3.3 to 3.8×10^9 yr ago. Wood (to be published) has recently chosen a similar approach.

A constant initial temperature of 1250 °C was used for model 3; a near-surface melting and fractionation event at the initial stage was considered for the outer

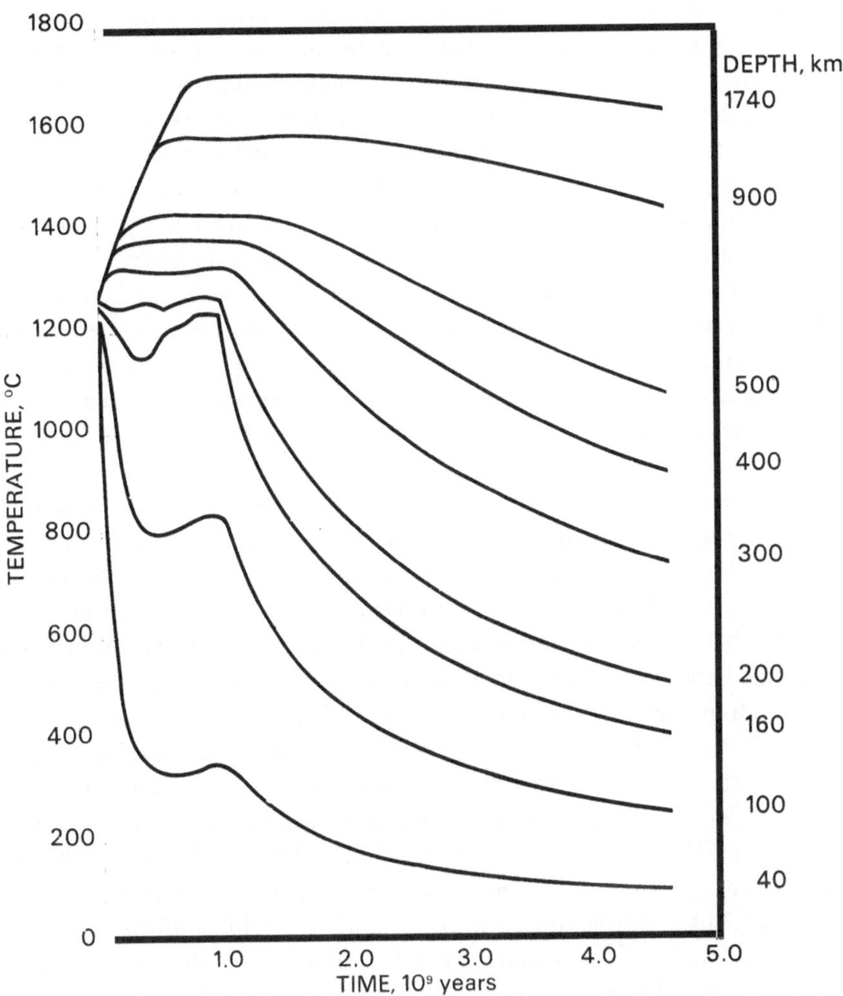

Fig. 3. Effects of melting and time-dependent fractionation of radionuclides on the temperature distribution within initially hot Moon having 'lunar' radioactivity (model 3).

200 km. If the internal temperatures reach the liquidus of basalt the radioactive heat sources are removed toward the surface. A residual fraction of 0.1, corresponding to 10% of the radioactive mass fraction, was employed. In this model, the temperature maximum at depths from 200 km to 700 km is reached during the first 1.5×10^9 yr and is then followed by a pronounced cooling period. In the outer 400 km, the present temperatures are below 900°C, that is less than in the 'cold' model 1. The surface heat flow is 18.5×10^{-7} J cm^{-2} s.

3. Discussion

The present thermal calculations confirm that a definite solution cannot be established on the basis of the known data and constraints. However, the inadequacy of two extreme types of thermal history models can be shown. The uniform model with low initial temperatures (Figure 1), even with a molten and fractionated outer layer of the order of 200 km, is not consistent with the reported occurrence of basaltic lava flows of internal origin during the period 0.8 to 1.3×10^9 yr after formation. To a lesser degree, an initially molten and completely fractionated model (Figure 2) meets with similar difficulties. An intermediate model characterized by relatively high initial temperatures close to the melting range and by time-dependent redistribution of radioactive heat sources (Figure 3) is more consistent with the actual evidence.

It is likely, in any event, that the bulk of the Moon has been affected by melting and fractionation. The occurrence of complete melting on a large scale would have been inhibited mainly by radioactive depletion of partially molten regions and by subsequent concentration of the radioactive heat sources in the near-surface layers.

Low initial temperatures have been advocated by Urey and MacDonald (to be published) in order to explain nonhydrostatic conditions as inferred for the lunar interior from the figure and the existence of mascons. However, the amount of structural strength within the interior is uncertain (Reynolds *et al.*, in press). It can also be noted that the present temperatures in the outer part of the Moon are lower for the intermediate model 3 (Figure 3) than for the uniform solid model 1 (Figure 1).

Clues regarding the distinction between the different possible types of thermal history models should be obtained by the planned lunar heat flow measurements. The present surface heat flow for the initially molten and fractionated model 2 as well as for model 3 with time-dependent fractionation is higher by a factor of 2 in comparison with the calculated heat flow of 9.1×10^{-7} J cm^{-2} s for the initially low-temperature, solid model 1, even when an initially molten outer layer of 200 km is considered. Thus a distinction between an initially 'cold' uniform model and an initially 'hot' or molten, fractionated model should be possible by means of heat flow determinations. The two types of fractionated models (models 2 and 3) could probably not be separated on this basis but additional new information from space missions should eventually permit to further delimit the number of possible solutions.

Acknowledgements

One of us (P. E. Fricker) acknowledges partial support by the Aerospace Institute Agreement between Ames Research Center and the University of Santa Clara.

References

Akimoto, S. I., Nishikawa, M., Nakamura, Y., Kushiro, I., and Katsura, T.: 1970, *Geochim. Cosmochim. Acta, Suppl. 1, Proceedings of the Apollo 11 Lunar Science Conference*, 129.

Albee, A., Burnett, D. S., Chodos, A. A., Eugster, O. J., Huneke, J. C., Papanastassiou, D. A., Podosek, F. A., Price Russ II, G., Sanz, H. G., Tera, F., and Wasserburg, G. W.: 1970, *Science* **167**, 463.

Albee, A., Burnett, D. S., Chodos, A. A., Haines, E., Huneke, J. C., Papanastassiou, D. A., Podosek, F. A., Price Russ II, G., Tera, F., and Wasserburg, G. W.: 1971, *1971 Lunar Science Conference Abstracts*, Houston, Texas, 56.

Cohen, L. H., Ito, K., and Kennedy, G. C.: 1967, *Am. J. Sci.* **265**, 475.

Compston, W., Chappell, B. W., Arriens, P. A., and Vernon, M. J.: 1970, *Geochim. Cosmochim. Acta, Suppl. 1, Proceedings of the Apollo 11 Lunar Science Conference*, 1007.

Fricker, P. E., Reynolds, R. T., and Summers, A. L.: 1967, *J. Geophys. Res.* **72**, 2649.

Fricker, P. E., Goldstein, J. I., and Summers, A. L.: 1970, *Geochim. Cosmochim. Acta* **34**, 475.

Goldschmidt, V. M.: 1954, *Geochemistry*, Oxford at the Clarendon Press, p. 730.

Kopal, Z.: 1962, 'Thermal History of the Moon and of the Terrestrial Planets: Numerical Results', TR 32-225, Jet Propulsion Laboratory, California Institute of Technology.

Kushiro, I., Yoder, H. S., Jr., and Nichikawa, M.: 1968, *Geol. Soc. Am. Bull.* **79**, 1685.

Levin, B. J.: 1962, in Z. Kopal and Z. K. Mikhailov (eds.), *The Moon*, Academic Press, New York, p. 157.

LSPET (Lunar Sample Preliminary Examination Team): 1969, *Science* **165**, 1222.

LSPET (Lunar Sample Preliminary Examination Team): 1970, *Science* **167**, 1325.

MacDonald, G. J. F.: 1959, *J. Geophys. Res.* **64**, 1967.

MacDonald, G. J. F.: 1962, *J. Geophys. Res.* **67**, 2945.

Maeva, S. V.: 1965, *Soviet Phys. Dokl.* **9**, 945.

McConnell, R. K., McClaine, L., Lee, D., Aaronson, J., and Allen, J.: 1967, *Rev. Geophys.* **5**, 121.

Phinney, R. A. and Anderson, D. L.: 1965, 'Internal Temperatures of the Moon', Minnesota University Rept., Tycho Meeting.

Reynolds, R. T. and Summers, A. L.: 1969, *J. Geophys. Res.* **74**, 2494.

Reynolds, R. T., Fricker, P. E., and Summers, A. L.: 1966, *J. Geophys. Res.* **71**, 573.

Reynolds, R. T., Fricker, P. E., and Summers, A. L.: in J. Lucas (ed.), *Lunar Thermal Characteristics*, AIAA Progress Series (in press).

Ringwood, A. E. and Essene, E.: 1970, *Geochim. Cosmochim. Acta, Suppl. 1, Proceedings of the Apollo 11 Lunar Science Conference*, 769.

Runcorn, S. K.: 1962, *Nature* **195**, 1150.

Sonett, C. P. and Colburn, D. S.: 1968, *Nature* **219**, 924.

Tatsumoto, M.: 1970, *Geochim. et Cosmochim. Acta, Suppl., Proceedings of the Apollo 11 Lunar Science Conference*, 1595.

Turner, G.: 1970, *Geochim. Cosmochim. Acta, Suppl. 1, Proceedings of the Apollo 11 Lunar Science Conference*, 1665.

Turner, G.: 1971, *1971 Lunar Science Conference Abstracts*, Houston, Texas, p. 63.

Urey, H. C.: 1952, *The Planets: Their Origin and Development*, Yale University Press, New Haven, Conn.

Urey, H. C.: 1962, in Z. Kopal (ed.), *Physics and Astronomy of the Moon*, Academic Press, New York, p. 481.

Urey, H. C. and MacDonald, G. J. F.: in Z. Kopal (ed.), *Physics and Astronomy of the Moon*, 2nd ed., Academic Press, New York (to be published).

Wakita, H., Schmitt, R. A., and Rey, P.: 1970, *Geochim. Cosmochim. Act,a Suppl. 1, Proceedings of the Apollo 11 Lunar Science Corference*, 1685.

Wasserburg, G. J., MacDonald, G. J. F., Hoyle, F., and Fowler, W. A.: 1964, *Science* **143**, 465.

Wood, J. A.: in G. Simmons (ed.), *The Geophysical Interpretation of the Moon*, preprint Sept. 1970 (to be published).

H. THE EVOLUTION OF THE MOON'S ORBIT

THE ROLE OF OCCULTATIONS IN THE IMPROVEMENT OF THE LUNAR EPHEMERIS

L. V. MORRISON

Royal Greenwich Observatory, England

Abstract. Analyses of occultation timings show that periodic correction terms with semi-amplitude as great as 0.″18 arise from corrections required to the empirical constants of the Brown/Eckert theory. Using the atomic time-scale, some of the occultation data have been used to determine a correction of $-30 \pm 16''/\text{cy}^2$ to Spencer Jones' value for the secular acceleration of the Moon. In the light of this correction, and previous determinations, attention is drawn to the possible weakness of Spencer Jones' value, which is not reflected in his quoted error of $\pm 1''/\text{cy}^2$. Further analyses of 50000 occultations observed since 1943 promise to reveal more accurately-determined corrections.

1. Introduction

Since the time of Ptolemy, occultations of stars have been used to provide a simple and effective method of monitoring the angular motion of the Moon relative to the stellar background. Naked-eye observations of a few bright stars made some two thousand years ago are used today in studying secular changes in the lunar motion. Nowadays small portable telescopes and a method of recording time to a precision of about 0.1 s are necessary to make a useful observation. From the star position, the time of the occultation to 0.1 s and the geodetic co-ordinates of the observer, we can (ideally) obtain a fix on a point on the limb of the Moon to ±0.″05. With two or more timings, the position of the centre of figure is known after allowing for the difference between the limb profile and a mean sphere using Watts' data (1963).

2. Accuracy of Occultation Observations

The standard error of the residuals ($\Delta\sigma$) formed by taking the difference between the observed position of the limb, deduced from the occultation time and star position, and the calculated position using the lunar ephemeris, is about ±0.″43 (s.e.). Analyses (Morrison, 1970) of occultations of the Pleiades group on 1969 March 23 have shown that this value is comprised of the following errors:

(1) Timing	±0.″20
(2) Star place (Robertson)	±0.26
(3) Lunar ephemeris ($j=2$)	±0.16
(4) Profile corrections (Watts)	±0.20
(5) Observer's position	±0.10

The error for the lunar ephemeris $j=2$ is an estimate based on the effect of the corrections to Brown's constants given in Table I (see later) and the results of comparisons with numerical integrations made by Garthwaite *et al.* (1970) which reveal the deficiencies of the ephemeris due to Brown's truncation of planetary terms.

Urey and Runcorn (eds.), The Moon, 395–401. All Rights Reserved.

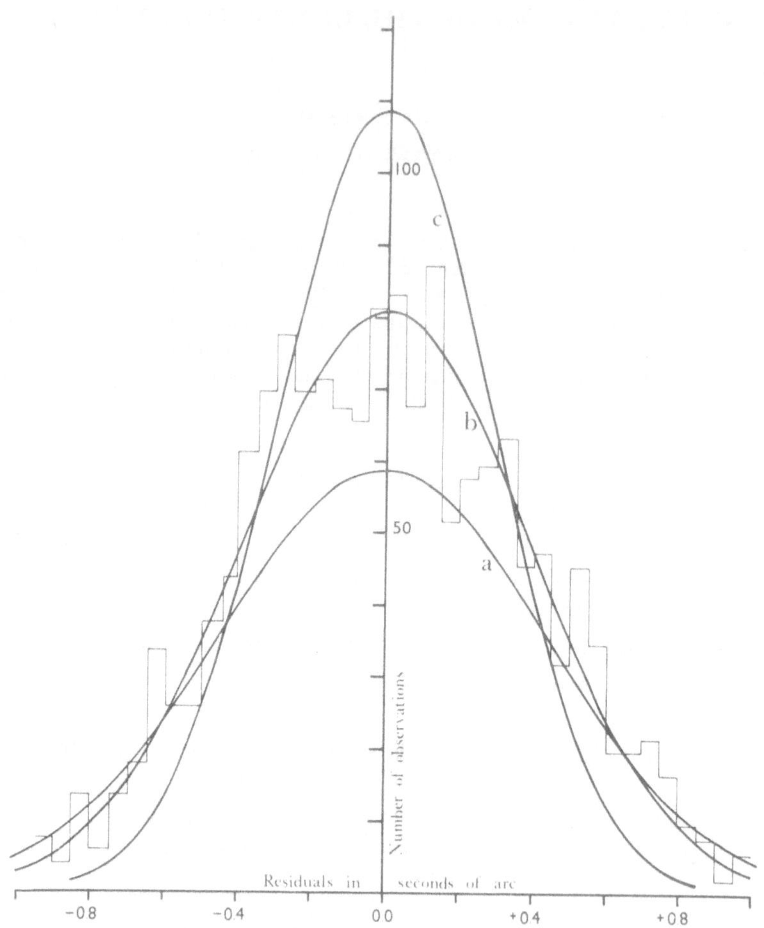

Fig. 1. Normal distribution curves for occultation residuals (see text).

Figure 1 shows three normal distribution curves corresponding to the standard deviations calculated from the residuals ($\Delta\sigma$) of occultations of the Pleiades group on 1969 March 23. The three curves are for the following cases:

(a) the residuals, uncorrected for errors 1 to 5 above s.e. $=0\overset{''}{.}43$

(b) the residuals, corrected for error 3 s.e. $=0.40$

(c) the residuals, corrected for errors 3 and 2 s.e. $=0.30$

The frequency distribution of the residuals for case (b) is also shown to indicate the reliability of a normal curve for this sample. The other two cases are equally reliable, but are not shown to avoid confusion in the figure.

The error due to the catalogue star places, which is the largest part of the standard error for one observation, is probably even greater for fainter stars. The desirability of improving the star places of Robertson's Zodiacal Catalogue (1940), which has a mean epoch of place around 1905, is therefore apparent. A programme of re-observation of these stars is now under way at the Royal Greenwich Observatory and the U.S. Naval Observatory.

3. Corrections to Brown's Constants from Occultations

Recently two independent analyses have been made of some of the occultation data which has been collated and coded at HM Nautical Almanac Office since 1943. Morrison and McBain Sadler (1969) analysed 10000 observations made during 1960–66, and Van Flandern (1971) analysed 7000 observations made during 1950–69. Their results for corrections to the adopted values of four of Brown's arbitrary constants (ILE, 1954), e, π, i, Ω (in elliptic motion, eccentricity, mean longitude of perigee, inclination, mean longitude of node) are given in Table I for epoch around 1960.0.

TABLE I

Summary of corrections to Brown's constants from occultations

Author	$\delta e \times 10^6$	$\delta\tilde{\omega}$	δi	$\delta\Omega$
Morrison and McBain Sadler	$+0.42$	$-1.''57$	$-0.''03$	$+2.''31$
	± 0.02	± 0.07	± 0.02	± 0.20
Van Flandern	$+0.24$	-0.7	-0.12	$+1.6$
	± 0.07	± 0.3	± 0.04	± 0.5

The differences between the corrections of Table I are probably due to two causes:

(1) Morrison and McBain Sadler's data only extends over seven years, thus, perhaps, not allowing complete separation of the unknowns in their solution; and

(2) the corrections given here are only four out of two different sets comprising 12 and 26 unknowns, respectively.

If we take mean values for the corrections to the constants in Table I this gives rise to the following correction terms in longitude and latitude to the ephemeris $j=2$ (IAU, 1967) with coefficients greater than 0."01:

$$\delta\lambda = +0.''14 \sin l + 0.''13 \cos l - 0.''03 \sin(l - 2D) - 0.''03 \cos(l - 2D)$$
$$\delta\beta = -0.''07 \sin F - 0.''18 \cos F,$$

where l, F and D have the usual meaning in Brown's notation.

The values in Table I do give an indication of the solution we might expect from a comprehensive analysis of about 50000 observations made since 1943 which is now under way. The unknowns to be determined will include the motions of the perigee and node, and also the secular acceleration in mean longitude.

4. Determination of the Secular Acceleration of the Moon from Recent Occultations

Much interest arises from the possibility of obtaining a reliable value for the secular acceleration in mean longitude, $\dot{n}_M$, using the occultation data after 1955.5 when a precise atomic-time scale is available to remove the effects of variations in the rate of rotation of the Earth from the universal time-scale. A range of 17 yr of data will be available in which to search for a possible correction to Spencer Jones' (1939)* value $-22''/\mathrm{cy}^2$ which is incorporated in the lunar ephemeris. The residuals in longitude after 17 yr resulting from a correction of, say, $10''/\mathrm{cy}^2$ would attain a value of 0".04, and the best straight line through the residuals would have a maximum departure from the postulated second degree curve of 0".02. Taking the standard error of an observation to be 0".40 [case (b) of Figure 1], and an average of about 1600 observations a year since 1955, we have 17 annual mean points, each with a standard error of 0".01. So the second degree correction will be detectable from the data, but only if it is not confused by other long-period corrections. For instance, the above correction is not large in comparison with the oscillatory variations (with a period of approximately 18 yr) in the residuals in longitude formed by the difference between a numerical integration ephemeris (JPL Lunar Ephemeris 16, Garthwaite *et al.*, 1970) and $j=2$ (Brown/Eckert theory). Confusion of this oscillation, which is not due to an erroneous secular acceleration, might lead to the doubtful significance of a correction derived from 17 yr of data. But if the oscillation is nearly sinusoidal with a period of 18 yr, then one might expect to separate the parabolic correction from this, provided that the observations are well distributed over the period.

5. Discussion of Several Determinations of $\dot{n}_M$

Using some of the occultation data in the period 1955.5–1969, Van Flandern (1970) has found $\dot{n}_M = -52 \pm 16''/\mathrm{cy}^2$, which is better in agreement with Newton's (1970) recent investigation of the 'ancient' solar eclipses where he finds $\dot{n}_M = -42 \pm 4''/\mathrm{cy}^2$, circa AD 0. Stephenson (1971) finds $\dot{n}_M = -32 \pm 5''/\mathrm{cy}^2$, circa 100 BC from a discussion of ancient total solar eclipses. Stephenson has rejected as unreliable some of the data included in Newton's solution, and has added new observations. In all, about 50% of the data is common to both discussions. Fotheringham (1920) does not actually give a value for $\dot{n}_M$ from solar eclipses alone, but a value of $-26''/\mathrm{cy}^2$ is consistent with his procedure. However, the triangular area of most probability in his famous diagram (p. 123) gives values of $-21''/\mathrm{cy}^2$ and $-33''/\mathrm{cy}^2$ for $\dot{n}_M$ at its extremities. De Sitter (1927) reconsidered Fotheringham's work and found $\dot{n}_M = -37''.7/\mathrm{cy}^2$ which, until recently, has usually been taken as the 'ancient' value. Thus, three different discussions of the ancient solar eclipses lead to negative values of $\dot{n}_M$ greater than $30''/\mathrm{cy}^2$. If one postulates that the secular acceleration of the Moon has remained nearly constant over the past 3000 yr, then, as has often been pointed out, the 'ancient' value is at variance with Spencer Jones' (1939) 'modern' value of $-22 \pm 1''/\mathrm{cy}^2$ based on observations of the position of the Moon, Sun, Mercury and Venus over the past 300 yr. The various values of $\dot{n}_M$ are collected together in Table II.

* Throughout this note the secular acceleration is the value of twice the coefficient of T^2 in the expression for the mean longitude of the Moon. In the lunar ephemeris the expression (due to Spencer Jones) is $-11.''22\,T^2$.

TABLE II

Values of $\dot{n}_M$ and their probable errors

Data	Author	$\dot{n}_M$ ("/cy^2)
From ancient total solar eclipses circa AD 0	Fotheringham	−26 (±5)
	de Sitter	−38 ±4
	Newton	−42 ±4
	Stephenson	−32 ±5
From modern meridian observations 1700–1930	Spencer Jones	−22 ±1
From recent occultations 1955–1969	Van Flandern	−52 ±16

6. The Secular Acceleration derived by Spencer Jones

Spencer Jones' solution for $\dot{n}_M$ is heavily dependent on the analysis of the transits of Mercury and meridian circle observations of the declination of the Sun. Observations of the Sun's declination taken together with the theory of its motion lead to the deduction of corrections to the Sun's longitude. Timing of the transits of Mercury across the face of the Sun also lead to corrections to the Sun's longitude and these corrections are interpreted as arising from the retardation of the Earth's rate of rotation. The combination of these observations with those of the longitude of the Moon made over the same period give a value of $\dot{n}_M$.

He took weighted mean values of the observed minus tabular declinations of the Sun over periods of 4 yr around 1930, extending to 10 yr at 1760. Thus, his quoted probable error for the value of $\dot{n}_M$ does not reflect the scatter of the original observations and would undoubtedly be greater than the value from the fit to the means. Moreover, it is well-known that large systematic errors are present in meridian circle observations of the Sun's position and the inevitable dependence of the T^2 term on the earlier observations, despite their lower weighting factors, makes one approach the solution with extreme caution. It might be valuable to repeat Spencer Jones' work by omitting the 18th-century observations of the Sun and extending the 20th-century observations to the present day, using the individual observations, rather than smoothed values, in deriving a solution. However, the solution from the Sun's declination observations agrees closely with the independent solution from Mercury's November transits*.

7. Discussion of the Values of $\dot{n}_M$ derived from Transits of Mercury

It is easier to inspect the solution from Mercury's transits than that from the Sun's declination. Here again, Spencer Jones preferred to use the weighted mean times of contacts II and III (contacts I and IV are very unreliable). He took these values from de Sitter's (loc. cit.) discussion, which was in turn based on the reduction of the times

* Transits can only occur in May and November when the Earth passes through the line of nodes of Mercury's orbit.

of transit given by Innes (1925). Williams (1939), later supplemented by Clemence (1943), published a thorough discussion, independently of Innes. I have summarized the values of $\dot{n}_M$ deduced from their analyses in Table III. The solutions and standard errors in Table III result from the different treatment of the observations rather than the differences in the times of contact used in the analyses.

Besides the unknown in T^2, a constant and linear term in T are introduced into the equations of condition when analysing the observations of Mercury's transits. Williams and Clemence also included five other unknowns: corrections to the longitude and motion of the node of Mercury's orbit; corrections to the adopted semi-diameters of the Sun and Mercury; and a correction to the mass of Venus. It is questionable whether all of these additional unknowns should now be included. The adopted inverse mass of Venus, 408 000, is relatively close to the recently determined value of $408\,522 \pm 3$. The difference of 522 leads to differences in the times of contact of usually less than 0''.1, which can certainly be ignored. From the duration of transits, Innes made preliminary solutions for the semi-diameters to be used in the analysis. He adopted values increasing with time to allow for the effect of the increasing optical power of telescopes, which reduced the effects of diffraction and irradiation on the observed diameters. Williams and Clemence's solution only permitted the derivation of mean values for the period. Whichever method is followed, the diameters are not well determined, but this is not of much consequence to the solution for the secular acceleration since the effects due to the diameters cancel out in the mean times of contacts II and III. When comparing the discordant solutions for the longitude and motion of Mercury's node from the transit and meridian observations, Clemence (loc. cit.) points to the weakness of the former which is largely dependent on the duration of the transits, and here there are inexplicable discrepancies in some of the observations, especially in the well-observed transit of 1940.

Provided that these corrections to the node and diameters are not closely correlated with the correction to the coefficient of T^2, it will be safe to allow them to be absorbed in the constant and linear term in T (with which they are correlated) in a solution which is primarily aimed at finding the coefficient of T^2.

Mainly to re-estimate the standard error of the solution made by Spencer Jones, I took Innes' reduced times of contact and weights and made a similar solution to Spencer Jones', but keeping the times of contact II and III separate, rather than taking weighted means. My solution for the November transits is shown in Table III.

TABLE III

Values of $\dot{n}_M$ derived from Transits of Mercury

	$\dot{n}_M$ (''/cy^2)	s.e.
Spencer Jones	−22.7	±1.6
Williams; Clemence	−17.9	±6.5
Morrison (Nov. only) unpublished	−21.1	±3.2

The value of $\dot{n}_M$ is not significantly different from Spencer Jones', but the standard error is doubled. The solution for the May transits gave an appreciably smaller value in accord with Spencer Jones' solution on page 554 of his paper. When I tried solutions by combining the November and May transits (with de Sitter's high weights for the latter drastically reduced) the residuals from the solutions differed systematically between the two sets of transits, and always gave an increased error for the solution of the secular acceleration. Spencer Jones arrived at his final value by taking a weighted mean of the separate solutions for November and May transits, which came close to the November value because of its appreciably smaller probable error. The combination of the November and May transits in the equations of condition probably accounts for the reduced value for $\dot{n}_M$ and its much greater standard error in the solution of Williams and Clemence.

8. Spencer Jones' Solution for $\dot{n}_M$ in Relation to the 'Ancient' Value

Unless one can reject the May transits – and there seems to be no obvious reason why one should – one should adopt Williams and Clemence's solution and standard error as being the most realistic solution for $\dot{n}_M$ from the transits of Mercury. Spencer Jones' value for $\dot{n}_M$ falls within Williams and Clemence's solution and standard error, but perhaps a figure of about $\pm 6''/\text{cy}^2$, rather than $\pm 2''/\text{cy}^2$ for the standard error should be borne in mind when comparing values of $\dot{n}_M$ from Mercury's transits with those from ancient eclipses, etc. Hence Spencer Jones' solution may be considered to be consistent with a value of, say $-30''/\text{cy}^2$, but not with a value as high as $-40''/\text{cy}^2$.

There still remains the independent solution from the Sun's declination, with its accompanying small probable error in Spencer Jones' analysis, but an extended re-analysis of the individual observations might well alter the position there also.

References

Clemence, G. M.: 1943, *Astron. Pap. Amer. Eph. Naut. Al.* **XI**, 53.
de Sitter, W.: 1927, *Bull. Astron. Inst. Neth.* **4**, 21.
Fotheringham, J. K.: 1920, *Monthly Notices Roy. Astron. Soc.* **81**, 104.
Garthwaite, K., Holdridge, D. B., and Mulholland, J. D.: 1970, *Astron. J.* **75**, 1133.
IAU, Report of Commission **4**, 1967. *Trans. IAU* **13B**, 49.
Improved Lunar Ephemeris, 1954. Suppl. to *Amer. Eph. Naut. Al.*, Washington D.C.
Innes, R. T.: 1925, Circular of Union Obs. South Africa, No. 65.
Morrison, L. V. and McBain Sadler, F. M.: 1969, *Monthly Notices Roy. Astron. Soc.* **144**, 129.
Morrison, L. V.: 1970, in C. de Jager (ed.), *Highlights of Astronomy*, p. 589.
Newton, R. R.: 1970, *Ancient Astronomical Observations and the Accelerations of the Earth and Moon*, Johns Hopkins Press, London.
Robertson, J.: 1940, *Astron. Pap., Wash.* **X**, Part II.
Spencer Jones, Sir H.: 1939, *Monthly Notices Roy. Astron. Soc.* **99**, 541.
Stephenson, F. R.: 1971, Ph. D. Thesis, Newcastle: result reported at this Symposium.
Van Flandern, T. C.: 1970, *Astron. J.* **75**, 657.
Van Flandern, T. C.: 1971, *Astron. J.* **76**, 81.
Watts, C. B.: 1963, *Astron. Pap., Wash.* **XVII**.
Williams, K. P.: 1939, Indiana U. Publ., Science Series No. 9, with Suppl., 1940.

ON THE INITIAL DISTANCE OF THE MOON FORMING IN THE CIRCUMTERRESTRIAL SWARM

E. L. RUSKOL

O. Yu. Schmidt Institute of Physics of the Earth, Academy of Sciences, Moscow, U.S.S.R.

Abstract. According to the Radzievskij-Artemjev hypothesis of the 'locked' revolution of the circumplanetary swarms around the Sun, the initial Moon-to-Earth distance and the angular momentum acquired by the Earth through the accretion of the inner part of the swarm can be evaluated. Depending on the concentration of the density to the centre of the swarm we obtain the initial distance for a single protomoon in the range 15–26 Earth radii R and for a system of 3-4 protomoons in the range 3–78 R, if the outer boundary of the swarm equals to the radius of the Hill's sphere (235 R). The total angular momentum acquired by the primitive Earth-Moon system through the accretion of the swarm particles is $\frac{1}{2}$–$\frac{2}{3}$ of its present value. The rest of it should be acquired from the direct accretion of interplanetary particles by the Earth. The contribution of satellite swarms into the rotation of other planets is relatively less.

The initial distance of the Moon forming in the circumterrestrial swarm of small Earth's satellites depends on the geocentric angular momentum of the swarm. The calculation of the value of that momentum was never done because of the great difficulty. The swarm may be imagined as a nearly spherical cluster of particles on initially chaotic orbits which gradually evolves into a regular disk-like system and then coagulates into one or several large satellites. It is of interest to evaluate the initial geocentric angular momentum of the Moon using the Radzievskij-Artemjev hypothesis of the 'locked' prograde revolution of the circumterrestrial swarm in respect to the Sun [1, 2]. It is based on a supposition that the peculiar components of the remnant velocities of the particles captured into the swarm, are symmetric in the frame rotating with the Earth around the Sun. In this case the total geocentric angular momentum of the swarm equals $+I\omega_c$, where I is the moment of inertia of the swarm relative to the axis going through the Earth and parallel to the axis of the ecliptics; ω_c – the angular velocity of the Earth's motion around the Sun. (Let us remind for a comparison that every flat eddy in the circumsolar cloud of particles moving on keplerian circular orbits has the angular momentum $+\frac{1}{4}I\omega_c$ relative to the axis going through the geometric center of the eddy and parallel to the axis of rotation of the cloud.)

Suppose that all new particles captured into the swarm on the distance l from the axis, in average have the specific angular momentum $l^2\omega_c$, corresponding to the 'locked' ('solid-body') rotation of the swarm on that distance. Inside the whole sphere of action of the Earth such a 'solid-body' momentum is much less than the geocentric angular momentum $\sqrt{(Gml)}$ of a satellite on a direct circular orbit in the equatorial plane of the swarm. Therefore the conversion of the 'solid-body' swarm into a satellite or a satellite system must be accompanied by the contraction and by fall out of its inner part onto the Earth.

Let us denote the boundaries of the inner and outer parts of the swarm correspondingly by (l_1, l_2), (l_2, l_3). Taking the dependence of the surface density of the swarm

Urey and Runcorn (eds.), The Moon, 402–404.

on the distance l in the form $\sigma(l)=\sigma_0(l/(l_1+1))^{-b}$, we can evaluate the ratio of the mass M_1 falling onto the Earth to the mass μ of the satellite:

$$\frac{M_1}{\mu}=\frac{\int_{l_1}^{l_2} 2\pi\sigma l\,\mathrm{d}l}{\int_{l_2}^{l_3} 2\pi\sigma l\,\mathrm{d}l}=\frac{l_2^{2-b}-l_1^{2-b}}{l_3^{2-b}-l_2^{2-b}}$$

and the ratio of the corresponding angular momenta:

$$\frac{K_{M_1}}{K_\mu}=\frac{\int_{l_1}^{l_2} \omega_c\cdot 2\pi\sigma l^3\,\mathrm{d}l}{\int_{l_2}^{l_3} \omega_c\cdot 2\pi\sigma l^3\,\mathrm{d}l}=\frac{l_2^{4-b}-l_1^{4-b}}{l_3^{4-b}-l_2^{4-b}}.$$

The specific angular momentum of the satellite matter will be:

$$k_\mu=\frac{K_\mu}{\mu}=\frac{\omega_c(l_3^{4-b}-l_2^{4-b})(2-b)}{(l_3^{2-b}-l_2^{2-b})(4-b)}.$$

Taking $\mu=\mu_☾$, we can evaluate K_{M_1}, the angular momentum transferred to the Earth from the swarm.

The inner boundary of the swarm may be taken on the axis of rotation ($l_1=0$), the outer boundary l_3 – the Hill's sphere radius $r_1=\mathfrak{R}(m/3M_\odot)^{1/3}$. For the present mass M of the Earth it equals 235 R, for $m=M/2$ – nearly 190 R, R being the present Earth's radius. The boundary between the inner and the outer portions of the swarm, l_2, may be put $\approx 100\ R$ from the condition that all circular orbits of prograde satellites should lie outside the Roche limit ($l'\gtrsim 3\ R$).

The distribution of masses and angular momenta between the inner and outer parts of the swarm bounded by the Hill's sphere of radius 235 R is represented in the Table 1 for different values of b. The change of b from 0 to 3 corresponds to a very significant change of concentration of the density to the center: from a uniform distribution of the surface density σ over the swarm to the variation of σ by 7 orders of magnitude within the swarm. This manifests in the great variation of the ratio M_1/μ. Meanwhile the specific angular momentum of the satellite matter k_μ changes only from 2.6×10^{15} to 1.9×10^{15} cgs which corresponds to the distance of a single protomoon in the interval 15–26 R. For a 3–4 protomoon system the distance interval is greater: from 3 to 78 R. The total momentum transferred to the Earth-Moon system by the swarm, $K_{M_1}+K_\mu$, is always less than the present one (3.45×10^{41} cgs); it reaches roughly $\frac{1}{2}$–$\frac{1}{3}$ of the present value. This result is in agreement with the R.T. Giuli's conclusion that a considerable part of the momentum of the Earth-Moon system (up to $\frac{1}{2}$) is brought to the Earth by particles from heliocentric orbits [3]. The initial distance of the Moon

TABLE I

The distribution of the mass and the angular momentum between the inner and the outer parts of the swarm inside the Hill's sphere (235 R)

b	0	1	2	2.5	3
M_1/μ	0.22	0.73	5.4	25.8	172
K_{M_1}/K_μ	0.033	0.083	0.22	0.38	0.73
k_{M_1}, cgs	4×10^{14}	2.7×10^{14}	8.7×10^{13}	3×10^{13}	8×10^{12}
k_μ, cgs	2.6×10^{15}	2.4×10^{15}	2.1×10^{15}	2×10^{15}	1.9×10^{15}
K_{M_1} at $\mu = \mu_☾$	6.5×10^{39}	1.4×10^{40}	3.4×10^{40}	5.6×10^{40}	1×10^{41}
K_μ at $\mu = \mu_☾$	1.9×10^{41}	1.75×10^{41}	1.55×10^{41}	1.46×10^{41}	1.4×10^{41}
$K_{M_1} + K_\mu$	2×10^{41}	1.9×10^{41}	1.9×10^{41}	2.1×10^{41}	2.4×10^{41}

The angular momentum of the Earth acquired by the direct accretion after Giuli [3] $K_{M_0} = 1.7\text{–}1.8 \times 10^{41}$.
Theoretical sum $K_{M_0} + K_{M_1} + K_\mu = 3.7 \times 10^{41}$.
The present value $K_{M+\mu} = 3.45 \times 10^{41}$ cgs.

depends considerably upon the choice of the outer boundary of the swarm. For the outer boundary 190 R, for example, all satellite orbits occur inside the distance 30 R, and the orbit of a single protomoon – at 10–15 R. These results do not contradict the conclusion by P. Goldreich that the most plausible zone of accumulation of the Moon is the interval 10–30 [4].

The plausible initial distances of the Moon may be considered as an argument in favor of the 'solid-body' swarm filling the Hill's sphere around the growing Earth.

References

[1] Artemjev, A. V. and Radzievskij, V. V.: 1965, *Astron. Zh.* **42**, 124.
[2] Artemjev, A. V.: 1963, *Ann. Faroslavl Pedagogical Inst., ser. Astronomia*, issue 56.
[3] Giuli, R. T.: 1968, *Icarus* **8**, 301.
[4] Goldreich, P.: 1966, *Rev. Geophys.* **4**, 411.

I. ORIGIN AND EVOLUTION OF THE MOON

FLUIDIZATION ON THE MOON AND PLANETS

A. A. MILLS

Depts. of Geology and Astronomy, Univ. of Leicester, England

Abstract. A dual origin of lunar formations by both exogenous and endogenous mechanisms is now generally accepted. These processes are normally equated with impact and volcanism respectively, but many aspects of the latter cannot be readily identified with present-day terrestrial volcanism. It is proposed that lunar volcanism involved large volumes of gases and volatiles, so that 'fluidization' of particulate systems occurred. Terrestrial fluidization structures and model studies are described and illustrated. Certain lunar rilles and transient phenomena are also suggested to result from degassing of the interior. Some possible applications of fluidization to Mars and the synthesis of electropolymers are mentioned.

When Galileo turned his telescope on the Moon he was able to distinguish broad, dark, uniform plains – the maria – and a multitude of ring-shaped structures which have come to be called 'craters'. Subsequent centuries of Earth-based telescopic study enabled maps to be drawn on a scale limited mainly by atmospheric aberrations [1]. In the last few years automatic and manned lunar probes have vastly increased our knowledge of lunar topography, [2, 3] and added direct measurements of a number of physical and chemical properties. [4] In particular, it has been shown that the far side of the Moon is extensively cratered, the maria being almost entirely confined to the Earth-facing hemisphere.

The Mariner 4 television pictures of 1965 demonstrated that Mars too is cratered; a discovery confirmed and amplified by Mariners 6 and 7. It is suspected that Mercury and the larger satellites of Jupiter are similarly marked, so it would seem that the cratering mechanism – whatever it may be – has been of widespread importance in the evolution of the Solar System. However, the proximity of the Moon naturally renders it the major source of data.

1. Impact vs. Volcanic Origins

At one time two major schools of thought could be distinguished for the origin of lunar craters: those supporting impact in some form on the one hand, and those favouring volcanism on the other [5]. However, even before the advent of space probes the boundary was becoming less distinct, each school having accepted a modicum of the claims of the other to produce a number of dualistic approaches. [1, 2, 6] None has met with universal acceptance, the chief point of dispute being the ratio between impact and volcanic structures.

Proponents of the impact hypothesis pointed to the expected ubiquity of debris and collision processes (especially in the youth of the Solar System) and drew attention to the circular characteristics resulting from the explosive impact of bodies colliding at cosmic velocities. [7] Considerable emphasis was placed on plots of crater diameter vs. depth – the interpretation of which has not passed unchallenged. [6, 8, 9]

Urey and Runcorn (eds.), The Moon, 407–425.

The rival school stressed the alignments and groupings that may be distinguished among the lunar craters as evidence for an endogenous mechanism, and drew analogies with terrestrial calderas and other volcanic formations. [10–14] However, those terrestrial calderas formed by destruction of the apex of a volcanic cone are elevated *above* the surrounding terrain, [15] while collapse calderas are not remarkable for their circularity. [16] And why are there no large 'intact' parent structures on the Moon? A more fundamental objection is the absence on the Moon of evidence for the active convection which has led to plate tectonics, continental drift and fold mountains on Earth, and provided an energy source for the restricted distribution of volcanic belts along the edges of moving plates. [17, 18] It would seem that if an endogenous mechanism has contributed to the shaping of the lunar surface, then it must be very different from that which is normally identified with present-day terrestrial volcanism. [19] Several authors have suggested that some form of 'gassy' endogenous process is responsible. [8, 15,18]

It is particularly desirable that any postulated mechanism be capable of the formation of both single and multiple craters by a comparatively quiet, gentle process extending over an appreciable period of time – as contrasted with the sudden destructive violence of impact or explosive evisceration. [1, 11, 20–23] The general rule that where lunar craters intersect the broken formation is the larger points to the available energy diminishing with time. [24]

2. A Return to First Principles

As neither impacts nor conventional volcanism satisfactorily explain all aspects of the morphology of the lunar surface, let us return to first principles to search for additional or alternative mechanisms. It is now many years since Urey promulgated the idea of a *cold* accretion process in the early stages of planetary evolution. [25] There now seems to be a very general agreement that such a process did indeed occur, although argument continues over mechanism, time-scale, and many other important factors.

The composition of the solid primordial material appears to be most closely approached by the Type I carbonaceous chondrites, but this would have been associated with a much greater mass of volatile material – water, carbon dioxide, ammonia, methane, etc. – in the form of ices. [26] These grains of 'dirty ice' are suspended in the hydrogen and helium of the primeval solar nebula. There are obvious parallels with Whipple's theory of the composition of cometary nuclei, [27] and with the presumed compositions of the Jovian planets. (Recently, it has been shown that under certain conditions even hydrogen may condense on interstellar grains, [28] and that comets are associated with enormously larger clouds of hydrogen, oxygen, and other volatiles than was previously suspected [29]).

Let us now suppose this icy material to accrete into protoplanets without passing through any intermediate high-temperature stage. It would seem likely that a substantial proportion of the volatiles would thereby be incorporated in the interiors of the growing bodies. [26] However, when a certain size is achieved, gravitational and radioactive heating result in warming-up and consequent degassing. [1, 8, 26, 30–33] This could be concomitant with the final stages of accretion – when impact heating and modification of the surface is also important – and would seem to lead to a situation where we have gasses rising through a deep, particulate, outer shell. Initially, the gases would be expected to be largely methane, ammonia, carbon dioxide and water vapour; but with increase of temperature at depth hydrogen and carbon monoxide would become dominant. [34] The evolution of only 10% volatiles from the mass of the Moon gives 19 tonnes/cm^2 to be lost from the lunar surface.

3. Fluidization

Now it has long been established by experiment and practice in quite different contexts that if a stream of gas is introduced beneath a bed of solid, particulate material, then a series of phenomena may be distinguished as the rate of flow is gradually increased: [35, 36]

(I) At very low rates the gas merely percolates through the stationary bed.

(II) As its velocity increases there comes a point at which the viscous drag on the particles becomes equal to their weight. The bed then begins to expand and to display many of the properties of a liquid – such as flow, wave, and buoyancy phenomena. It is than said to be 'fluidized.'

(III) With somewhat greater gas flow the fluidized system becomes more expanded, less dense, and flows even more easily.

(IV) With still greater gas flow 'bubbles' appear in the fluidized bed, giving the appearance of boiling. But, of course, the bubbling bed need not really be hot at all; and the 'bubbles' show many features which render them quite distinct from bubbles in true liquids. [37]

(V) Finally, with still greater gas flow, this phase passes into one of pneumatic transport of the particles in the gas stream.

The turbulent motion and intimate contact between solid and gaseous phases make fluidized beds very efficient in the exchange of heat and mass, and in the promotion of chemical change. They are therefore finding wide application in technology today. [38] Industrial practice and theoretical investigation commonly employ spherical particles of narrow size range, and such irregularities as agglomeration, channelling and slugging in the fluidized bed are minimized by careful engineering design.

It has been suggested [39] that conditions for widespread fluidization were temporarily achieved on the young Moon and other bodies destined to become the terrestrial planets: this is the sought-for 'gassy endogenous process' mentioned above. The duration of intense activity would be comparatively short, but it is expected that regionalised, and later localised and sporadic, escape of gas from the interiors would persist over a long period – indeed, up to the present time (see below).

This hypothesis provides a necessary mechanism for the effective removal of heat from the interiors of the evolving bodies, [40] while exposure to the hot reducing gases would result in extensive and efficient chemical alteration of the original material, including reduction of iron and transport or loss of a proportion of volatile elements and those forming volatile hydrides or halides. As the gas flow diminishes so the depth which can be held in the fluidized state decreases, and the resulting static levels can be indurated by sintering, pressure, and chemical alteration.

This model leads to the accumulation of massive primordial atmospheres around the larger proto-planets, [41] making it necessary to appeal to the enhanced solar wind from a hotter, younger, Sun passing through a T-Tauri stage to drive them off. [26, 42] This also removes heavy volatiles like mercury, which are hard to lose by a Jeans mechanism.

It is generally agreed that the present atmosphere and hydrosphere of the Earth are (with the exception of free oxygen) derived by secondary degassing processes. [43, 44] The Moon is insufficiently massive to retain a permanent atmosphere. [45]

4. Fluidization on the Earth

Any structures formed by the primary degassing of the proto-Earth could not be expected to survive the turmoil of further accretion and the subsequent convective history. Present-day examples must necessarily be on a restricted scale.

The best-known natural fluidized systems are *nuées ardentes*: remarkable forms of volcanism where particulate material is suspended in, and transported by, the associated hot gases. [46] They can be extremely destructive – as was the type example which erupted from Mt. Pelée to overwhelm the town of St. Pierre in Martinique in 1902. [47] *Nuées ardentes* are infrequent in modern volcanism, and the few recent examples have fortunately been confined to sparsely-inhabited areas. However, they appear to have been much more common in the geological past, apparently erupting from fissures to lay down vast deposits in the American Middle West, South America and New Zealand. [48] With loss of the fluidizing gas the mass settles and compacts: the hotter central region of the flow commonly undergoes sintering and thermal alteration to produce hard, crystalline rocks now known as welded tuffs or ignimbrites, but at one time mistaken for true lavas. There is, understandably, a paucity of information on the mechanics of *nuée ardente* flow; but important observations were made by Perret, [47] while more recently McTaggart [49,50] and Brown [51] have contributed laboratory experiments and calculations. Salient points are

(a) Compositions range from nearly basic to acid.

(b) Temperatures of emplacement range from 700–1000 °C.

(c) The presence of fine material markedly increases the mobility, whereas addition of quite large amounts of coarse material to a bed of fine particles produces only a small change in characteristics. Surprisingly low rates of gas supply are sufficient for fluidization if fine material is present.

(d) There is little lateral sorting according to size: enormous blocks may be 'rafted' for miles. Perret [47] shows photographs of both rounded and angular examples, as well as of the block-littered surface of the spent flow. The fine material may be seen to contact the entire perimeter of a block – there is no trace of the concave depression that would be produced by its fall as an aerial projectile.

(e) Once a thick bed of particles is fluidized, considerable time is required for it to deflate completely. (A problem familiar to the manufacturers and packers of numerous powdered products.)

(f) True lavas may be found co-existing with ignimbrite deposits, having flowed over the surface or been intercalated between layers.

Doris Reynolds [52] first promoted the wider application of fluidization in geology, suggesting that certain intrusive granites were emplaced by this mechanism. Her ideas were taken up and extended by Holmes [53], who demonstrates the applicability of fluidization to the Swabian tuffisite pipes, kimberlite diamond pipes, and other diatremes. Of particular significance in the present context is Holmes' discussion of what he terms 'fluidization craters' – flat-floored ring structures with low ramparts found in the Rift Valley region of Africa. Holmes illustrates his textbook with a photograph of Lake Nyamununka – a ring crater about 2 miles in diameter – for which he believes fluidization to offer the only possible genesis. (For location of this crater see ref. 54.) The border between 'fluidization craters' and the gas-explosion structures called 'maars' is not well-defined: research is urgently required on the entire spectrum of types. Ollier [55] points out that maars are usually associated with basaltic rather than acidic igneous activity, and gives cross-sections of various examples, some with igneous masses (tholoids) extruded within the crater. The Wangoom maar in Australia would be a fluidization crater in Holmes' terminology. [56]

King and Sutherland [57] believe fluidized transport occurred in the carbonatite complexes of eastern Uganda. A review of fluidization as a volcanological agent was presented by Reynolds [58] at a recent meeting of the Geological Society of London devoted to this topic.

5. Fluidization on the Moon

The first description of a fluidized bed (although not by this name) appears in Robert Hooke's 'Micrographia' of 1665 [59], where he describes the appearance and properties of a bowl of 'boyling alabaster' (hot calcium sulphate dihydrate losing its water of hydration). Later in the same book this remarkable man compares the structures left on the surface of the alabaster with the craters of the Moon! It would seem that the application of fluidization to the origin of lunar surface features does, in fact, antedate both the volcanic and impact theories! The idea lay fallow for three hundred years, until independently resurrected by a number of workers [39, 60–64] in the 1960's as a result of growing interest in space exploration. However, two earlier papers which contain the germ of this idea should not be overlooked. [65, 66]

Particularly important is the work by O'Keefe and Adams [67], which includes a mathematical treatment of proposed lunar ash flows. They show that the amount of gas required to induce fluidization is inversely proportional to the square of the gravitational attraction at the site: lunar flows therefore require only $\frac{1}{36}$ of the gas content of their terrestrial counterparts. It seems reasonable to expect the terrestrial fluidization crater Lake Nyamununka mentioned above to be equivalent to a 10–12 mile diameter ring structure on the Moon.

The only study of the effect of reduced pressures on fluidization appears to be that of Miller and King. [68] They found that the effect of vacuum is to cause the top surface of the bed to erupt into a dilute phase and produce extensive elutriation. They suggest that any lunar flows must therefore have been accompanied by an extensive dilute phase, soon falling back after the end of the eruption. A sufficiently elevated temperature would presumably result in melting, as individuals, of a proportion of the suspended grains; surface tension then pulling them into the glassy spheres found in fly-ash and the returned lunar samples. [69–71]

Experimental craters have been produced in the laboratory by Emmons [62], Mills [72] and Schumm. [73] Mills found that ring structures were generated at the surface of a slowly-collapsing, aggregative, inhomogeneous fluidized bed. In such an environment the gas does not escape uniformly, but instead finds and maintains

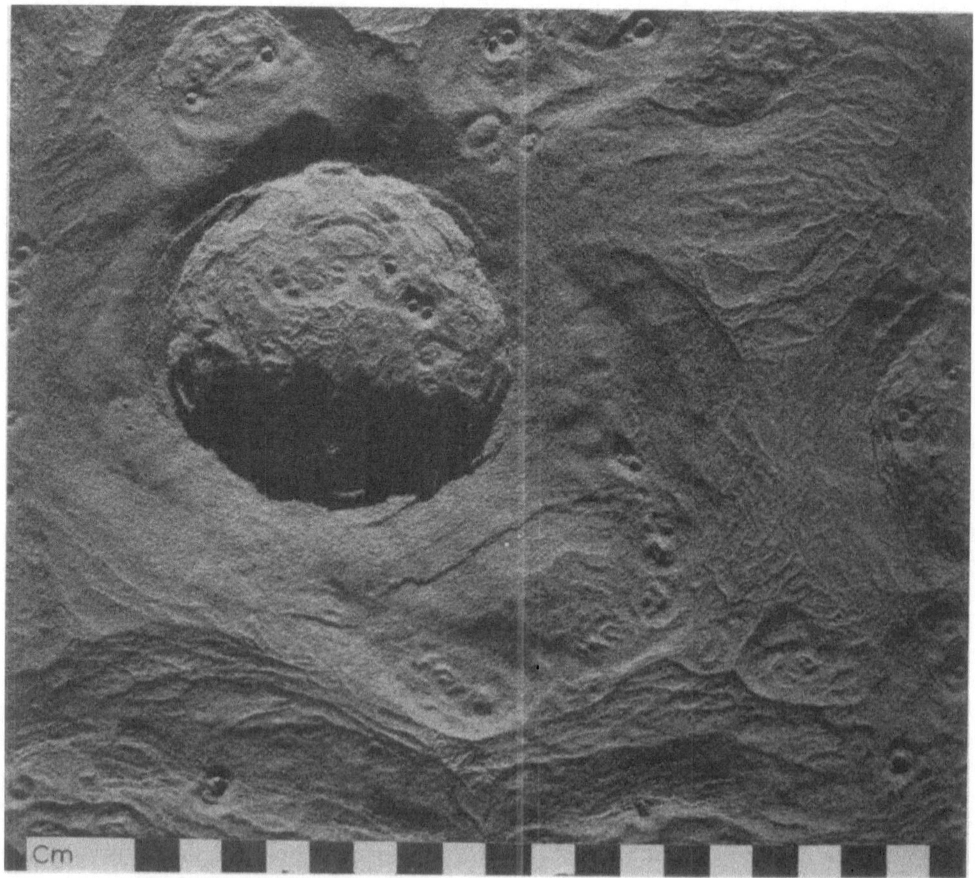

Fig. 1. Model single crater. Note the concentric slump terraces and the flow structures on the right.

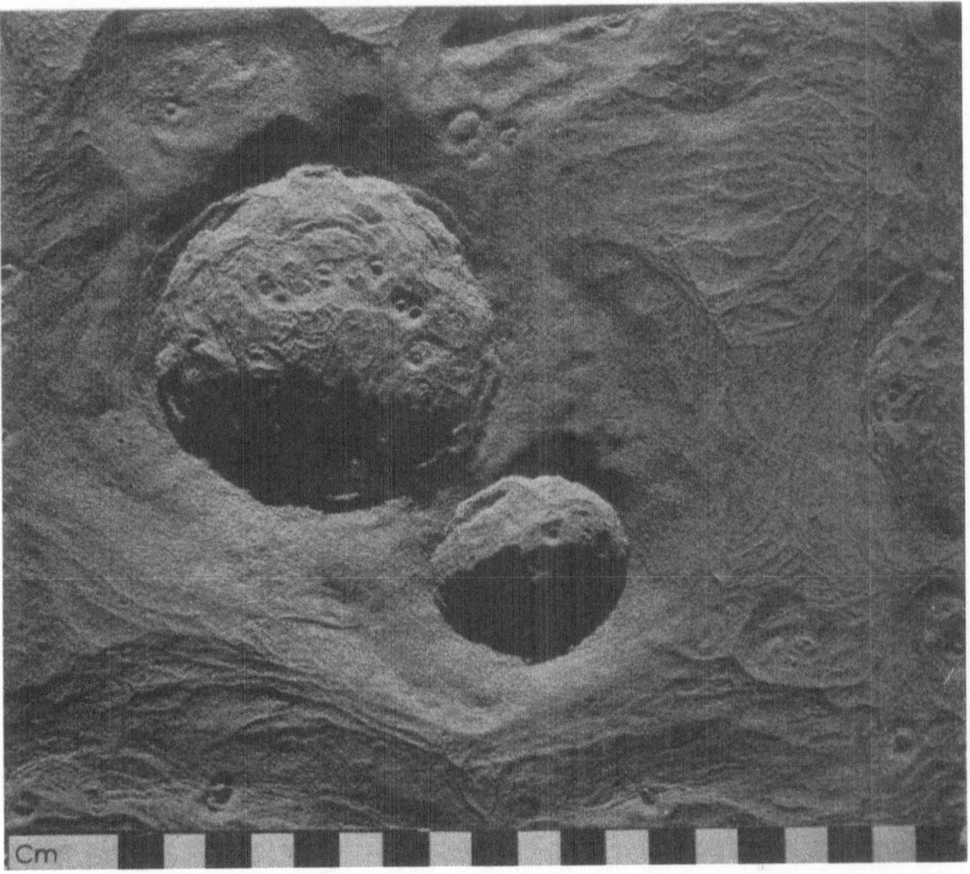

Fig. 2. Model craters. A second, smaller, crater has been produced at a short distance from the single crater shown in Figure 1.

certain channels of preferential escape. [74] (Discontinuities are inherent consequences of the fundamental equations of fluidization [75]) It was observed that

(a) A structure with a strong tendency towards a circular plan is produced independently of the shape of the area of enhanced gas release, and its diameter may exceed the depth of the bed.

(b) Formations may be abruptly intercepted by the walls of the apparatus.

(c) A large, shallow, depression in the original bed may be transformed into a flat-floored, raised-rim crater by fluidization of its interior.

(d) The height of the structures could be small compared with their diameter, strong side-lighting being required to delineate them clearly. It was later found that the relief is controlled by the degree of overall (or 'background') fluidization. A high level of mobility 'dissolved' any structure, a moderate level produced craters with shallow slopes, and an immobile background resulted in craters of funnel-like section.

Fig. 3. Model craters. A second crater has been produced nearer an original large single crater, resulting in penetration. Note how the smaller crater intrudes the larger, producing a talus slope (but otherwise little destruction) within it.

(e) The process of formation is comparatively gentle, and may operate over an extended period of time with quiescent intervals.

(f) The extent to which two or more structures interact with one another depends on their distance apart and sequence of formation. Figures 1–4 illustrate this.

(g) Concentric slump terraces appear within the ring as the gas supply diminishes and bed contraction ensues.

(h) The terminal stage is commonly the formation of small craterlets within the ring. Tapping the apparatus at this stage often results in transient activity confined to these craterlets.

(i) The lobate flows apparent in the illustrations might be termed the results of *nuées froides* – they were neither true liquids nor hot.

(j) The effect of bed depth is illustrated by Figures 5–7. A shallow bed results in a large number of 'blowholes', but as the depth increases these tend to be replaced by

Fig. 4. Model craters. Dimultaneous production of craters may result in either a straight dividing wall or coalescence of the two structures, depending on the amounts of gas involved. Here we see a coalescent structure bearing resemblances with, for example, the lunar crater Fauth.

zones of fluidization and a smaller number of escape channels. Certain areas on the Moon exhibiting excess crater numbers (e.g. the lunar 'playas') might well be associated with terminal gas release through a thin particulate cover.

The concept of 'regional fluidization' is shown diagrammatically in Figure 8. It is essential that this scheme be clearly distinguished from any hypothesis involving bubbles in viscous liquids. [5] Not only is it impossible to form very large bubbles, [76] but the liquid phase would flow back level again! [77] However, the property of the moving cavities ('bubbles') in boiling fluidized beds of bringing up patches of underlying material [37] may be relevent to the genesis of lunar dark-halo craters.

Although bearing many visible resemblances with the surface of the Moon, [78] it is freely conceded that these laboratory simulations are not true-scale models, and

Fig. 5. Result of passage of gas through a bed of rock powder ½″ deep.

that the greatest caution is required in extrapolating them to any natural situation. Similar objections apply to cratering experiments with accelerated dust-size particles. All these experiments are perhaps best considered as aids to the conception of hypotheses to be tested in the field – in the manner so successfully used by Kuenen [79] in his pioneering work on marine turbidity currents.

Murray, Spiegel and Theys [80] have published a preliminary attempt to quantify aspects of fluidization on the Moon. Unfortunately they employed data derived from the *secondary* degassing of the Earth, thereby arriving at an unacceptably low value for the period during which large-scale fluidization could be simultaneously supported over the entire surface of the Moon. Substituting the above-mentioned value of 19 tonnes/cm^2 for degassed volatiles gives the more realistic figure of 10^3 yr.

It must also be reiterated that it is not suggested that fluidization is the only mechanism shaping the lunar surface. Impact craters occur on Earth, and must surely be represented on the Moon. The constitution of the impacting bodies was at one time generally considered to be identical with that represented by the meteorites in our museums, but in view of the surprisingly low concentration of such meteoritic material

Fig. 6. Result of passage of gas through a bed of rock powder 1″ deep.

in returned lunar samples [81–83] there is now an increasing body of opinion that weakly-cohesive or icy masses were important, and might even have predominated. The possibility of base surges [84] and fluidization induced by their impact [85] should be considered. [86, 87]

6. The Deeper Structure of the Moon

Loss of all but traces of its volatiles leads to the following picture of the present-day Moon

I – A fragmented and particulate surface layer, containing much fine material and up to several metres deep. [88–90]

II – A sintered, heterogeneous, vesiculate breccia crust some 3–5 km thick, gradually passing into –

III – Increasingly compacted rubble and breccia making up a thick mantle.

IV – A small core, probably with no distinct boundary.

This model appears to be in accord with the low density, seismic observations, [91–93] and measurements of the electrical characteristics of the Moon. [94, 95]

Fig. 7. Result of passage of gas through a bed of rock powder 2″ deep.

The crystalline material in the returned lunar samples indicates that actual melting has occurred within or below the maria (perhaps with the assistance of extra energy on the Earth-facing hemisphere) to allow true lavas and igneous differentiation products to form. Lava may flow out at the surface (wrinkle ridges?) or be injected between other deposits (giving domes as 'quelkuppen' [46]). However, the returned lunar basalts cannot represent the bulk composition of the Moon, for transformation to eclogites at depth would give too high a density. [96] It is considered more likely that local melting beneath the maria has occurred, rather than that the Moon has ever been molten on a global scale. [97, 98] Mackin's [64] scheme for the origin of maria allows material to be brought into the basin from the edges, thus giving rise to mascons.

7. Lunar Rilles

Two distinct types of rille are immediately obvious – the straight and the sinuous – although intermediate and other forms may be found on further study. Straight rilles (Figure 9) extend for tremendous distances over the lunar surface, maintaining their linearity without regard to surface topography. [2, 3] Craters are divided with no sign

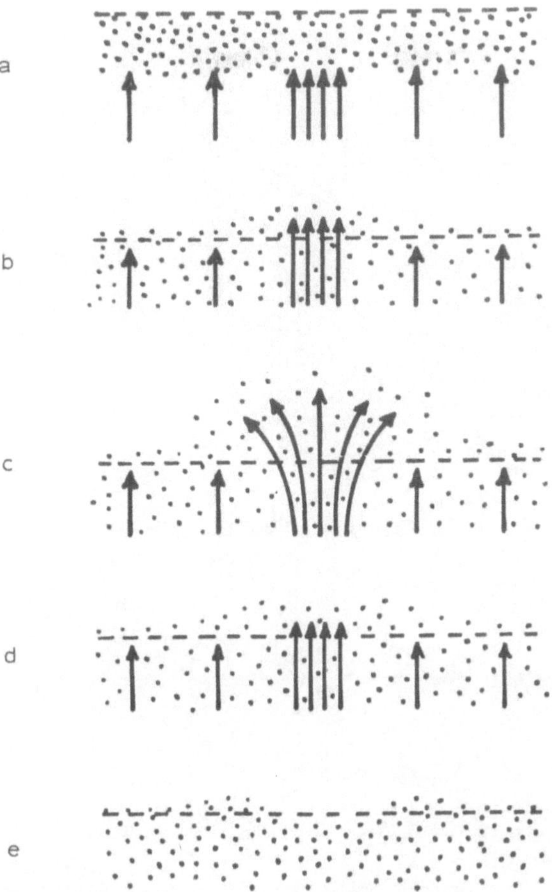

Fig. 8. Successive stages of regional fluidization. a, Release of gas from depth; b, bed expansion, with doming over channel of preferential escape; C, lateral flow of basal fluidized system beneath a disperse cloud; d, gas flow diminishing-activity subsiding within the ring structure; bed contraction is accompanied by ring slumping; e, cessation of gas flow and general consolidation results in an approximately circular depressed area surrounded by a raised wall.

of diversion by cone sheets, feeder pipes, or dykes of strong igneous rock. The depression of the floor of a crater-crossing rille is similar in both mare and ramparts. [99] The orientation of straight rilles seems to follow a pattern – the lunar grid – indicative of an endogenous origin. [2, 100]

In 1885 Osborne Reynolds [101, 102] coined the term 'dilatancy' for the property of close-packed granular masses of expanding in bulk with change of shape, due to the increase of space between individual particles as they change their relative positions. Reynolds' interest in the phenomenon was confined to its application in his mechanistic theory of the ether: the more practical role of dilatancy in geology had to await the investigations of W. J. Mead. [103]

Fig. 9. Part of the Sirsalis rille: an example of a long, straight rille. It is named from the 32 km Sirsalis at upper right. (NASA/National Space Science Data Center; Orbiter 4 – 161H).

In a modification of one of Reynolds' experiments, Mead shows that if sand is placed in an impervious elastic container (itself of negligible strength, such as a toy balloon) and the air pumped out, then

(a) The mass becomes exceedingly rigid, and
(b) If the mass is stressed to failure it responds as a *brittle solid*, failing along definite shear planes.

i. e. granular masses free to dilate deform by flow, but when dilation is prevented or restricted by relatively slight confining pressure deformation causes *fracture*. This manner of failure requires a minimum increase in volume, involving dilation only in the shear zone.

It is suggested that the straight rilles are simply an analogous expression of tensional failure in the weakly-cohesive lunar crust. Whether the tension is a result of cooling or doming is not yet clear. The relief of pressure and extra space created along these deep fractures (or set of closely-spaced fractures) permitted dilatancy and plastic flow to occur, giving the present depressed flat floors of the rilles.

Besides providing an escape route for gases from below, Mead himself pointed out

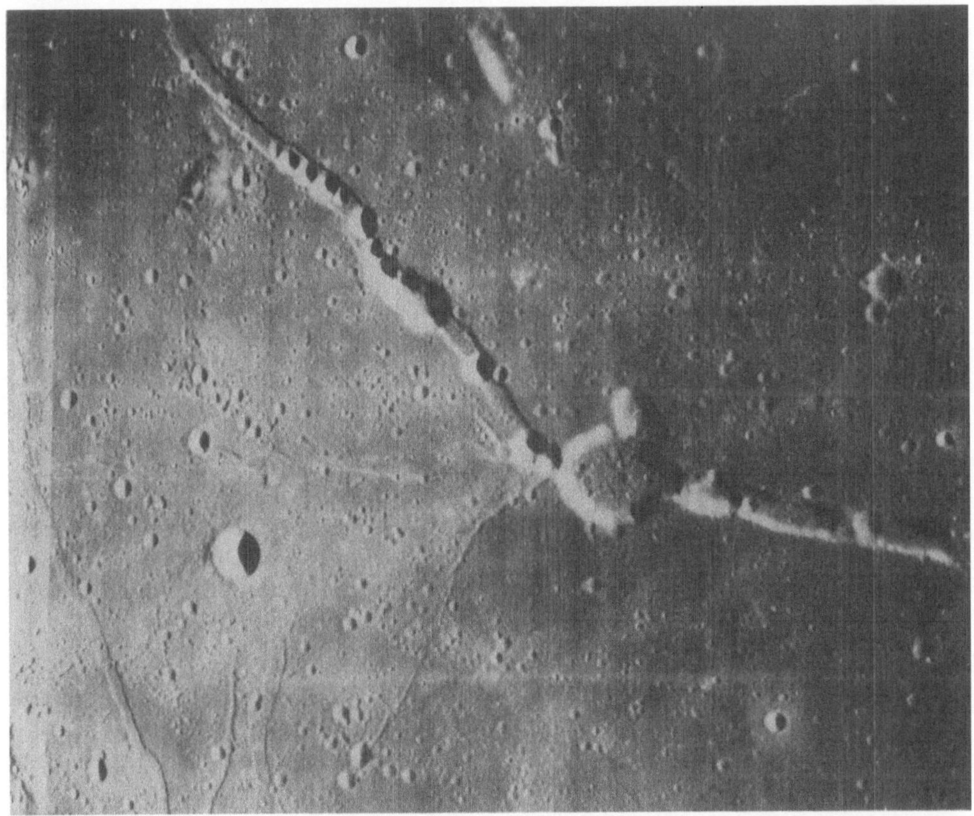

Fig. 10. The Hyginus rille. The largest crater, Hyginus itself, is about 6.5 km in diameter. (NASA/ National Space Science Data Center; Orbiter 5 – 96M).

that this mechanism leads to lateral migration of any fluids contained in the surroundings towards the zone of fracture and dilation. Movement and escape of gases could lead to fluidization or maar-like excavations within the rille, depending on quantity, rate, and point of release. The Hyginus rille (Figure 10) is put forward as an example of a rille modified by such an escape of gas.

The form of the sinuous rilles has suggested to various workers that they have been created by actual flow of some material – water, lava (in channels or tubes) or nuées ardentes. [104] This is difficult to reconcile with the observable penetration of topographic highs, and the occasional breaks in continuity. [105, 106] It is considered more likely that fine underlying tensional cracks are responsible, allowing gas release with

(a) The formation of closely-spaced craters which have coalesced, leaving only their cuspate walls.

(b) Fluidized flow in those sections of a rille where the topography is appropriate. These flows need be neither continuous nor hot.

In the absence of gas release the finest fractures remain masked by the superincumbent particulate material. Experimental investigations along these lines have been conducted by Schumm [73] and Mills (unpublished).

8. Transient Lunar Phenomena

That a small amount of gas is even today being evolved from the Moon is suggested by the temporary obscurations and light emissions known as transient lunar phenomena (TLP's). Gas venting is also considered a likely candidate for the fixed-location seismic signals observed on the Moon, [107] while the cold-cathode gauge left behind by the astronauts has yielded some interesting preliminary results.

TLP's continue to be observed and catalogued [108]: correlations with perigee [109] and seismic events [110] have been reported. However, postulated origins in luminescence or thermolominescence have proved inadequate [111, 112], leaving electrostatic glow discharges (generated by friction in dust-laden gas discharges) as almost the only contender. [113, 114] This subject will be discussed at greater length elsewhere.

9. Application of Fluidization to Other Systems

The surface of Mars displays eroded craters of enormous size, rilles, 'chaotic' and 'featureless' ground. [115] There is some evidence for a Martian grid system, with its implications for the participation of endogenous processess. [116] Fluidization has been suggested as a possible cause of the featureless terrain of Hellas. [117]

The possibilities of electrostatically-charged fluidized systems for the abiogenic generation of electropolymers and other necessary precursors of life have recently been discussed by Sylvester-Bradley. [118] Carbonaceous chondrites may have experienced a fluidized stage. [119]

It is concluded that fluidization has been of importance in many aspects of the evolution of planets and satellites.

References

[1] Kopal, Z.: 1966, *An Introduction to the Study of the Moon*, D. Reidel Publ. Co., Dordrecht-Holland.
[2] Lowman, P. D.: 1969, *Lunar Panorama*, Weltflugbild, Zurich.
[3] Kosofsky, L. J. and El-Baz, F.: 1970, *The Moon as Viewed by Lunar Orbiter*, NASA SP-200.
[4] Levinson, A. A.: 1970, *Proceedings of the Apollo 11 Lunar Science Conference*, Pergamon.
[5] Green, J.: 1967, in *Encyclopedia of Atmospheric Sciences and Astrogeology* (ed. by R. W. Fairbridge) Reinhold.
[6] Mutch, T. A.: 1970, *Geology of the Moon*, Princeton Univ. Press.
[7] Baldwin, R. B.: 1963, *The Measure of the Moon*, Univ. of Chicago Press.
[8] Green, J. and Poldervaart, A.: *Report of the 21st Int. Geol. Congress 1960*, Part XXI, pp. 15–33.
[9] Steinberg, G. S.: 1968, *J. Geophys. Res.* **73**, 6125.
[10] Von Bülow, K.: 1965, *Ann. N. Y. Acad. Sci.* **123**, 528.
[11] McCall, G. J. H.: 1965, *Ann. N. Y. Acad. Sci.* **123**, 843.
[12] Fielder, G.: 1965, *Lunar Geology*, Lutterworth.
[13] Moore, P. and Cattermole, P. J.: 1967, *The Craters of the Moon*, Lutterworth.
[14] McCall, G. J. H.: 1969, *J. Brit. Astron. Assn.* **80**, 19 and further parts in this series.
[15] Miyamoto, S.: 1965, *Ann. N. Y. Acad. Sci.* **123**, 776.
[16] Smith, R. L. and Bailey, R. A.: 1968, in *Studies in Volcanology* (ed. by R. R. Coats, R. L. Hay and C. A. Anderson) Geol. Soc. Amer. pp. 613–62.
[17] Miyamoto, S.: 1964, *Icarus* **3**, 486.
[18] Dietz, R. S. and Holden, J. C.: 1965, *Ann. N.Y. Acad. Sci.* **123**, 631.
[19] Emmons, R. C.: 1965, *Tectonophysics* **1**, 541.
[20] McCall, G. J. H.: 1963, *Nature* **197**, 273.
[21] Moore, P.: 1963, *Nature* **197**, 273.
[22] Moore, P.: 1966, *J. Brit. Astron. Assn* **76**, 256.
[23] McCall, G. J. H.: 1969, *Nature* **223**, 275.
[24] Moore, P.: 1967, *J. Brit. Astron. Assn.* **77**, 119.
[25] Urey, H. C.: 1952, *The Planets*, Yale Univ. Press.
[26] Ringwood, A. E.: 1966, in *Advances in Earth Science* (ed. by P. M. Hurley), MIT Press.
[27] Whipple, F.: 1950, *Astrophys. J.* **111**, 375.
[28] Lee, T. J., Gowland, L., and Reddish, V. C.: 1971, *Nature Phys. Sci.* **231**, 193.
[29] Biermann, L.: 1971, *Nature* **230**, 156.
[30] Bernal, J. D.: 1961, *Times Science Review*, No. 40, pp. 3–4.
[31] Green, J.: 1965, *Ann. N.Y. Acad. Sci.* **123**, 403.
[32] Kopal, Z.: 1968, *Exploration of the Moon by Spacecraft*, Oliver and Boyd.
[33] Nagy, B. *et al.*: 1971, *Nature* **232**, 94.
[34] Ringwood, A. E.: 1972, Private communication.
[35] Lewis, J. B.: 1961, *Times Science Review*, No. 40, pp. 15–8.
[36] Davidson, J. F. and Harrison, D.: 1963, *Fluidised Particles*, Cambridge Univ. Press.
[37] Rowe, P. N.: 1965, *Sci. J.* **1**, 59.
[38] Lewis, J. B. and Partridge, B. A.: 1967, *Nature* **216**, 124.
[39] Mills, A. A.: 1969, *Abstr. Meeting on Fluid Mechanics in Relation to Natural Phenomena*, Newcastle.
[40] Anderson, D. L. and Phinney, R. A.: 1967, in S. K. Runcorn (ed.), *Mantles of the Earth and Terrestrial Planets*, Wiley.
[41] Urey, H. C.: 1954, *Astrophys. J. Suppl.* **1**, 147.
[42] Keays, R. R., Ganapathy, R., Laul, J. C., Anders, E., Herzog, G. F., and Jeffrey, P. M.: 1970, *Science* **167**, 490.
[43] Brancazio, P. J. and Cameron, A. G. W.: 1964, *The Origin and Evolution of Atmospheres and Oceans*, Wiley.
[44] Condie, K. C.: 1968, *J. Geophys. Res.* **73**, 5466.
[45] Öpik, E. J.: 1969, *Sci. J.* **5**, 67.
[46] Rittmann, A.: 1962, *Volcanoes and Their Activity* (transl. by E. A. Vincent), Wiley.
[47] Perret, F. A.: 1935, *Carnegie Inst. Wash. Pub.* **458**.
[48] Ross, C. S. and Smith, R. L.: 1961, *U.S. Geol. Survey Prof. Paper* **366**.

[49] McTaggart, K. C.: 1960, *Amer. J. Sci.* **258**, 369.
[50] McTaggart, K. C.: 1962, *Amer. J. Sci.* **260**, 470.
[51] Brown, M. C.: 1962, *Amer. J. Sci.* **260**, 467.
[52] Reynolds, D. L : 1954, *Amer. J. Sci.* **252**, 577.
[53] Holmes, A.: 1965, *Principles of Physical Geology*, Nelson.
[54] Holmes, A.: 1951, *Geol. Mag.* **88**, 73.
[55] Ollier, C. D.: 1967, *Bull. Volc.* **31**, 45
[56] Ollier, C. D.: 1969, *Volcanoes*, MIT Press.
[57] King, B. C. and Sutherland, D. S.: 1966, in O. F. Tuttle and J. Gittins (eds.), *Carbonatites*, Interscience.
[58] Reynolds, D. L.: 1969, *Proc. Geol. Soc. Lond.*, No. 1655 pp. 110–5.
[59] Hooke, R.: 1665, *Micrographia*, Royal Society (Dover facsimile, 1961).
[60] Shoemaker, E. M.: 1962, in Z. Kopal (ed.), *Physics and Astronomy of the Moon*, Academic Press.
[61] Cameron, W. S.: 1964, *J. Geophys. Res.* **69**, 2423.
[62] Emmons, R. C.: 1965, *Tectonophysics* **2**, 83.
[63] Elston, W. E.: 1965, *Ann. N.Y. Acad. Sci.* **123**, 817.
[64] Mackin, J. H.: 1969, *Bull. Geol. Soc. Amer.* **80**, 735.
[65] Macqueen, P. O.: 1936, *Trans. Amer. Geophys. Union*, Part 1, pp. 85–89.
[66] Gaydon, A. G. and Learner, R. C. M.: 1959, *Nature* **183**, 37.
[67] O'Keefe, J. A. and Adams, E. W.: 1965, *J. Geophys. Res.* **70**, 3819.
[68] Miller, G. H. and King, I. R.: 1966, *NASA Contractor Report CR-627.*
[69] Hart, A. B. and Raask, E.: 1969, *Nature* **223**, 762.
[70] Agrell, S. O. *et al.*: 1970, *Science* **167**, 583.
[71] Tolansky, S.: 1970, *Science* **167**, 742.
[72] Mills, A. A.: 1969, *Nature* **224**, 863.
[73] Schumm, S. A.: 1970, *Bull. Geol. Soc. Amer.* **81**, 2539.
[74] Whitehead, A. B. and Young, A. D.: 1967, in H. H. Drinkenburg (ed.), *Proc. Int. Symp. on Fluidization, Eindhoven, 1967*, Netherlands Univ. Press.
[75] Murray, J. D.: 1965, *J. Fluid Mech.* **21**, 465.
[76] Krejci-Graf, K.: 1965, *Ann. N.Y. Acad. Sci.* **123**, 751.
[77] Burns, G. J.: 1902, *J. Brit. Astron. Assn.* **12**, 334.
[78] Apollo 12 commentary from the Moon, Nov. 1969.
[79] Kuenen, P. H. and Migliorini, C. I.: 1950, *J. Geol.* **58**, 91.
[80] Murray, J. D., Spiegel, E. A., and Theys, J.: 1969, *Comm. Astrophys. Space Phys.* **1**, 165.
[81] Ganapathy, R., Keays, R. R., Laul, I. C., and Anders, E.: 1970, in A. A. Levinson (ed.), *Proc. Apollo 11 Lunar Science Conference* **2**, 1117, Pergamon.
[82] Keil, K., Prinz, M., and Bunch, T. E.: 1970, *Science* **167**, 597.
[83] Arrhenius, G. *et al.*: 1970, *Science* **167**, 659.
[84] Fisher, R. V. and Waters, A. C.: 1969, *Science* **165**, 1349.
[85] Hartman, W. K. and Yale, F. G.: 1968, *Arizona Univ. Lunar and Planetary Lab. Commun.* **7**, 131.
[86] Van Dorn, W. G.: 1969, *Science* **165**, 693.
[87] McKay, D. S., Greenwood, W. R., and Morrison, D. A.: 1970, *Science* **167**, 654.
[88] Bastin, J. A., Clegg, P. E., and Fielder, G.: 1970, *Science* **167**, 728.
[89] Fryxell, R. *et al.*: 1970, *Science* **167**, 734.
[90] Quaide, W. L. and Oberbeck, V. R.: 1968, *J. Geophys. Res.* **73**, 5247.
[91] Report in *Nature* **228**, 906 (1970).
[92] Latham, G., Ewing, M., and Dorman, I.: 1970, *Science* **170**, 620.
[93] Gold, T. and Soter, S.: 1970, *Science* **169**, 1071.
[94] Sonett, C. P. *et al.*: 1971, *Nature* **230**, 359.
[95] Kuckes, A. F.: 1971, *Nature* **232**, 249.
[96] Ringwood, A. E. and Essene, E.: 1970, *Science* **167**, 607.
[97] Gilvarry, J. J.: 1970, *Nature* **225**, 623.
[98] Urey, H. C.: 1971, *Science* **172**, 403.
[99] Leatherbarrow, W. J.: 1969, *Brit. Ast. Assn. Lunar Section Circular* **4**, 77.
[100] Fielder, G.: 1967, in S. K. Runcorn (ed.), *Mantles of the Earth and Terrestrial Planets*, Interscience

[101] Reynolds, O.: 1885, *Phil. Mag.* **20**, 469.
[102] Reynolds, O.: 1886, *Nature* **33**, 429.
[103] Mead, W. G. 1925, *J. Geol.* **33**, 692.
[104] Greeley, R.: 1971, *Science* **172**, 722.
[105] Schumm, S. A. and Simons, D. B.: 1969, *Science* **165**, 201.
[106] McCall, G. J. H. 1970, *Nature* **225**, 714.
[107] Report in *Nature* **229**, 521 (1971).
[108] Moore, P. A.: 1971, *J. Brit. Astron. Assn.* **81**, 365.
[109] Middlehurst, B. M. and Moore, P. A.: 1967, *Science* **155**, 449.
[110] Ewing, M.: 1970, Address to the IAU General Assembly, Brighton, England.
[111] Blair, I. M. and Edgington, J. A.: 1970, in A. A. Levinson (ed.), *Proc. Apollo 11 Lunar Science Conf.* Vol. 3, 2001, Pergamon.
[112] Nash, D. B. and Greer, R. T.: 1970, in A. A. Levinson (ed.), *Proc. Apollo 11 Lunar Science Conf.* Vol. 3, 2341, Pergamon.
[113] Mills, A. A.: 1970, *Nature* **225**, 929.
[114] O'Keefe, J. A.: 1970, *Science* **168**, 1209.
[115] Leighton, R. B., Horowitz, N. H., Murray, B. C., Scharp, R. P., Herriman, A. H., Young, A. T., Smith, B. A., Davies, M. E., and Leovy, C. B.: 1969, *Science* **166**, 49.
[116] Wells, R. A.: 1969, *Geophys. J. Roy. Astron. Soc.* **17**, 209.
[117] Sharp, R. P., Soderblom, L. A., Murray, B. C., and Cutts, J. A.: 1971, *J. Geophys. Res.* **76**, 331.
[118] Sylvester-Bradley, P. C.: 1971, *Proc. Geol. Assn.* **82**, Part. 1, pp. 87–136.
[119] Mills, A. A.: 1968, *Nature* **220**, 1113.

ON THE POSSIBLE DIFFERENCES IN THE BULK CHEMICAL COMPOSITION OF THE EARTH AND THE MOON FORMING IN THE CIRCUMTERRESTRIAL SWARM

E. L. RUSKOL

O. Yu. Schmidt Institute of Physics of the Earth, Academy of Sciences, Moscow, U.S.S.R.

Abstract. The model of the origin of the Moon in the circumterrestrial swarm during the active stage of the Earth's growth [1, 2] receives now a support (see e.g. [3]). In this model the feeding substance is the same for both bodies. However two different processes may be mentioned of chemical fractionation between the Earth and the Moon.

1. The Partial Loss of Volatile Elements from the Circumterrestrial Swarm

The active formation of the swarm should take place when the mass of the growing Earth reached nearly a half of its present value. In this period the typical velocities of mutual collisions of particles captured into the swarm, were of the order of 3–7 km $\sec^{-1}$ in the whole sphere of action of the Earth. In subsequent collisions inside the swarm the velocities decreased. The collisions with the high energy (10^{10}–10^{11} erg g^{-1} etc.) led to multiple breaking up and both complete and partial evaporations of the particles' material, with the subsequent condensation of vapours on the solid particles. Different atoms and molecules could then escape into space under the action of the solar wind. Predominantly escaped the volatile substances (the water, the elements with the low melting temperatures Pb, Bi, Tl, etc.).

It can be shown that the preplanetary cloud of a mass equal to the mass of terrestrial planets was rather transparent in radial direction ($\tau \lesssim 1$) at the Earth's distance from the Sun, if the distribution of radii of the preplanetary bodies followed the law $\mathrm{d}N(r) \sim r^{-n}\,\mathrm{d}r$, where $n \leqslant 3.5$ and $r_{max} \approx 10^8$ cm. Thus, on the periphery of the circumterrestrial swarm also a transparency may be expected. In this region every free atom of lead, with the cross-section $\approx 10^{-15}$ cm^2, had a chance to collide with a high-energy proton and to be ejected from the swarm once in two months, if the intensity of the solar wind was 2×10^8 protons cm^{-2} $\sec^{-1}$, as at the present time. The condensation of that atom on a solid particle at the optical depth $\tau \approx 1$, took a time comparable with the period of revolution around the Earth, that is several months on the periphery of the swarm. If the intensity of the solar wind in the past was much higher than at present time, then the ejection of volatile elements from the periphery of the swarm would be more effective than their condensation on solid particles. On the contrary, at the accumulation of the Earth itself the volatile substances do not dissipate. It can be expected that the volatiles were retained also in the inner, more dense and obscure part of the swarm, which presumably has fallen onto the Earth. This fractionation may account for the deficiency, described by E. Anders and others, of volatile elements in lunar basalts and the enrichment of the latter in refractory ones in

Urey and Runcorn (eds.), The Moon, 426–428.

comparison with the terrestrial basalts [4]. It can be assumed that the primitive Moon was considerably depleted in water in comparison with the Earth. This strengthens the arguments against early lunar hydrosphere and atmosphere which were put forward earlier in the assumption of equal primitive composition of the Earth and the Moon [6].

The explanation of the deficiency of volatiles given by E. Anders seems to us less plausible, because he suggests the coincidence in time of a very short period of condensation of volatile elements at the earliest stage of the solar nebula (first 10^5–10^6 yr) with the terminating period of growth of the Earth and the Moon. We believe that the time-scale of their growth was of the order of 10^8 yr [7], but not much less. Ringwood [5] explaines the deficiency of volatiles in the Moon by the solar wind. However his model of the origin of the Moon from a primitive massive iron-silicate Earth's atmosphere needs much more mechanical foundations (too short a time-scale, unclear mechanics of the formation of a satellite from an atmosphere etc.).

2. The Selection of the Finest Particles at Their Capture into the Circumterrestrial Swarm and Resulting Enrichment of the Primitive Lunar Matter in Silicates

If the Earth's core consists from iron, the Earth should contain nearly 35% of iron by weight. Meanwhile, the Moon contains only 14% of iron [10, 11].

We are tempting to explain this difference by the peculiarities of the formation of the circumterrestrial swarm from the preplanetary cloud, if in that cloud either (a) separately existed the iron and the silicate particles which had systematic differences in size due to some differences in physico-mechanical properties; or (b) there were no pure iron or silicate particles but the iron content increased with the particles dimension because the iron-rich bodies had more chance to survive in collisions and to grow up.

As Orowan has pointed out [9], the silicate particles are brittle and give fine debris in collisions, meanwhile the iron particles are plastic-ductile even at low temperatures and can coalesce at collisions. Altshuler and Sharipdjanov [11] also discuss some reasons stimulating the advanced growth of the iron particles in the preplanetary cloud.

As we have shown [2, part III], in the process of capture of particles into the circumterrestrial swarm, their size distribution changes even in the case without the breaking-up. If to approximate these distributions by power functions in the form $dN(r) \sim r^{-n}\,dr$, then the power index n_1 for the interplanetary particles transforms into n_2 immediately after the capture of them into the swarm according the formula:

$$n_2 = 2n_1 - 3.$$

In particular, the index $n_1 = 3.5$ found for the preplanetary bodies [7, 8] transforms into $n_2 = 4$, which signifies the predominant capture of the finest component of the interplanetary assembly of particles and the permanent enrichment of the swarm by this component. The fragmentation of captured particles strenthens this effect. This selection does not affect the size distribution in the whole Earth's zone, if the mass of the swarm is small relative to the mass of all matter in the zone. At the same time the growing Earth accretes all particles without choice.

In both cases (a) and (b) mentioned above the enrichment of the swarm by the finest component signifies also the enrichment in silicates relative to iron. The case (b) may be illustrated by a numerical example.

Suppose the increase in Fe: Si ratio in bodies from the smallest particles to the greatest preplanetary bodies corresponds to the change of their density in the form:

$$\delta(r) = \delta(r_{\min})[1 + a \log(r/r_{\min})].$$

The ratio $r_{\max}/r_{\min}$ can be taken equal 10^{12}–10^{13}. Suppose that $\delta(r_{\min}) = 2.5$ g cm^{-3}, roughly the density of the lightest silicates; and that $\delta(r_{\max}) = 4.5$ g cm^{-3}, roughly the density of the matter of the Earth at zero pressure. Then the average density for the assembly of particles of the size distribution with $n_1 = 3.5$ (preplanetary matter) will be $\bar{\delta}_1 = 4.4$ g cm^{-3}, and for that with $n_2 = 4.0$ (the swarm matter) $\bar{\delta}_2 = 3.5$ *g* cm^{-3}. We see that the values of $\bar{\delta}_1$ and $\bar{\delta}_2$ only slightly differ from the average densities correspondingly of the Earth and of the Moon. Thus the model of the circumterrestrial swarm permits to explain the difference in Fe:Si ratio in the bulk composition of the Earth and the Moon.

References

[1] Schmidt, O. Yu.: 1959, *A Theory of the Origin of the Earth,* Lawrence and Wishart, London.

[2] Ruskol, E. L.: 1960, *Sov. Astron. AJ* **4**, 657 (I); *ibid.* **7**, 221, 1963 (II); 1971, *Astron. Zh.* **48**, 819 (III).

[3] Kaula, W. M.: 1971, *Rev. Geophys. Space Phys.* **9**, 217–38.

[4] Ganapathy, R., Keays, R. R., Laul, I. C., and Anders, E.: 1970, *Proc. Apollo 11 Lunar Sci. Conf.* **2**, 1117.

[5] Ringwood, A. E.: 1970, *Earth Planetary Sci. Letters* **8**, 131.

[6] Safronov, V. S. and Ruskol, E. L.: 1962, *Proc. XIII Internat. Astronaut. Congr., Varna*, 42.

[7] Safronov, V. S.: 1969, *The Evolution of the Preplanetary Cloud and the Formation of the Earth and Planets* (in russian), Nauka, Moscow.

[8] Dohnany, I. S.: 1967, in J. L. Weinberg (ed.), 'Collisional Model of Meteoroids', *Proc. Symp. Honolulu, Hawaii.*

[9] Orowan, E.: 1969, *Nature* **222**, 867.

[10] Reynolds, R. T. and Summers, A. L.: 1969, *J. Geophys. Res.* **74**, 2494.

[11] Altschuler, L. V. and Sharipdjanov, I. I.: 1971, *Izv. An S.S.S.R., Fiz. Zemlii*, No. 4, 3.

THE ORIGIN OF THE MOON AND SOLAR SYSTEM

HAROLD C. UREY

University of California at San Diego, La Jolla, California, U.S.A.

Abstract. The Rb^{87}-Sr^{87} ages of many of the lunar rocks suggest that the fundamental differentiation took place 4.5×10^9 yr ago and that remelting occurred without exchange of the rubidium and strontium with the surroundings. The Apollo A rocks are an exception to this. They appear to have acquired rubidium without all of the Sr^{87} produced during the first aeon. Also in the remelting about half of the radiogenic leads were lost to the surroundings probably the soil by a vaporization process. We interpret these results to mean that remelting occurred in a system that was nearly closed to the surroundings and that the high early concentrations of radioactive elements in highly insulating surroundings made this possible.

1. Introduction

The problem of the origin of the solar system is a very old one, and many very competent people have made suggestions in regard to the origin. The problem involves physics, astronomy and chemistry. Most problems of the stars involve the properties of gases only, but the problem of the origin of planet Earth and its satellite, as well as the other planets, involves such low temperatures eventually that liquid and solids appear. With this, complicated problems of chemistry must be considered. Today, there is no unanimity of opinions in regard to this problem.

The general facts in regard to the system are well known. The system has very nearly a planar structure; the orbits of the planets and satellites are nearly circular; the division into terrestrial planets, major planets, asteroids and comets are very obvious, and the increased spacing of the planetary orbits roughly represented by Bode's law have been recognized for many years. The early theories of origin have been reviewed very well by Jastrow and Cameron (1963) and the chapter by Cameron and ter Haar is especially pertinent. Also, a review by Woolfson (1969) and a number of papers edited by Marsden and Cameron (1966) give excellent summaries of the many ideas in regard to this problem that have been presented during the past years, decades and centuries. Most theories have involved highly complex mathematical discussions, and these certainly are necessary, but the complexity of the subject is so great that these can hardly be adequate. Much observational data must be secured, and I, personally, have tried to concentrate my efforts in this direction. In particular, I have naturally looked for chemical evidence while trying not to violate physical laws and physical and astronomical evidence. The physical facts are also very complicated involving magnetic effects, strength of materials etc., in addition to the gravitational fields.

2. Some Chemical and Physical Properties of the Sun and Planets

The isotopic composition of the elements of higher atomic weights are within the

Urey and Runcorn (eds.), The Moon, 429–440.

limits of error – the same in the Earth, meteorites and Moon. There are small differences in these abundances in the case of the lightest elements, e.g. H, Li, C, N, O, but these variations are generally explained as due to chemical fractionation of the isotopes. This means that these objects have been formed from the same nuclear synthetic events, and, hence, that the meteorites, Moon and Earth must be derived from the same fundamental mixture of elements and that appropriate fractionation processes must have occurred to produce any observed differences in composition. We cannot be certain that the Sun belongs to the same synthetic process, since exact measurements of abundances of the isotopes cannot be made. However, it is generally assumed that this is true. Thus, the assumption of two sources of material, i.e., one for the planets and another for the Sun, is generally assumed not to be true. This conclusion applies to the following discussion, and we believe that planetary objects have been derived from solar-type material.

Hydrogen and helium make up about 97 or 98% of the Sun by mass, conventional solids about 0.25%, and the remainder consists of carbon, nitrogen, oxygen and inert gases other than helium of which neon is the most abundant. Thus, there is one gram of solids in 400 g of gas, and, hence, the Earth and its component of solar gases would have somewhat more mass than Jupiter. One cubic centimeter of solids must be separated from about 10 m^3 of gas at STP. In order to produce the terrestrial planets, an immense settling process is needed, which must have occurred at such temperatures that terrestrial type, rocky materials were present as liquids or solids. Though high temperatures were probably present during the condensation of the Sun and Planets, lower temperatures must have been present at some stages of the process. Also, gases have been lost to some extent from the major planets, though this loss has not been so very nearly complete as is the case for the terrestrial planets. The problem of the extent of loss of these substances from the major planets is quite complex, but we can conclude that Uranus and Neptune must have lost much gaseous material while little may have been lost from Jupiter and Saturn. The gases which have been lost almost completely from the terrestrial planets are H_2, He, C as CH_4, N_2 or NH_3, H_2O, Ne, other inert gases and possibly some of the more volatile elements which are condensed into rocky materials at lower temperatures. These gaseous substances were probably lost before planets of the mass of the Earth or Venus had accumulated because of the difficulty of removing gases of higher atomic weight than helium from these planets due to their high gravitational fields. It is just possible that strong magnetic fields sweeping over these planets would remove ionized atoms and molecules from their atmospheres. In this case, ionization by far ultraviolet light from the Sun is required for many of the constituents, and it is not certain that this would be very effective. It should be noted that ionization potentials of the abundant gases, e.g. H_2, N_2, H_2O, and Xe, are very high, i.e., 15.427, 15.576, 12.60 and 12.127 respectively, and will require very short wavelengths of light to produce ionization. Even small amounts of interplanetary gases will absorb this light. It is far from certain that the primitive Sun emitted intense light of the required wavelengths and that this light would reach planetary atmospheres sufficiently to permit magnetic fields to sweep them away. The problem is very complex involving the solar spectrum, the amounts of gas between the Sun and planet, the ionization of molecules, the recombinations of ions and electrons, the efficiency of removal by a magnetic field of unknown intensity and probably other difficult variables. Perhaps I am lazy and mother nature is not, but I prefer the loss of gases before strong gravitational fields develop. It seems likely to me that gases were lost from the region of the terrestrial planets and from Uranus and Neptune before the great gravitational fields had developed, that the planets accumulated from smaller objects from which gases had been lost to space, and that the terrestrial planets had accumulated by relatively small solid objects.

Considerable variation in chemical composition with respect to the more non-volatile consituents exists in the terrestrial planets and the meteorites. The densities of the terrestrial planets and the Moon estimated for low temperatures and pressures vary considerably, and we suppose that this is due to a variation in the proportion of the abundant high density element of iron either in its elemental or combined state. The estimated iron to silicon atomic ratios in various objects are presented in Table I. The

TABLE I

	Fe/Si
Achondrites	~ 0.0–0.3
Sun (uncorrected)	0.12
Sun (corrected)	0.25
Moon	0.3
L chondrites	0.59
Mars	~ 0.6
H chondrites	0.81
Type I Carb. chondrites	0.89
Earth	1.0
Venus	~ 1.0
Mercury probably	$\lesssim$ 3

iron ratios are those generally accepted some years ago. More recently, the oscillator strengths for iron lines have been revised downward, and, thus, the solar abundance has been raised until the estimated iron to silicon ratio is approximately unity. However another study by my colleague, Brueckner (1971), indicates that for some of the solar iron lines, the damping constant used is in error, and that the abundance of iron must be lowered again until the sun (corrected) value may be about correct. Cowley (1970) has insisted upon the importance of this. These calculations apply only to the strong lines, but abundances from weak lines may be in error due to blends with other weak lines. This makes an error in the observed abundances which is always on the high side. (This has been called to my attention by John Ross.) Whatever the final conclusion may be, considerable fractionation of the chemical constituents has occurred with the Moon and Mercury being at opposite extremes for planetary objects and with great variations in the meteoritic compositions also being present. Low density silicate material must have been lost during the accumulation of Mercury, and if the higher abundance of iron, i.e., an iron silicon ratio of unity, is correct, then high density iron must have been lost from the Moon or the interior of the Moon contains some percentage of low density materials such as water or carbonaceous substances. In view of the low water and carbon content of the lunar surface, I am not enthusiastic about this interpretation, but I cannot definitely exclude it as a possibility.

It seems most likely that some large objects of approximately lunar size were present during the accumulation of the planets. If the planets accumulated from small objects or from gases, one would expect that their axes of rotation would all be nearly perpendicular to the ecliptic plane or more probably to the planes of their orbits, and the orbits would all lie nearly in one plane. This is only true for Jupiter. The Earth, Mars, Saturn and Neptune have axes tilted relative to the verticle to their orbits by 23° to 29° and Uranus by 98°. Venus rotates in the reverse direction to the orbital rotations. Mercury's rotation has been modified by tidal action from the Sun. Collisions of fairly large objects of about the mass of the moon are an obvious explanation of these facts. Objects colliding with terrestrial planets should have lost their atmospheres while those which collided with the major planets may have retained all or part of their gaseous atmospheres. Incidentally, at this point in our discussion, such objects might make it possible for one of them to be captured by the earth and thus supply a very unusual satellite moving in an orbit about its primary whose plane apparently could never have coincided with that of the Earth's equator.

3. One Model for Solar System Origin

I have adopted a model for the origin of the solar system which starts with a Sun probably of greater diameter than it has at present, and a flat nebula extending out to the farthest planet. It is assumed that the contraction of the original rotating mass of gas threw off a disk of gas and dust when it contracted to form the Sun. Solids would settle to the median plane of such a nebula if turbulence did not persist. Gravitational instability would occur in such a layer of gas, and it should break up into spheres of gas. Kuiper (1951) first used such a model to make protoplanets which were of planetary size plus the solar component of gas. I argued against this since it appeared difficult to construct a very few large objects of this kind, and also especially because of the difficulty of dissipating such enormous masses of gas from the planets, e.g., a mass of Jupiter approximately from the Earth's or Venus' protoplanet. Mostly, it appears that this type of nebula is assumed by others though the processes of further development, i.e., the gravitational instability mechanisms, are generally not used. I have argued that such instability may have produced approximately lunar sized objects, i.e., objects having the mass of the Moon plus its proportion of solar gases, and, thus, total masses in the neighborhood of $2.2–3.7 \times 10^{28}$ g. (See Urey, 1966). The escape of gases from objects of this mass should be considerably more plausible than from the Earth with its component of solar gases. Others have assumed that accumulation of sizeable solid objects in the solar nebular should be possible and that the gases were dissipated leaving these objects which then accumulated into the planets. (See Hartmann (1971) for example.) Indeed, it seems likely that both mechanisms may have occurred, and also that much dust and fragmented material may have escaped to space with the gases that must have been lost to space in order that the sun could lose its excessive angular momentum. The dissipation of gases to space probably occurred by the rotating primitive intense magnetic dipole field of the Sun. Thus, processes capable of producing considerable numbers of objects of lunar mass and smaller can reasonably be postulated for the earlier development process of the solar system.

4. Recent Data From the Moon

(1) The surface of the Moon has been fashioned physically by the collision of many objects which have produced many craters, possibly, in fact, many layers of craters and the great circular maria. Volcanic craters of small size are certainly present, and great flows of lava or volcanic ash have filled many craters and the shallow maria. I, personally, doubt the existence of great calderas, but many students interpret some craters in this way. This great bombardment of the Moon must have occurred before the oldest rocks of the Earth were laid down either as sediments or as molten silicates, since these rocks show little signs of having been formed from fragments produced by great collisions, and it would be impossible to bombard the Moon without bombarding the Earth at the same time. Thus, the intense bombardment of the Moon occurred earlier than 3.5 aeons ago. This conclusion is confirmed by the ages of the rocks which will be discussed later in this dissertation. (See Gilbert, 1893.)

(2) In 1968, Muller and Sjogren (1968) showed from studies of the movement of the orbiters that very marked mass concentrations exist below the surface of the Moon mostly in the circular maria regions. These were detected by observing the velocities of the orbiters as they passed over the visible side of the Moon. As the orbiter approached Mare Imbrium, for example, it was accelerated. As it left this region, it was decelerated. By very careful and detailed calculations, they were able to estimate the excess mass over what should have been present if the Moon's density did not vary with angular position, i.e., with latitude and longitude, and if the excess mass was near the surface. Thus, for Mare Imbrium, the excess mass is 1.6×10^{21} g. If this mass had a density 10% greater than that of the surrounding material, its mass would be $1.6 \times \times 10^{22}$ g, and, with a total density of 3.7 g cm^{-3}, it would cover Great Britain to a depth of 21 km or the State of California to a depth of 12 km. These masses are truly very large. If due to volcanic effects, the transports of material required to produce them would compare with the largest volcanic effects on Earth. Furthermore, they would not be supported by the outer rocks of the Earth. A positive gravity anomaly exists on the Island of Hawaii, but it stands in a depression on the ocean floor. Thus, the Earth's surface is sinking under this load. If the anomaly was measured from about 100 k above the Earth's surface, it is probable that no anomaly would be observed. Other anomalies, both positive and negative, exist on the Earth, but they are probably due to the great convection cells in the Earth's mantle which also move the continents about the Earth's surface and push up the great mountain chains as

well. Convection cells within the Moon are also postulated, but these have no relation to the positions of the circular maria. The outer parts of the Moon must be very rigid and must have been very rigid throughout all of lunar history, since the maria were formed probably approximately 4.5 aeons ago.

If we assume that the objects that fell on the Moon to produce the circular maria were moving with about the escape velocity of the Moon, i.e., they were moving along with the Moon about the Sun at about the same distance, and if we use formulae for the relation of the required energy to produce these maria with their observed diameters, we find that these calculated masses agree fairly well with the masses estimated by Muller and Sjogren (1968). Several of us have proposed this as their origin. Others propose complicated volcanic effects to produce them, and I definitely do not agree with these models. However, the important point to be made at this time is that in order to support these mascons, the Moon must be sufficiently rigid, and, hence, be at a rather low temperature in its outer parts, and must have been so throughout its history.

(3) The evidence from the Turkevich *et al.* (1967) analyses on Surveyors 5, 6 and 7 showed that the lunar surface was highly differentiated, and the chemical analyses of samples returned by Apollos 11, 12 and 14, as well as the spectrographic studies of McCord (1969), completely confirmed this data and added much detailed information in regard to this question. Basalts, anorthosites and very acidic rocks have been recognized, and the indications are that these very highly differentiated rocks must have been produced from a partial melting of some parent silicate rocks or to crystallization of a melted pool of silicate melt. The concentrations of minor elements show that the basalt rocks and soil must be only a small fraction, i.e., a few percent, of the original parent material, if it was of approximately meteoritic composition. The composition of these rocks are in all cases, it seems, somewhat different from terrestrial rocks, though the differentiation process must have involved fractionation between liquid and crystalline phases of silicon, aluminum, magnesium, calcium, iron etc. and oxide melts. If the parent material was of approximately meteoritic composition, very large increases in the concentrations of many minor elements have occurred. Also, the more volatile elements are missing to a marked degree as are also the siderophile elements. Thus, at some time, these volatile elements must have been lost by a high temperature process, i.e., possibly 1500 °C, and liquid iron must have trickled through some liquid material removing the siderophile elements. In fact, the lunar surface, at least, must have been melted at some time.

(4) The lunar seismic data are most interesting and puzzling in many ways. The seismologists suggest that the lunar surface consists of highly fragmented materials to a depth of possibly 20 km with no indication of discontinuities. Lava flows in the shallow maria may have been broken up by the lesser collisions that have occurred in great numbers in the maria. Possibly this problem would be solved by information from the great circular maria where deep layers of lava should exist, i.e., if the smooth material is indeed lava and not finely fragmented material of the general composition of lava flows. The seismic data indicates that the surface is highly fragmented to

considerable depths as is to be expected if the great craters are due to collisions.

(5) Various ages of the lunar surface have been determined. The mixing times of the soils as determined by the effects of cosmic rays give very little information in regard to the early history of the solar system which is the principle theme of the present discussion, though the effects of cosmic rays may produce great difficulty in understanding the data bearing on this early history. The Rb^{87}-Sr^{87}, U^{238}-Pb^{206}, U^{235}-Pb^{207} and Th^{232}-Pb^{208} data are not complicated by the cosmic ray effects, but lead under the highly reducing conditions represented by the presence of liquid iron (at the melting point of iron (1535°C) or somewhat lower than this temperature) is fairly volatile and may move from one phase to another as seems to have occurred. The K^{40}-A^{40} ages are plagued by the loss of A^{40}, if even moderate heating for long periods of time has occurred.

Two types of ages have been determined and discussed. The fundamental equation for the time elapsed during which the concentrations of the parent and daughter radioactive elements have not changed is

$$\left(\frac{Sr^{87}}{Sr^{86}}\right) = \left(\frac{Sr^{87}}{Sr^{86}}\right)_i + \left(\frac{Rb^{87}}{Sr^{87}}\right)(e^{\lambda t} - 1)$$

where the ratios without subscripts are the present measured ratios, the subscript, i, indicates the initial ratio, λ is the decay constant, and t is the time elapsed. Similar equations for other parent/daughter relations can be written. If the initial ratio is known or thought to be known, as for example from meteoritic studies, t can be calculated, and this is known as the *model age*. If the initial rubidium to strontium ratio is not known, but variable ratios of rubidium to strontium occur in different crystalline masses in a rock, the variable values of the measured ratio (Sr^{87}/Sr^{86}) can be plotted against the measured ratio (Rb^{87}/Sr^{87}), and then the slope of the curve, which should be a straight line, is equal to $(e^{\lambda t} - 1)$, the value of t can be calculated, and the intercept on the (Sr^{87}/Sr^{86}) axis is the value of the $(Rb^{87}/Sr^{87})_i$. The age calculated in this way is known as the *isochron age*.

(6) The isochron ages of the Apollo 11 rocks by the Rb^{87}-Sr^{87} method are close to 3.65 aeons, and of the Apollo 12 rocks about 3.3 aeons. The 12013 rock has an isochron age of 4.0 aeons. The model ages run near 4.5 aeons. The model ages by the uranium-thorium lead ages are about 4.65 aeons, but the isochron ages vary rather badly and do not agree with the Rb^{87}-Sr^{87} ages. In another paper presented at this symposium, my colleagues and I find that consistent results can be secured if about one-half of the lead isotopes produced during the time between the initial melting and the second melting, about one aeon later, was lost by evaporation into the soil. The initial melting time by the U-Pb method then becomes 4.5 aeons instead of 4.65 aeons. The K^{40}-A^{40} ages, when corrections for loss of A^{40} are made, give very similar results. For our present purposes, we note that the times of the first and second meltings are about 4.5 and 3.3–4.0 aeons ago respectively. These times for the first melting are close to those for the ages of the meteorites and the earth, and they substantiate the initial arguments for a great age for the Moon. We must conclude that an early melting

occurred, followed by a cooling and freezing of the surface in order to support the great collisional craters and mascons, and then a remelting occurred about one aeon later. It is necessary to conclude that that first melting occurred when the radioactive heating was greater than it was an aeon later by about 40%. This is an important problem in devising a history for the Moon.

Meteorites contain rare gases which indicate that they had cooled to sufficiently low temperatures to retain xenon gas before certain radioactive elements had decayed to such low concentrations that their daughters could not be detected. Thus, I^{129}, with a half-life of 17 million yr, decays to Xe^{129} which appears in excess in some meteorites. Excess Xe^{129} has not been observed in lunar samples. If it did, this would indicate that the Moon, as a non-degasing body, would be older than the Earth since terrestrial Xe shows no excess of Xe^{129}.

Fission of uranium and atoms of higher atomic weight fission spontaneously to atoms of lower atomic weight, and among these are Xe^{134} and Xe^{136}. Slight excesses of these have been observed by Marti (1971), but they are so slight that no confidence can be placed in the observations. At present, we have no evidence from these xenon observations that these elements were present in lunar material.

(7) No general magnetic dipole field associated with the Moon has been detected. However, the rocks which were last melted some 3.5 aeons ago do have detectable magnetic moments. Also, local magnetic fields up to 100 gamma have been detected. This means that these surface rocks were in magnetic fields as they cooled through the Curie point, and that local areas of some kilometers in dimensions at least do retain magnetic fields. This means that at some time, when these rocks cooled through the Curie points (770 °C for metallic iron) a magnetic field was present and estimated to have had an intensity of about 1000 gammas or 0.01 Oe. It seems that no field of such intensity exists on the Moon, and, hence, the field that did exist has disappeared.

5. Model for the Moon's Early History

The rigid Moon required by the mascons, and the melted Moon required by the chemical differentiation indicate that the Moon was accumulated at low temperatures, and that its surface was melted by some external source of heat while the interior remained relatively cold. Then the surface cooled down sufficiently to support the craters which were produced by collisions following this cooling off period. A rather high thermal conductivity of crystalline rocks made this possible even though radioactive heating was at a maximum at this time. The mascons are probably supported on the interior rigid material. The collisional processes produced a highly fragmented layer with a low thermal conductivity which made it possible for the decreased amounts of radioactive elements to produce melting about one aeon later. The differentiation of the lunar surface rocks occurred during the initial melting, and the subsequent cooling process, i.e., the differentiation of the surface materials, occurred during the very early processes on the Moon. Subsequent processes changed the chemical compositions only in a minor way. Reduction of some of the iron occurred in this first melting, and it settled into a lower layer and carried the siderophile elements with it. This metallic layer is very probably the highly conducting layer which has been detected by Sonett *et al.* (1971) from the flow of solar winds over the lunar surface. This iron layer may have been cooled below the Curie point of 770°C during the early cooling period, and, hence, it would have remained highly magnetized if an intense solar dipole field was present at this early time. This may have caused the magnetization of the rocks an aeon later, and heating since then may have raised this metal layer above the Curie point so that the dipole field has disappeared though the surface rocks retain their magnetized characteristics.

The melting processes of 3.2–4.0 aeons ago may be due to collision melting during the smaller collisions on the Moon as suggested by Gold, or to collision melting due to the mare collisions, though in this case the masses that produced the maria must be stored somewhere for an aeon. Possibly a catastrophic event in the asteroidal belt produced objects which fell on the Moon to some extent.

My colleagues and I suggest a possible remelting about an aeon after the first melting and show that this is possible, providing the initial melted layer was about 200 km in thickness. Sonett *et al.* estimate the depth of the highly conducting layer as about 250 km. It is assumed that the highly radioactive basalt and highly silicic and radioactive rock 12013 material lies in layers about 20 km below the surface, and that the collisions which followed the initial solidification broke up the surface layers producing a highly insulating layer. (The details of this work were presented at the Newcastle conference, and will be published in the Apollo 12 Conference Report by Urey *et al.*, 1971.) Radioactive heating in the course of about one aeon produced a flood of lava, or more probably ash, which covered the shallow maria and left rocks with isochron ages of 3.2–4.0 aeons.

6. The Origin of the Moon

Up to the present, it seems that no one has changed his mind in regard to the origin of the Moon, and we can only conclude that all evidence from the space program is indecisive in regard to this question. The escape from the Earth requires that a primitive Earth with high angular momentum accumulated and then separated into the present Earth and Moon. This has been discussed by many authors since George Darwin proposed this at the end of the 19th century. No really satisfactory solution of the problem has been secured. However, it seems that we cannot exclude this origin.

Various suggestions of the double planet hypothesis have been made. This hypothesis meets with many difficulties. For example, the growth of two bodies moving about each other requires a very nice balance of centrifugal and gravitational forces during the period of growth. Ringwood (1970) has suggested a modification of this origin that the Earth formed at very high temperatures with an extensive atmosphere of volatilized, rocky material including the elements found in the Moon's surface. This atmosphere is supposed to have condensed and the rocks so formed to have accumulated into the Moon. This is far from being a simple model. The condensing vapor must have been more than 2.89 Earth radii above the Earth, since this is the Roche limit, and no accumulation into the Moon could have occurred unless the solid objects condensed from the vapor beyond this limit. What the chemical composition of both Earth and Moon might be would be difficult to estimate. Possibly the loss of volatiles might result from this model in some way if an escape mechanism for vapors could be devised. However, to me the greatest objection to this model is that Venus, a planet of comparable mass and distance from the Sun, not only does not have a Moon but indeed rotates slowly in the opposite sense to the Earth. Thus, the two planets must have had very different specific angular momentum. If the terrestrial planets all had Moons similar to that of the Earth, it would have been recognized by everyone, including Galileo, that some double planet mechanism, escape from the primary or condensation near the primary was such a probable origin that there would have been no arguments against some such origin. It seems most improbable that Ringwood's mechanism could be true for the Earth/Moon system, and that no approximation to this would be true for Venus.

The capture hypothesis presents serious difficulties for capture of the Moon by the Earth requires the loss of the energy of capture in some way. Gerstenkorn (1967) has presented a vary carefully calculated process of capture in an initially reverse orbit. Urey and MacDonald (1971) suggest capture by collision with other objects moving in orbits near the Earth. Both mechanisms are very special, and one could hardly expect this to be a regular event. It does require that many lunar type objects were present in order that the probability of capture by one planet would be reasonable. Also, capture by Venus of such an object may account for the reverse rotation of Venus, as suggested by Singer (1970), and the tilts of the axes of the planets could be produced by such objects. Thus, a limited number of massive primitive objects could account for this irratic characteristic of the planets. However, the Moon must have accumulated elsewhere, and a problem is presented by the curious low density of the Moon which indicates that the abundance of some high density element, i.e., iron, is lower than it is in the terrestrial planets. Small amounts of water or carbonaceous materials in the lunar interior or a lower abundance of iron in the primitive solar composition is required. For many years, the iron-silicon ratio in the Sun was estimated to be about $\frac{1}{10}$ to $\frac{1}{5}$, but, recently, this has been revised to about 1. However, this result is disputed by some experts in the field, and a certain conclusion cannot be reached at present.

7. The Moon and the Origin of the Solar System

The review of data on the Moon is reviewed because it may have a special importance relative of the origin of the entire system. Thus, if the Moon has been captured by the Earth, it is a more primitive object than the Earth and other planets, and being near the Earth, the thorough investigation of this primitive object is possible. Of course, it may be that Ceres and other asteroids are the primitive objects, and, if so, the intense investigation of these objects should be made even though it will require many years to do so. It may be that both the Moon and larger asteroids are primitive objects, and, of course, it may be that neither are such objects.

The investigation of Mars will be very interesting, especially if life is present or has existed in the past on this planet. However, we can expect little evidence in regard to the early history of the system, since it seems probable that water has been present on Mars. The planet is red, probably due to trivalent iron which has been oxidized from the ferrous state by water. This means that sedimentary rocks will be present with all the attendant difficulties in understanding early planetary history that this entails. It will be interesting to determine, if we can, whether the planet was formed in an extremely hot condition or at a more moderate temperature, and whether the planet has a core. However, the many detailed questions that we ask in connection with Moon's ancient history for the most part cannot be answered by detailed investigation just as is true for studies of the Earth.

The recent space studies have added many details in regard to the structure of the major planets and their satellites, but little in regard to their origin. Jupiter looks like a miniature solar system with its axis of rotation tilted only three degrees from the verticle of its orbital plane and a system of satellites with orbits lying closely to the equatorial plane. The other planets do not resemble closely this character of Jupiter at all. The marked tilts of the axes relative to the planes of their orbits again suggest the presence of some massive objects which accumulated into these planets.

Many factors of immense complexity exist in regard to the origin of the system, and many observational details are needed before we can reach definite conclusions in regard to these problems. Possibly no general agreement is possible, and, in a way, we may never know!

References

A. REFERENCES TO THE TEXT

Brueckner, K.: 1971, *Astrophys. J.*, in press.

Cowley, C.: 1970, *Astrophys. Letters* **5**, 149.

Gerstenkorn, H.: 1967, in S. K. Runcorn (ed.), *Mantles of the Earth and Terrestrial Planets*, Wiley, Interscience, New York, pp. 228.

Gilbert, G. K.: 1893, *Bull. Amer. Phil. Soc., Washington* **12**, 241.

Goldreich, P.: 1966, *Rev. Geophys* **4**, 411, 439.

Hartmann, W. K.: 1971, Preprint.

Kuiper, G. P.: 1951, in J. A. Hynek (ed.), *Astrophysics*, McGraw-Hill Book Co., New York, Chap. VIII.

Marti, K.: 1971, Private Communication.

McCord, T. B.: 1969, *J. Geophys. Res.* **74**, 3131.

Muller, P. M. and Sjogren, W. L.: 1968, *Science* **161**, 680.

Ringwood, A. E. and Essene, E.: 1970, *Proc. Apollo 11 Lunar Science Conference* **1**, 769.

Sonnett, C. P., Smith, B. F., Colburn, D. S., Schubert, G., Schwartz, K., Dyal, P., and Parkin, C. W.: 1971, *Second Lunar Science Conference* (unpublished proceedings).

Singer, S. F.: 1970, *J. American Geophys. Union* **51**, 637.

Turkevich, A. L., Franzgrote, E. J., and Patterson, J. H.: *Science* **158**, 635 (1967); *Science* **160**, 1108 and preprint submitted to *Science* (1968); *Science* **162**, 117 (1969a); *Science* **165**, 277 (1969b).

Urey, H. C.: 1966, *Monthly Notices Roy. Astron. Soc.* **131**, 212.

Urey, H. C. and MacDonald, G. J. F.: in Z. Kopal (ed.), *Physics and Astronomy of the Moon*, 2nd edition, Academic Press, New York and London, Chapter 6.

Urey, H. C., Marti, K., Hawkins, J. W., and Liu, M. K.: 1971, *Proc. of the Second Lunar Science Conference* **2**, 987.

B. GENERAL REFERENCES

General Reviews:

Baldwin, R.: 1963, *The Measure of the Moon*, University of Chicago Press.

Jastrow, R. and Cameron, A. G. W.: 1963, *Origin of the Solar System*, Academic Press.

Kopal, Z.: 1962, 1971, *Physics and Astronomy of the Moon*, 1st and 2nd editions, Academic Press, New York and London.

Marsden, B. G. and Cameron, A. G. W.: 1966, *The Earth-Moon System*, Plenum Press, New York.

Woolfson, M. M.: 1969, *Evolution of the Solar System, Reports on Progress in Physics* **32**, Part 1, pp. 135–186.

For much observational data from the Apollo landings:

Mason, B.: 1970, *The Lunar Rocks*, Willey-Interscience.

Proceedings of the Apollo 11 Lunar Science Conference, Pergamon Press (1970).

Proceedings of the Apollo 12, Lunar Science Conference, Pergamon Press, in press.

EVOLUTION OF THE MOON: RECENT MODIFICATION OF PREVIOUS IDEAS

B. J. LEVIN

O. Schmidt Inst. of Physics of the Earth, U.S.S.R. Acadamy of Sciences, Moscow

New data on the Moon obtained from its study by space probes and analyses of returned lunar samples give new 'boundary conditions' for a study of its origin and evolution.

1. Proofs for a Hot Moon

First of all, the new data have brought to the end the dispute between proponents of a hot and a cold Moon. As lunar maria are flooded by basalt – a product of magmatic differentiation – the lunar interior must be hot and, at least sometimes, – partially molten. Earlier the partial melting of lunar interior and its present day hot state was obtained in calculations of thermal history for conductive models with chondritic radioactivity (Urey, 1952, 1962; MacDonald, 1959; Levin and Majeva, 1960; Levin, 1962, 1966a, b; Fricker *et al.*, 1967; McConnell *et al.*, 1967; Majeva, 1971). Some authors tried to modify the model regarding the Moon to be a throughout solid body. The others combined these results with the interpretation of lunar maria and flatbottom craters as lava-filled basins and predicted that the Moon indeed is hot and has a semi-molten interior.

Urey and other proponents of a throughout solid Moon regarded a hot Moon to be incompatible with its desequilibrium shape. One must remind, however, that the concept of hydrostatic equilibrium can be applied only to an isothermal body or a body in which isothermal surfaces coincide with equipotential ones. In the Moon the decrease of surface temperature from the equatorial zone toward the poles produce a flattening of isothermal surfaces for several tens of kilometers which load to a flattening of its figure for about 1 km (Levin, 1964b, 1966a, b, 1967; Safronov, 1967; Volkov, 1967). Indeed such flattening is the main deviation of the Moon from the hydrostatic equilibrium. As to the small ellipticity of the lunar equator, the stresses it produce, as well those produced by mascons, must be supported by the rigidity of the outer cold solid layer of the Moon.

2. Evidence from Magnetic Experiments

Few years ago the idea of a hot, semi-molten lunar interior seemed to be in conflict with low values of its electric conductivity derived by Ness from measurements of magnetic field perurbations in cislunar space. Later Ward (1969) stated that both cold and hot models of the Moon are consistent with evidence from magnetic experiments aboard of Explorer 35 lunar orbiting satellite available at that time. Recently Sill (1971) combining measurements of the lunar surface magnetometer and the magneto-

Urey and Runcorn (eds.), The Moon, 441–449. *All Rights Reserved.*

meter on Explorer 35, dedused that the electric conductivity increase up to a depth of 200–300 km and beyond stabilise at the value which for an olivinlike composition corresponds to temperatures of beginning of melting. This is in accord with previous calculations for initially cold or warm Moon (Fricker *et al.*, 1967; Majeva, 1971) and with new calculations for initially hot Moon (Wood, 1971) provided the redistribution radioactive elements toward the surface in the course of magnetic differentiation proceeded at sufficient rate.

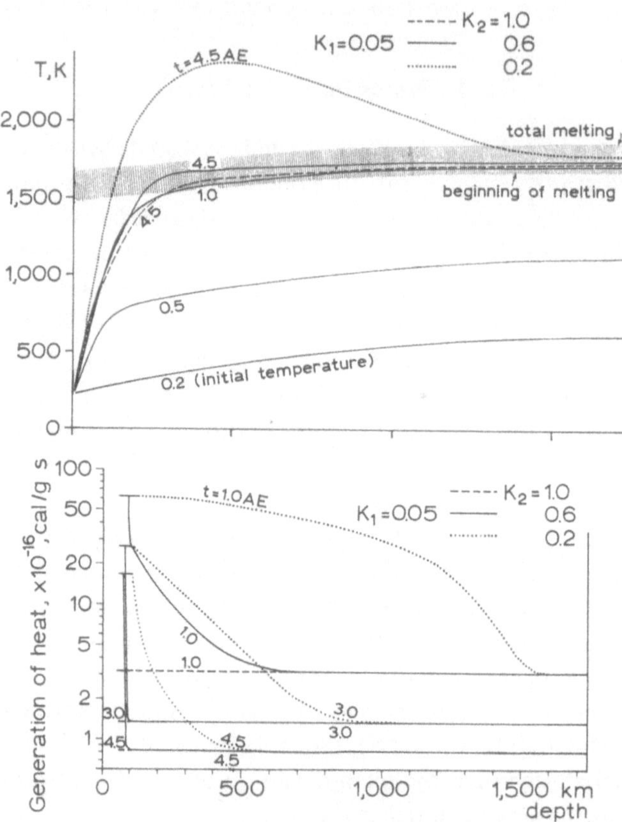

Fig. 1. Distribution of temperature (above) and generation of heat (below) in the lunar interior for different rates of redistribution of radioactive elements (variant with initial temperature giving melting about 3.5 AE ago).

Results of calculations for an initially warm Moon with simplified initial temperature profile are shown on Figure 1 for a case when the redistribution of radioactive elements toward the surface decrease 20 times their concentration in the interior ($k_1 = 0.05$) and for different rates of redistribution ($k_2 = 1.0$; 0.6; 0.2). For rapid redistribution ($k_2 = 1.0$ and $k_2 = 0.6$) the interior is semi-molten and the thickness of the outer solid layer is about 400 and about 300 km respectively. But for slow redistribution ($k_2 =$ 0.2) the interior, except the central part, is molten and the outer solid layer is only 150 km thick. The lower part of Figure 1 show profiles of generation of heat for three moments of time for different values of k_2. For $k_2=1.0$ and 0.6 the present day temperature profiles are in accord with the results by Sill (1971).

However Sonett *et al.* (1971) obtained from the same measurements as Sile a more complicated profile of electrical conductivity implying a layered structure of the upper 300–400 km. Assuming below 400 km a poorly conducting matter they obtained a temperature of about 1000 K only, while the melting temperature is few hundreds degree higher. A calculation based on more conducting matter would depress the computed temperatures. Thus there is a moderate but important disagreement in the interpretation of magnetic experiments.

3. Early Melting of the Moon

The convincing evidence in favour of the hot Moon was the immediate result of the presens of basalts in the returned lunar samples. Their further study showed that magmatic differentiation of the Moon began very early – nearly at the end of the accumulation of the Moon (Papanastassiou and Wasserburg, 1970, 1971). while according to previous calculations of lunar thermal history, melting had to begin only $1–2\times10^9$ yr after the formation of the Moon. The only reasonable way to account for such early melting is to assume a high initial temperature of the Moon. (Wood, 1971; Wood *et al.*, 1971; Papanastassiou and Wasserburg, 1971).

Besides evidence for early melting there is evidence that this early melting was restricted to outer parts of the Moon while melting of deep interior occurred about 10^9 yr later and produced dark basaltic lavas filling lunar maria. Such thermal history requires initially hot outer parts of the Moon and initially cool interior.

Calculations of the thermal history for several variants of such type of initial temperature profile were done by Wood (1971). His results clearly show that such initial profile can give a necessary course of lunar thermal history.

4. Composition of Highlands

Evidence for early melting and differentiation of the outer part of the Moon eliminates the problem of the fate of the primordial outer layer of the Moon and of the composition of highlands. The highlands are so densly covered by impact craters (on some places it is a saturated density (Marcus, 1970)) that their origin must be connected with intence bombardment by big bodies at the last stage of lunar accumulation. Previously when calculations of lunar thermal history had given that melting and consequently – magmatic differentiation began only $1–2\times10^9$ yr after the formation of the Moon, it seemed impossible to regard highlands to be composed of products of magmatic differentiation. It followed from these calculations that highlands, if composed of products of differentiation, should be formed after beginning of melting. But by that time intense bombardment of the Moon has to cease. Therefore some astronomers regarded the highlands as preserved parts of the initial outer layer (Kuiper, 1954, 1959; Levin, 1962). However it seemed strange why blocks of solid primordial lunar material which had to be denser than the molten differentiated magma, had not sink into the latter (Urey, 1955). Now when it is proved that melting and magmatic differentiation of the Moon occurred already near the end of its formation, it is no more difficulties to combine the igneous origin of highlands with the impact origin of craters on them. After the effusion of magma on the surface its cooling and solidification could be sufficiently rapid to permit the surface to be densely cratered during the abatement of lunar accumulation. According to the hypothesis by Wood *et al.* (1970a, b, 1971) the highlands are composed of anorthosites.

5. Origin of the Moon

To discuss possible sources of energy for early heating of the Moon at least some general idea of the origin of the Moon is required. We shall rest upon the idea of its accumulation from a circumterrestrial swarm of particles, which had to exist when the Earth in its turn accumulated from bodies and particles forming a circumsolar swarm. This hypothesis of origin of the Moon proposed 20 yr ago by Prof. O. Schmidt (1950) seems to be the most promising. It is similar to the so-called sediment ring hypothesis. The continuing discussion of the hypothesis of separation of the Moon from the Earth seems to the present author to be a curious episode in the history of science. Darwin's mechanical foundation of this hypothesis is since long refuted by Liapunov (1906–1914), Cartran and others (see Lyttleton, 1953, 1954) while recent attempts by Wyse and O'Keefe to save Darvin's scheme cannot withstand a critic. As to the capture hypothesis, it studied only the past evolution of the lunar orbit, the Moon being treated as a material point. But it never studied the origin of the Moon as a cosmic body. It simply shifted this problem away from the Earth – to some other place in the solar system. One can hope that further discussion of the capture hypothesis will stop because now all its supporter agree that when the evolution of the lunar orbit is extrapolated far into the past, this leads to a deep penetration of the Moon inside Roche's limit. This is possible for a material point or a small body but not for a body of lunar size.

6. Possible Sources of Energy for Early Heating of the Moon

As it was already said at the end of Section 3, a source of energy is needed which can not only produce an early heating, but can heat preferentially the outer parts of the Moon retaining a cool central part. If to accept the accumulation of the Moon from a circumterrestrial swarm, than the following ways of its early heating are in principle possible:

A. HEATING BY GRAVITATIONAL ENERGY LIBERATED AT ACCRETION

The gravitational energy if totally retained, could increase the mean temperature of the Moon for $\sim 1800°$. But the energy was liberated at impacts of accreting bodies and particles on the surface, causing local heating. Therefore most of heat was radiated into space. A substantial part of released gravitational energy could be retained in case of a rapid accumulation of the Moon from a pre-existing circum-terrestrial swarm. But it had to be a gradual accumulation from a swarm, which continued to be replenished in the course of accumulation of the Earth. Thus the duration of accumulation of the Moon was nearly the same as that of the Earth, namely about 10^8 yr.

As the specific energy (energy per unit of infalling mass) increased with increasing mass of the Moon, Wood (1971), assuming a retention of all accretional energy (or its partial retention proportional to specific energy) concluded that this source can produce the required increase of initial temperature toward the surface. However radiation losses are proportional to T^4 and besides the intensity of bombardment of the Moon had to decline to the end of the process due to exhaustion of accretable material.

Perhaps the gravitational energy could be retained and produce a preferential heating of the outer parts of the Moon if the latter by the end of the accumulation process posessed a temperary opaque atmosphere. It is possible that such atmosphere existed at that time due to degasing of infalling planetesimals at impacts.

The suggested existence of such temporary atmosphere seems to give the only possibility to retain a major part of gravitational energy. Without such atmosphere this energy was able to produce neither the sufficient heating of the Moon, nor the required temperature profile.

B. HEATING BY SHORT-LIVED RADIOACTIVITIES

It requires that the Moon should be formed no more than few million years after the nucleosynthesis of short-lived radioactive isotopes. The most promising isotope – Al^{26}, has a half-live of 0.74×10^6 yr and cannot survive in significant quantity more than $3–5 \times 10^6$ yr. Few years ago the formation of Al^{26} was ascribed to additional nucleosynthesis supposed to occur in the early solar system. But even then the time-interval of less than 10^7 yr between this nucleosynthesis and the formation of the Moon seemed uncomfortably small.

Recently all variants of additional nucleosynthesis were abandoned and the situation became much worse, because it is known that 30–200 m.y. elapsed between the last galactic nucleo-synthetic event in the solar matter and the condensation of solid protometeoritic particles. No anomalies in the isotope ratio Mg^{26}/Mg^{24} (Mg^{26} is the decay product of Al^{26}) were found in meteorites, as well as in lunar and terrestrial samples (Schramm *et al.*, 1970). Thus, there is neither theoretical, nor experimental evidense for the presence of Al^{26} in the early solar system.

C. HEATING BY ELECTRIC CURRENTS INDUCED BY THE EARLY INTENCE SOLAR WIND

The possibility of such heating is based on a premiss that the early Sun passed through a T-Tauri-stage when it had to have a corpuscular radiation 10^6–10^7 times more intence than the present-day solar wind. Corpuscular radiation of T-Tauri stars decrease e times in about 10^7 yr. Calculations confirming the efficiency of electric heating for the Moon, published by Sonett *et al.* (1968) are based on the implicit assumption that the Moon was formed nearly simultaneously with the Sun*. This is obviously incorrect. Only if for our Sun the decrease of intensity of the early solar wind occurred at much smaller rate than in typical case, the electric heating could play an important role for the Moon whose accumulation lasted about 10^8 yr. However calculations by Sonett *et al.* (1970) for small bodies show no preferential heating of the upper parts.

D. ACCUMULATION OF HOT PARTICLES

Analyses of returned lunar samples revealed a depletion of several trace elements in lunar basalts as compared with terrestrial ones (Keays *et al.*, 1970; Ganapathy *et al.*, 1970; Anders *et al.*, 1971). Some of these elements are largely concentrated in the crust and therefore their depletion is extrapolated by Anders and his co-workers on the whole Moon. Observed depletions they ascribe to accumulation of the Moon in the source of cooling of the solar nebula and consequtive condensation of more and more volatile elements and compounds. From analyses of Apollo 11 and 12 samples they conclude that the mean accretion temperature was about 620 K. (Anders *et al.*, 1971). However, the basic idea does not give an unambigious interpretation of chemical evidence while its astronomical aspect is in conflict with existing theories of the early evolution of the solar system. The cooling of the solar nebula had to last no more than 10^5–10^6 yr what is by 2 or 3 orders of magnitude shorter than the accumulation time for the Earth and Moon.

The mean accretion temperature estimated by Anders *et al.* (1971) is too low to explain the early melting of the Moon. Besides accretion during cooling would give a decreasing temperature profile with central parts of the Moon hotter than the outer parts (like that on Figure 1).

E. HEATING BY TIDAL DEFORMATIONS

Probably the newly formed Moon was in the state of free rotation which was rapidly decellerated by dissipation of energy of tidal deformations. At this stage the Moon had to be heated on the expense of energy of its rotation. When a free rotation changed into synchronous one (or if it was such from a very beginning) the Moon would continue to be heated due to changes of tidal deformations when it pass from apogeum to perigeum, and back. At this stage the Moon had to be heated on the expense of energy of the Earth's rotation.

A rapid recession of the Moon from the Earth had to cause a rapid decrease of tidal heating, so that the latter was practically restricted to the early stage of the existence of the Moon. A proper choice of inelastic characteristics of lunar and terrestrial globes probably will permit to attain a necessary heating.

* Recently Sonett *et al.* (1971) wrote: "...the accretional time-span could not be more than 5000 yr and possibly much less". What origin of the Moon they have in mind is not explained.

The heating of the synchroneously rotating Moon caused by the eccentricity of its orbit was studied by Kaula (1963, 1964). I am not sure that his result that the main heating had to occur in the centrum is correct. It seems possible that the outer parts should be preferentially heated. One can hope that further studies of tidal effects will provide for early heating of the Moon and the necessary temperature profile.

7. Post-mare volcanic activity

The post-mare volcanic activity on the Moon is revealed by a study of high resolution pictures of Tycho, Aristarchus and Copernicus, taken by Lunar Orbiters (Hartmann, 1968; Strom and Fielder 1970). Manifestations of volcanic processes in Tycho are especially significant for two reasons: it is one of the youngest post-mare craters and it is situated not on a mare but on a highland. Lava lakes and lava flows are seen on the rim, slopes and floor of Tycho. It becomes more and more clear that impact formation of large craters on many, if not all, occasions was accompanied by volcanic processes of different intensity. Most aspects of these volcanic processes remain unknown. Even for circular maria not all students of the Moon agree that their flooding was trigged by impacts that formed maria basins and occurred shortly or even immediately after impacts (Levin, 1966a). Some authors suggest that a large interval of time elapsed between the formation of a mare basin and its flooding by lava (Shoemaker, 1964; Baldwin, 1970). The same question upon the interval of time between impact and the beginning of flooding by lava and other volcanic processes, remain unsolved for craters. For maria it is clear that lava had to come directly or indirectly form a semi-molten lunar interior. For young craters on highlands, like Tycho, the same deep origin of lava seems probable, although the generation of lava by impact cannot be excluded.

Unfortunately we have yet no reliable observations of present day volcanic activity on the Moon. It seems that good example of recent volcanic process on the Moon represent dark-halo craters photographed by Ranger 9 on the floor of Alfonsus. Their youth is manifested by a low density of small craters on the dark blanket as compared with surrounding terrain.

Few years ago it was thought that magmatic differentiation had to last a limited interval of time of the order of 10^9 yr and thus the volcanic activity on lunar surface abated about 2×10^9 yr ago. As the differentiation had to embrass the whole interior of the Moon up to the centrum, later epochs of lunar cooling were reagarded as unsuitable for volcanic activity. Now from the Orbiter pictures we see that, in spite of early beginning, volcanic activity was not completed long ago but lasted up to relatively recent time or even continues up to now.

8. A general picture of lunar evolution

New data obtained by cosmic studies and analyses of returned lunar samples required a rather serious modification of earlier ideas on the evolution of lunar interior.

But the idea of three stages of bombardment of lunar surface, including the relatively late impacts of big bodies which produced circular mare basins (Levin, 1964a), remain valid.

A modern picture of the evolution of the Moon was recently summarized by Baldwin (1970) in the form of 13 items. In essence it coincides with a picture discussed in the previous sections and will not be repeated here.

The evolution of lunar interior was to a large extent determined by its thermal history. The early melting of the Moon means the early concentration of radioactive elements to the surface and the early change of heating of the Moon into its cooling. The cooling affected mainly the outer parts and lead to an increase of thickness of the outer solid layer.

For lunar interior composed of different minerals there is no definit melting point but some interval of temperature within which melting occures. Probably the present-day temperature in the Moon is in the lower part of this interval.

At the present time we need new studies of the initial temperature of the Moon. We need new calculations of lunar thermal history – for improved models. We need data on the present-day generation of radiogenic heat in the Moon that can be obtained from measurements of the heat flow on its surface. We need better present-day temperature profiles that can be obtained from magnetic measurement. Probably it is not too long to wait when cosmic studies will supply us with these data.

Acknowledgement

The author is grateful to Dr. S. V. Majeva for calculations of the tentative variant of lunar thermal history represented on Figure 1.

References

Anders, E., Ganapathy, R., Keays, R. R., Laul, J. C., and Morgan, W.: 1971, *Proc. Apollo 12 Lunar Sci. Conf.* (in press).
Baldwin, R. B.: 1970, *Science* **170**, 1297.
Fricker, P. E., Reynolds, R., and Summers, A. L.: 1967, *J. Geophys. Res.* **72**, 2649.
Ganapathy, R., Keays, R. R., Laul, J. C., and Anders, E.: 1970, *Proc. Apollo 11 Lunar Sci. Conf.* **2**, 1117.
Hartmann, W. K.: 1968, *Comm. Lunar Planet. Lab.* **7**, 145.
Kaula, W. M.: 1963, *J. Geophys. Res.* **68**, 4959.
Kaula, W. M.: 1964, *Rev. Geophys.* **2**, 661.
Keays, R. R., Ganapathy, R., Laul, J. C., Anders, E., Herzog, G. F., and Jeffery, P. M.: 1970, *Science* **167**, 490.
Kuiper, G. P.: 1954, *Proc. U.S. Natl. Acad. Sci.* **40**, 1096.
Kuiper, G. P.: 1959, *J. Geophys. Res.* **64**, 1713.
Levin, B. J. and Majeva, S. V.: 1960, *Dokl. Akad. Nauk S.S.S.R.* **133**, 44.
Levin, B. J.: 1962, in Z. Kopal and Z. K. Mikhailov (eds.), 'The Moon', *IAU Symp.* **14**, 157.
Levin, B. J.: 1964a, *Proc. 13th Intern. Astronaut. Congr.* (Varna, 1962), 11.
Levin, B. J.: 1964b, *Nature* **202**, 1201.
Levin, B. J.: 1966a, in *The Nature of the Lunar Surface, Proc. of the 1966 IAU-NASA Symp.*, 267.
Levin, B. J.: 1966b, *Proc. Caltech-JPL Lunar and Planet. Conf.* (Pasadena), 61.
Levin, B. J.: 1967, *Proc. Roy. Soc. Lond.* **A 296**, 266.
Lyttleton, R. A.: 1953, *The Stability of Rotating Liquid Masses*, Cambridge Univ. Press, p. 150.

Lyttleton, R. A.: 1954, *Trans. IAU* **8**, 717.
McConnell, R. K., McClaine, L. A., Lee, D. W., Aronson, J. R., and Allen, R. V.: 1967, *Rev. Geophys.* **5**, 121.
Majeva, S. V.: 1971, *Izv. Akad. Nauk SSSR, Fizika Zemli*, No. 3, 3.
Marcus, A. H.: 1970, *J. Geophys. Res.* **75**, 4977.
Papanastassiou, D. A. and Wasserburg, G. J.: 1971, *Earth Planetary Sci. Letters* **11**, 37.
Safronov, V. S.: 1967, *Icarus* **7**, 275.
Schmidt, O.: 1950, *Four Lectures on the Origin of the Earth*, Russian sec. ed., p. 65–66. O. Schmidt: *A Theory of the Origin of the Earth. Four lectures*, p. 58–59 (Foreign Lang. Publ. House, Moscow, 1958; Lawrence and Wishart, London, 1959).
Schoemaker, E. H.: 1964, *Sci. Amer.* **211**, 38.
Schramm, D. N., Terra, F., and Wasserburg, G. J.: 1970, *Earth Planetary Sci. Letters* **10**, 44.
Sill, W. R.: 1971, *J. Geophys. Res.* **76**, 251.
Sonett, C. P., Colburn, D. S., and Schwartz, K.: 1968, *Nature* **219**, 924.
Sonett, C. P., Colburn, D. S., Schwartz, K., and Keil, K.: 1970, *Astrophys. Space Sci.* **7**, 446.
Sonett, C. P., Colburn, D. S., Dyal, P., Parkin, C. W., Smith, B. F., Schubert, G., and Schwartz, K.: 1971, *Nature* **230**, 359.
Strom, R. G. and Fielder, G.: 1970, *Comm. Lunar Planet. Lab.* **8**, 235.
Urey, H. C.: 1952, *The Planets, Their Origin and Development*, Yale Univ. Press, New Haven.
Urey, H. C.: 1955, *Proc. Nat. Acad. Sci.* **41**, 127.
Urey, H. C.: 1962, in Z. Kopal (ed.), *Physics and Astronomy of the Moon*, Acad. Press, New-York, London.
Volkov, M. S.: 1967, *Bull. Inst. Teor. Astron. (Leningrad)* **11**, 262.
Ward, S. H.: 1969, *Radio Sci.* **4**, 117.
Wood, J. A.: 1971, in G. Simmons (ed.), *The Geophysical Interpretation of the Moon* (in press).
Wood, J. A., Dickey, J. S., Marvin, U. B., and Powell, B. N.: 1970a, *Science* **167**, 602.
Wood, J. A., Dickey, J. S., Marvin, U. B., and Powell, B. N.: 1970b, *Proc. Apollo 11 Lunar Sci. Conf.* **1**, 965.
Wood, J. A., Marvin, U. B., Reid, J. B., Taylor, G. J., Bower, J. F., Powell, B. N., and Dickey, J. S.: 1971, *Smithsonian Astrophys. Obs.*, Spec. Rep. 333.

LUNAR TIDAL PHENOMENA AND THE LUNAR RILLE SYSTEM

BARBARA M. MIDDLEHURST
Encyclopaedia Britannica, Chicago, U.S.A.

Abstract. Two types of tide-linked lunar phenomena now exist: the so-called lunar *transient events*, short-lived changes in brightness or colour and obscurations in small areas of the Moon (the reported duration is typically from a few seconds to a few hours and the areas involved are usually a few kilometres or less in diameter); and the *A*- and *B*-type *seismic signals* relayed back from the Apollo 12 passive seismic experiment. The frequency diagrams of both the lunar transient events and the seismic signals show strong peaks at perigee when the Earth is closest to the Moon, with a smaller peak at apogee (transient events and *B*-type signal only) and both have been attributed to endogenous causes.

Both sets of data appear to be linked to areas where cracks and rilles exist. The most likely interpretation of the two sets of observations is that they are complementary and that some sort of gas release and excitation (in the cases of the glows and colour changes) is involved.

Association of the lunar events with craters with seamed and cracked floors and with other crack systems is described. Only a few associated areas of the Moon-wide lineament systems have been noted, but the Apollo 12 seismometer signals seem likely to originate from the nearby system of parallel features in the Fra Mauro area. The case for possible gas volcanism is examined.

1. Tidal Analysis

On the Moon, the gravitational tide-raising potential is much simpler than that causing the terrestrial tides. Because of its greater mass, the Earth's contribution dominates; also, the much slower rotation of the Moon and the absence of surface water with the associated tidal lag and distortion allows us to consider body tides under free oscillation only. The terms introduced by the distortion of the Moon's shape are very small and are neglected. However, it is necessary to consider both the changes in distance of the Earth and the librations of the Moon. The radial component of the tidal force at a given point on the Moon, that is, the vertical tidal term, F_r, in the gravity field can then be expressed in quite simple terms. Neglecting terms in $(r/R)^2$ and higher in an expansion in terms of zonal harmonics, we have

$$F_r = -\frac{GM}{2R^2}\frac{r}{R}(3\cos^2 z - 1) + \text{a similar term for the Sun}$$

$$= -\frac{GM}{2R^2}\cdot\frac{r}{R}(3\cos 2z + 1) + \cdots. \qquad (1)$$

The solar term is only of the order of 1% to 3% of the term due to the Earth but it has been taken into account in the computer-generation of tides undertaken at Houston under the direction of William Chapman of NASA. Results of this programme are used in the following analysis.

In expression (1), the two variables are the Earth's distance and the angular distance, z, of the point on the Moon to be considered from the sub-Earth point. The quantity R varies through the anomalistic month but the amount, or amplitude, of this change also changes from month to month over longer cycles due to perturbations of the orbit. The value of z varies because the position of the sub-Earth point varies with the

Urey and Runcorn (eds.), The Moon, 450–457.

librations of the Moon. The value of F_r then goes through a modified sinusoidal cycle in the course of an anomalistic month and with passage of time, the amplitudes of the cycle are further modified.

The anomalistic month is on the average 27½ days, while the tropical month from New Moon to New Moon is 29½ days, but as Knopoff (1970) has shown, the 6-month perturbation of the lunar orbit by the Sun produces cyclical changes in the length of the anomalistid month; in consequence, perigee is more likely to occur at New or at Full Moon than at the quarter phases. If the observer pattern was such that lunar visual observations tended to take place at New or Full Moon preferentially, then the perigee correlation would have to be adjusted. The pattern of observations up to 1964, used in the frequency diagram for lunar events, did not show such features; most observations are made when shadows are favourable for the enhancement of detail in the object being studied. Except at eclipse times very few observers look at the Moon when it is full and the observations were well distributed with phase. Both the transient events and the seismic *A*- and *B*-type signals occur most frequently at times of tidal maxima and minima and now appear to originate in similar ways. Of course, no adjustment of the seismic data for phase need be considered.

2. Characteristics of the Transient Events

Throughout the lunar literature of the 18th and 19th centuries, with some few reports from earlier sources – the first listed was dated 1540 – descriptions occur of the disappearance of detail for short periods of minutes or hours (clear before and after) and of temporary glows or increases in brightness and colour changes (the so-called 'red spots', 'violet glares' and so on). The best known of modern observations are the reports of colour changes in the lunar crater Alphonsus by N. Kozyrev who took spectra in 1956, the observations of red spots in Aristarchus at Flagstaff, Arizona by the lunar mappers, J. C. Greenacre and E. Barr in October and November, 1963, and the description of a lunar 'volcano' noted by Sir William Herschel on May 4, 1783, which strongly resembled that of the Flagstaff phenomenon except that it was seen on the dark side of the Moon, while Aristarchus was not illuminated by the Sun.

In addition to about 600 visual reports, around 20 records of a more permanent nature, often obtained by chance, are known. The following report was sent to the present author by Dr V. P. Dzhapiashvili of Abastumani Observatory, Georgian S.S.R. in 1970 and has not appeared before in western literature.

Though the phenomenon was registered nearly 19 years ago, the recollection is vivid even today. On the night of July 3, 1952, Tamara G. Negrelishvili and I were carrying out a routine programme of electro-polarimetric observations of the Moon with the help of a 33-cm mirror telescope. We were studying the polarization properties of lunar surface formations. The results were published in 1957....

The sky was absolutely clear that night. The author was at the telescope directing it at the object to be measured, guiding and manipulating the analyser.... Reflected light readings for each position of the analyser were repeated to reduce the error of pointing....

We measured two objects – the craters Aristoteles and Eudoxus. It is worth noting that the readings of both series for the respective analyser positions and for each crater separately coincided.

However, when the telescope was directed at the crater Posidonius and when we began to take its readings, we witnessed a totally different picture. The pointer of the galvanometer swung to and fro and the readings for the second series of measurements failed to coincide with those of the first. To rule out the possibility of faults in the apparatus we switched back to the craters Aristoteles and Eudoxus. However, we only confirmed our earlier results. We also obtained good results for the crater Aristillus which we observed immediately after the crater Posidonius. So we made the following entry in the record book: "During the observations of the crater Posidonius, the pointer of the galvanometer for some reason swung to and fro. The other craters showed normal readings...." [Later] the author observed Posidonius with the help of the above method in 40 different phases but he failed to witness any similar phenomenon again.

Previous events have been noted as occurring in Posidonius in 1821, 1849, and 1890, and in each case changes in the shadow of the crater were described. Only visual reports were made.

The association of lunar events with perigee and apogee times first led to the conclusion that they were endogenous. This deduction has since been supported by the

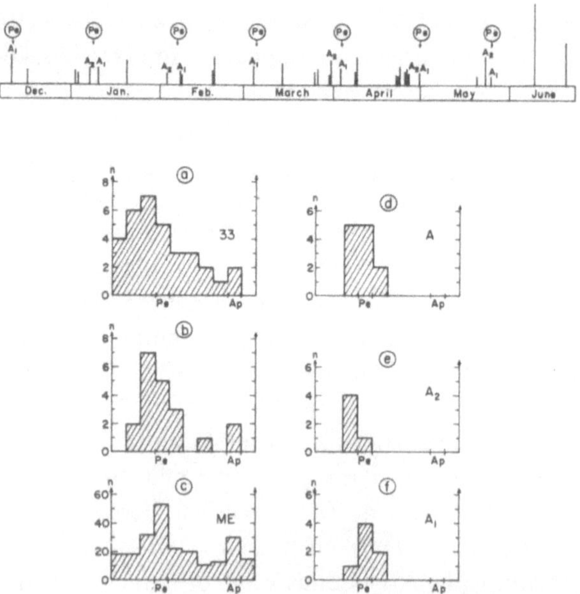

Fig. 1. Frequency distribution for different classes of events with respect to the lunar anomalistic month. (a) 33 of the strongest seismic signals; (b) A- and B-type seismic signals and those with similar frequency spectra; (c) dated lunar transient events before 1964; (d) A-type seismic signals; (e) A_1-type seismic signals; (f) A_2-type seismic signals. Symbols: Pg = perigee; Ap = apogee; ME = Middlehurst Effect. *(Courtesy of R. Meissner.)*

similarity of the frequency diagram to that found for the *A*- and *B*-type Apollo 12 seismic signals (Figure 1) and also by the following characteristics.

The distribution of the event sites over the lunar surface is not random but is associated with the borders of the regular maria, and with types of crater containing extensive crack systems, and/or central peaks; in addition to those mentioned already Taruntius, Sabine, Copernicus and Kepler are examples. Most of the larger ray-craters have been reported as sites of activity. With the exception of the ring-plain site such as Plato, also often reported as showing an anomalous appearance, all craters that have been listed as event sites have previously been classified as young formations. Other correlations are with the areas where the 'rolling stones' have been found (these objects appeared in Orbiter photographs at the ends of tracks that seemed to have been produced across the lunar surface by gentle movement of the stones themselves; they are mysterious as, in a few cases, the tracks travel uphill, no motive force has been satisfactorily found and there is no evidence to show how the tracks could have been set in motion). Black-haloed craters, and mascons have both been located in active area.

3. Tides at Aristarchus and Gassendi

A correlation of increased activity with increases in the amplitude of local tides was noted at two sites for the period from 1963 and 1968.

The crater Aristarchus, named by Hevelius *Mons Porphyrites*, the burning mountain or red hill, has, with Schröter's Valley and other features of the immediately area – the Aristarchus Uplift – accounted for fully one-third of all the reports of lunar activity. Nevertheless, the times are not random. The usual perigee peak is shown but, in addition, an increase in reported activity was noted by William Chapman (1967) to have occurred when the local tidal amplitudes (see expression (1)) was greatest. In the last few years, since 1963, it has been possible to compare the changes with those at another crater, Gassendi. In 1966, after a long period of quiescence, Gassendi burst into activity. A whole series of reported colour changes, glows, and obscurations were seen by many observers at this time. This spurt died down in 1967. Aristarchus which had been more active during 1963–64 before the Gassendi phenomena began, decreased its activity during the Gassendi peak period (see Figures 2). When the tides for Gassendi and Aristarchus were computed and compared, it was found that in each case the greatest activity was shown at each site at the times of maximum amplitudes of the local tides. The relative positions of the two craters were such that the tides were out of phase with each other.

Observational selection can probably be ruled out. The general morphology of the two craters is similar. Both are large (Aristarchus about 40 km and Gassendi about 120 km in diameter) and easily distinguished. The floors of each are seamed with cracks and both have central peaks. Aristarchus, though smaller is brighter. Sunrise occurs one day earlier at Gassendi (10 days after New Moon) than at Aristarchus, and both craters are at about the same distance from the mean centre of the Moon's disc, so that the range and mean value of the tidal amplitudes is the same. Any knowledge of the present analysis among the observers that might have led, even unconsciously, to bias in the observations can be ruled out as the analysis for Gassendi was not begun until most of the data used had been assembled. It is also of interest that an infrared survey made of the lunar surface during a lunar eclipse in 1967 by Salisbury *et al.* confirmed an earlier survey by Saari and Shorthill in 1964 but noted one additional infrared anomaly or hot spot in Gassendi.

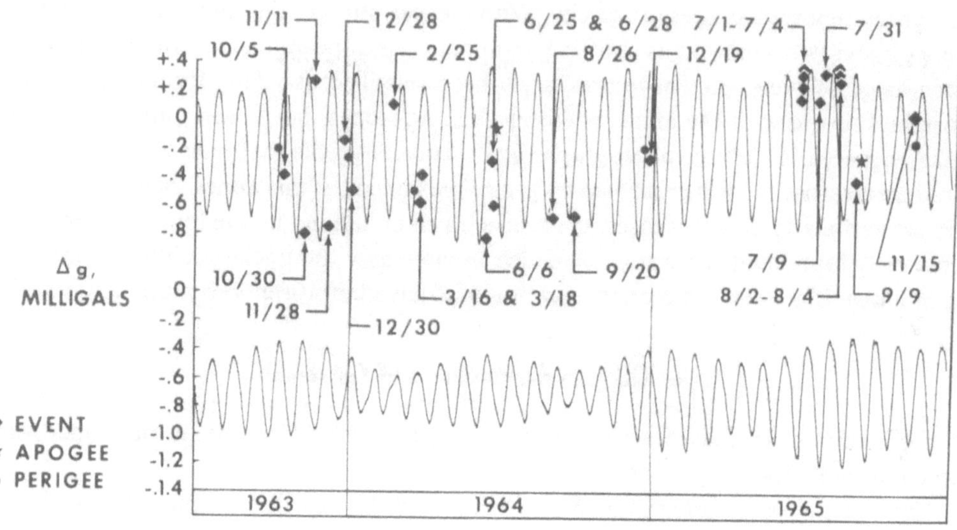

Fig. 2a.

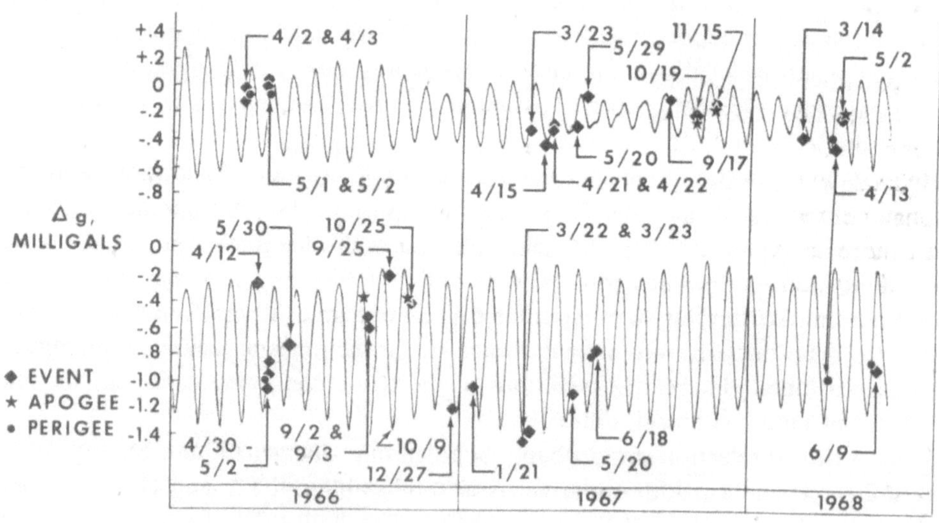

Fig. 2b.

Fig. 2a–b. Tidal-Gravity change. Local tide-raising vertical force change per gram of lunar material (a) for July 1963 to December 1965 and (b) for January 1966 to June 1968. Upper tides at Aristarchus and lower tides at Gassendi. Events at local Aristarchus tidal-gravity values for given date when observed. Each apparent event triggering perigee and apogee is indicated.

4. The Passive Seismic Experiment of Apollo 12

With the exception of Aristarchus and its environs, activity in most sites has been sporadic. The passive seismic experiment of Apollo 12 has produced continuous records of moonquakes and meteorite and other impacts since November 19, 1969 up to

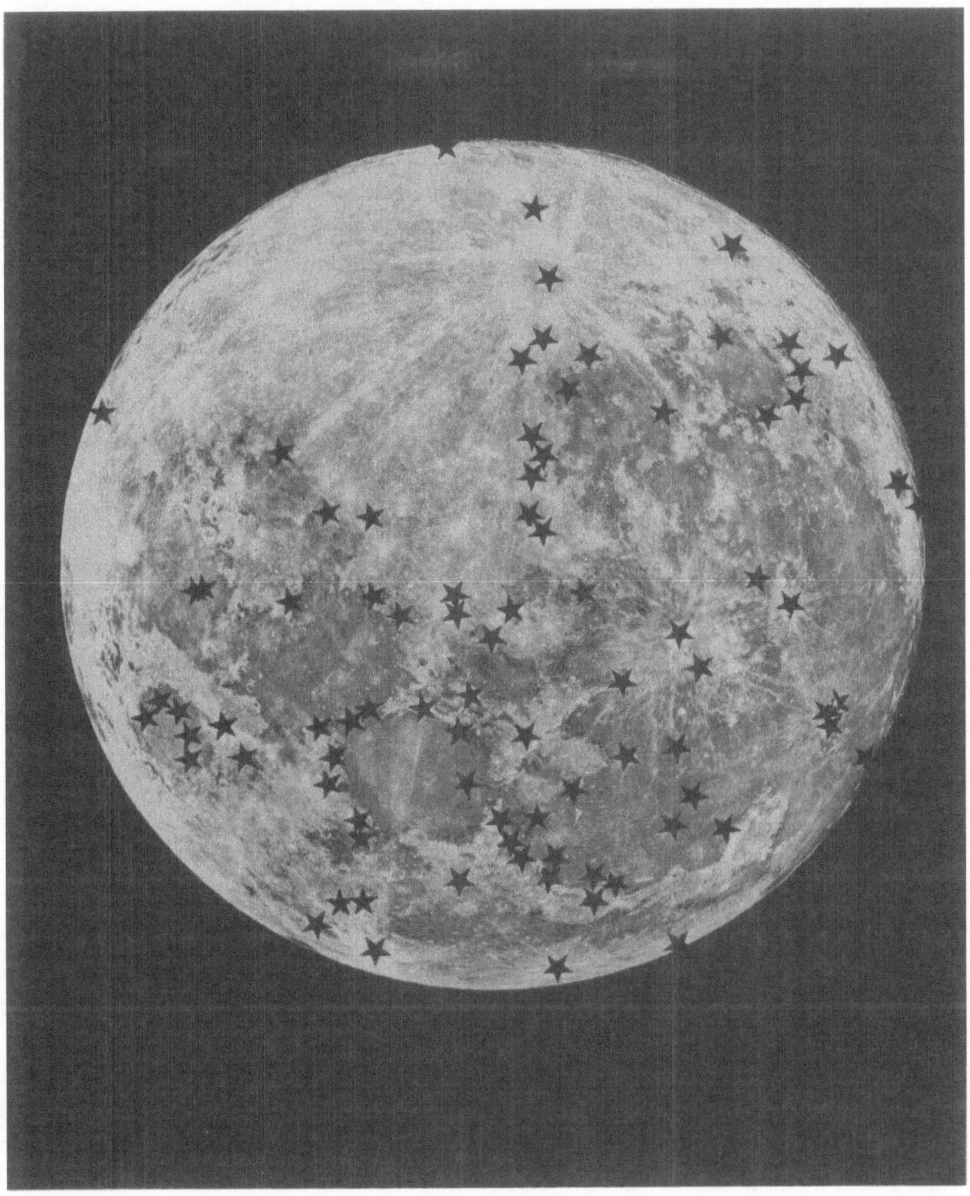

Fig. 3. Sites of lunar transient events for dates before June, 1964. Note that the sites are not distributed at random but occur preferentially around the regular maria, in ray craters, and down the central meridian following the Alphonsus crater chain.

August 1970. During that time, 208 events, or 23 per month were recorded. The moonquakes are strongly concentrated near perigee and less strongly, as for the transient changes, near apogee. The region of greatest activity has been tentatively located in the vicinity of the well developed set of parallel rilles and lineaments near Fra Mauro and Parry. Figure 1 shows the distribution over the orbit for the first 6 months.

These rilles are directed preferentially from S60°W to N60°E and at the time of the records were directed towards the sub-Earth point. At such periods, the horizontal component is directed across the rille and has maximum disruptive force. The times of perigee occur within one day of tidal minimum gravity.

Mr Chapman has run tides at Fra Mauro and has shown that the repetitive moonquakes occur at local minimum tidal gravity. Tidal triggering of moonquakes or of transient events might be explained by the release of subsurface fluids as has been previously suggested for the transients alons by Green (1965) and by Middlehurst (1966). Periods of minimum gravity offer periods of decompression and the widest opening of pore spaces when the release of gases is most likely, all other factors for the moment being considered equal.

What is somewhat puzzling is that successive perigees do not correspond to exactly Earth-Moon distances so that a simple correlation with the tidal gravity change is not the whole explanation. In other words, events tend to occur at successive perigees or at successive minima, not at successive times of equal tidal force. This is apparently true of the moonquakes, just as it was earlier of the transient events. More deepseated disturbances, possibly affecting large areas of the Moon may be involved. A second possibility is that either rates of change of stress, or cumulative effects are more potent in triggering off the events and these possibilities are being investigated.

5. The Lineament System and Other Lunar Rilles and Faults

In addition to the lunar craters, other features are of structural importance. Three moon-wide systems of lineaments were investigated and plotted by Fielder (1963) and by Strom (1964). They probably represent surface expression of very deep-seated disturbance of the lunar material. The general directions of these lineament systems, which include straight rilles, graben, wrinkle ridges, sections of (polygonal) crater walls and crater chains, are N-S, ENE-WSW, and WNW-ESE, and they continue on the far side of the Moon.

Not many transients are associated with these systems. An apparent fault, the Straight Wall runs N-S near the border of Mare Nubium and is the site of an event reported for 1956 (an obscuration). A N-S chain of craters runs from W. Herschel through Alphonsus to Clavius, along which many events have been reported; this area is apparently depressed below the surrounding level.

Sinuous rilles, of which examples are Schröter's Valley, the Hadley Rille and the Hyginus Rille, all of which are event sites, occur in many areas and may be associated with fluid flow. Schröter's Valley has features strongly resembling terrestrial river meanders and appears to originate in Herodotus, but there are many puzzling features in this area and the Orbiter photographs unhappily do not show the whole valley. The Hadley Rille has been compared to a collapsed lava tube. The Hyginus rille has crater chains along it and these seem to be structurally related to it, but clearly have nothing to do with meteorite impact.

No reports of transient events are known for the Fra Mauro-Parry area but the crater forms are quite similar to others that have been reported as sites.

6. Gas Volcanism

Many of the processes that lead to volcanism on the Earth are still imperfectly understood. The lunar environment is, however, so different that in any case, it seems unlikely that the phenomena of terrestrial volcanism would be reproduced there. The seismic traces are reported to be quite similar to those occuring in volcanic-triggered seisms on the Earth and passing through porous lava-type rock. Under external conditions of near-perfect vacuum, many substances, liquid under standard temperatures and pressures, would soon form pockets of gas that could escape through fissures and produce disturbances of the rock detectable as lowlevel seismic signals of the kind recorded by the Apollo 12 seismometer. The transient event descriptions are entirely consistent with this possibility, and the hypothesis has been further strengthened by Dr Mills' elegant production of not one but many types of crater forms by fluidization of dust through the passage of gas through it in the laboratory. See chapter by Mills, p. 407.)

In summary, The mechanism of production of both the *A*- and *B*-type seismic traces and the transient events is likely to be tide-triggered disturbance of existing cracks, not necessarily in progressive faulting, but by repeated opening and closing of the pores so that gas within the Moon can alternately be built up and released.

References

Burley, J. M. and Middlehurst, B. M.: 1966, *Proc. Nat. Acad. Sci.* **55**, 1007.
Chapman, W. B.: 1967, *Geophys. Res.* **72**, 6293.
Chapman, W. B. and Middlehurst, B. M.: 1971, (abstract), *Trans. Am. Geophys. Union* **52**, 4.
Chapman, W. B. and Middlehurst, B. M.: 1972, (abstract), *Trans. Am. Geophys. Union* **53**, 4.
Fielder, G.: 1963, *Quart. J. Geol. Soc. London*, No. 119, pp. 65.
Green, J.: 1965, *Ann. N.Y. Acad. Sci.* **123**, 433.
Knopoff, L.: 1970, *Moon* **2**, 143.
Latham, G. and Meissner, R., *et al.*: NASA Prelim. Sci. Rep. for Apollo 12 (SP 235).
Meissner, R., Sutton, G., and Duennebier, F.: 1970, *Mondbeben*, Umschau, Heft 4, 111.
Middlehurst, B. M.: 1967, *Rev. Geophys.* **5**, 173.
Middlehurst, B. M. and Chapman, W. B.: 1971, *Strolling Astronomer* **23**, 17.
Strom, R.: 1964, *Comm. Lunar Planet. Lab.* **2**, 205.

INDEX OF NAMES

The page numbers are typed in normal type in ascending numerical order, followed by the reference numbers (these only appear in a few papers) in brackets. The numbers underlined refer to the pages on which the references are actually listed.

INDEX OF SUBJECTS